Tableau Strategies

Solving Real, Practical Problems with Data Analytics

Ann Jackson and Luke Stanke

Beijing • Boston • Farnham • Sebastopol • Tokyo

Tableau Strategies

by Ann Jackson and Luke Stanke

Published by O'Reilly Media, Inc., 1005 Gravenstein Highway North, Sebastopol, CA 95472.

O'Reilly books may be purchased for educational, business, or sales promotional use. Online editions are also available for most titles (*http://oreilly.com*). For more information, contact our corporate/institutional sales department: 800-998-9938 or *corporate@oreilly.com*.

Acquisitions Editor: Michelle Smith
Development Editor: Jill Leonard
Production Editor: Caitlin Ghegan
Copyeditor: Sharon Wilkey
Proofreader: Stephanie English
Indexer: Cheryl Lenser
Interior Designer: David Futato
Cover Designer: Karen Montgomery
Illustrator: Kate Dullea

October 2021: First Edition

Revision History for the First Edition

2021-07-27: First Release
2021-10-08: Second Release

See *http://oreilly.com/catalog/errata.csp?isbn=9781492080084* for release details.

978-1-492-08008-4

[LSI]

Table of Contents

Preface

If you've picked up this book, chances are we already have something in common. What is it, you may ask? It's that we love data—working with it, analyzing it, and presenting it. The Tableau portfolio of products, most specifically Tableau Desktop, provides us data enthusiasts the opportunity to get hands-on with our data, interrogate it, and create beautiful charts and outputs.

Tableau as a visualization tool is unique. It takes an analysis-first approach by offering an open canvas where creators can start from scratch and build any chart they imagine. It also provides a fantastic toolkit to create both interactive dashboards and pixel-perfect data displays.

But since we're all professionals here, we know the truth: it takes more than just connecting a dataset to Tableau to derive insight and value. That's where this book comes in. Throughout these pages, you'll find an exhaustive collection of use cases and scenarios designed to provide both contextual commentary and technical know-how. Whether you're a data beginner (we've got you covered) or a seasoned Tableau professional, the chapters in this book are designed to guide you through *how* and *why* charts are made.

And one of the most exciting parts about this book? It cultivates examples from many industries and includes many types of data. We all know that both the type of data and the contents of a dataset define the direction of analysis, but often in your day job, you're stuck with just one type. Here, you'll be able to draw inspiration and gain practical experience working with data that you may be completely unfamiliar with.

In the nearly 20 years that Tableau has been around, neither of your authors has found a visualization tool that provides as much depth, flexibility, and innovation as Tableau Desktop has. Add on top of that a world-class community and free social sharing platform called Tableau Public, and you have the perfect recipe for why we both continue to invest our careers in Tableau and to teach it to learners and classrooms of all sizes and types.

Who Should Read This Book

This book is designed for anyone who is familiar with Tableau Desktop and creating charts, visualizations, and dashboards with the application. We've designed this book for someone who already knows how to connect to data, but is looking for inspiration and enhanced application of knowledge.

If you're a data analyst, this book will resonate with you as it provides use cases and hands-on exercises designed from our own experiences. Each strategy is framed with stakeholder questions and context, and we take you from start to finish of how to build a visualization, why to build it, and lessons along the way.

If you're an analytics leader, this book will help you navigate the myriad requests you and your team are bound to receive as you use Tableau to analyze and present data. With our strategies, you can avoid common mistakes, learn how to create the right data product for your audiences, and spend less time maintaining the products you do build.

If you're a business intelligence developer or data engineer, we hope this book serves as a bridge between the technical realm of data and the more nuanced side of satisfying the business with analytics. You'll see firsthand how the agile nature of using Tableau can provide direction on what type of data to capture, how to store it, and most importantly, how to share it.

Finally, if you're new to the world of data and analytics, this book will serve as a comprehensive road map of the most common requests you're bound to receive. This book does what no other before has done: takes real use cases and presents them in a format that any reader can learn and implement. As you delve into these chapters, know you're reading the book we both wish we had as we started in our careers.

Why We Wrote This Book

We wrote this book to demystify the process of using Tableau to create data products and visualizations. In a sea of information about Tableau, we both felt there was a gap to be filled—one that outlines successful strategies for creating Tableau visualizations in any type of organization. This book is unique in that every strategy is something one or both of us has actually implemented and is accompanied by the thoughts and perspectives associated with the strategy.

Navigating This Book

The first 10 chapters of this book are designed to take you through specific types of analysis and the data products you can generate using Tableau. Within each of these chapters, you will find hands-on exercises called *strategies*, and accompanying datasets (*https://tableaustrategies.com*), to build solutions.

You'll start with the fundamentals in the first three chapters of the book, beginning with categorical data. First you'll learn how to build the most basic visualization, the bar chart. From there, strategies will continue to build and will include tips on how to customize each chart to fit your audience.

Chapters 4 through 7 guide you through more specific types of visual analysis. You'll work with time and learn the ins and outs of constructing perfect date calculations, communicating key metrics to executive audiences (Chapter 5), and building maps (Chapter 7).

Chapters 8, 9, and 10 will test your mastery of Tableau in a big way. You'll see our most challenging strategies and learn how to bend Tableau well beyond its defaults.

The final four chapters of the book take a step back from exercises to explore broader concepts. You'll learn how to bring it all together in Chapters 11 and 12, as you create interactive, beautiful, and impactful dashboards and data products. And in Chapters 13 and 14, we'll leave you with a commentary on analytics, including how to create a full analytics platform and a discussion on frameworks from the most popular industries.

Conventions Used in This Book

The following typographical conventions are used in this book:

Italic
: Indicates new terms, URLs, email addresses, filenames, and file extensions.

`Constant width`
: Used for program listings, as well as within paragraphs to refer to program elements such as variable or function names, databases, data types, environment variables, statements, and keywords.

`Constant width bold`
: Shows commands or other text that should be typed literally by the user.

This element signifies a general note.

This element signifies a tip or suggestion.

This element indicates a warning or caution.

O'Reilly Online Learning

For more than 40 years, *O'Reilly Media* has provided technology and business training, knowledge, and insight to help companies succeed.

Our unique network of experts and innovators share their knowledge and expertise through books, articles, and our online learning platform. O'Reilly's online learning platform gives you on-demand access to live training courses, in-depth learning paths, interactive coding environments, and a vast collection of text and video from O'Reilly and 200+ other publishers. For more information, visit *http://oreilly.com*.

How to Contact Us

Please address comments and questions concerning this book to the publisher:

O'Reilly Media, Inc.
1005 Gravenstein Highway North
Sebastopol, CA 95472
800-998-9938 (in the United States or Canada)
707-829-0515 (international or local)
707-829-0104 (fax)

We have a web page for this book, where we list errata, examples, and any additional information. You can access this page at *https://oreil.ly/TableauStrategies*.

Email *bookquestions@oreilly.com* to comment or ask technical questions about this book.

For news and information about our books and courses, visit *http://www.oreilly.com*.

Find us on Facebook: *http://facebook.com/oreilly*

Follow us on Twitter: *http://twitter.com/oreillymedia*

Watch us on YouTube: *http://www.youtube.com/oreillymedia*

Acknowledgments

Collectively, we both would like to thank every single member of the Tableau community, past, present, and future. The knowledge, questions, and discussions that you continue to bring are the reason we do our jobs. More specifically, we'd like to thank Andy Kriebel and Emma Whyte, two individuals who took the idea of technical challenges in Tableau and framed them as a weekly challenge called Workout Wednesday. Without the foundation of this community initiative, we wouldn't have been able to build up a large portfolio of examples to choose from or recognize the need for this type of book.

We'd also collectively like to thank Rody Zakovich, Curtis Harris, Lorna Brown, Sean Miller, and Candra McRae. Each has supported Workout Wednesday throughout the years and provided a sounding board for challenges and knowledge to us. You've made each of us better as a result of your collaboration.

And finally, collectively we want to thank the technical reviewers of this book, Jami Delagrange, Alicia Bembenek, Christopher Gardner, Allen Hillery, and Bonny McClain, who put in many hours of work to ensure that the final product met our satisfaction. We appreciate every comment, every correction, and every suggestion you each shared throughout the process.

Ann would like to thank the following individuals:

Jami Delagrange
: Thank you for being more than a technical reviewer. I appreciate your friendship and perspective.

Patti and John Provost
: Thank you for believing in and supporting my goals. Thank you for answering every text message and phone call of panic with love, understanding, and humor.

Josh Jackson
: Thank you for being my life partner and husband. Thank you for supporting my dreams and guiding us together when I don't quite see the light. I'm so thankful for the life we've created together and that I get to share my days with you.

Luke would like to thank the following:

My coauthor
Earlier in our careers, Ann and I worked for the same company, but we didn't get a chance to meet until Tableau Conference in 2017. Since then, we've maintained a strong professional relationship, which led to us working together on this book. Thank you, Ann, for the time you've committed to our friendship over the years.

My wife and parents
Thank you for taking the time during a global pandemic to provide child care support so that I could write a technical book on nights and weekends.

The team at Tessellation
You are a dream team and one of my everyday sources of motivation and inspiration. Watching each of you blossom over the years has been a gift. I also want to lend additional kudos to Alicia Bembenek, who joined the team two years ago and jumped at the opportunity to act as a technical reviewer. Thank you for taking extra time out of your life to help review and provide feedback on the book.

The analytics community
Members of the analytics community have not-so-secretly provided distributed input through social media. Thanks for reacting to all of my random thoughts.

CHAPTER 1

Categorical Analysis

Categorical analysis is the foundation of data visualization. It is the first and most frequent type of data visualization that data analysts use. Categorical analysis takes a dimension (for example, [Regions]) and breaks it apart by a measure (for example, [Sales]). A *dimension* is typically a categorical value; these do not get aggregated. They are likely used to create data headers or to generate filters. A *measure* is a (usually numerical) value that can be aggregated using mathematical functions (like sum, average, or median). Measures create unbroken axes, those that extend from one end of a range to the other.

This type of analysis aids in answering common business questions such as these:

- How does *A* compare to *B*?
- How is *X* measure distributed across *Y* categories?
- How much do *A*, *B*, and *C* contribute to the total?
- How does *X* measure change over time (where time is the dimension)?

Categorical analysis is usually presented as bar charts. *Bar charts* use height or length as visual encoding to express a measure. *Visual encoding* refers to the techniques used to display data in charts; Figure 1-1 shows some examples. Encoding data in bar charts is effective because humans can quickly analyze the variation among the size of the bars; they are also easy to understand and label.

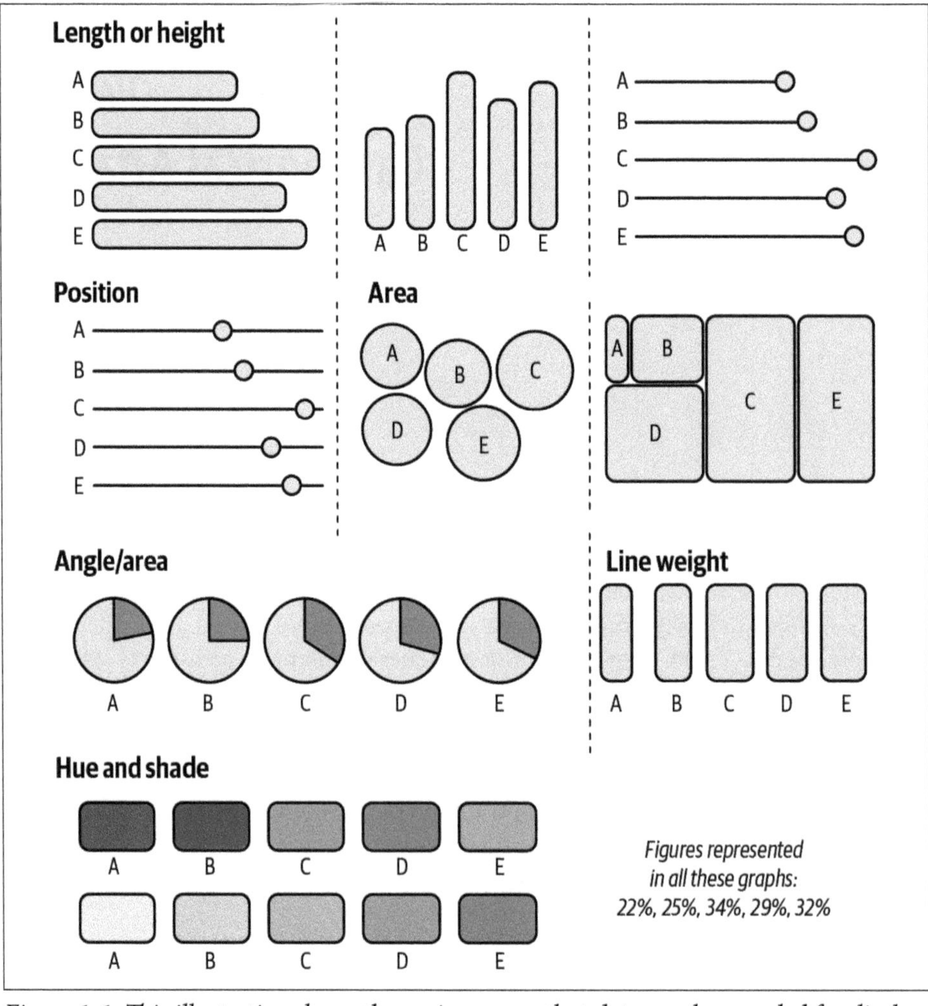

Figure 1-1. This illustration shows the various ways that data can be encoded for display, and aligns them to a comprehension scale indicating how precisely the human eye can discern differences

In our first use case, we will explore how to make effective *bar charts*. You'll play the role of a large financial institution that wants to understand which merchant categories make up the majority of transactional spending in order to drive marketing efforts and partnerships and better serve customers' interests. We will also expand from the defaults and learn two additional methods for making bar charts that demonstrate the most important information.

In the second case study, you'll learn about working with many dimensions. While bar charts are very useful, you'll need other data visualization tools when doing categorical analysis. When a dimension has many members, displaying each one as a bar chart sometimes becomes problematic. When this happens, you can use alternative chart forms to conserve space but still display all members. The most useful chart for this scenario is a treemap. We'll explore treemaps through this case study about a nonprofit organization.

In our final use case of this chapter, you'll learn how to use pie charts and donut charts to visualize whole relationships. This case study involves conducting a survey about IT professionals and mental health. Pie charts are often the first type of data visualization you learn in school, but we like to use them sparingly and as an alternative option. By the time we get to this use case, you'll see how properly executed pie charts can be great tools to craft and share data with your audience.

In This Chapter

To build the visualizations in this chapter, you'll use these datasets: Financial Institution Transactions, IT Survey Data, and Nonprofit Grant Data. In this chapter, you'll learn how to do the following:

- Create compelling bar charts that work dynamically to display top contributor information and that can automatically group together dimensions of small values.
- Understand when to utilize bar charts versus treemaps when faced with a dimension of several members. Utilize drill-down features within treemaps to explore tiered dimensions. Leverage additional data encoding by way of color to express alternative information.
- Utilize pie charts to demonstrate part-to-whole relationships. Turn pie charts into donut charts that communicate multiple data points. Utilize small multiple charts for dual-dimension comparisons.

Bar Charts: Banco de Tableau Case Study

Our first case study involves a large financial institution that is trying to understand consumer behaviors. We'll call it *Banco de Tableau* (*BoT* for short).

The data team at BoT is working to understand how and where consumers spend their money. This objective is fundamental to the organization's success because it will drive the direction of marketing efforts, partnerships, and product promotions. It will also provide insight into profiling customers and may even unearth opportunities to grow the customer population. What kind of chart should the team use to present its results?

Bar charts should be the first visualization type you try when exploring categorical analysis. Because they use length and height as visual encoding, they make it easy to interpret and compare members.

To solve the bank data team's problem, you're going to start with a basic bar chart. It will help you compare types of merchants by how much consumers spend. It will also serve as the first step in understanding the data.

Strategy: Build a Bar Chart in Tableau

To build your first bar chart, you'll use the Financial Institution Transactions dataset as you follow along in Tableau. Here are the steps:

1. Drag the [Merchant Category] dimension to the Rows shelf.
2. Drag the [Transaction Amount] measure to the Columns shelf as `SUM([Transaction Amount])`.
3. From the toolbar, sort the merchants in descending order by Transaction Amount (aka spending).

Congratulations—you've built your first bar chart. Now this may not seem revolutionary, but it is the first step in finding out where consumer spending is focused. This simple chart, shown in Figure 1-2, dispels any intuition-based theories and presents us with the facts.

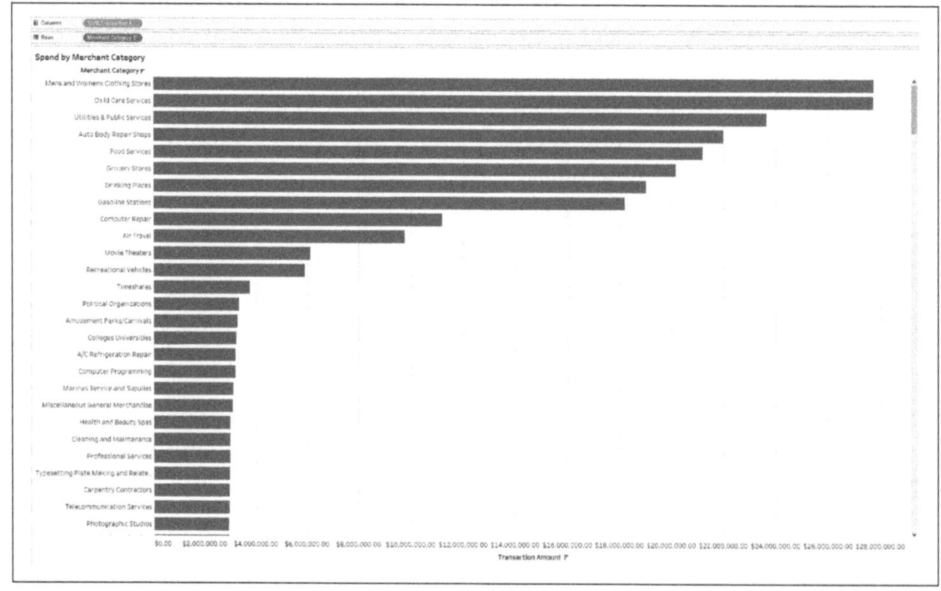

Figure 1-2. A bar chart showing the merchant categories sorted in descending order by transaction amount

This is a great starting point. We've now visualized the data and can see that the largest merchant category is Men's and Women's Clothing Stores. We can also see that several small merchant categories aren't responsible for a lot of spending, like Cleaning and Maintenance. Knowing that there are many types of merchant categories, and that some are much larger than others, is additional insight we can act on to improve our visualization.

A good incremental way to do this is to limit the number of visible merchant categories by using a *Top N filter* with a parameter. This filter limits the chart by *N*, the number defined, in order to show only the top members in the chart. A *parameter* is a dynamic entry field defined by the end user.

In this scenario, we will create a parameter that allows the user to dynamically define the number of categories they want to see. Adding the parameter not only gives our audience more control over the visualization, but also provides a more conversational way to understand the chart's content. A Top 10 illustration is a much more tangible and bite-size takeaway than a long list of bars.

In the upcoming strategy, you'll be working with a parameter for the first time. We use the mnemonic *ABC* to remember the most common steps to building a parameter:

- A: Add a new parameter.
- B: Bring the parameter control onto your worksheet (show the control).
- C: Include the parameter in a calculated field.

Strategy: Create a Top N Bar Chart

To create the Top N bar chart, follow these steps:

1. Drag [Merchant Category] to the Filters shelf.
2. Navigate to the Top section.
3. From the drop-down list, select Create a New Parameter.
4. Name the parameter **`[Top N]`** and save it.
5. Right-click the parameter at the lower left of the Data pane to expose the parameter control.

Notice in Figure 1-3 that we've put the parameter value in the title of the chart. Now, when the user makes a dynamic change, the number will update in response. With this small act, we've created a portfolio of charts that can be customized to suit the audience's needs. The change also makes the chart feel responsive to the audience; their actions change the visualization.

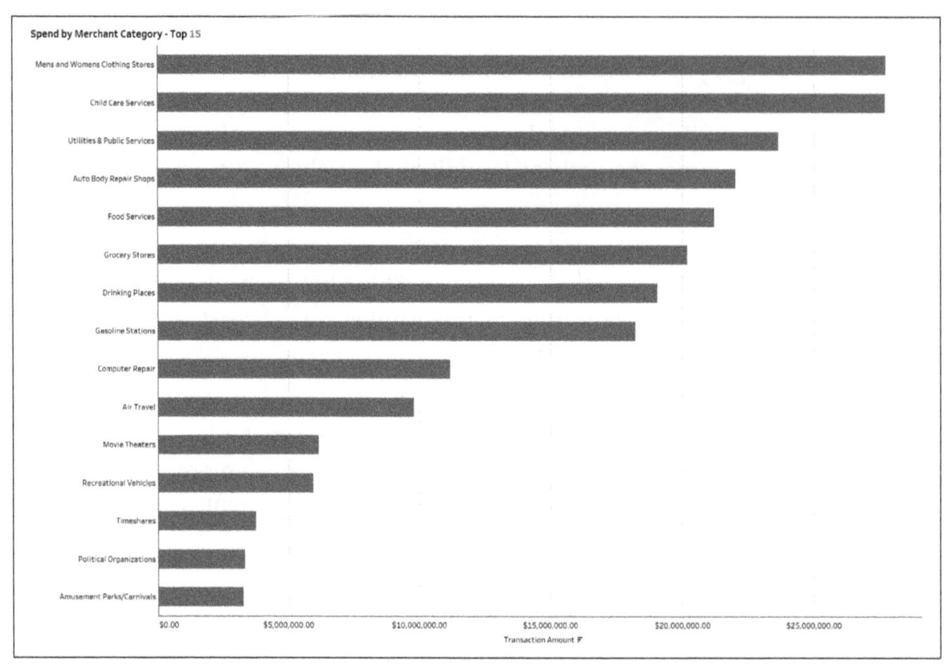

Figure 1-3. A bar chart with Top N filtering and a parameter applied

So how can we take this one step further? We're guessing that you've been wondering what percentage of the total each of these categories represents. Although comparing them is useful as we try to synthesize our thinking further, a natural inclination would be to change the commentary that "Men's and Women's Clothing Stores makes up $27 million in customer spending" to "Close to 4% of all customer spending is attributed to Men's and Women's Clothing Stores."

How do we approach presenting this information? Well first, we can change the measure from a direct measure to a percent of the total. But we're now left with the lingering notion that if we limit our chart to a Top 10, we'll lose the context of how much spending is in all the other categories.

To work around this constraint, we can allow the audience to utilize a parameter to define the proportion of customer spending that they want to see broken out into the merchant categories, and to automatically group together all other categories. They still have dynamic control over the chart, but are left with a full picture of the data, as shown in Figure 1-4.

Notice that you'll need to construct a calculation that is equivalent to the percentage-of-total calculation you made. To do this, you can utilize *level-of-detail (LOD) expressions*. An LOD expression lets you define the aggregation of the calculation, independent of the dimensions used in the visualization. This calculation takes the

SUM([Transaction Amount]) and divides it by the total SUM([Transaction Amount]) for the entire dataset.

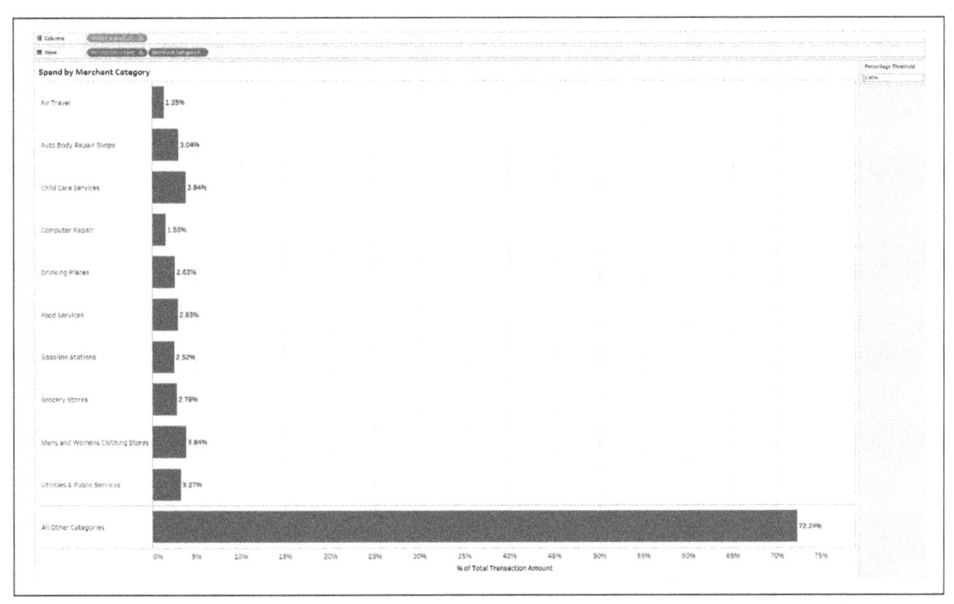

Figure 1-4. The updated bar chart, now with All Other Categories grouped at the bottom

Strategy: Dynamically Group Other Dimensions

To update our chart to dynamically group other dimensions and show the percent of the total, follow these steps, using the worksheet from the previous strategy:

1. Remove the [Merchant Category] filter from the Filters shelf by right-clicking and selecting "Remove" or by dragging it into the gray space beneath the Marks card.
2. Change the measure to a percentage of the total by using Quick Table Calculations. Right-click `SUM([Transaction Amount])` and choose Quick Table Calculations → Percentage of Total.
3. Create a parameter called **`[Percentage Threshold]`**. Set the data type to a float with 0.01 as the current value. Display the number format as a percentage.
4. Create a set based on the [Merchant Category] dimension. This will be a formula set based on the calculation that the percent of the total is greater than or equal to the parameter. Right-click [Merchant Category] in the Data pane and choose Create → Set.
5. In the Create Set dialog box, select "Use all" and then navigate to Condition tab and enter the following in the "By formula" text box:

```
SUM([Transaction Amount])/MAX({SUM([Transaction Amount])})
        >= [Percentage Threshold]
```

6. Create a calculated dimension called **[Merchant Category to Display]**:

```
//Merchant Category to Display
IF [Merchant Category Set] THEN [Merchant Category]
ELSE "All Other Categories" END
```

7. Drag the new [Merchant Category to Display] dimension on top of [Merchant Category] on the Rows shelf.
8. Drag [Merchant Category Set] to the left of [Merchant Category to Display]. This will organize the way the categories are listed. Right-click and hide the header.
9. Finish up the visualization by right-clicking and adding the [Percentage Threshold] parameter to the sheet. Also right-click and hide the field header for [Merchant Category to Display]. Add labels by clicking Label on the Marks card and selecting the "Show mark labels" check-box.

The updated analysis is much more flexible to the audience's preferences. Now, they have contextual information about the percentage of the total, and input to determine how much data is shown. This visualization is a step ahead of the bar chart with the sum of spending, because we are no longer sacrificing knowing the total distribution of data.

If you've reached this point and still want more, you can introduce additional items to add even more context and feedback. Similar to our original parameter for Top N, these additional techniques will provide feedback to the audience as the chart reacts to their input, and will help enhance their trust in the chart.

Strategy: Enhance Your Bar Chart with Color

One addition you can make to your bar chart is color. Follow these steps for this enhancement:

1. You can use the parameter as a reference line to reinforce the concept of dynamic entry. Right-click the [% of Total Transaction Amount] axis and choose Add Reference Line. Set the Scope to Entire Table, the Value to Percentage Threshold, and the Label to Value. Click OK.
2. Now adjust the Percentage Threshold to 0.75% (0.0075). Notice that additional categories display, but none are less than 0.75%.
3. You can also further encode the target large merchant categories by utilizing our set for color. Drag [Merchant Category Set] onto Color. Those merchants in the set will appear as one color, while those not in it and part of "All Other Categories" will be another color.

Figure 1-5 shows the result.

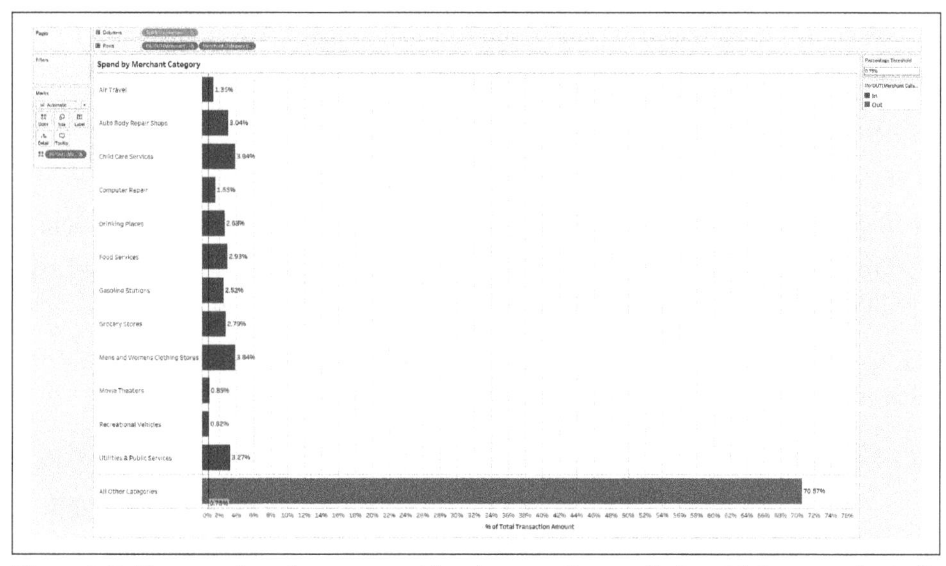

Figure 1-5. The same bar chart, now with color encoding to distinguish between the individual categories and the grouped category

Well done! We love bar charts—they are critical tools for any analysis. While they can start out very simple, you can take a bar chart from basic to amazing through abstracted metrics and dynamic entries.

Sometimes the value is in the way you format text on and around those bar charts. In this section, we discuss three more ways to spice up your bar charts with formatting that makes them pop. You'll continue using the Financial Institution Transactions dataset to build these.

Strategy: Left-align Text

Starting with a new worksheet, you'll use Network to create this formatted bar chart:

1. Add [Network] to the Rows shelf and `SUM([Transactions])` to the Columns shelf. Set the view to Entire View and use the sort icon on the axis to sort the networks in descending order by `SUM([Transactions])`.
2. Add [Network] and `SUM([Transactions])` to Label on the Marks card by holding down the Ctrl key and dragging each one from their respective Rows/Columns shelf.
3. Right-click [Network] on the Rows shelf and deselect Show Header.

4. To edit the label, click Label on the Marks card. In the dialog box that opens, click the ellipsis next to the Text option. Customize the label to read **`<Network> // <SUM(Transactions)>`** and to be left-aligned. We recommend setting the font size of the dimension to be about 1.5 times larger than the measure's font. You can do this by setting <Network> to 12 and <SUM(Transactions)> to 10.
5. Click Label again and adjust the horizontal alignment to be Left.
6. Right-click the axis and deselect Show Header to hide the axis for [Transactions].
7. From the toolbar, choose Format → Lines. Remove both the grid lines and zero lines by setting them to None. Set the rows' Axis Rulers to solid black.

The result is a visualization with labels that contain the dimension name and the value associated with the measure (Figure 1-6).

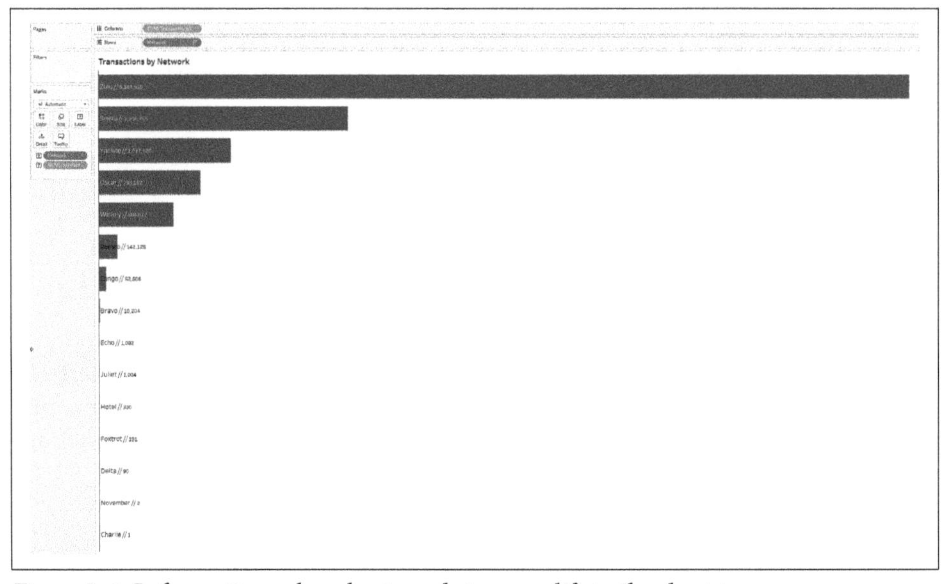

Figure 1-6. Reformatting a bar chart can bring new life to the chart type

Strategy: Create Bars with Labels on Top

This strategy also places labels and the values directly above the bars. The trick about creating this bar chart is that we are not going to use the Bar mark type at all:

1. Right-click the worksheet from the previous strategy and duplicate it.
2. Create a calculation called **`[Baseline]`**. This calculation will be used to establish the baseline location of the bar charts:

```
//Baseline
MIN(0.0)
```

3. Add [Baseline] to the Columns shelf. Right-click and make a dual axis. Synchronize the axis and then change the mark type to Gantt. Hide the axis by right-clicking and deselecting Show Header. You may have to change the mark type of `SUM([Transactions])` back to Bar. While you're doing that, also uncheck "Show mark labels."
4. On the Marks card for [Baseline], edit the width of the Gantt chart by clicking Size and adjusting the slider so the Gantt bars are as wide as possible. Change the color to white and set the opacity to zero.
5. Adjust the label alignment to be right-justified horizontally and at the top vertically.
6. Now you can adjust the size of the `SUM([Transactions])` bar to be smaller, giving the appearance of the labels from the Gantt mark sitting on top of the bar. Figure 1-7 shows the reformatted chart.

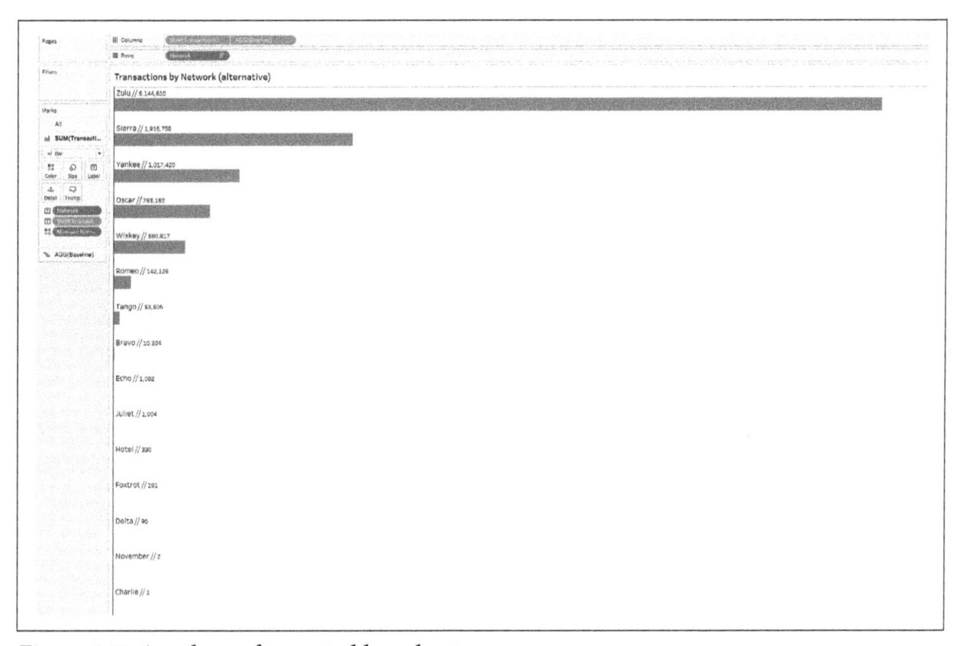

Figure 1-7. Another reformatted bar chart

Strategy: Create a Percent-of-Maximum Bar Chart

For this example, you'll build another bar chart (Figure 1-8), but this time the background of the bar will also be highlighted with color:

1. Create a calculation called **`[Total Bar]`**:

   ```
   // Total Bar
   MIN(1.0)
   ```

   We're going to use this calculation to represent 100% of the transaction amount within the visualization.

2. Add [Network] to the Rows shelf.
3. Add [Total Bar] to the Columns shelf. Fix the axis to start at 0 and stop at 1.1. You can do this by right-clicking the axis and selecting Edit Axis. In the dialog box, set the Range to Fixed and use 0 as the fixed start and 1.1 as the fixed end.
4. Set the color opacity to 40%. Drag `SUM([Transactions])` to Label.
5. Create a measure called **`[Percent of Maximum]`**:

   ```
   // Percent of Maximum Transactions
   SUM([Transactions])/WINDOW_MAX(SUM([Transactions]))
   ```

6. Add this new measure to the Columns shelf. Create a dual axis and synchronize the axes. You may have to change your mark types back to Bar. Remove [Measure Names] from Color on all Marks cards. Remove the label from the [Percent of Maximum Transactions] Marks card.
7. Sort the networks by clicking the axis and selecting descending order. Then hide both axis headers. Adjust the sizing of the bar charts to be at the center tick mark on Size.

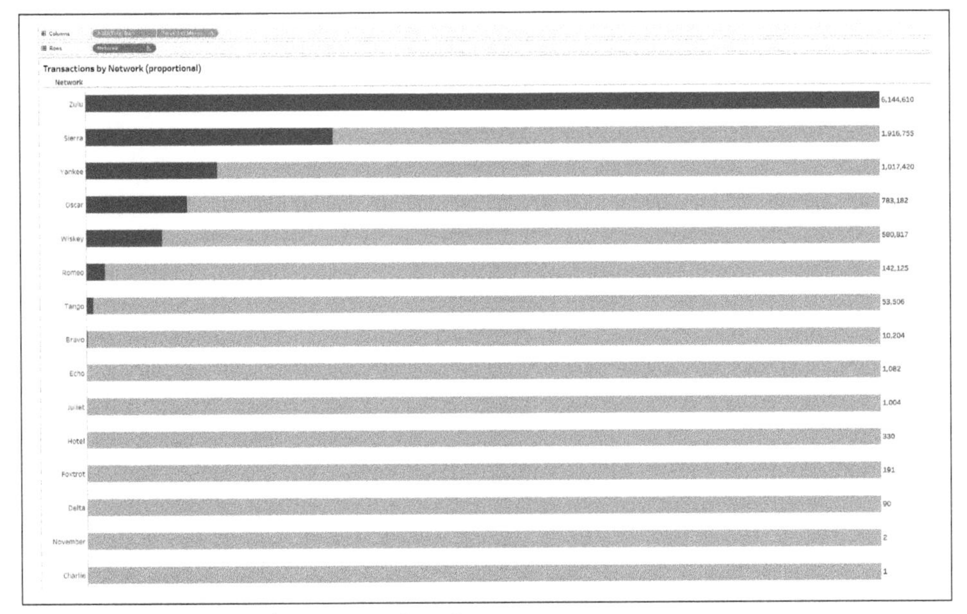

Figure 1-8. A percent-of-maximum bar chart

Bar-on-Bar Charts: Amplify Performance Case Study

Our next case study looks at a nonprofit organization, Amplify Performance (AP), that controls and awards grant money for creative, performing, and cultural arts programs and initiatives for the state of New York. Grant money is broken into two category types: one is related to the organization's budget and the other more directly categorizes the programs' initiatives. The AP data team tried to show both types with side-by-side bar charts, but the results were confusing. What kind of visualization would work better?

Not all categorical comparisons are going to require a simple bar chart. Sometimes the comparisons are more complex. For example, you might have to compare groups on a single metric but across two different time periods. Your audience will want to understand changes across members, but also how individual groups have changed over time.

Novice developers' first instinct in these situations is often to use a side-by-side bar chart. These can be effective, but they take up a lot of space. When doing this type of analysis, we prefer to use a bar-on-bar chart instead. In Figure 1-9, you can see the total grant sizes for 2018 and 2019 for each category in a side-by-side bar chart.

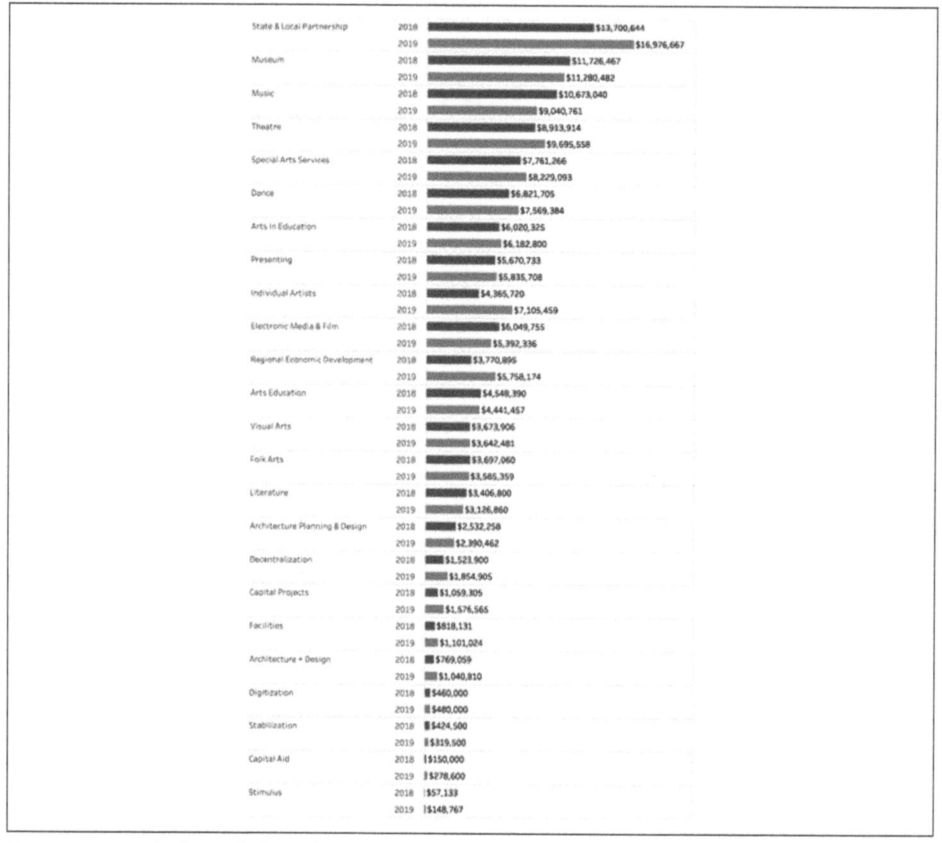

Figure 1-9. Side-by-side bar charts are effective but take up too much space

It looks like you have all the information you really want: there are bars for two years, and you can quickly compare. But you'll notice that the data is not sorted on the total for 2019; rather, it is sorted on the total *across* the two years.

Additionally, any comparisons we make within a group (for instance, Arts Education) requires the audience to do mental math to understand the magnitude of the change from 2018 to 2019. It would be great to have that information directly in the visualization. It also might be helpful if the audience could quickly note which categories increased from the prior year and which decreased.

You can do all of this using a bar-on bar chart, as shown in Figure 1-10. Follow along to create this using the Nonprofit Grant Data dataset.

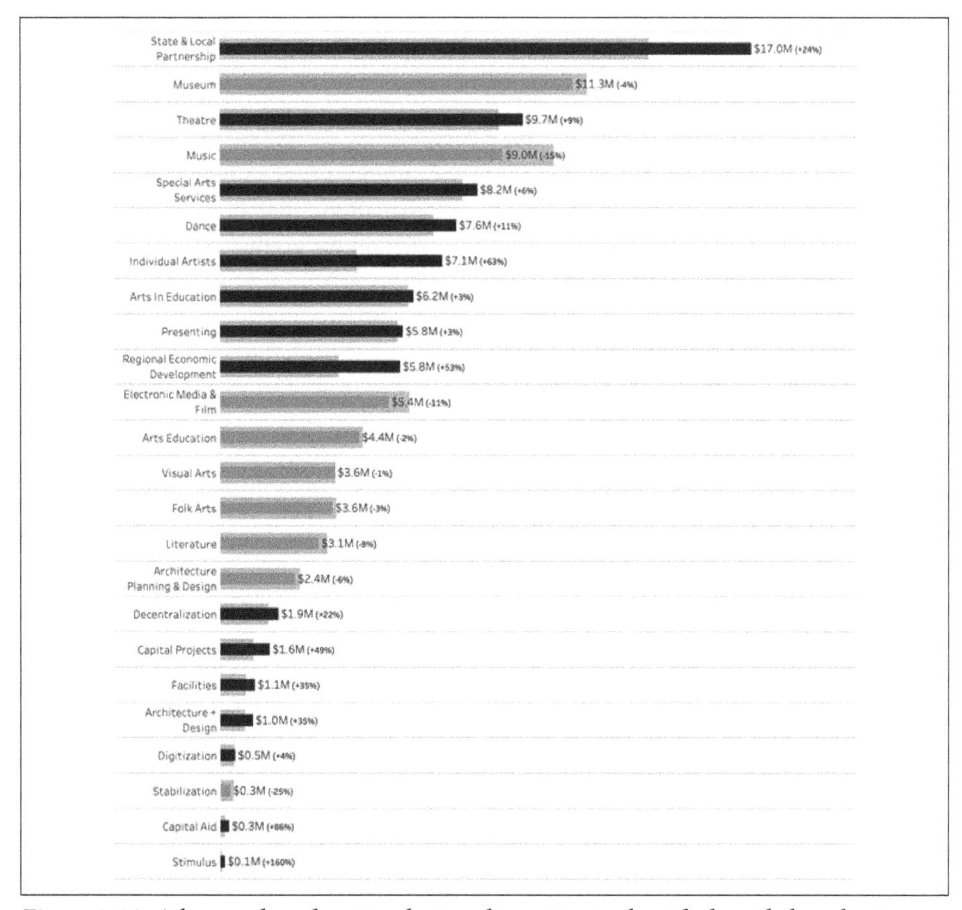

Figure 1-10. A bar-on-bar chart is a better alternative to the side-by-side bar chart because it takes up less space while displaying the same information

Strategy: Create a Bar-on-Bar Chart

In Figure 1-10, we've placed 2019 grant totals over the 2018 grant totals for each category. Here, your audience can still quickly compare groups across the base year, in this case 2019, as well as how that group performed versus the previous year.

We aided in the year-to-year comparison by adding color—we didn't choose two distinct colors, but two colors with the same hue. Totals that decreased from the prior year are represented with a brighter, less saturated version of the color used to indicate totals that increased from the prior year.

Finally, we added the year-over-year change as a percentage next to the total for 2019. The result in Figure 1-10 is a chart that consolidates three comparisons: total grant dollars across groups for 2019, changes in total grant dollars from 2018 to 2019 for each group, and changes in magnitude from 2018 to 2019.

So how do you create this chart? Follow these steps:

1. Create your measures. Instead of using a date dimension to partition your data, it's more effective to create two separate calculations that filter to the relevant data inside the calculation. Let's create a calculation for grant amounts in 2018:

   ```
   // Grant Amount | 2018
   SUM(
    IF YEAR([Date]) = 2018
    THEN [Grant Amount]
    END
   )
   ```

   And a calculation for grant amounts in 2019:

   ```
   // Grant Amount | 2019
   SUM(
    IF YEAR([Date]) = 2019
    THEN [Grant Amount]
    END
   )
   ```

   Generally speaking, avoid hardcoding anything inside calculations. In this case, we'd normally use calculations or parameters to automate change as the data is updated. (We'll discuss this more in Chapter 4.)

2. Create the base visualization by adding [Budget Category] to Rows and both [Grant Amount | 2018] and [Grant amount | 2019] to Rows.

   a. Create a synchronized dual-axis chart with both mark types as Bars. Be sure to place 2018 as the leftmost dimension in the dual axis.

      Change the 2019 bar size to be narrower than the 2018 bars. You might have to adjust both to get your bars in a happy place.

   b. Be sure to remove [Measure Names] from both Marks cards. (You didn't add this; Tableau did this automatically when you created a dual-axis chart.)

   c. Set the color on the outer bar to a light gray that is still distinguishable from the background, as shown in Figure 1-11.

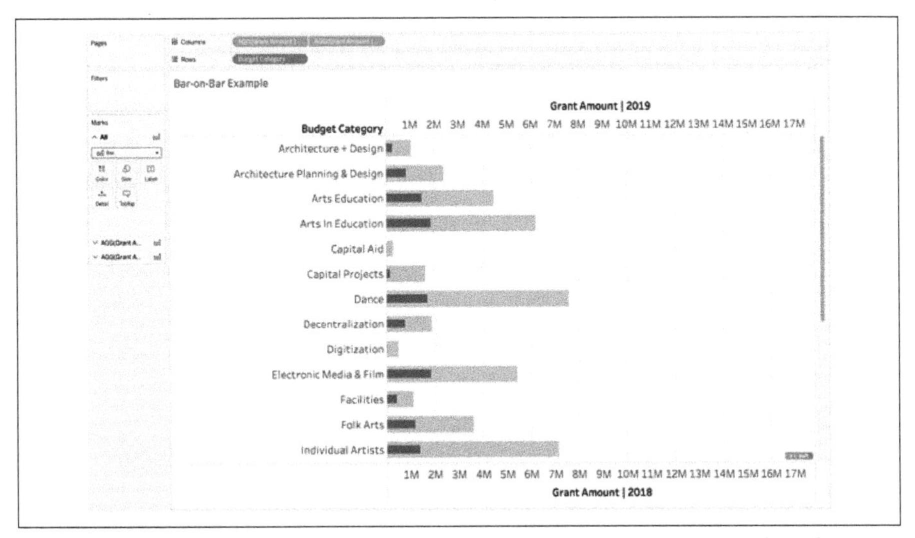

Figure 1-11. Use a dual axis with custom calculations to create bar-on-bar charts

We prefer to make the width of our outer bars (in Figure 1-11, 2018) equal to the width of whitespace between bars. For our inner bars, we look to have the width between 50% and 75% of the outer bars' width.

3. We could rely on the axes for comparisons, but because we are using a horizontal bar chart, it makes sense to add labels:

   a. On the [Grant Amount | 2019] Marks card, click and drag [Grant Amount | 2019] to Labels.

   b. Create a new calculation called **`[Grant Amount | % Change]`** for the percent change from 2018 to 2019:

   ```
   // Grant Amount | % Change
   ([Grant Amount | 2019] - [Grant Amount | 2018]) / [Grant Amount | 2018]
   ```

   c. After you create the measure, right-click it and change the default settings of the number format, as shown in Figure 1-12.

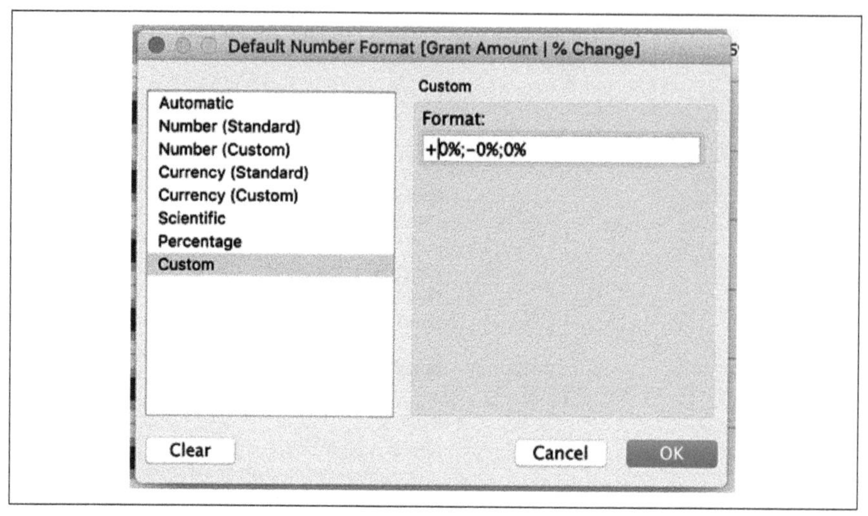

Figure 1-12. Use custom formatting to get your percentage displays just right

This will display a plus sign in front of the positive values, a minus sign in front of the negative values, and no sign when there is no change in the direction.

d. Add this calculation to Label as well. Now edit the text of the label. Format [Grant Amount | 2019] to be both larger and a darker shade than the [Grant Amount | % Change] measure. Your chart should look like Figure 1-13.

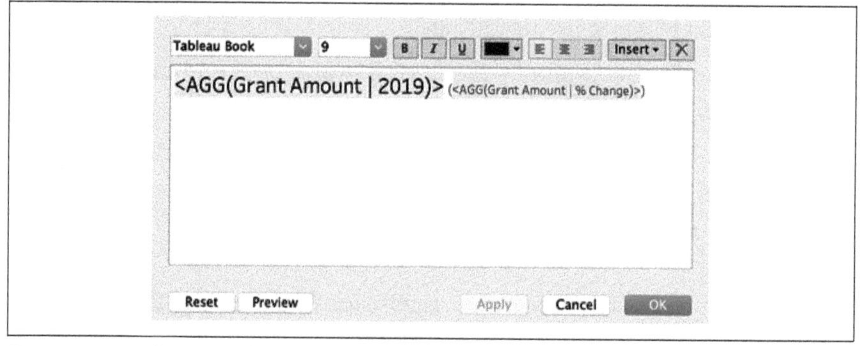

Figure 1-13. The text editor showing how you should format the text labels on the visualization

For this example, we're using size 15 and size 9 fonts, respectively. Text colors are #000000 (black) and #555555 (dark gray), respectively. Additionally, we've added [Grant Amount | % Change] between parentheses. Your chart should look like Figure 1-14.

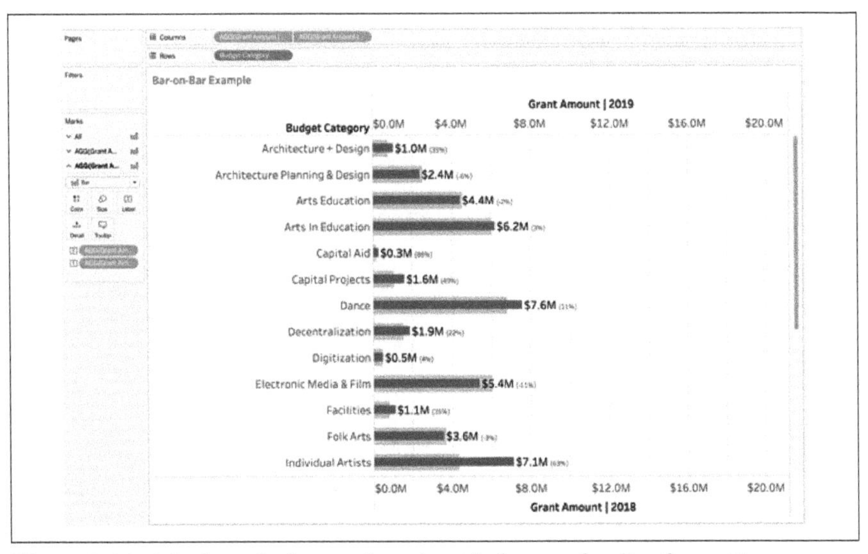

Figure 1-14. A look at the bar-on-bar chart before we finalize formatting

4. To add color, create a simple Boolean called **`[Color]`** that compares 2019 to 2018 and then add it to the [Grant Amount | 2019] Marks card:

   ```
   // Color

   [Grant Amount | 2019] > [Grant Amount | 2018]
   ```

   You can edit the color and select two colors that start with the same color: for instance, the base hex color #19626B for values that are True and a second color, #84B6BC, that is brighter and less saturated.

5. Add finishing touches:

   a. Sort your categories by total grant amounts in 2019.

   b. Hide your axes and row header labels. Remove all extra lines.

   c. Remove your vertical divider. Keep your horizontal divider, but make sure it separates each member.

   Your result should look like Figure 1-15.

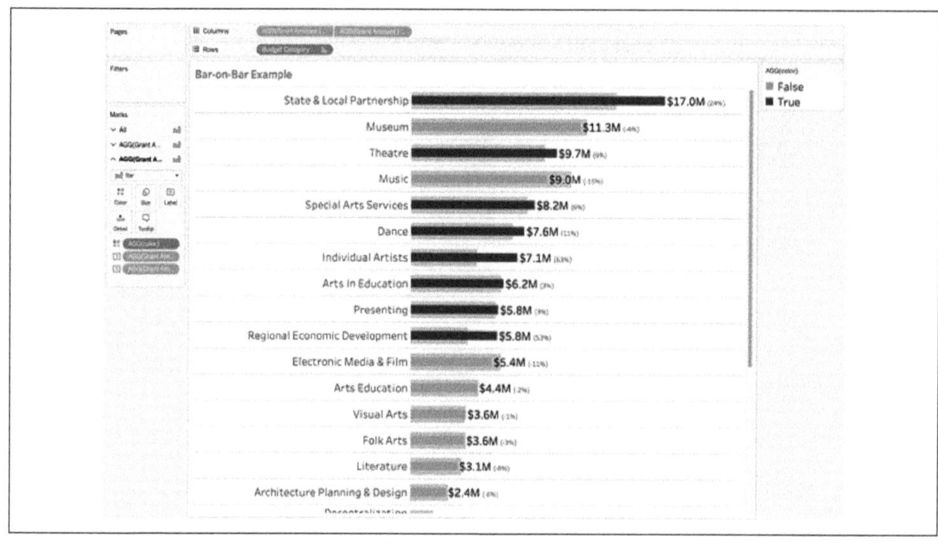

Figure 1-15. The bar-on-bar chart after adding color

Treemaps: Amplify Performance Case Study

A *treemap* is similar to a bar chart, but uses the area of a rectangle relative to the height or length to encode data. If you have many members of a dimension and must show all members, a treemap is a great alternative to a bar chart.

Area is a less precise measure, but often when working with treemaps, the goal is not to be completely precise but rather to display all the members of a category in a single, compact visualization that is sorted from largest to smallest.

One of the main benefits we'll explore with treemaps is being able to use color to represent a measure or a dimension. In the example with a drillable treemap, we'll use color to represent both budget categories and the program categories within them. We'll then show how to display additional detailed members to the audience on demand. With this feature, the audience is free to explore multiple facets without being overwhelmed. We'll also represent color as a measure, both directly and indirectly.

Next, we'll be focusing on what to do when you need to show all members of a dimension in a single visualization. You've already seen the problem of a long scroll when using a bar chart, so what chart type can you employ to get around this barrier?

Imagine you are working with the nonprofit organization introduced previously, AP, that controls and awards grant money for creative, performing, and cultural arts programs and initiatives for the state of New York. Grant money is broken into two category types: one focused on the organization's budget and the other on the programs' initiatives.

In this scenario, you can't sacrifice small members. Having visibility into some of the smaller categories is crucial, in order to provide insight into where additional grant money should go.

If you're facing a similar scenario, we recommend a treemap. While you'll lose some precision in comparing your chosen measure, you will get a well-ordered and compact visualization that will show all the members of your dimension.

Strategy: Create a Basic Treemap

Create a new worksheet using the Nonprofit Grant dataset and then follow these steps:

1. Drag [Budget Category] to Text.
2. Drag `SUM([Grant Amount])` to Size, and Ctrl-drag this to Color as well.
3. Ensure that the mark type is set to Square.

   The result is shown in Figure 1-16.

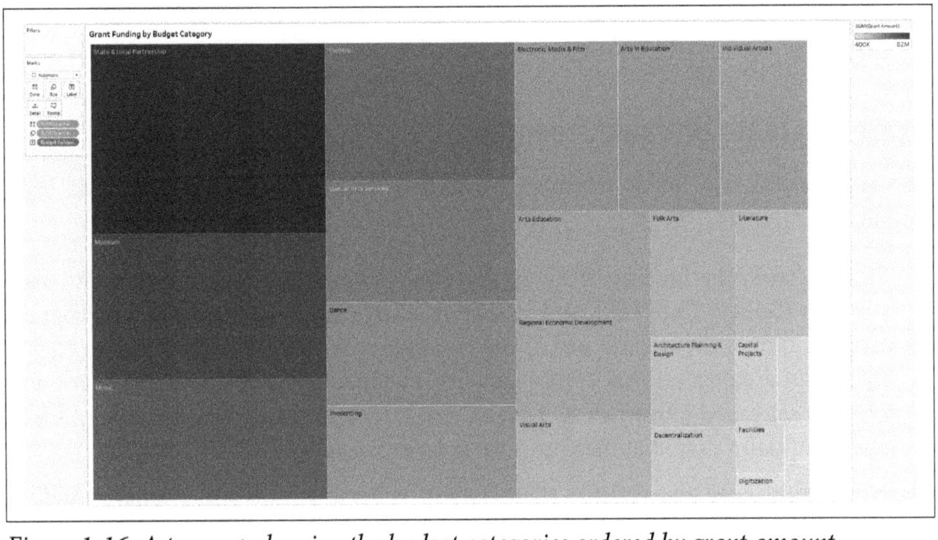

Figure 1-16. A treemap showing the budget categories ordered by grant amount

This will give you a full picture of how AP's grants are distributed. No single budget category has a significant majority of funding, but some smaller categories take up less than 1%.

The standard convention when working with treemaps is to double-encode a measure by using size and color. This helps to further distinguish the pieces and members. But it is not a requirement. As an alternative, you could consider encoding the categories on color. However, we recommend caution: there are many members and, in this case, you would be encoding redundant information.

This treemap is a great start, but takes us only halfway to providing specifics about the grant data. Dollars are not only divided among budgets, but also assigned to program categories. There are 55 unique program categories, a significant additional level of detail that could be overwhelming. We really need to know only which program categories that budget dollars are tied to within one given budget category at a time.

To solve the next-level question of program categories within budgets, the AP team can create a *drillable treemap*: an interactive treemap in which the audience can click a specific budget category to see further information.

In the following strategy, you're going to use set actions. Sets and set actions allow your end users to interact with visualizations to assign dimensions to a set. Here, once a dimension is part of a set, more detail will show in the drilled section.

Strategy: Create Drillable Treemaps

Let's try creating a drillable treemap, continuing to use the Nonprofit Grant Data dataset:

1. Starting from the finished treemap from the previous strategy, create a set based on both [Budget Category] and [Program Category] by first dragging [Program Category] onto the Marks card of the treemap view. This will allow you to click a mark and create a set that combines both dimensions. Right-click any mark and create a set called **`[Program & Budget Set]`**. It doesn't matter what values are in the set initially, only that there are two columns, one for each dimension.
2. Now create a calculated field called **`[Label Program]`**. This will evaluate whether something is part of the set and return the program if it is:

```
//Label Program
IF [Program & Budget Set] THEN [Program Category] END
```

3. Drag this calculated field on top of [Program Category] on the Marks card.
4. Create the drill-down functionality. Choose Worksheet → Actions → Add Action → Change Set Values.

5. Call it **`Drill Down to Program`**. It will be run on Select. The Target Set is [Program & Budget Set]. The action you want when clearing the selection is "Remove all values from set."
6. Drag [Budget Category] onto Color, replacing `SUM([Grant Amount])`.
7. Now click Dance, and the treemap rectangle will drill in to show all the programs that comprise the Dance budget.

Notice in Figure 1-17 that you've changed how color is leveraged here. Instead of tying color to the repetition of the grant spending, you are using it to distinguish the budget categories.

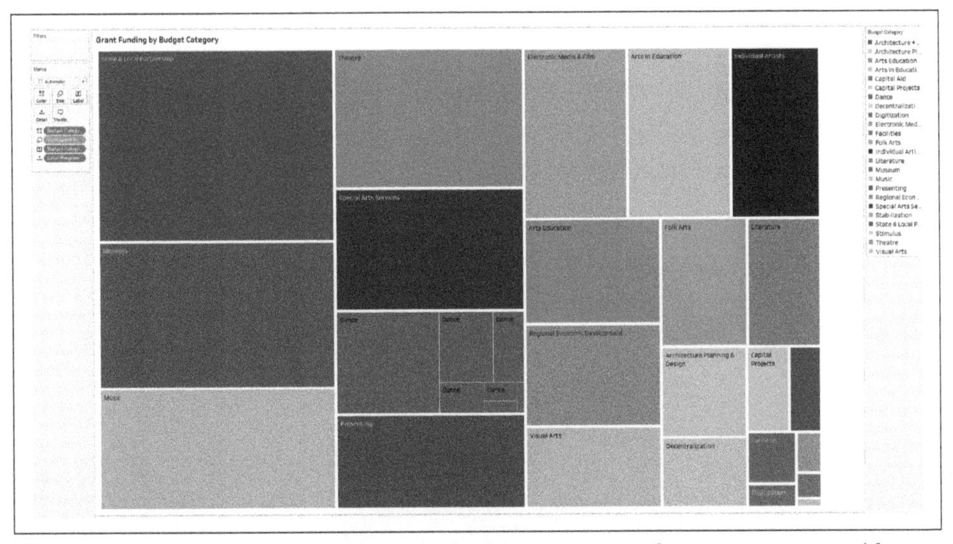

Figure 1-17. A treemap showing both the budget category and program category (the treemap is colored by budget category, and the size of the rectangle represents the grant amount)

Let's go back to our treemap example one more time, and take color encoding in one more direction. In this scenario, you're going to start with program categories to create the treemap. This time, the AP team is trying to ensure not only that different program types are getting a sufficient distribution of funding, but also that diversity exists in the types of programs that are funded and supported.

Strategy: Encode a Continuous Measure with Color

For this visualization, you'll use color encoding to spot opportunities for programs to be revitalized. You'll also use it to highlight another continuous measure: days since the most recent grant was funded.

A *continuous measure* is one that spans an infinite range, typically on a number line or timeline:

1. On a new worksheet, create a treemap of [Program Category] and [Grant Amount].
2. Create a calculated field that evaluates how long it has been since a grant was awarded in a category:

    ```
    //Days Since Last Grant
    DATEDIFF( 'day',MAX([Date]),TODAY())
    ```

3. When you use this calculation in the view, it will evaluate the maximum or most recent date per [Program Category] and then calculate the number of days since today.
4. Put this measure on Color and change the Palette to Blue-Green Sequential reversed.

    You can see the result in Figure 1-18.

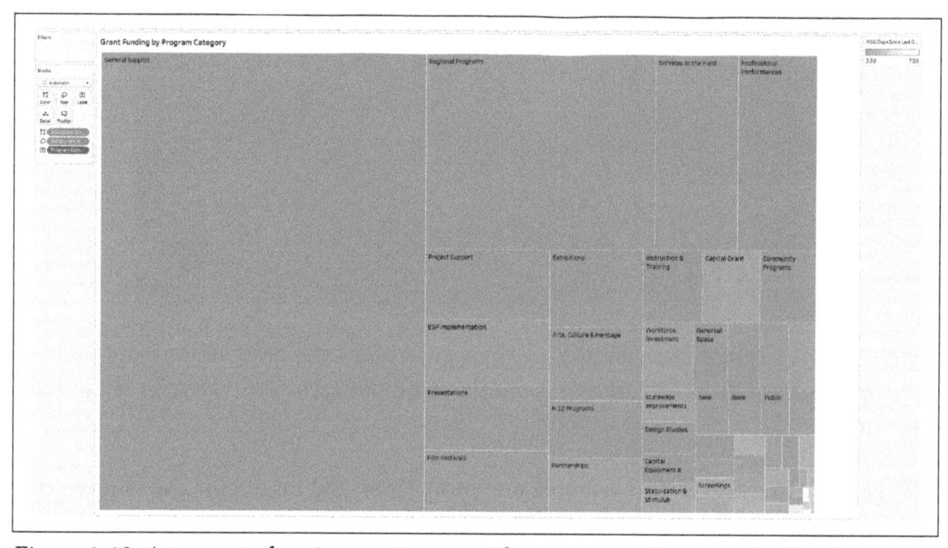

Figure 1-18. A treemap showing grant amount by program category; color has been encoded to show the number of days since the program was last funded

You've now seen three approaches to treemaps, utilizing different color-encoding techniques and dynamic elements that allow the audience to dig into a category and explore it in even more depth.

Pie and Donut Charts: IT Employee Wellness Project Case Study

Our last case study in this chapter involves the IT Employee Wellness Project, an initiative that conducts surveys of IT professionals and their employers. The project investigates how IT work affects employees' mental health, as well as employers' attitudes about mental health.

As with most surveys, there are many questions to analyze to determine attitudes, but a core task is to provide a demographic overview of respondents. The project's data analysts' goal is to show gender distribution among the survey respondents by profession. They would also like to compare that distribution across several professions at the same time. What chart types would help them convey this most effectively to the project's board members?

When you're looking at a parts-to-whole relationship, a natural place to start is a pie chart. A *pie chart* divides a circle into slices by members of a dimension, and each piece represents a proportion of the whole. Some chart lovers cringe at the thought of a pie chart, but pie charts are familiar to most people. They also use space efficiently and can serve as color legends or interactive filters. They're not the best choice in every situation, but they do have their place when used correctly.

In our final strategy for the chapter, you'll take your donut charts one step further by making *small multiples*, or repeated versions of the same chart, separated out by profession. This will let the data team compare gender distribution among several professions at the same time.

Strategy: Build a Basic Pie Chart

Let's start with the basic pie chart. You'll use the IT Survey Data dataset to follow along:

1. Drag [Gender] onto Color.
2. Change the mark type to Pie.
3. Create a calculated field called **`[# Respondents]`**.

   ```
   //# Respondents
   COUNTD([Respondent ID])
   ```

4. Drag [# Respondents] onto Angle. Change the colors to your choosing (we're using the Summer palette and a white border). Drag [Gender] onto Label along with [# Respondents]. Format the Label to match the mark colors.

   Figure 1-19 shows the result.

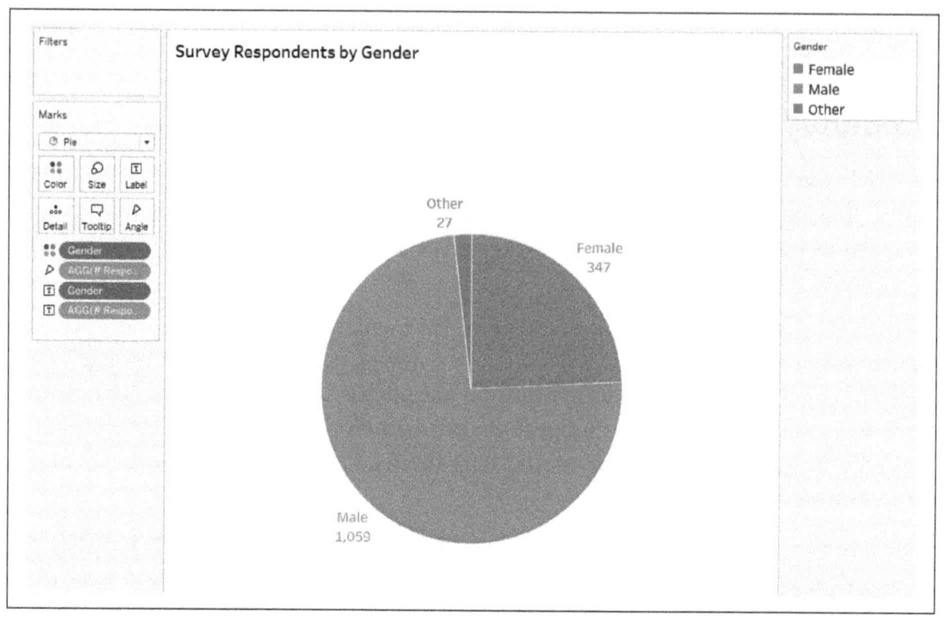

Figure 1-19. A pie chart showing the distribution of gender for survey respondents

Why does this pie chart work? First, it has only three slices; this pie doesn't have many pieces, so it's pretty easy to compare the differences. Next, you've gone the extra mile to directly label and put the percentage next to each slice. This makes it easy for the audience to comprehend. If you had many more values, or several slices of relatively the same size, our recommendation would be a bar chart, but here that's clearly not the case.

Strategy: Build a Donut Chart

Now let's turn this pie chart into a donut chart by adding a hole to its middle. This hole allows you to communicate an additional piece of information. In this case, it allows you to represent two concepts in a single chart—the number of respondents and the distribution of gender:

1. Create a dummy measure **`[MIN(1)]`**; this will be used as multiple measures for a dual-axis chart:

   ```
   //Dummy
   MIN(1)
   ```

2. Drag it onto Rows twice, right-click, and make a dual-axis chart. You can use the visualization from the previous strategy as a starting point.
3. Make the size of the first measure the right tick on recommended size.

4. Right-click [# Respondents] and change it to a Percent of Total Quick Table Calculation. Then format the percentage to have no decimals by right-clicking the field and selecting Format.
5. Click the second measure and remove all fields, except [# Respondents], which should be on Label. Add the text **Respondents** in 8 pt. font beneath the field.
6. Make the size of the second measure the left tick on recommended size.
7. Align the label middle and center, and set the color of the mark to white.
8. Hide the axes and remove all lines from the chart.

   You can see the result in Figure 1-20.

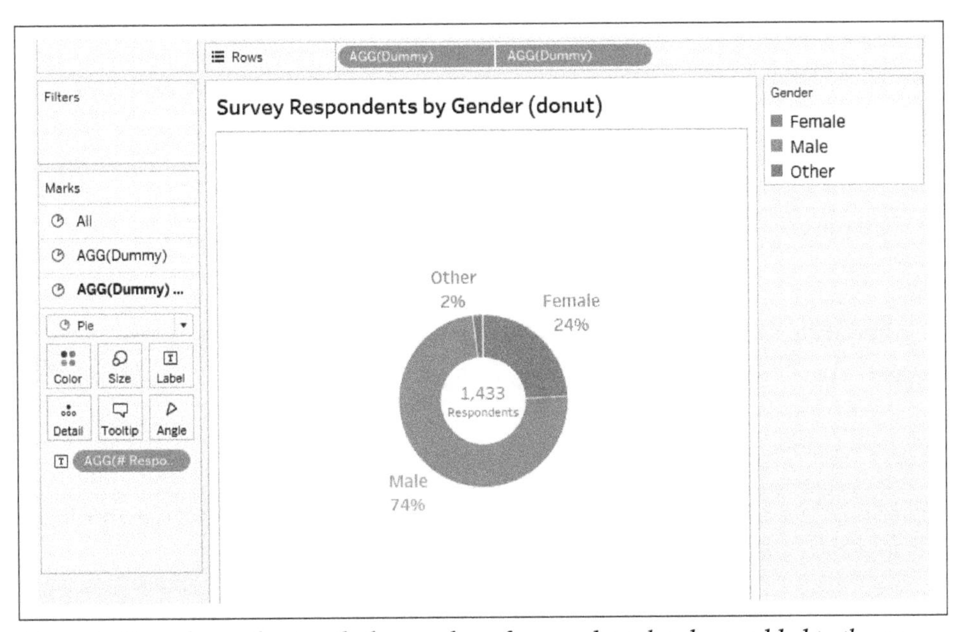

Figure 1-20. A donut chart with the number of respondents has been added to the center; the slices represent the distribution of survey respondents by gender

Now you have a donut chart that serves two purposes and is fantastic at generating insightful descriptions. You can say "24% of respondents were female" and immediately know the survey's sample size, which will be necessary for the audience to make decisions from the results.

To complete the donut chart, you could utilize it as an interactive filter; when a user clicks a slice, Tableau filters in subsequent visualizations. It could also serve the purpose of a color legend—one that does more than just say that green means male.

Strategy: Create Small Multiples

Since we're working with dessert charts, there's one last visualization we'd like to introduce: the *small multiple,* which is any chart repeated multiple times in a smaller format. With pie and donut charts, small multiples become pretty powerful. You can take a dimension with more members and use that to create repetitive charts for comparisons.

We know the project board wants to see how gender distribution changes among professions. Small multiples can help you separate out the answer quickly. Yes, you could create a filter to select the role, but you'll get more insight at one glance when you create small multiples that can show the distribution and still provide the sample size in context.

Starting with the donut chart you just made, you need to make only a few tweaks:

1. Remove the word "Respondents" from the label to save space.
2. Drag [Professional Role] onto the Columns shelf.

   You can see the result in Figure 1-21.

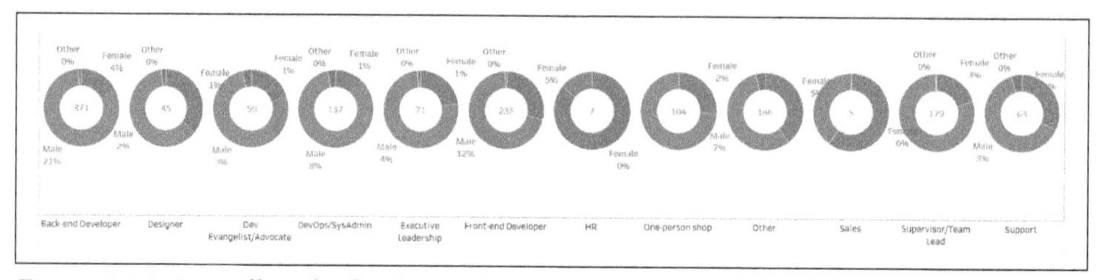

Figure 1-21. A small-multiples donut chart, which includes separating out the distribution by gender among different professions

Now that you've seen some examples of pie charts, we hope you'll recognize when it is appropriate to utilize them—and more importantly, when utilizing them can enhance the data presentation for your audience.

Conclusion

You've now had a chance to see different chart types used for categorical analysis. You started with a basic horizontal bar chart and quickly learned how to create a Top N chart to limit the data for your audience to the most relevant information.

From there, you explored dynamically grouping small categories into an Other bar and allowed your audience to define the scope of analysis with a parameter, allowing them to define the percentage contribution a category must have to be shown in the chart.

You also explored using color encoding with bar charts to further accentuate categorical analysis and learn some of our favorite formatting techniques to make bar charts pop.

Next, we moved onto our case study for Amplify Performance and used a treemap to ensure that all members of a dimension (even really small ones!) were represented in a visualization. You learned how to make this chart type even more dynamic by allowing drill-down to further split the data. And you also used color encoding and dipped a toe into date calculations to show AP executives which areas may not be getting funding (if this was fun, wait until Chapter 4).

Finally, we wrapped up the chapter by exploring dessert charts, first the pie and then the donut chart. You learned when to effectively use a pie chart (not too many slices) and how to take advantage of the hole inside a donut chart to display two pieces of information at one time.

With all of these new techniques, we are confident you will be able to create flexible, compelling, and insightful visualizations that let your audience explore questions and analyses dynamically.

In the next chapter, we'll introduce you to quantitative analysis—all charts focused on different ways to plot numerical fields and to use statistics. You'll be taking the foundations you learned through creating charts and putting them into practice with a variety of chart types (bye bye, bar chart).

CHAPTER 2

Quantitative Analysis

This chapter is about crafting visualizations that show distribution, spread, and plenty of nuance. If you've ever worked with data, you know it can be a challenge to communicate detailed, complex information in clear and accessible ways. We'll offer techniques to help you overcome this challenge.

The visualizations in this chapter focus on measures and statistics. Using numeric fields, we'll show you how different charts can provide different insights through numerical representations of data values. *Measures* are numeric fields to which we can apply functions, like `SUM()`. *Statistics* is, of course, a branch of mathematics; in Tableau, we use its concepts in specialized functions, working with averages, medians, means, standard deviations (SDs), and more. (Don't worry if you never took statistics in school—we'll walk you through it!) These functions don't necessarily produce a summative value, like total sales; instead, they usually describe how the data points are distributed: their highs, their lows, where data points are concentrated, and where the outliers are.

Every dataset has a unique data shape. Most have some common features once you explore them. Some data shapes can be hard to work with, so we'll also explore techniques to work with those.

In this chapter, we'll explore the following types of visualizations:

- Histogram
- Dot plot
- Jitterplot
- Ranged dot plot
- Box plot
- Line chart
- Pareto chart

Because we know you're using these visualizations to do your job, we'll keep using case studies to help you see how these chart types function in real-world business communications that involve real people (like your boss). Through our advice to analysts for an office-supply retailer, a delivery and logistics company, and a large bank, we'll give you a general overview of each type, including what it looks like, its advantages and disadvantages, and the kinds of information it's best for. We'll also spend time working with some core concepts from statistics that you can use to bring your data insights to the next level.

> **In This Chapter**
>
> For this chapter, you'll need to download the Sample – Superstore dataset and the Logistics dataset from the code repository (see the Preface for details). In this chapter, you'll learn how to do the following:
>
> - Build a histogram to show customer purchase frequency
> - Make different variations of dot plots—including the jitterplot, which is useful when you have many data points
> - Construct charts that summarize and display statistics for your audience

Histograms: Office Essentials Case Study

Our first case study involves a large retail chain that sells office supplies. We'll call it *Office Essentials* (*OE*). The data team is getting ready to make a big presentation to the OE C-suite and board. The team has been asked to do an analysis of customers' purchasing behaviors. The board wants to know who is placing orders, how often, and how many.

What metrics should the team members be trying to tease out of this data? And what's the best way for them to present what they find, knowing that they might have only a few seconds to communicate the most important insights from the data? The team members know that they need to be thinking about order frequency, for one thing. How many orders do OE's customers place? Are most of them frequent return shoppers, or do they purchase only once?

To answer this question, the best chart to turn to is the histogram. Before we dive into making one, let's give you some background on this popular chart type.

A *histogram* is a specialized bar chart used to visualize *distributions*, or spreads, of data points that have a numerical value. In this case, how are OE's sales distributed across its customers? You might decide to look at the number of customers who spent $100 to $200 last year at one of OE's retail stores, or the number of times a certain customer purchases a specific product.

Unlike a standard bar chart, the y-axis of a histogram (that's the vertical one) has a numerical value, range, or interval (often referred to as a *bin*), represented as a header, and its x-axis (the horizontal one) represents a frequency, like the number of orders per customer per year. A histogram is a bar chart with the frequency on the x-axis and the number of customers on the y-axis.

Strategy: Make a Simple Histogram of Purchasing Behavior

For this type of chart, we will utilize a LOD expression instead of bins to construct histograms. As noted in Chapter 1, Tableau's LOD expressions allow you to define the *aggregation* of a measure (such as sum, average, or median). By default, Tableau bases the aggregation of any field on the detail or fields in the view or worksheet, but LODs let you create customized aggregated fields. Let's try it out:

1. Create a field called **`[# Orders per Customer]`**. This tells Tableau to count the distinct number of orders by each customer name:

   ```
   //Orders per Customer
   {FIXED [Customer Name]: COUNTD([Order ID])}
   ```

2. Right-click the measure and choose Create → Bins. A dialog box opens.
3. Tableau will automatically suggest a bin size after reviewing the various values for your measure. You are free to adjust these to suit your needs. For this analysis, it doesn't make sense for the number of orders per customer to be a decimal, so change the bin size to 1.
4. Once you create the bin, it shows up as a discrete dimension in the Data pane. This is an important point, because it will give you insight into how the field will behave when we use it in a chart.

5. Now, drag [Orders per Customer (bin)] onto Columns. As expected, Tableau has created discrete headers.
6. Now, drag [# of Customers], which is the COUNTD([Customer Name]) measure, onto Rows:

```
//# Customers
COUNTD([Customer Name])
```

You should end up with something like Figure 2-1. Let's stop and digest this visualization.

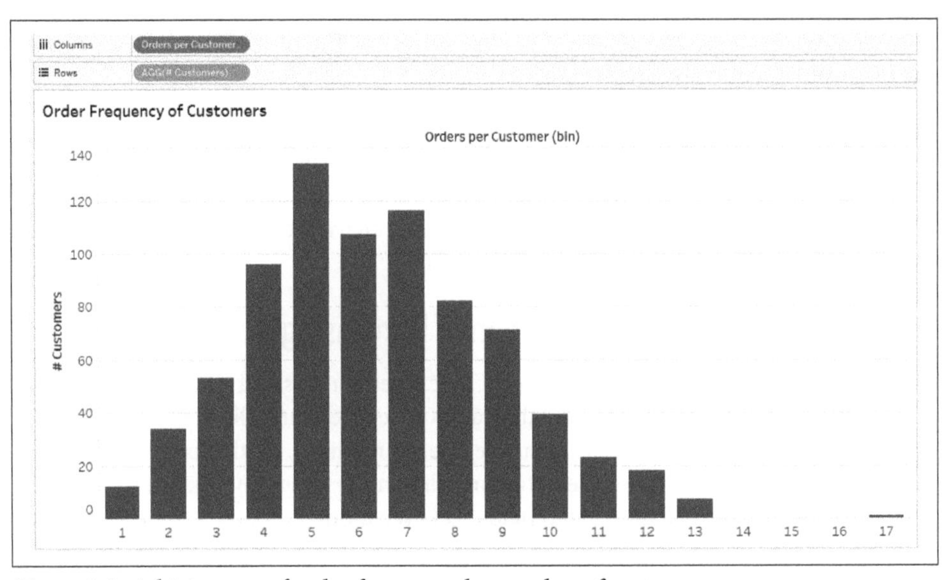

Figure 2-1. A histogram of order frequency by number of customers

Even if you haven't taken a statistics class, you're probably familiar with one of the most common visuals in statistics: the normal distribution, also widely known as a *bell curve* (Figure 2-2). Whenever you're looking at a histogram, the normal distribution should come to mind.

While the histogram you created just now is *not* a normal distribution, comparing it to the normal distribution will prove useful in trying to explain the shape of the data. In a *normal distribution*, data is represented such that the top of the curve represents the mean, or average. The curve is symmetrical: exactly half of the data is to the left, and half is to the right of the mean.

When you are working with numbers, *standard deviation* (*SD*) refers to how the numbers in your dataset spread out from "normal," or the average. In practice, the larger your SD, the more variation exists among your data points; conversely, the smaller your SD, the smaller the difference, or data spread. In a normal distribution, a majority of the data is within one SD of the mean; 95% of the data is two SDs from the average.

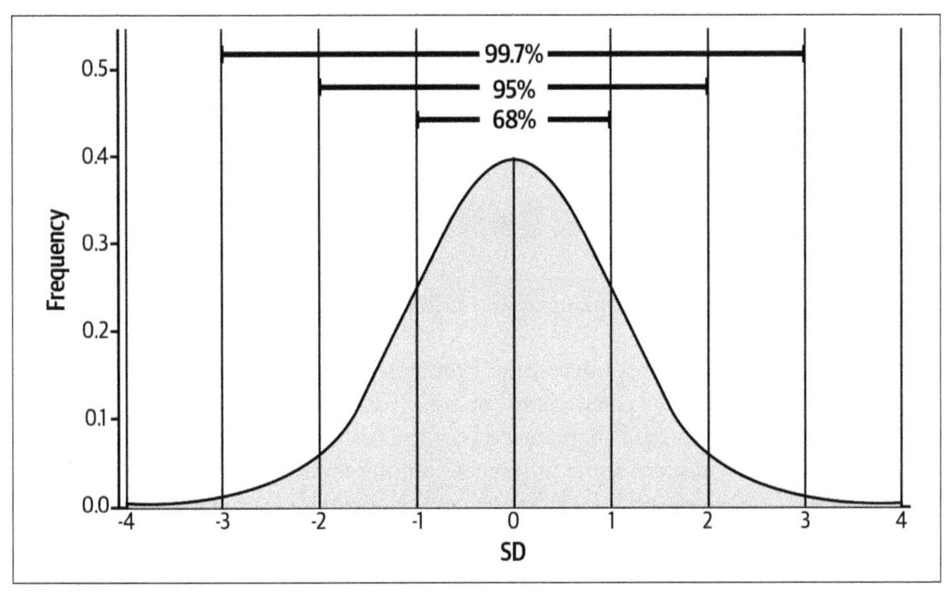

Figure 2-2. A normal distribution with percentages for each standard deviation

You can see from the chart you made (Figure 2-1) that the order frequency is definitely centered on a range of four to eight orders, and that a few customers order much more frequently (we'll call them outliers).

We can create even more immediate value out of the histogram by cutting it into small multiples and comparing the distribution, or spread, of the data among a dimension such as [Segment], as shown in Figure 2-3. We'll go into this chart type more in Chapter 3, but for now, just know that a small multiples chart (also known as a *trellis chart*) is actually a series of charts that use the exact same axes, most often ordered in a grid.

Try dragging [Segment] onto Rows and also onto Color.

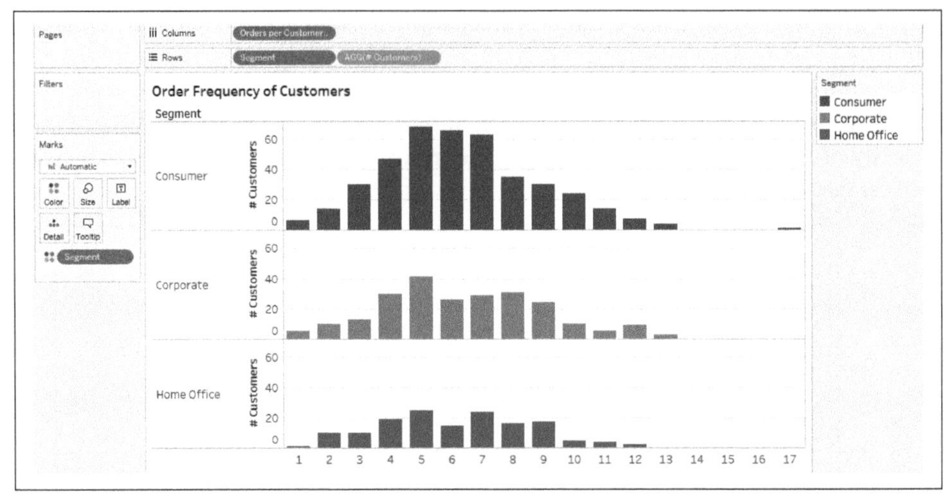

Figure 2-3. Histograms of order frequency by customer separated by segment

Now here is where things get interesting. Previously, we saw the overall distribution of order frequency among customers, but now we've separated it out by business unit/segment. While the data shape of each of the histograms is *similar* to the original, we can definitely discern some differences among them:

- The overall number of customers purchasing from the Consumer segment dominates the other two segments.
- Those outlier customers exist only within the Consumer segment.
- The order frequency for Home Office looks relatively stable, generally ranging from 4 to 9; the amounts are much closer together for this segment than for others.

For more precision, try changing the y-axis to be independent for each row, as shown in Figure 2-4. When you make this change, though, do so with caution! It should be very clear what the bar lengths represent and that they are unique for each of the segments.

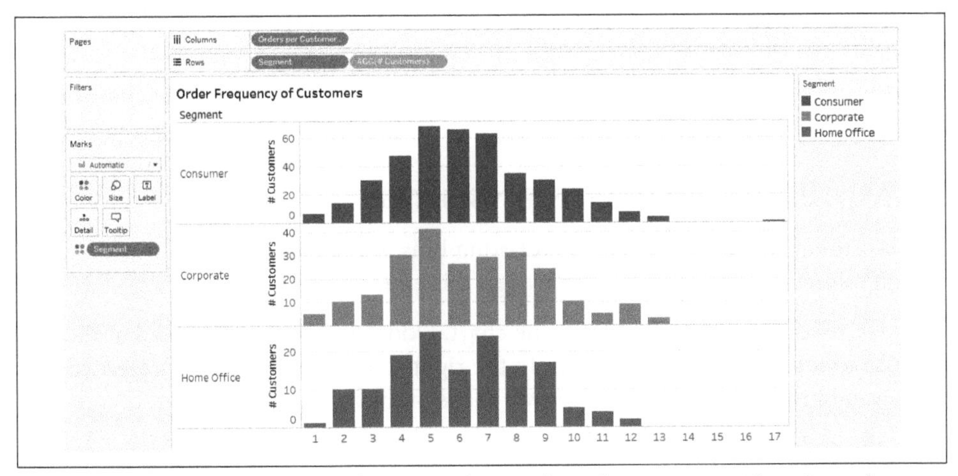

Figure 2-4. Histograms of order frequency by number of customers with independent y-axes

Let's look at a few other techniques you can utilize to make even more-effective histograms.

Strategy: Create a Histogram with a Continuous Bin

In mathematics and statistics books, histograms usually show the bars without any space between them, as seen in Figure 2-5. To do this, change your bin from discrete to continuous by right-clicking the field.

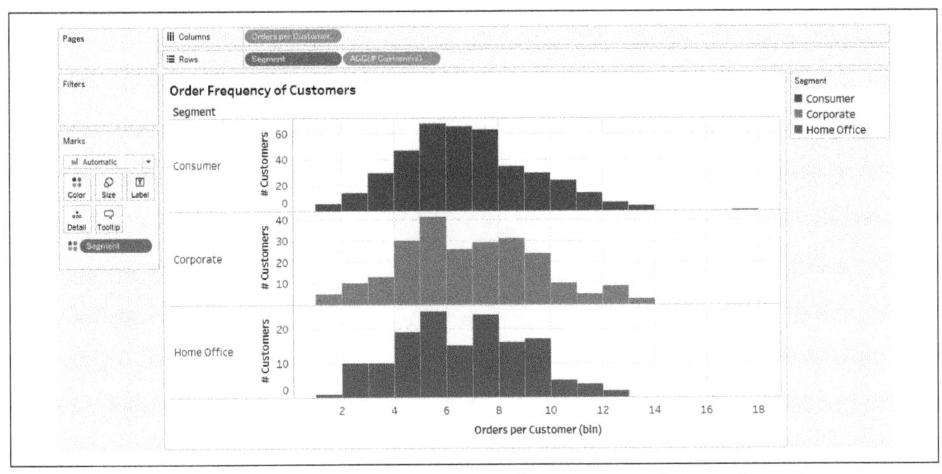

Figure 2-5. In these continuous histograms of order frequency by number of customers, note the lack of space between bars

Recall, from Chapter 1, that continuous fields will draw axes: this means that instead of having a header for each order bin, we will have an unbroken number line. Additionally, the width of the bars will automatically change to the size of the bins—in our case, 1.

Strategy: Use LOD Expressions in Histograms

The last thing we can do is take the existing bins and create a bin for anything of a certain number or greater. Why go through this extra effort? Two reasons.

First, we reduce the overall noise in the chart. Odds are, when you looked at the initial histogram, your eyes were drawn to the right. This may have even prevented you from accurately remembering the overall shape of the histogram.

Next, the chart becomes easier to read aloud, which will be helpful in your presentation. Since there are so few members in the 11 through 17 group, bucketing them together allows us to assign a single number to that grouping. Figure 2-6 shows the visualization that results from these steps:

1. Create the same histogram as in Figure 2-1 (which shows the order frequency by customer).
2. Create a calculated field called **`[Orders per Customer Grouping]`**. This calculation will combine orders per customer that are greater than 11 into a single section. This will get rid of the empty bins and group everything into one final bar:

```
//Orders per Customer Grouping
IF [Orders per Customer] <=10 THEN [Orders per Customer]
ELSE 11
END
```

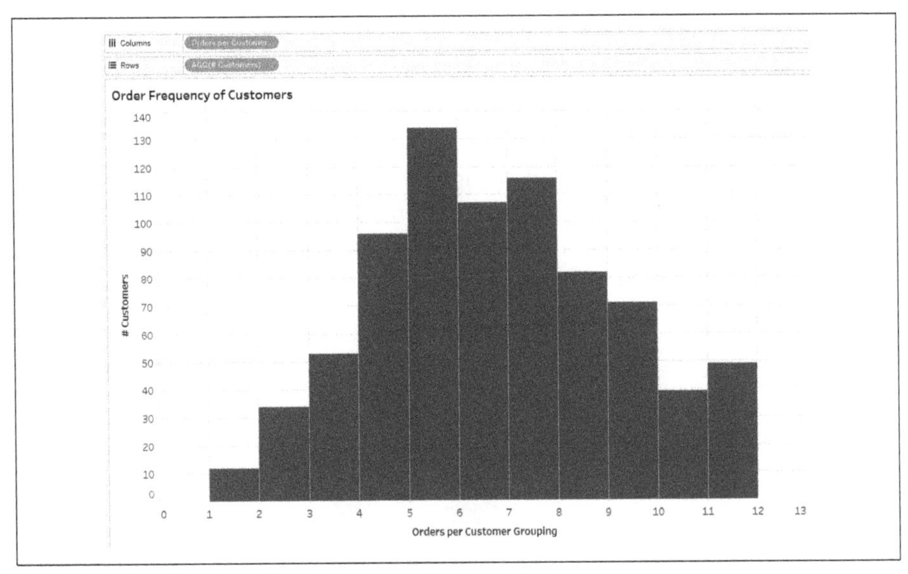

Figure 2-6. Orders per Customer grouping

3. Drag and drop your newly calculated field on top of [Orders per Customer] on the Columns shelf. You should right-click and drag this into the visualization so that you can quickly drop it as a dimension, as opposed to an aggregated measure (see Figure 2-7).

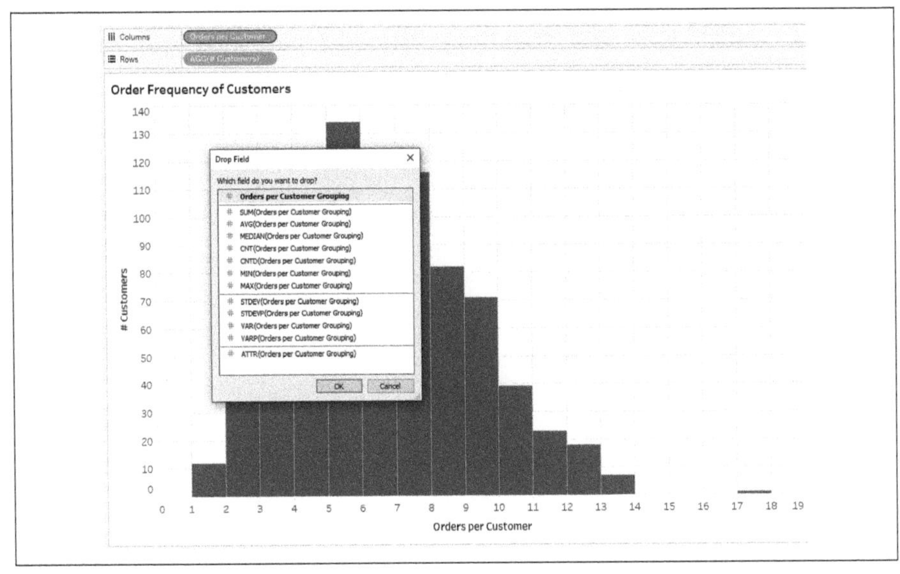

Figure 2-7. Dropping a field

Now your chart has everything over 10 in a single bar. Consider this an intermediary step to our solution. We would *never* suggest that you should share this histogram with your audience: the final bar would be confusing and potentially misleading.

4. Instead, we will make one last change for the sake of accuracy. First we'll change the field to discrete by right-clicking it on the Columns shelf. Next we will *alias* (that is, assign an alias to) the value of 11 to 11+. You can alias this value by right-clicking 11 and selecting Edit Alias. Figure 2-8 shows the result.

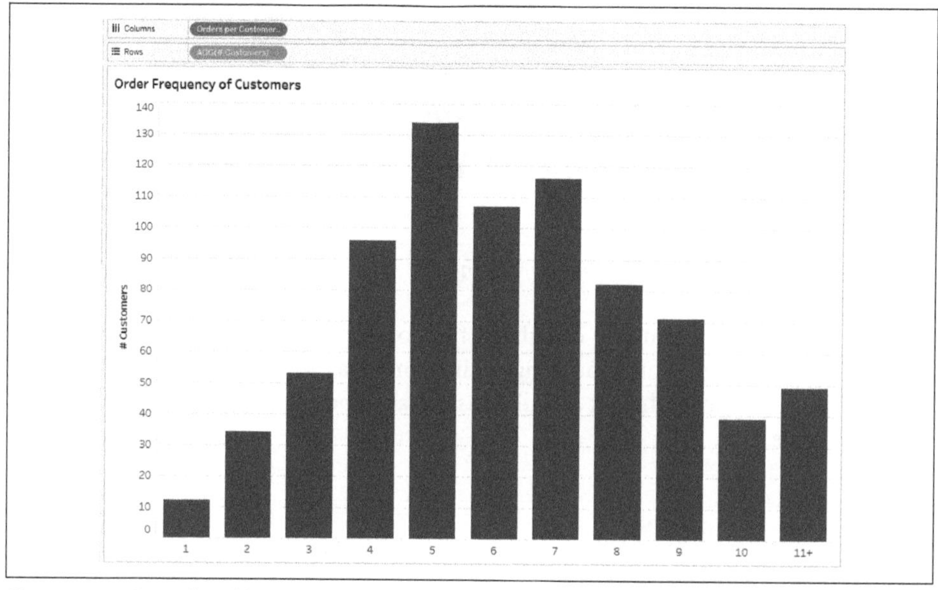

Figure 2-8. Completed histogram

Now we've neatly dealt with the right "tail" of the data, and the 11+ clearly communicates to the audience that the final bar represents everything greater than or equal to 11. Good luck with your presentation!

Dot Plots and Jitterplots: Office Essentials Case Study

The senior management team at Office Essentials now wants to know, on average, how much customers spend on each order. What kind of chart should the data team use?

Histograms alone won't be enough to communicate all of your insights. You need a strategy that incorporates other chart types. This section introduces you to two types of charts you can use to display quantitative information: dot plots and jitterplots.

If anyone ever asks you to provide the average of a metric, proceed with caution. An *average* is a very aggregated value; whenever you present only that level of information, you risk oversimplifying the underlying dataset. What's more, your audience might rightly ask: "But what if there are outliers that skew the average?"

If you have only the aggregate, responding might be difficult. To avoid that, we suggest starting with the detailed data and *then* presenting the aggregated average. Including this additional detail immediately provides you with a response and allows the audience to be more analytical in their thought process.

Strategy: Create a Basic Dot Plot

A *dot plot* is a one-dimensional chart representing data as circles (dots). Each dot represents a single data point. This type of chart is effective when you have lots of data points or members within a dimension, or when the distribution of those data points is an important insight. We also like working with dot plots because they take up very little visual screen space. Let's go ahead and build one out (Figure 2-9) and then take a closer look:

1. Drag [Order ID] onto Detail on the Marks card. This will define the level at which you will display your data.
2. Drag `SUM([Sales])` to Columns.
3. Change your mark type to Circle.
4. Since we have such dense data toward smaller values, add Transparency by changing the Opacity of Color to 60%.
5. To ensure that you answer the original question, drag an Average Line from the Analytics pane across the entire Table.

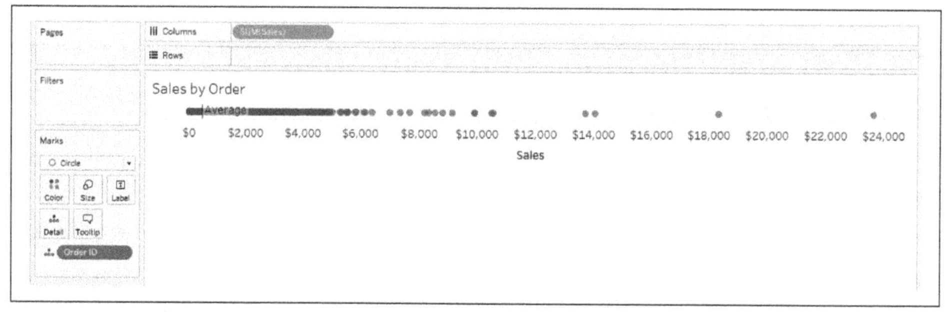

Figure 2-9. A dot plot showing the sales amount per order with an average

Now that you have the underlying picture, what conclusions can you draw? We would conclude that a lot of data density certainly exists around smaller values, and most likely around the average, but we can't neglect the obvious outliers.

The problem is, we now have a lot of data points that overlap, clumping together at a shared numerical value—which makes it hard to see the individual data points. To clean up this view, you need a way to expand all those data points that are collecting on top of one another. The technical term for this is *jittering* the data—that is, turning the dot plot into a jitterplot.

Strategy: Create a Jitterplot

When you create a jitterplot, you get the advantage of a dot plot, but with the members spread out randomly among the y-axis. The data points no longer overlap as much. A *jitterplot* is almost identical to a dot plot, but includes a numeric value on the y-axis. That value has no analytical value and should not be shown; its only purpose is to spread out the data and make it less dense, so it's easier to read.

If you are using a Tableau Extract or Published Data Source, you will have access to a built-in function called `RANDOM()` that returns a random decimal number from 0 to 1. We highly recommend that you take advantage of this function if you can: it will ensure that your jitter is natural and not influenced by any other metrics. (Other methods involve indexing your data, requiring you to impose some type of ordering on the data.)

Follow these steps to create your jitterplot:

1. Double-click into Rows and type **`RANDOM()`**, and then press Enter. You will notice that the aggregation will be `SUM(RANDOM())`. Change this so it doesn't overaggregate and return values greater than 1.

2. Change the aggregation to `MIN(RANDOM())` by right-clicking `SUM(RANDOM())` and changing the Measure to Minimum (see Figure 2-10).
3. Right-click the y-axis (rows) and uncheck "Show Header."

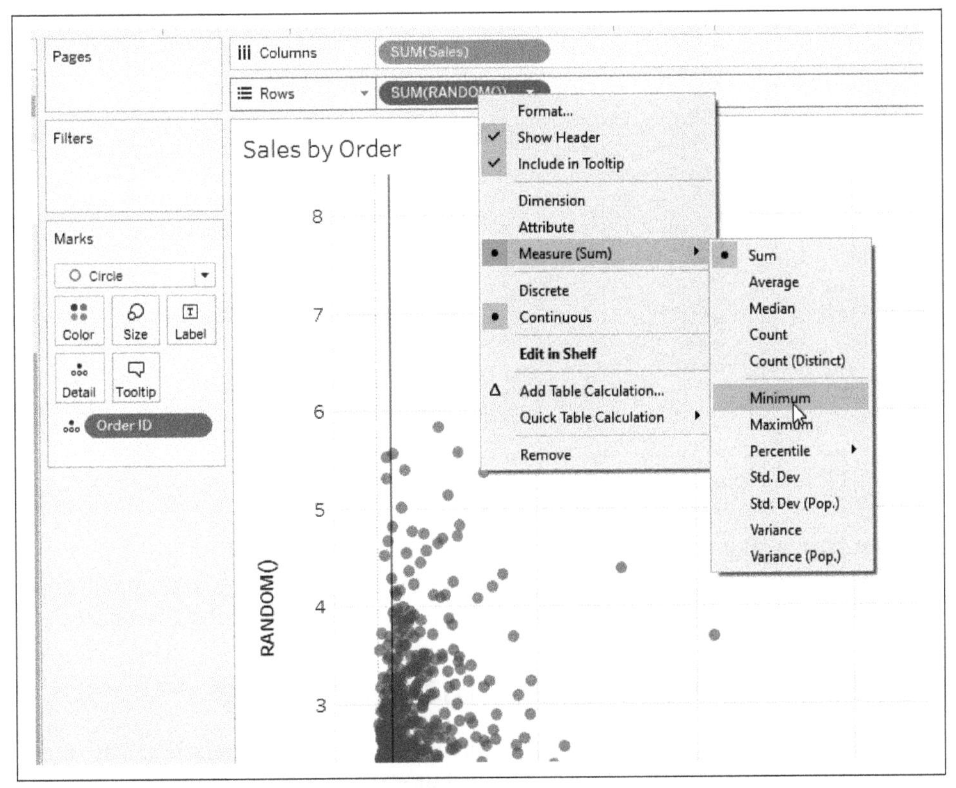

Figure 2-10. Changing the measure to minimum

Let's look at the resulting chart in Figure 2-11.

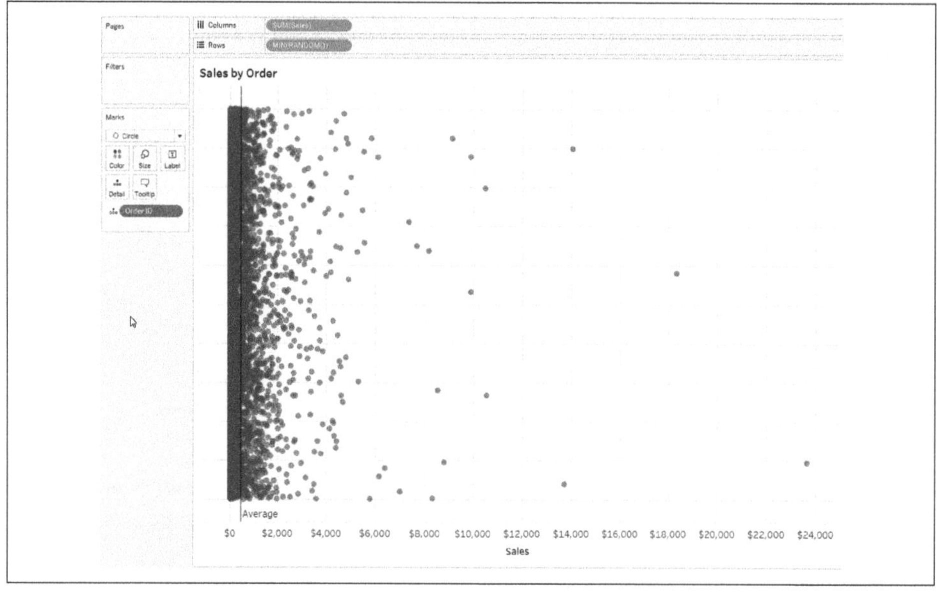

Figure 2-11. A completed jitterplot

Now we have a much clearer picture of the data. When we looked at the previous chart, it was one-dimensional, so we really had no idea how much density there was toward the left side of the chart. This should give management the insight it needs.

Ranged Dot Plots: Office Essentials Case Study

The jitterplot we created for OE has too much detail—it's overwhelming. The team's visualization will be shown as a slide during a presentation, so it needs to convey the point quickly. The senior managers want a soundbite—information they can easily remember and tell others.

So you can't convince your executives to embrace the jitterplot; it's just too much data for them to consume. We've been in your shoes! How can we simplify this chart without losing important information? When we've encountered this situation, our first response has been to suggest a happy medium: show only the aggregate, but include the detail in a secondary place, like the tooltip.

If that's still too much data, though, we have another solution: a *ranged dot plot*. It's a distilled version of the dot plot that still demonstrates some key features of the data shape. Ranged dot plots represent aggregate values like the average, minimum, maximum, or median, instead of all the individual data points. They can be great visual representations of the overall scope of a dataset.

Since your data is about the average amount customers spend, the key features you'll want to focus on are the average (of course), the minimum, the maximum, and the range between the minimum and maximum.

Strategy: Create a Ranged Dot Plot

Use these steps to create your ranged dot plot:

1. Make sure there is a measure that represents the total sales per Order ID. In this dataset, that means you will create an LOD expression. This calculation is telling Tableau to find the `SUM()` of Sales for each Order ID, because your final visualization will not have an Order ID detail:

   ```
   //Sales by Order ID
   {FIXED [Order ID]: SUM([Sales])}
   ```

2. Right-click and drag [Sales by Order ID] onto the Columns shelf to select the aggregation, and choose `AVG()`. Set the mark type to Circle and choose a color of your liking.
3. Make a calculated field called **`[Range of Order Sales]`** to find the difference between the maximum and the minimum:

   ```
   //Range of Order Sales
   MAX([Sales by Order ID])-MIN([Sales by Order ID])
   ```

4. Right-click and drag [Sales by Order ID] to the Columns shelf again, but this time select Minimum. (You need a starting point for your range, and what better starting point than the minimum?) Make this a dual axis and synchronize your axes. Also make sure the marks for this new axis are in the back by right-clicking and selecting "Move marks to back."
5. Change the mark type to Gantt Bar. This will allow you to put a measure—namely, our [Range of Order Sales]—onto Size.
6. Drag [Range of Order Sales] onto Size.
7. Drag [Sub-Category] onto Rows so you can compare the range and averages among the product categories.
8. Adjust formatting and sizing until you like the final visual (ours is in Figure 2-12).

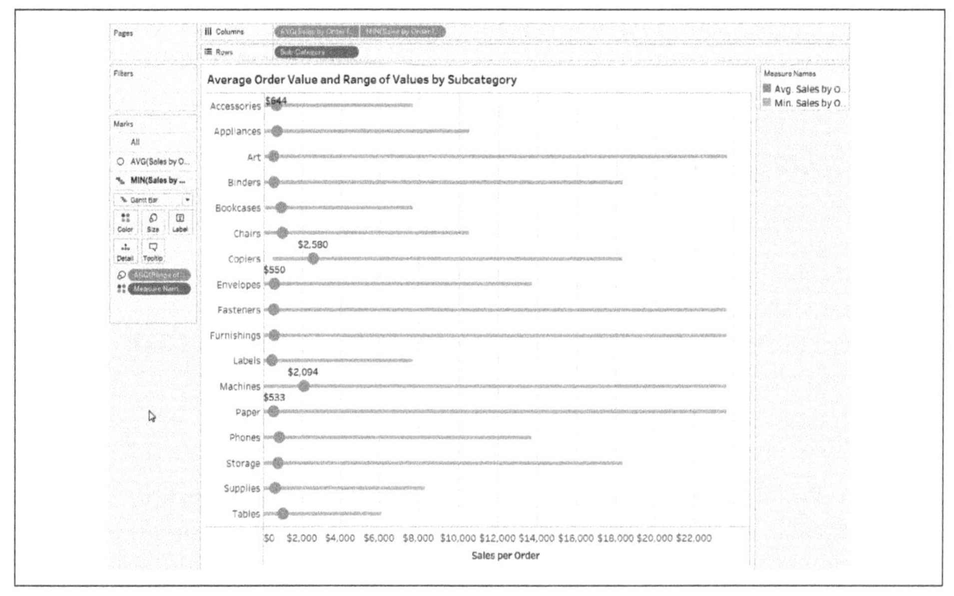

Figure 2-12. Ranged dot plot

This visualization tells a much more compelling story. At a glance, your audience can see many elements, like the average per sub-category and the range, and do some natural visual comparisons. They can see, for example, that Envelopes orders average $550, and that the range isn't too extreme compared to similar subcategories.

If you like this visualization, you can take it further as follows:

- If you find that you have heavy skew or outliers—as your data does—try changing the average to the median.
- If your audience can't decide whether they want to see the median or the average, build in a parameter that allows them to toggle between the two.
- You can also leverage percentiles instead of the minimum and maximum range.

If you see that one or two extremes significantly pull the data shape, your strategy should be to make this known, but provide a solution. Let's try percentiles and see what we come up with.

Strategy: Use Median and Leveraged Percentiles

If you imagine your data spread out, *percentiles* allow you to cut up your data points into chunks, ranging from 0% to 100%. The most common percentiles are quantiles and deciles. *Quantiles* break the data into quarters: four equal portions, each 25% of the dataset. *Deciles* break the data into 10 equal pieces, each 10%.

Percentiles are also used in statistical scenarios that relate back to the normal distribution. Remember that in a normal distribution, 95% of the data is two SDs from the average—so it might be appropriate to limit the range to between 2.5% and 97.5%, or to say that "95% of the data" is within a specific range.

Beginning with the chart from the previous strategy, follow these steps to create a range:

1. Start by changing the aggregation of the average from the previous strategy to the median with a simple right-click. Now you can construct the percentile calculations:

   ```
   //2.5% of Data
   PERCENTILE([Sales by Order ID],.025)
   ```

2. You can repeat this calculation for 97.5%, or 0.975. Then you can create a new Range that is nearly identical to our original. Keep going to complete the chart in the exact same fashion:

   ```
   //97.5% of Data
   PERCENTILE([Sales by Order ID],.975)

   //Percentile Range
   [97.5% of Data]-[2.5% of Data]
   ```

3. Replace the second axis, MIN([Sales by Order ID]) with [2.5% of Data], and replace [Range of Order Sales] with [Percentile Range]. Figure 2-13 shows the updated view.

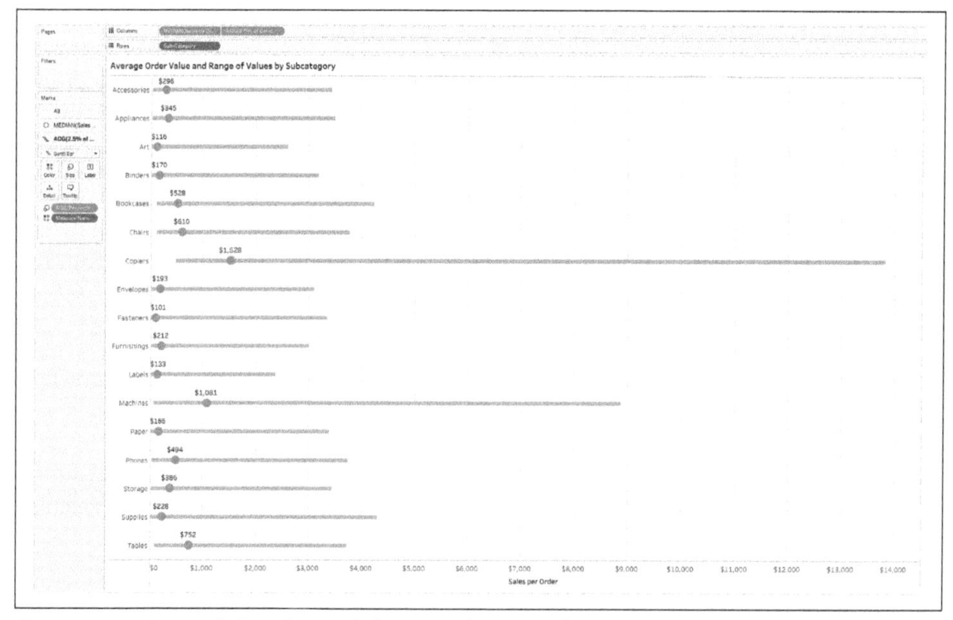

Figure 2-13. Ranged dot plot with leveraged percentiles

So what makes this view special? Sure, we've reduced the range, but we've also reduced the amount of noise in the view.

Anytime you make a chart, ask yourself: What insights do you want the audience to take away? What insights are relevant to their needs?

If the OE executives need an average or median order value and a soundbite, this ranged dot plot will fulfill that need. Now they have an easy way to convey information about each segment, such as this: "Aside from some outliers, most of our copier orders averaged about $1,500."

Box Plots: Spear-Tukey Shipping Case Study

Our next case study takes place at Spear-Tukey Shipping (STS), a transportation and logistics company.[1] STS ships products all over the world, on incredibly tight deadlines, so processes have to be precise. If a problem arises, the company needs to know about it.

---

1 Not its real name. Instead, we've used the names of Mary Eleanor Spear, who created the range bar, and John Tukey, who developed the box plot and whiskers plot (*https://en.m.wikipedia.org/wiki/Box_plot*). By the time we're done answering your question, you'll understand why we chose them.

When a customer ships with STS, they have to select a shipping method. There are four methods. They're all different, and each one has its own delivery deadline. The delivery *must* happen within the specified number of days.

Recently, STS has been hearing complaints about late deliveries, but executives aren't sure how big the problem is (or even if it's really a problem). The data team has collected data and is working on analyzing it. The team plans to present the results to the CEO, who wants to know how often the company meets its required deadlines per method.

Table 2-1 lists STS's methods and deadlines.

Table 2-1. Shipping methods and deadlines

Method	Days to delivery
Same Day Air	1
Priority Air	2
Priority Ground	3
Ground	5

Before we get too deep, let's think about expectations. In the OE case study, we came in without an expected answer; we may have had some institutional or intuitive guesses, but we were using visualizations to *find* answers. Here, though, we have a process that should be controlled; the objective is to find *examples* where that process isn't working as designed. We also have lots of data for each method, so we need to figure out how to visualize it all in one chart without the information getting overly complex and confusing.

With any highly controlled process, we find it best to lean on statistics to communicate performance. For this use case, we will combine a few concepts into one visualization, the *box plot* (Figure 2-14). Box plots are a little more complicated than the other charts we've shown so far, so let's look at how they're constructed.

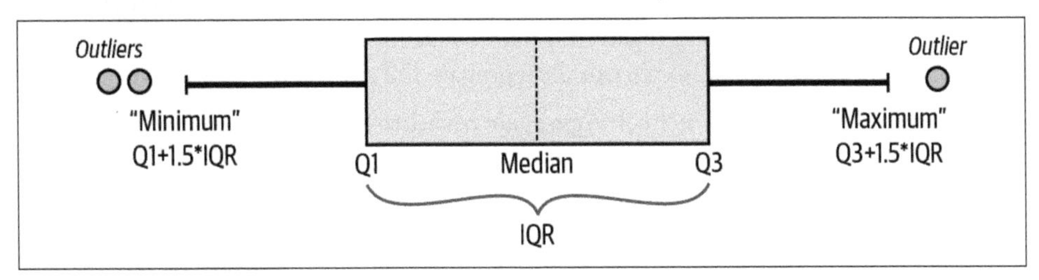

Figure 2-14. Anatomy of a box plot

A box plot comprises the following elements:

Median
: The *median*, or the midpoint the of data, is represented as a line inside a box.

Quartiles
: Q1 and Q3 are quartiles; you'll also see these labeled as *hinges* in Tableau. Q1 is at 25% of the data, and Q3 is at the 75% mark.

Inner quartile range (IQR)
: The range between Q1 and Q3.

Whiskers
: These are the horizontal lines that extend out from the box. They can be constructed in two ways:

  - The *minimum* whisker can extend to Q1 minus 1.5 times the IQR. The *maximum* whisker can extend to Q3 plus 1.5 times the IQR.
  - Alternatively, the whiskers can extend to the minimum and maximum of the dataset.

Outliers
: Any data points that extend beyond the whiskers and are shown as dots.

Strategy: Create a Basic Box Plot

Now that you have a sense of what we're building, follow along and let's try it:

1. Create the **`[Days to Delivery]`** field. You'll construct this calculated field to compute the days between the shipping date and the delivery date:

   ```
   //Days to Delivery
   [Delivery Date]-[Ship Date]
   ```

2. Drag the [Days to Delivery] field onto the Rows shelf. Notice that when you drag the field, Tableau will aggregate the data by using a measure. We don't want to look at aggregated data, so we can disaggregate the data in one of two ways:
   - Choose Analysis → Uncheck Aggregate Measures.
   - Drag [Manifest ID] onto Detail. This is the lowest level of detail in the dataset and will construct an aggregate that is equal to the row-level value.
3. Include the shipping method. Also drag [Ship Method] to Columns.

4. Change your mark type to Circle and reduce opacity to 60%. Notice that you again have something very similar to a dot plot, as shown in Figure 2-15.

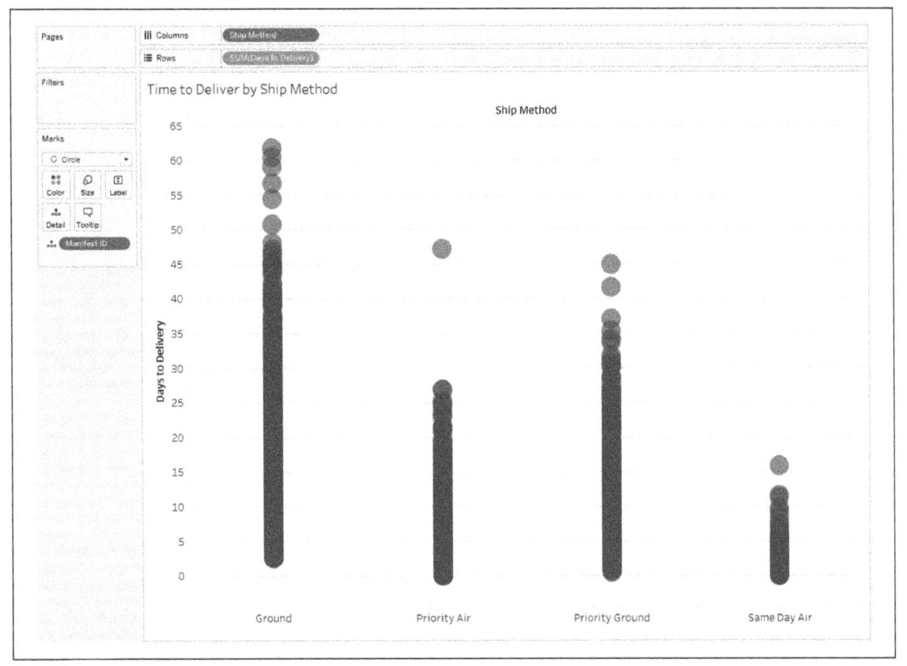

Figure 2-15. Each dot is the number of days it took for a delivery

5. To turn this into a box plot, go to the Analytics pane and drag Box Plot onto the worksheet. When you drag it on, drop it on Cell. Tableau has now drawn a box plot on top of your data.

When you hover over the box plot, the summary tooltip will show the values for the median, the quartiles, and the whiskers (Figure 2-16).

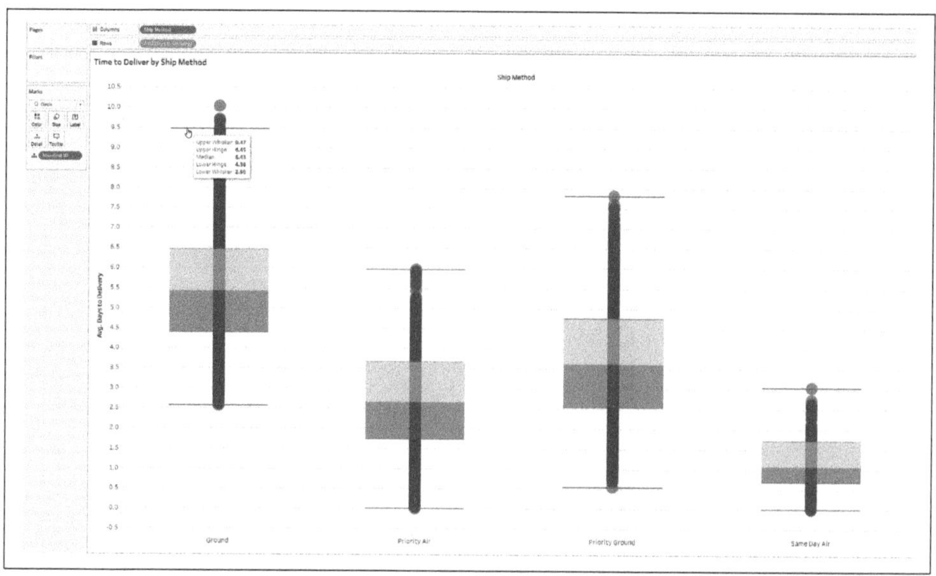

Figure 2-16. Finished box plot

At this point, you're probably wondering how to interpret this box plot—and how to explain it to the CEO. First, narrow the data. Let's zoom in on the Same Day Air method (Figure 2-17) and try interpreting just that part first.

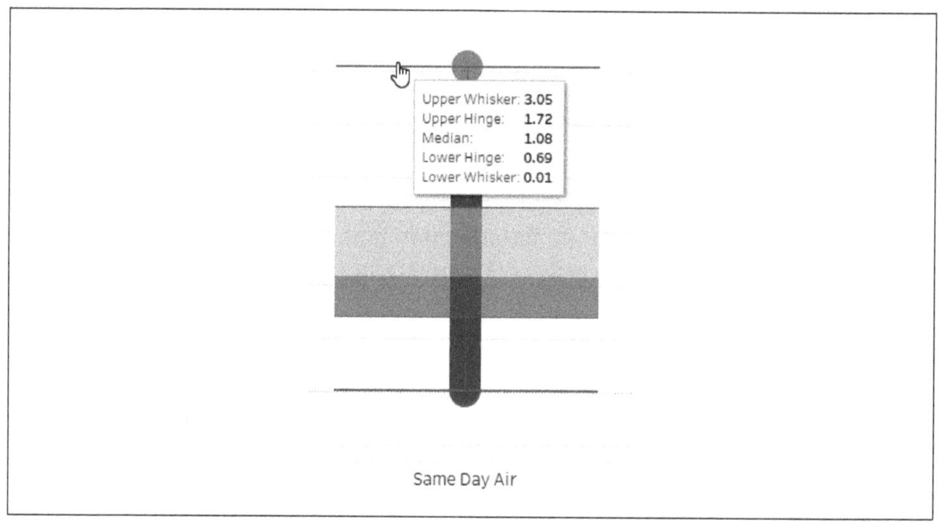

Figure 2-17. Interpreting a box plot

We can see in the third row that the median time for a Same Day Air delivery is 1.08 days. You could phrase this as "50% of our Same Day Air deliveries arrive within 1.08

days." Looking at the hinges (IQR), you can add that "25% of deliveries arrive in less than one day (0.69 days), and 75% arrive in less than 2 days (1.725)."

It looks like the maximum is at 3.05 days. This means we can confidently say that 100% of STS's Same Day Air deliveries arrive within three days—not great for something that should be same day!

In Tableau, we can customize this visualization further, by right-click editing. There's a lot of information here. We can reduce the total number of marks on the screen, leaving only the box plot and any outliers.

As we mentioned, we can extend whiskers in two ways. You'll see both as options when you right-click to edit. Click the checkbox to hide the underlying marks within the box plot. You can see the finished version in Figure 2-18.

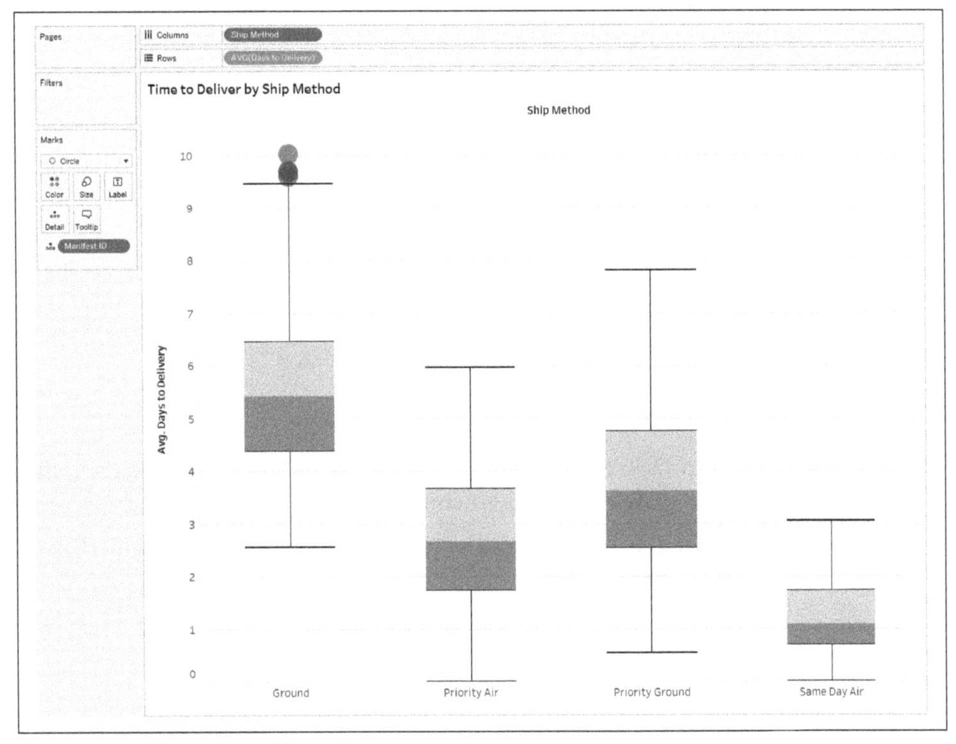

Figure 2-18. Customized box plot

If this box plot will work for your audience, great! But there's a lot going on here, and sometimes box plots can intimidate audiences. Knowing that, we usually pair the box plot with a less intimidating chart, like the histogram, to help communicate the contents and ensure that everyone understands the key takeaways.

Strategy: Combine a Box Plot and Histogram

To create this chart, we'll start with the box plot you just built. This time, instead of disaggregating your measures, you will need to aggregate them. If instead you added [Manifest ID] to Detail, you're good to go!

1. Modify the box plot so that the whiskers range from minimum to maximum.
2. Now that you have this view, you can bring on other measures that will need to be aggregated—specifically, as a dual axis. Create a calculated field, specifically an LOD called **`[Days to Delivery by Manifest ID]`**:

   ```
   //Days to Delivery by Manifest ID
   {FIXED [Manifest ID]: AVG([Days to Delivery])}
   ```

3. Create bins, using the same method that you used when constructing the histogram at the beginning of this chapter. Make the bins at every 0.5 days. Right-click the field in the Data pane and convert it to continuous.
4. Toggle the layout of the rows and columns by clicking the Swap Rows and Columns icon in the top toolbar (next to the sorting icons).
5. Drag the new bin calculation, [Days to Delivery by Manifest ID (bin)], to Columns, to the right of AVG([Days to Delivery]).
6. On the Marks card for this visual, change the mark type to Bars. Remove [Manifest ID] from Detail.
7. Create a calculated field called **`[# Shipments]`** and drag it to Rows:

   ```
   // # Shipments
   COUNTD([Manifest ID])
   ```

8. Right-click [Days to Delivery by Manifest ID (bin)] and select Dual Axis; right-click the newly created dual axis and synchronize axes.
9. Remove [Measure Names] from Color on the All Marks card, and right-click to hide the extra axis on the top.

   Figure 2-19 shows our result—it's easier to read, isn't it? It's also easier to explain the data.

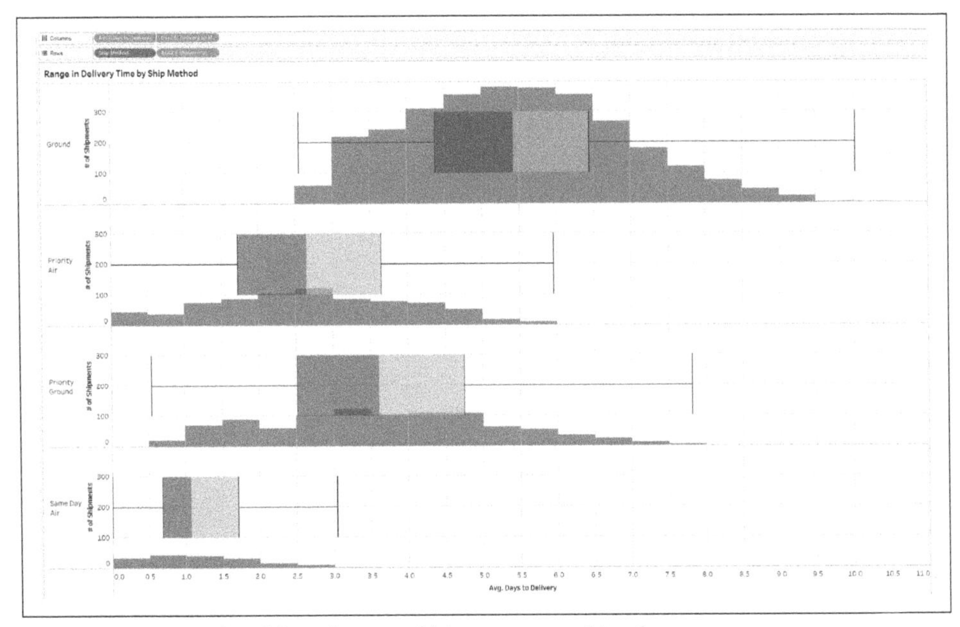

Figure 2-19. Completed box chart and histogram combined

Now the STS CEO can see how the construction of the box plot relates to the underlying data. The executives can see the frequency constructing each of the bars and understand how the data shape presents itself in box plot form. They can also see both the frequency and the density of the data at one time.

Line Charts: Office Essentials Case Study

Let's go back to OE, where the data team now needs to show the company's overall sales numbers for the last few years, broken down by month, using a sample of the data. The idea is that a good view of those numbers will help predict future months' sales and identify trends, especially if there's anything alarming the company should act on quickly. Showing sales by month is easy enough, but how do you indicate what's normal and what's not?

One thing we've learned over the years is that whenever someone asks about trends, in their mind they've drawn a mental picture of a line chart. Like bar charts, *line charts* are a popular tool in data visualization. They take a metric and display it over time. Here we won't dive too deep into working with time (we'll save that for Chapter 4), but we do want to introduce some ways you can use statistical values and derivative measures to make really impactful line charts. It sounds like OE's leadership is interested in becoming more data-driven, and these techniques will help.

Strategy: Build a Line Chart

Let's start with a basic line chart showing OE's Sales by Month (Figure 2-20). To build this, right-click and drag [Order Date] onto Columns, selecting Month toward the bottom. Drag `SUM([Sales])` to Rows.

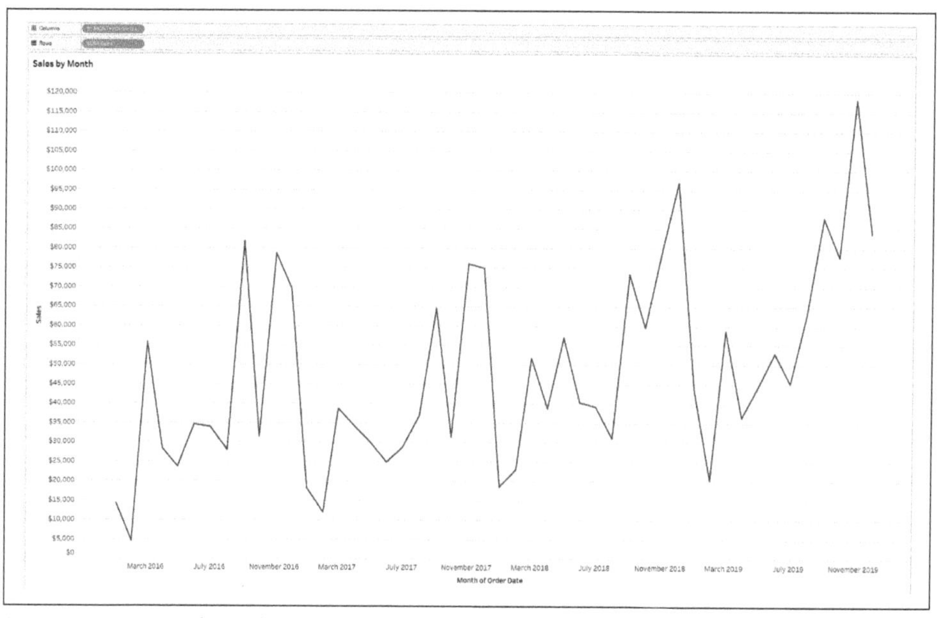

Figure 2-20. Line chart showing sales by month

This provides some immediate insight. We can tell there is seasonality in the data: every year spikes occur in September, November, and December. And in general, it looks like sales are going up over time.

But we can get more sophisticated in the story that we're telling by including statistical measures. We'll start with tools that are built in and easy to utilize in Tableau. In Chapter 1, we explored adding a reference line for an average; this time, let's go a step further and add *reference distributions*: a set of bands on the chart that indicate a range, so the audience can see where the lines fall into that range.

Strategy: Add Reference Distributions to a Line Chart

Remember we warned you that averages (and other aggregate values) don't always communicate enough on their own? This is a great way to mitigate that by showing exactly how sure you are about the averages you're providing.

The reference distribution we'll include here shows the range of a confidence interval. A *confidence interval* is a range of values computed from a random sample that is likely to contain the true value of a chosen measure. It's a function of two items of your data: the *sample size* (or number of data points) and the overall *confidence level.* As you increase your sample size, your confidence interval is likely to narrow, because there are more observations. The confidence level has the inverse effect: the more confident you need to be with your data (say, if you need to be 99% confident rather than 95%), the wider the band.

To add the confidence interval, follow these steps:

1. From the Analytics pane, drag an Average line onto the view, and then drop it on Table.
2. Focusing on the middle section, add a confidence interval in addition to your line, and specify the level of confidence. As you're doing this, change the Label option to Value instead of the computation.

   The chart (Figure 2-21) now shows average monthly sales, with a gray band for the confidence interval.

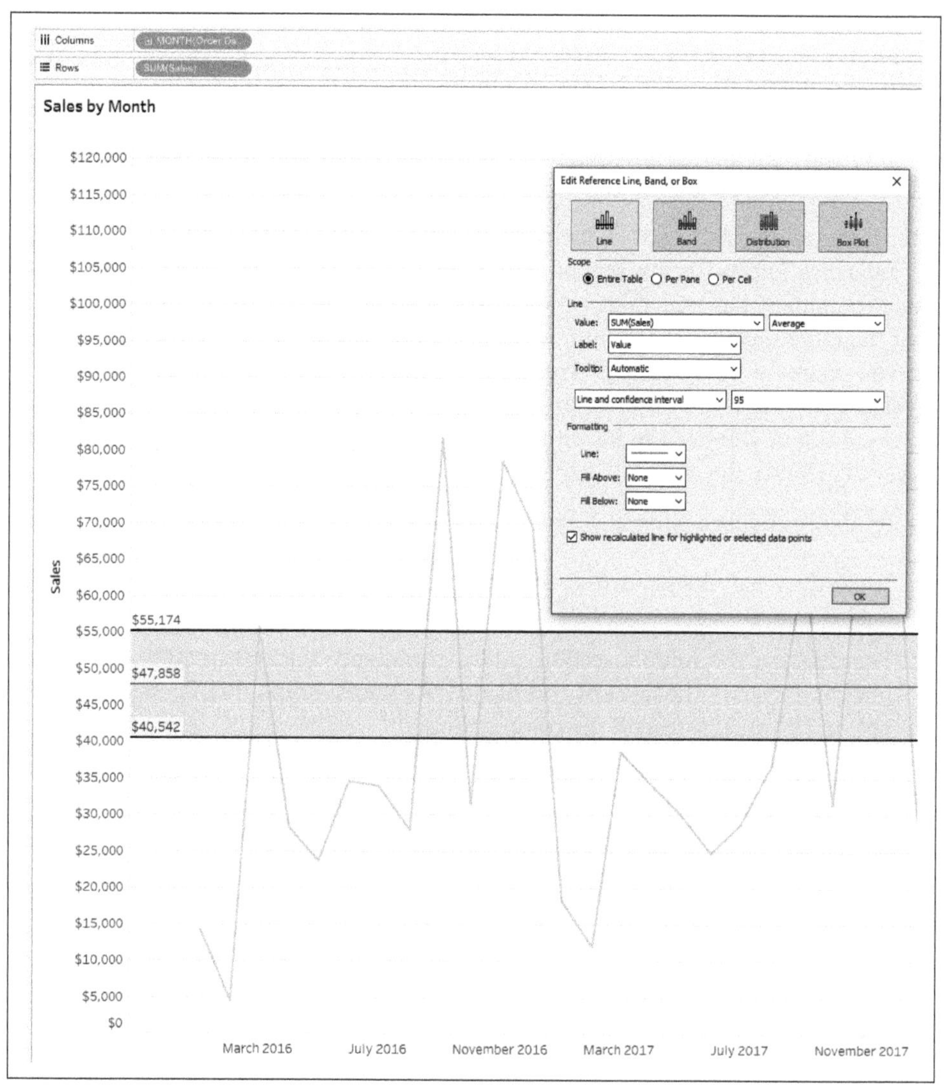

Figure 2-21. Average monthly sales line chart with confidence interval

Why add this type of context to charts? For this presentation, it's to add detail for the audience, the OE executives and board, who are relying on the average value that we present. Now they can easily see, thanks to that gray band, which months had sales that were below or above average—exactly the kind of information they're looking for.

While the chart we just created shows a computed average of $47,858 (based on the sample data provided), a *margin of error* of about $7,000 exists on either side to show what the average *could* be. The margin of error is a way of saying that we can state with 95% confidence that if we had *all* the values of monthly sales, the average would fall between $40,542 and $55,174.

Strategy: Add a Standard Deviation to a Line Chart

You can also use distribution bands and reference lines in Tableau to represent percentiles or SDs. We'll add SDs next:

1. Right-click the average line and click Edit.
2. Change back to Line Only.
3. To create another reference line, right-click the axis and this time choose Distribution.
4. Change the Computation Value to Standard Deviation and leave the factors at –1 and 1. Similarly, leave the Sample option selected, since we don't have *all* the data, just a subset.
5. Change the Label option to Value and click OK (see Figure 2-22).

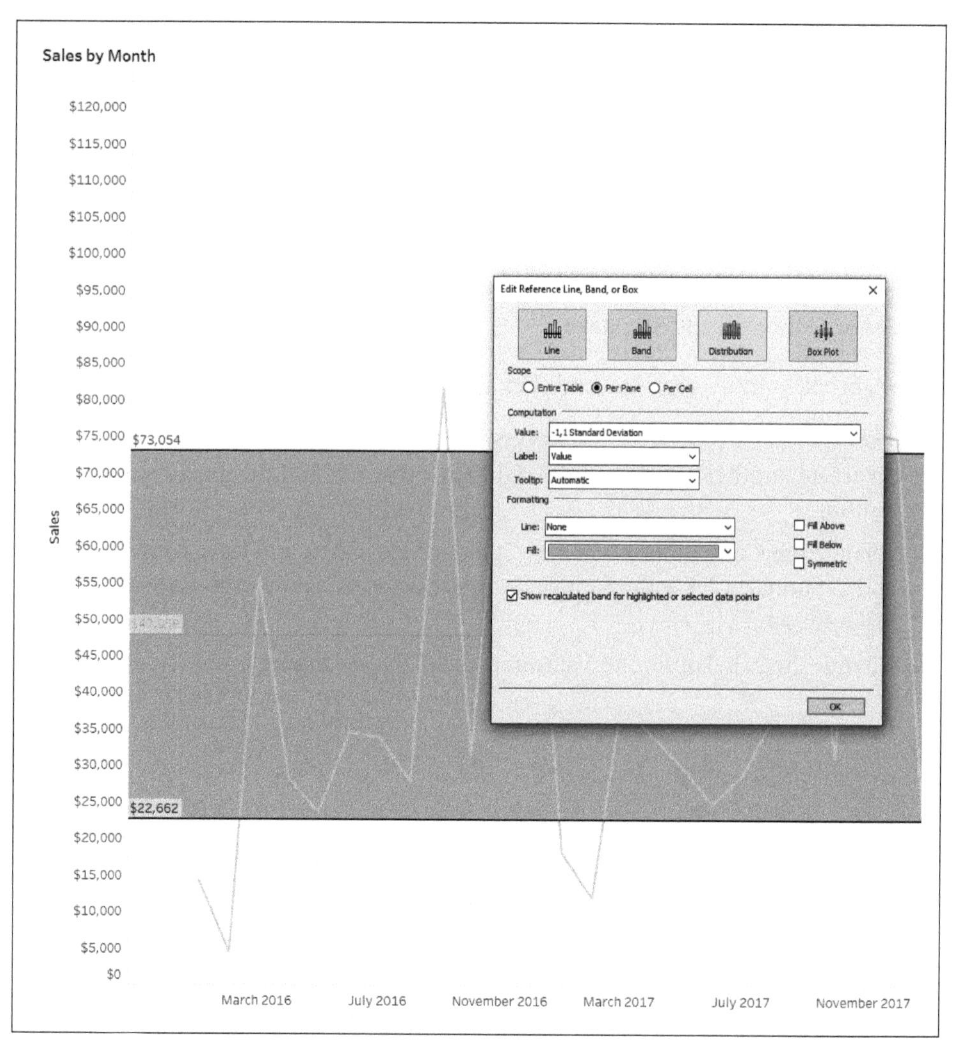

Figure 2-22. Adding a SD to a line chart

So what can we conclude from this resulting chart? For one thing, there is a *lot* of spread from the average. For any given month, the sales numbers might be as far as $25,000 from the average. That's nearly half the computed monthly average—not so great.

What should the OE executives take away from that? It's important for them to realize that a lot of variability exists in the monthly sales data. At this point in time, it probably isn't appropriate to use the data to forecast future values.

Our audience also wants to know what's normal and what isn't. When should they worry?

We're going to try another way of working with SDs—but instead of showing them as reference distributions, we'll use them as an alerting tactic. It's an unbiased way to review your dataset and see where meaningful statistical variation truly exists. (In statistics, it's common practice to say that something is "out of control" when the reported value is outside a certain number of SDs from the average.)

Strategy: Use Reference Distributions as an Alerting Tactic

Our next chart will have monthly sales and an average line, as it did before. This time, though, we'll add red dots to months that are more than one SD above or below the average:

1. Create a new calculated field called **`[SD Alert Dot]`**. This calculation will compare the monthly sales amount against the average of the view plus 1 SD, and likewise for minus 1 SD. If the condition is true, it will return the sales amount. We will then turn that sales amount into a dot.

    ```
     //SD Alert Dot
    IF SUM([Sales])>=(WINDOW_AVG(SUM([Sales]))+WINDOW_STDEV(SUM([Sales])))
    OR
    SUM([Sales])<=(WINDOW_AVG(SUM([Sales]))-WINDOW_STDEV(SUM([Sales])))

    THEN SUM([Sales])
    ELSE NULL
    END
    ```

2. When you read this calculated field, it looks like a lot is going on, but let's simplify the various components:

    - `SUM([Sales])` is just the sales amount, which will be drawn by month because that's the level of detail you have in your view.
    - `WINDOW_AVG(SUM([Sales]))` is exactly the same as the Average Reference line: it is the average of all the data points in the view (in the "window").
    - `WINDOW_STDEV(SUM([Sales]))` is the same concept as the average, but this time it's the SD.

    The rest of the conditional is determining whether the Sales value is above or below those points. And if it is, you want to return the sales; otherwise, you can null it out.

3. Drag the [SD Alert Dot] field onto your Rows shelf.
4. Right-click and make it a dual axis and then synchronize your axes.
5. Change your mark type for the secondary axis to Circle. Make sure your first mark is still a line.
6. Color the measures appropriately and right-click the null indicator in the bottom right to hide it.

7. You can also drag the reference distribution from the visualization, by clicking where the visual line for the top or bottom SD is.

   Figure 2-23 shows the resulting line chart.

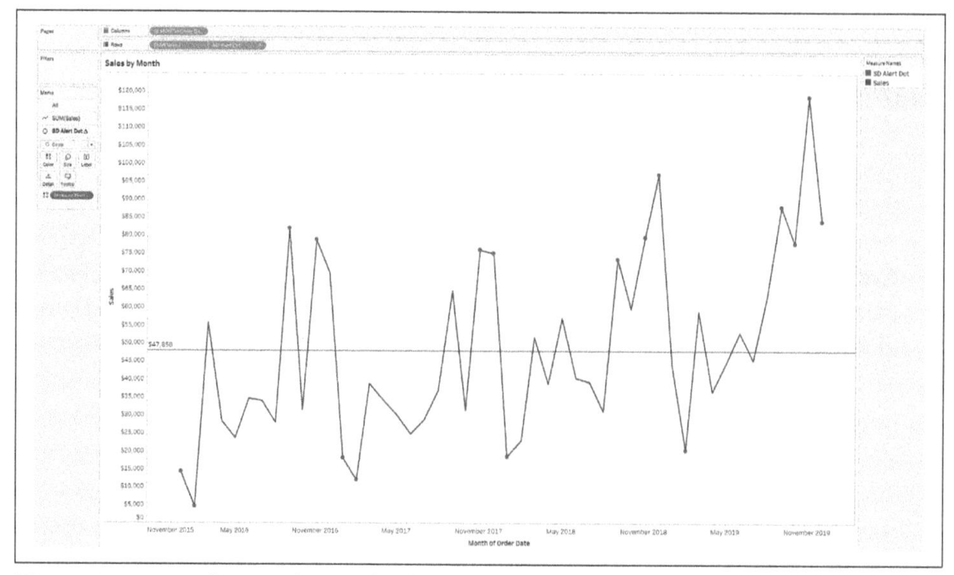

Figure 2-23. Line chart with months dotted in red when they are ±1 SD from the average

So what's the difference between this chart and the previous example with reference bands? Both show the trended metric and SD, but the key difference is in *how the audience will perceive them*. With the reference bands, the team will present the executives with the SD spread—but the alert dots invite them to take action.

That's the difference between showing information with charts and being truly data driven. A strong, data-driven organization defines alerts or thresholds with metrics, so that its leaders know when to take action.

Our next case study has us looking at another way to take a statistical output and turn it into a dimension.

Calculated Dimensions Based on Statistics: Banco de Tableau Case Study

Our next study is at the Banco de Tableau (BoT), a large (fictitious) financial institution you met in Chapter 1. The bank is finding new ways to leverage its merchant-category data to better understand its customers. The data has a lot of categories, and the BoT data team needs a way to group and score them based on their total spending.

The Top N bar chart from Chapter 1 is great, but now the team needs a more holistic way to communicate results. It should include *every* category in the dataset, not just the top 10. Team members want to compare the lowest spenders to the top spenders, for instance, and drill down for even more pointed insights.

Deploying statistics is a great way to deepen your data visualizations. We'll try two approaches here. The first is to assign a summary statistical value to the data point based on what data is in the view. The second is to use a Pareto chart (stand by for more about those).

Strategy: Assign a Summary Statistical Value to Data Points

Here are the steps to assign a summary statistical value to data points:

1. Start with a new horizontal bar chart that has the total transaction amount for each [Merchant Category]. Sort the bars in descending order.
2. Build out a calculated field called **`[Merchant Category Level]`** that will allow you to categorize merchant categories into three cohorts: one for the top 10%, one for the bottom 10%, and one that contains the rest of the data. Drag the calculated field onto Color. Remember that here, `window_percentile` will be the equivalent percentile based on the data in our view:

   ```
   //Merchant Category Level
   IF SUM([Transaction Amount])
     <=WINDOW_PERCENTILE(SUM([Transaction Amount]),.1)
   THEN 'Bottom 10%'
   ELSEIF SUM([Transaction Amount])
     >=WINDOW_PERCENTILE(SUM([Transaction Amount]),.9)
   THEN 'Top 10%'
   ELSE 'Middle'
   END
   ```

   This calculation really evaluates only one essential thing: whether the transaction amount is above or below a specific percentile.
3. Add the calculated field as a header to the view. This will allow you to create new aggregates, like the average, for the various cohorts. You may have to manually reorder the headers.

There are a few reasons to assign a summary statistical value. Now, with mathematical analysis, we have separated out our merchant categories evenly. Doing this also lets us speak in aggregate about those new categorizations. If we want to draw even more pointed insights, we can add back a reference line that addresses only merchants within a certain percentile.

As an example, in our chart (Figure 2-24), we can see that the average spending in the top 10% of merchant categories is $10.1 million. Compare that to those in the bottom 10% of merchant categories, whose average spending is a mere $1.75 million.

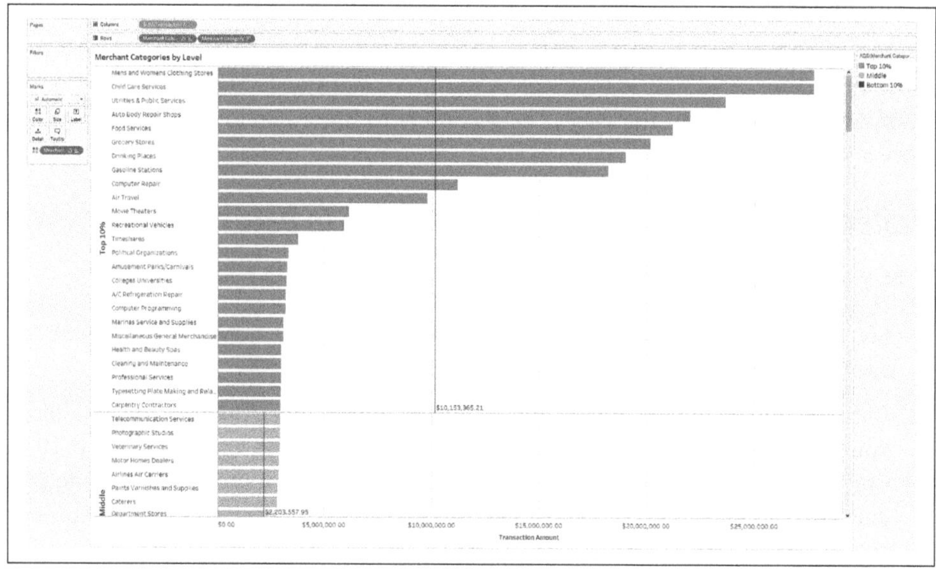

Figure 2-24. Assigning summary statistical value to data points

Using statistics to create the various groupings also makes the chart dynamic, allowing for the deep dives our bankers are hoping for. If our scope of analysis changes when we filter our data, the subsequent visualization will also update. We can filter the data to look at transactions completed during a certain time period or even in a specific region, and the visual will respond.

Up to this point, we have talked about how to show categorical data and some of its spread, but we haven't gone deep. Let's do that now.

Pareto Charts: Amplify Performance (AP) Case Study

We're back analyzing data for Amplify Performance. This time, the executives want to better understand the distribution of grant dollars. They've seen the treemap from Chapter 1 and know that a few program categories seem to take up the majority of funds, but they want to be able to communicate this to a broader audience. What kind of visualization can we build to satisfy this request?

Strategy: Use Pareto Charts to Show Categorical Data

Have you ever heard of the *80/20 rule*? Often called the *Pareto principle*, it suggests that 80% of a metric or outcome is related to 20% of the data.

To see that more directly in a visualization, we can construct a *Pareto chart*: a dual-axis chart that takes the initial sorted bar chart and adds one line representing a running percent of the total. The running percent of the total is what illustrates the 80/20 rule. More specifically, this chart is designed to present a distribution of numerical values that relate back to categorical data.

The Pareto principle holds that 80% of consequences come from 20% of causes. We adapt this to data visualization by saying that 80% of metrics can be traced back to 20% of the data. It was developed by the pioneering quality manager Joseph M. Juran (*https://oreil.ly/govqe*), who named it after Vilfredo Pareto (*https://oreil.ly/r8HS0*) (1848–1923), an Italian economist who famously proved that 80% of the land in Italy was owned by just 20% of the population.

Pareto charts are great when you need to help your audience focus on what is important and to easily communicate about a distribution. We'll start by building a basic bar chart for AP, this time vertical. (If you don't remember how to do this, revisit Chapter 1.) Then we will turn it into a Pareto chart.

Will you find that 20% of program categories make up 80% of grant dollars? Here are the steps:

1. Drag [Program Category] onto Columns.
2. Drag [Grant Amount] onto Rows.
3. Use the toolbar to sort the program categories in descending order by [Grant Amount], as shown in Figure 2-25.

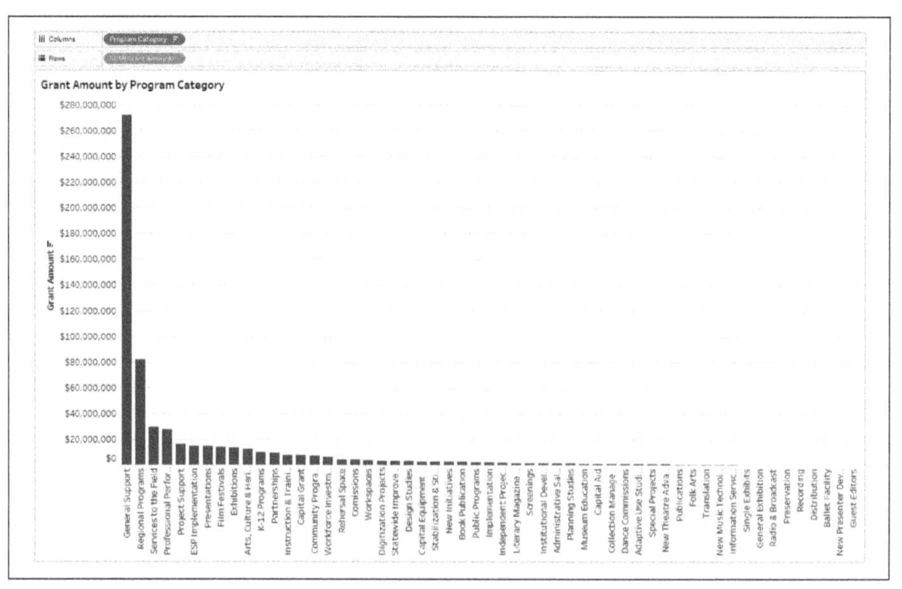

Figure 2-25. Sorting program categories in a vertical bar chart

4. Hold down your Ctrl key and drag `SUM([Grant Amount])` as a copy onto your Rows shelf.
5. Right-click the new measure and create a Quick Table Calculation; use the Running Total.
6. Right-click the Table Calculation again and Edit it. This time click "Add secondary calculation" and select Percent of Total. This takes the running total and presents it as a cumulative total percentage (Figure 2-26).

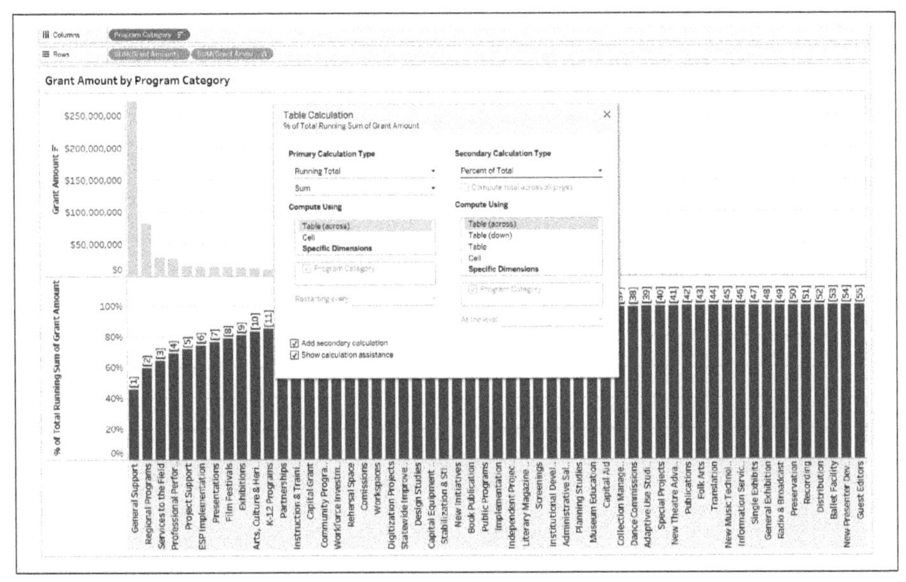

Figure 2-26. Table calculation for a Pareto chart

7. Change your mark type for the second metric to Line. Then right-click it to make it a dual-axis chart, and turn on labels for the line chart (Figure 2-27).

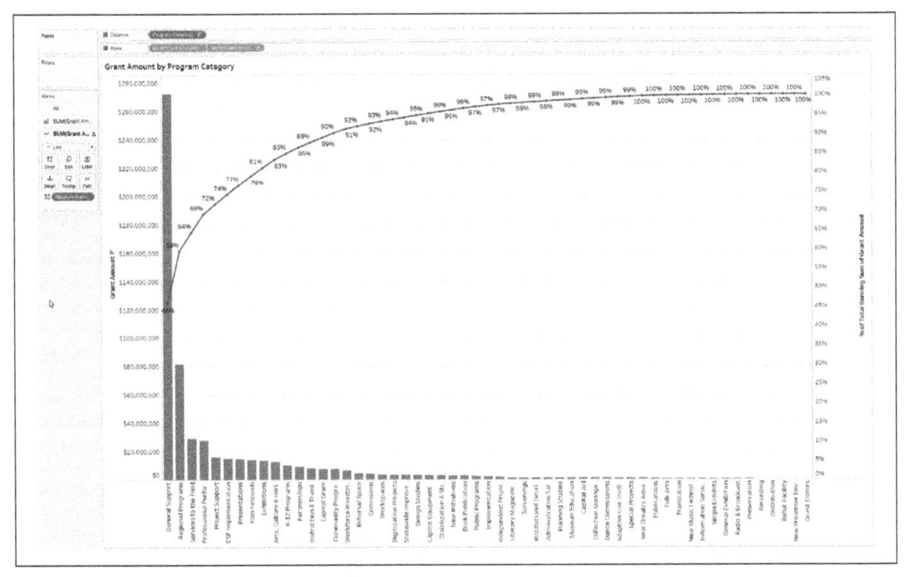

Figure 2-27. Dual-axis Pareto chart

8. Clean up the view by editing your axis labels. Ann likes to label the secondary axis as **`[Cumulative %]`**.

9. You can also add a parameter or constant reference line to highlight exactly where 80% of the data is. Now let's take a look at the finished product (Figure 2-28).

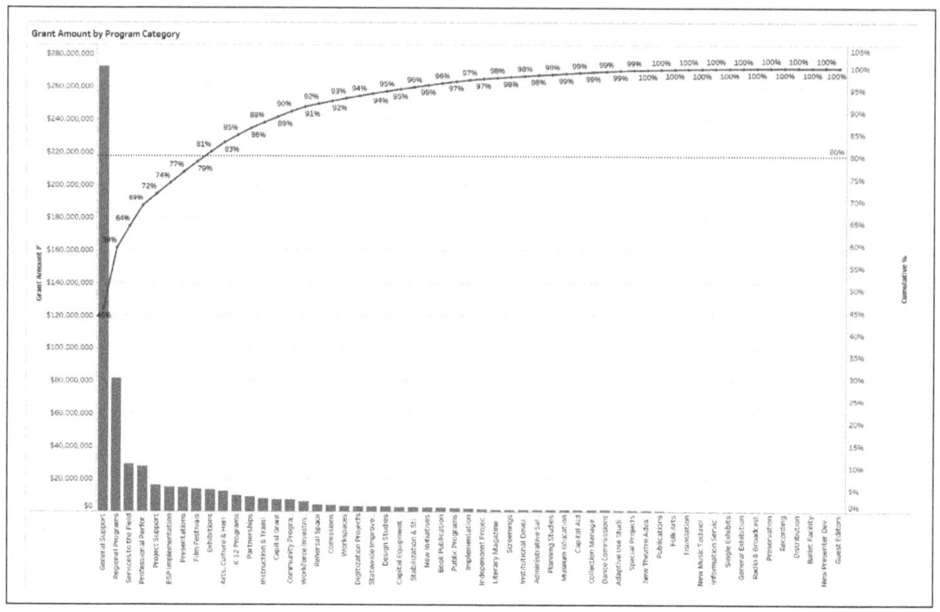

Figure 2-28. Completed Pareto chart

Does the funding at AP follow the 80/20 rule? It does: from the chart, we can see that 9 program categories, out of 55 total categories, take up 81% of the funding. In other words 16% of program categories consume 81% of funding. This means that AP's spending is actually slightly more concentrated than a Pareto distribution! We're sure that seeing funding distributed in this way will give the executive team a lot to think about.

Conclusion

Throughout this chapter, you've seen several ways to work with numeric data in the real world. We've explored ways to view the spread of data with histograms, dot plots, and jitterplots, the former allowing your audience to see summarized data, and the latter allowing them to see the individual data points that make up the total picture.

You've also worked with applied statistical concepts to create box plots that show important summary statistics (median and the IQR are your friends), and learned how you can persuade your audience to embrace this chart with the inclusion of a histogram.

And you've seen how built-in statistical calculations, like SD, can easily be added to line charts to help express the nuances in your data. Moreover, you've explored using statistical thresholds to help make your organization data driven, allowing your audience to quickly see data points that are "out of control" and guide in decision making.

And in the last two sections, you explored how even with categorical data, new statistical dimensions can be calculated to describe your data (and to test the all-important 80/20 rule).

With this knowledge, you'll be primed to take on the next chapter on comparisons, where we show you strategies for comparing both categorical data (dimensions) and measures in new chart types.

CHAPTER 3

Making Comparisons

Comparisons are at the heart of data visualization. Our audiences are often laser-focused on comparing two or more values. Our job as practitioners is to simplify the comparison of each set of values.

Some comparisons are easier than others. If you're comparing two groups on a single value, your comparison is straightforward: use a bar chart. Most times, you should just use a bar chart simply because bar charts are the most versatile and common chart type used for comparisons.

But every bar chart presents one challenge: it's a bar chart! Your audience has seen this chart type hundreds, thousands, or millions of times. In this chapter, we'll offer you a few alternatives, such as Cleveland dot plots and lollipop charts, as well as various ways to format your bar charts.

When creating these alternative charts, really think about whether you're seeking a bar chart for your audience's needs or for your own. If it's for you, keep it a bar chart. If it's for your audience, perfect—use the alternative.

Not all comparisons are as straightforward as bar charts. Sometimes you are comparing across members of a dimension and at different time points within that member. Other times you're comparing multiple members against multiple dimensions. Sometimes you're worried about the actual metric, sometimes the rank of the value. Sometimes you're interested in comparing values not against other members, but against the overall group.

In this chapter, we'll provide other methods for comparison including the bar-on-bar chart, displaying the change in rank between two time periods, bump charts, barbell plots, trellis, and parallel coordinates.

In This Chapter

You will work with two data sources: the nonprofit dataset (Nonprofit Grant Data) and the superstore retail sales dataset (Sample – Superstore). In this chapter, you'll learn how to do the following:

- Format bar charts so they have a fresh or unique look and feel
- Build a lollipop chart (and understand some of its drawbacks)
- Build a Cleveland dot plot
- Show change from year to year in a single bar chart
- Build a bump chart
- Build a barbell plot
- Build a trellis chart or create small multiples
- Create a parallel coordinates chart

Bar Charts and Alternatives: Amplify Performance Case Study

As we've already discussed, data visualization practitioners create a lot of bar charts—justifiably so. As you learned in Chapter 1, a bar chart compares two or more groups across a single metric. Bar charts are easy to use because you have to compare only the length or height of one bar against another. With a quick glance, your audience understands the magnitude of the difference.

A nonprofit in the performing arts space, AP is trying to better understand its grant program. Its executives want to know how the average grant size compares to the type of grant. They have shown the same chart to funders for the last four years and are looking for different ways to spice it up! How can they increase the visual impact of this data?

In the grant data, you will use a group called [Program] to compare the data, and you will create a measure called [Avg Grant Amount]. You will use this group and this measure for five examples.

Strategy: Compare with a Basic Bar Chart

First, let's start with a basic bar chart:

1. Connect to the Nonprofit Grant Data dataset.
2. Create a calculated field called **`[Avg Grant Amount]`**.
3. Use existing fields and write the following formula:

```
// Avg Grant Amount
SUM([Grant Amount])/COUNT([Record ID])
```

4. Add [Program] to Columns and [Avg Grant Amount] to Rows.
5. Change the mark type to Bar.
6. Rotate the labels on the Columns shelf so the text is vertical.

We started by creating this vertical bar chart because almost everyone learns how to make a bar chart by plotting it vertically. This is often reinforced when you're using document editors or spreadsheet tools. Most people can't help that the default settings are vertical bar charts. The alternative: a horizontal bar chart, which you'll make next.

But first, let's dissect this vertical bar chart. Here, [Program] is on Columns and [Avg Grant Amount] is on Rows, as you can see in Figure 3-1.

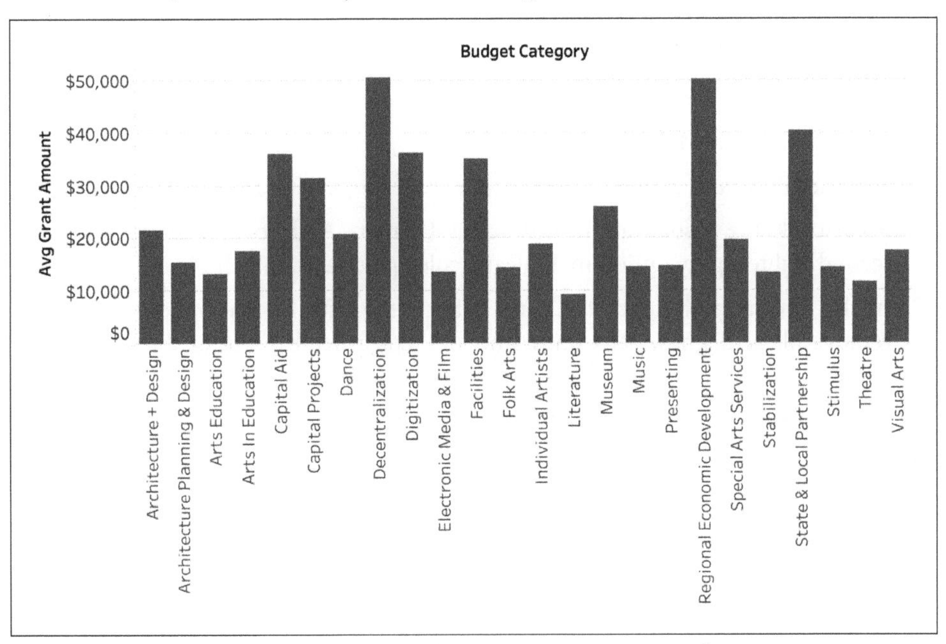

Figure 3-1. Vertical bar chart depicting the average grant amount by category

Let's look at this chart as if you were someone from your audience. At first glance, it looks pretty good. You can easily identify Decentralization and Regional Economic Development funds as two categories with the largest average grant amounts, for example.

When you start to dig into the comparisons, however, determining the order of the groups is more difficult. Although alphabetical order is great, it's not that meaningful. Additionally, to read the labels, you need to bend your neck slightly. If only there was a better way.

We almost always recommend horizontal bar charts over vertical bar charts. It's completely OK if you've made a lot of vertical bar charts before reading this. It's not your fault that your grade school teachers made you create vertical bar charts for all those math and science projects. Many of us read from left to right, especially if we grew up speaking a left-to-right language, so it's easy for us to process text and information the same way. This means choosing vertical over horizontal bars. (Remember, that doesn't mean you have to give them up.)

Strategy: Convert a Vertical Bar Chart to Horizontal

For this next example, you are going to make a horizontal bar chart:

1. Duplicate the sheet from Blueprint 3-1.
2. Set the sheet to fill the entire view.
3. Click the Swap Axes button to swap rows and columns.
4. If necessary, rotate the labels on rows.
5. Sort the [Program] group in descending order by [Avg Grant Amount].
6. Right-click the visualization and select Format. Add row dividers, using the second-lightest gray option in Tableau's color palette, to match Figure 3-2.
7. In the Format pane, click the Line button (the fifth of five options) and then remove all grid lines, zero lines, and the axis ruler by setting those values to None.
8. Size the bars to be about 50% of the maximum value.
9. Click Label on the Marks card and select "Show marks label."
10. Right-click the [Avg Grant Amount] axis and deselect Show Header to hide the header.

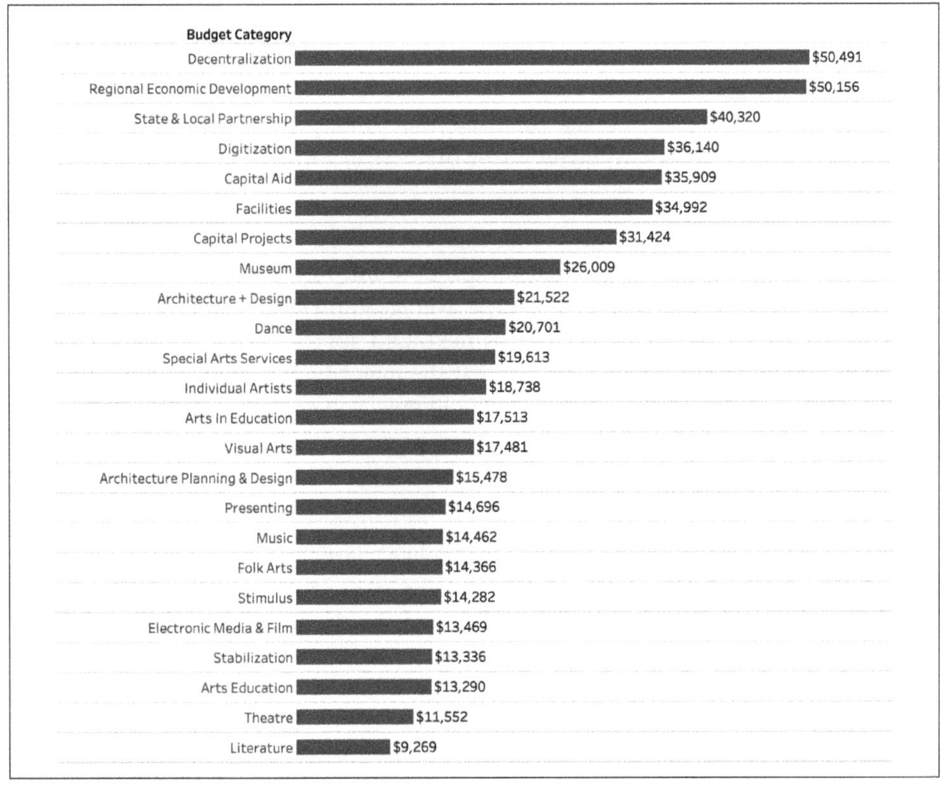

Figure 3-2. Horizontal bar charts are easier for your audience to interpret because text can be read from left to right

Horizontal bar charts open up opportunities for the labels of your bars. You could include labels to add more context to the bars, indicating a change compared to a previous period, a percent of the total, or the rank versus other members.

If you choose to add labels, you can hide the axis. Axes and labels accomplish the same goal: they give scale to the bar size. Including both means including redundant information, and as a practitioner, you want to minimize redundancy. (However, if you do find your audience often misinterpreting information, you may want to include redundancies, provide additional context in text form, or choose a different visualization to communicate.)

For this horizontal bar chart example, we've chosen to add labels, so we had you hide your axis and remove grid lines, grid rulers, and all other line types included in Tableau's Format pane. We also recommend you sort your values on the metric. It's Luke's personal preference to also increase the whitespace between each bar (by decreasing the bar width) and include row dividers in formatting options, as shown in Figure 3-2.

Now let's pretend again that you are an individual from our intended audience. As before, you can quickly see that Decentralization and Regional Economic Development are the two categories with the largest average grant amounts. But now, you can also see that State & Local Partnership is next, and approximately $10,000 lower in terms of average grant size. You can see that the majority of categories have an average grant size between $14,000 and $20,000. Finally, you can see that Arts Education, Theater, and Literature have the smallest average grant size. Of course, all this information is available in the vertical bar charts, but with our horizontal bar chart, we can get to these insights faster.

Now that we've made our pitch for horizontal bar charts, let's talk about alternatives that help your audience break up the routine of seeing bar charts. Next, we'll show you how to create lollipop charts and single-point, or Cleveland, dot plots.

Strategy: Create a Lollipop Chart

One of the easiest ways to break up the monotony of bar charts is to create a *lollipop chart*. This is simply a bar chart with a circle at the end of each bar (Figure 3-3). For whatever reason—perhaps it's the rounded endpoints—audiences and developers love to use this chart type.

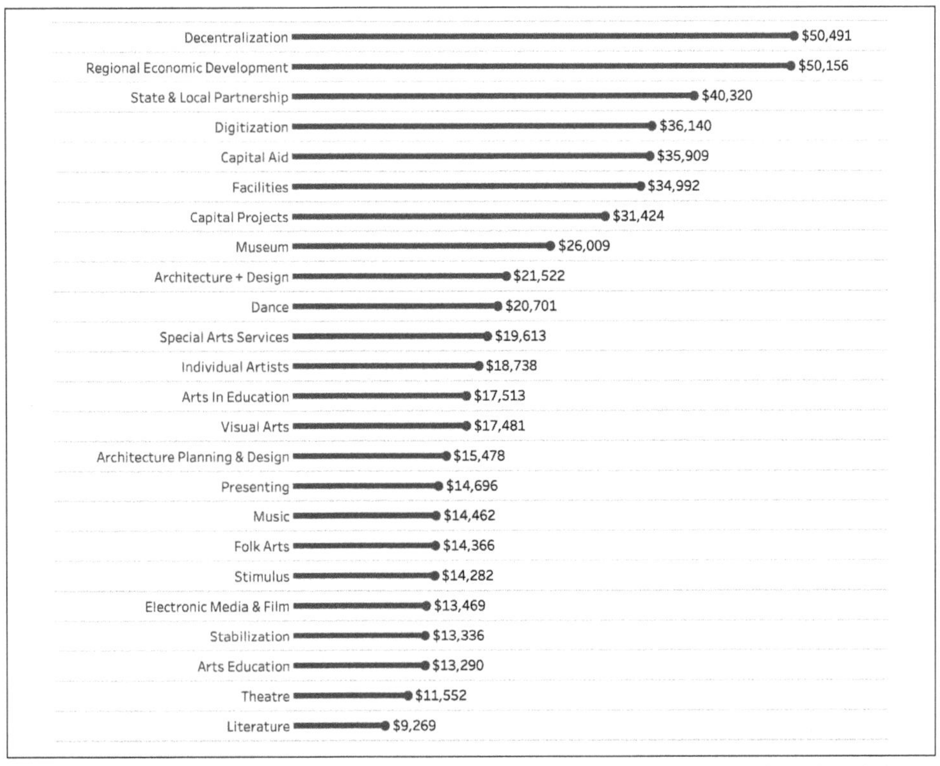

Figure 3-3. A lollipop chart is a quick way for you or your audience to take a break from looking at bar charts—even though a lollipop chart is a bar chart

Follow these steps to create your lollipop chart:

1. Create a new sheet.
2. Create a dual-axis chart using the same dimension—in this example, using [Avg Grant Amount]. This will give you two Marks cards you can control. If you haven't made a dual-axis chart, you can do this by adding [Avg Grant Amount] to the columns axis twice. Then right-click the rightmost measure on Columns and select Dual Axis. On the top axis on the view, right-click and select Synchronize Axis. The result is shown in Figure 3-4.

Synchronize Axis does exactly what it sounds like: it makes sure that the unit of measure is identical across a dual-axis chart. If you're using the same metric in both axes, you'll want to ensure that this option is enabled.

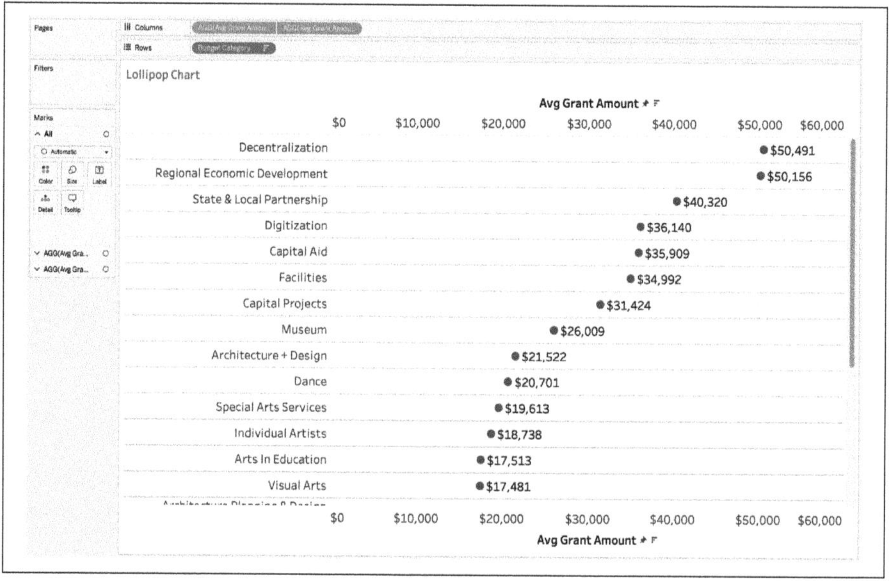

Figure 3-4. The first step to a lollipop chart

3. For Marks card 1, click the leftmost [Avg Grant Amount] on Columns. Then make these changes:
   a. Change the mark type to Bar.
   b. Change the size of the bar to be smaller than the circle.
4. For Marks card 2, click the rightmost [Avg Grant Amount] measure on Columns. Then make these changes:
   a. Change the mark type from Automatic to Circle.
   b. Show the mark labels on this Marks card and change the alignment to be right-middle.
   c. Adjust the circle size if necessary. See Figure 3-5 for the result.

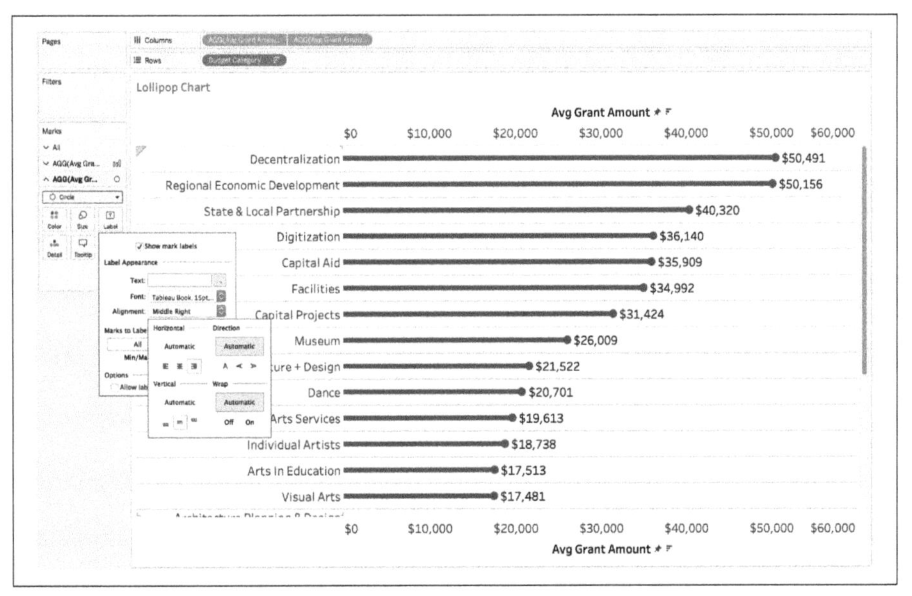

Figure 3-5. Create a synchronized dual axis with a bar and a circle to make a dual-axis chart. Be sure to add labels to the circle and not the bar.

5. Format the visualization as follows:
   a. Hide your headers by right-clicking and unchecking Show Header on the axes.
   b. Remove grid lines, zero lines, and axis rulers.
   c. Add row dividers.

   This produces the visualization in Figure 3-6.

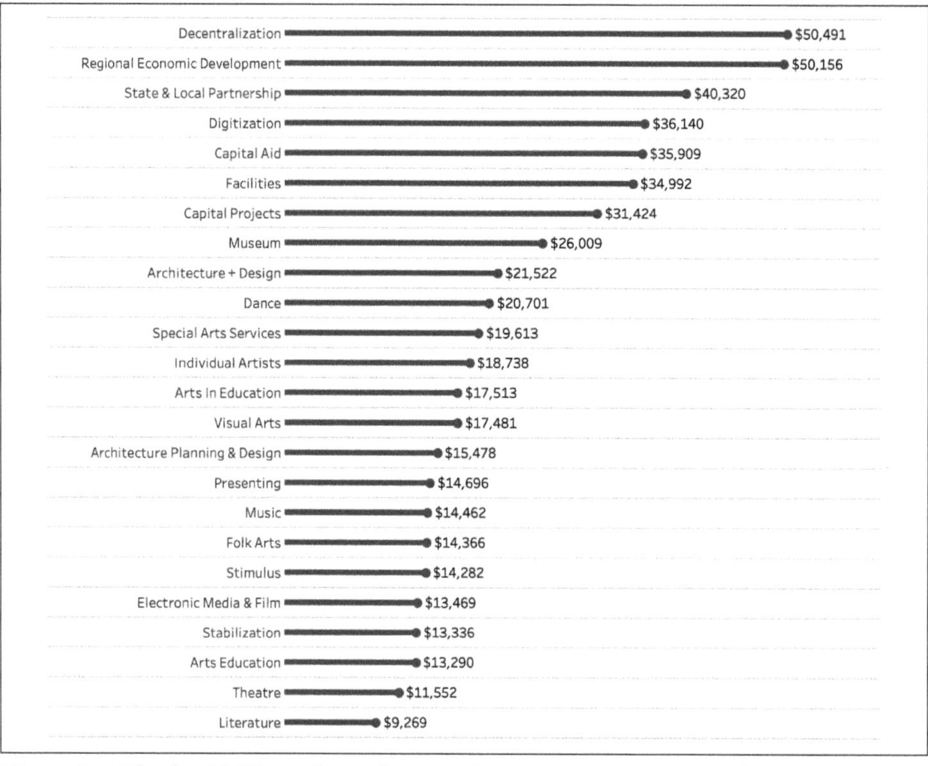

Figure 3-6. The final lollipop chart, showing the average grant amount by category

Audiences and developers love lollipop charts. They provide a subtle way to take a break from bar charts, and building a lollipop chart doesn't take a lot of work.

One drawback of lollipop charts: they are not a 100% accurate representation of the data. A bar chart is an accurate representation of the data, but then we tack a circle onto the end of each bar. The center of the circle is aligned with the center of the bar. This means that half of the circle misrepresents the data (Figure 3-7).

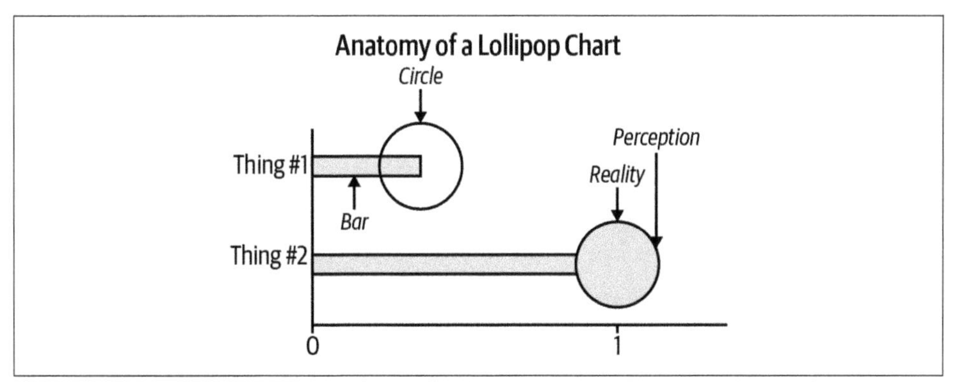

Figure 3-7. Lollipop charts are not pixel-perfect when it comes to creating a 100% accurate depiction of the visualization, but that's OK

Here's our take on this inaccuracy: it is our job to create visualizations that approximate the data. Most chart types are already approximations of the data. If this level of abstraction would change readers' interpretation of the data, don't use it. If it's still a close approximation, feel free to use it. Just be ready to justify your decision!

Strategy: Create a Basic Cleveland Dot Plot

As we mentioned at the beginning of the chapter, our audiences are looking for alternatives to bar charts. Luke's personal favorite is the *Cleveland dot plot*. He likes to use it because it increases the spacing of the data elements and de-emphasizes nondata elements while still providing the same information as a bar chart.

The Cleveland dot plot (*https://www.jstor.org/stable/2288400?seq=1*) is not from the city of Cleveland, Ohio. Rather, it's a design published by William Cleveland and Robert McGill in 1984. With the dot plot, we show our values with circles along an axis. We then display the value of each group at the end of the axis, as shown in Figure 3-8.

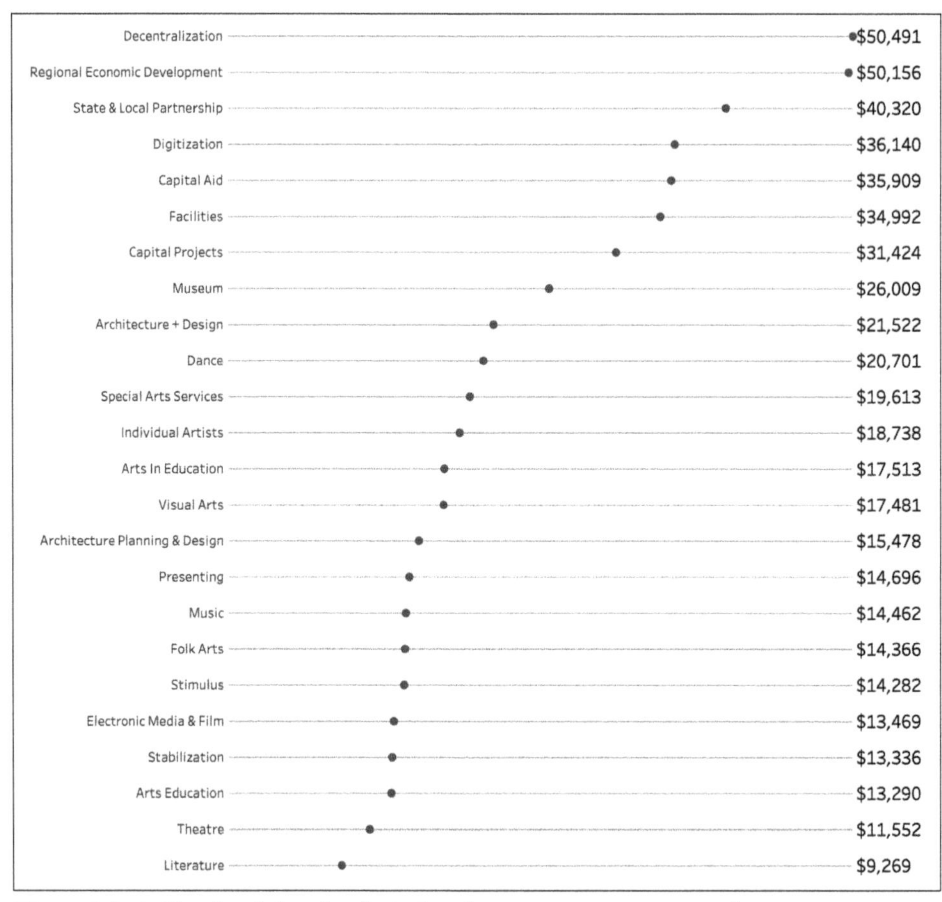

Figure 3-8. A Cleveland dot plot depicting the average grant amount by category

What we like about the dot plot over the bar chart is that we can increase our overall whitespace, maintain existing comparisons of the values, and align the text/values in a single column. We can also use the amount labels as references to quickly understand the range of our data (recall the dot plots from Chapter 2).

Follow these steps to build a basic Cleveland dot plot:

1. To build the visualization, create a new sheet.
   a. Add [Program] to Rows.
   b. As you did in the lollipop chart, add [Avg Grant Amount] to Columns twice and create a synchronized dual axis.

2. On the leftmost [Avg Grant Amount] measure on Columns, double-click and edit the calculation. Wrap [Avg Grant Amount] in the `WINDOW_MAX()` function (Figure 3-9).

   The `WINDOW_MAX()` calculation is a table calculation that will return the maximum value across all [Avg Grant Amount]. Edit the table calculation by right-clicking and setting Compute Using to "Table (down)."

> Table calculations allow you to do analysis on top of the data shown in the visualization you're creating. While we won't go into too much detail on *how* they function, you can check out Tableau's own Top 10 table calculations blog (*https://oreil.ly/jHVud*) for more insight.

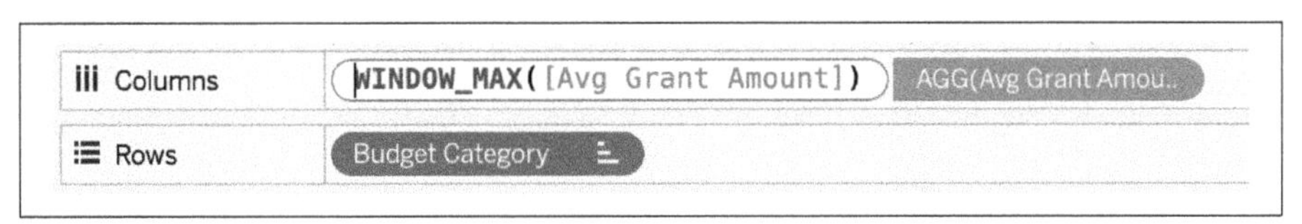

Figure 3-9. Creating an ad hoc `WINDOW_MAX()` calculation on columns

3. Customize your bars as follows:

   a. Change the mark type from Automatic to Bar. Change the size of the bar to the smallest value possible. We don't want the bar to stand out too much, so change its color to a light gray.

   b. Add [Avg Grant Amount] to Label.

   c. If you want to make your background bar size even narrower, you can click Color on the Marks card and set the border color to match the background. After this step, your visualization should resemble Figure 3-10.

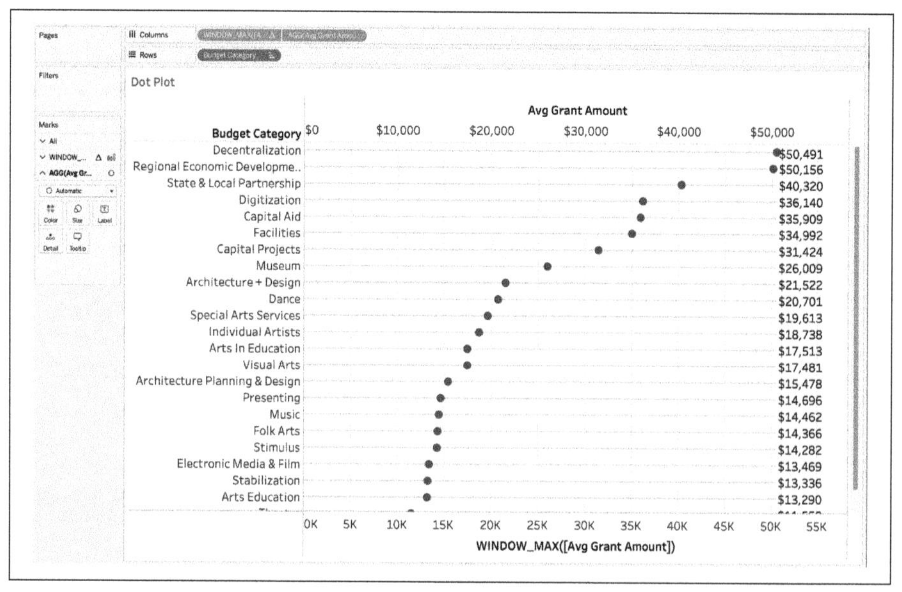

Figure 3-10. A sneak-peek into the Cleveland dot plot prior to adjusting formatting

4. Adjust the circles as follows:
   a. Click the rightmost [Avg Grant Amount] on Columns. Change the mark type from Automatic to Circle.
   b. Change the size of the circle. We prefer a smaller circle; that is, less than 50% of the distance between the lines that we created with `WINDOW_MAX()`. Sizing this mark must be done manually, so do your best to eye the size to something that seems appropriate.
5. Finalize the visualization (Figure 3-11):
   a. Adjust the color to something that stands out from the background lines.
   b. Hide the axes.
   c. Remove all lines and dividers in the Format pane.

Category	Average grant
Decentralization	$50,491
Regional Economic Development	$50,156
State & Local Partnership	$40,320
Digitization	$36,140
Capital Aid	$35,909
Facilities	$34,992
Capital Projects	$31,424
Museum	$26,009
Architecture + Design	$21,522
Dance	$20,701
Special Arts Services	$19,613
Individual Artists	$18,738
Arts In Education	$17,513
Visual Arts	$17,481
Architecture Planning & Design	$15,478
Presenting	$14,696
Music	$14,462
Folk Arts	$14,366
Stimulus	$14,282
Electronic Media & Film	$13,469
Stabilization	$13,336
Arts Education	$13,290
Theatre	$11,552
Literature	$9,269

Figure 3-11. The final results: a Cleveland dot plot depicting the average grant amount by category

As we mentioned earlier, the dot plot provides us with the same information as a bar chart, while increasing whitespace, retaining the ease of comparison, and vertically aligning the labels. We could get picky around right-aligning labels or increasing whitespace between the line and the labels, but for most audiences, the basic dot plot here will more than suffice.

Bar Charts for Rank Changes: AP Case Study

Not all comparisons require you to examine the magnitude across or within a measure. Sometimes your audience is merely interested in comparing how different groups change in rank and change in rank over periods. Tableau provides many ways to show ranks of values. We will start with a straightforward approach: showing the current rank in total grant dollars and the change in rank from one year to the next.

Now that we have a better understanding of the monetary amounts for the grants that were distributed, we want to compare grants by group and year. Let's explore how to show changes in rank order.

AP is taking a closer look at its books, and is trying to better understand funding trends on an annual basis. Executives want to understand how much funding has been given to each industry segment they serve and whether an increase or decrease occurred in each segment from 2018 to 2019. How would you build a visual that shows the relationship between these variables?

Strategy: Show Rank and Change of Rank on a Bar Chart

In this example, we'll again display the total grant dollars for 2019. This time, we'll display the rank relative to all other groups in 2019—but also the total change in rankings from 2018 to 2019 on the same metric. Figure 3-12 shows a preview.

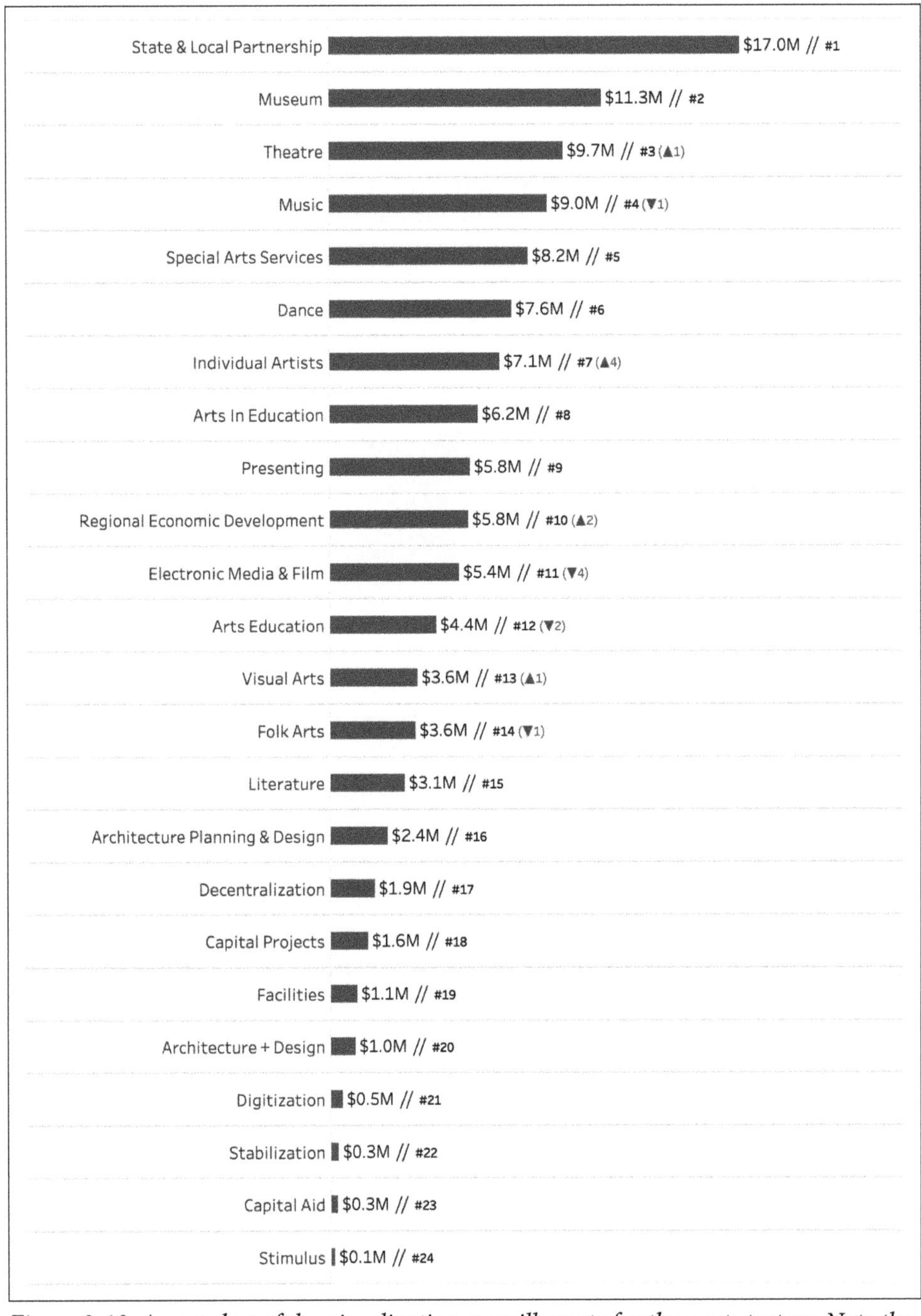

Figure 3-12. A snapshot of the visualization we will create for the next strategy. Note the rank values—and change in rank to the right side of the // dividers.

The reason for using rank in charts is simple: we humans can process only so much information. And understanding the change between two measurements is sometimes easier to understand when comparing the changes in rankings. Here are the steps for this strategy:

1. Create the base bar chart by adding [Program] to Rows, and [Grant Amount | 2019] to Columns as well as to Label on the Marks card (Figure 3-13).

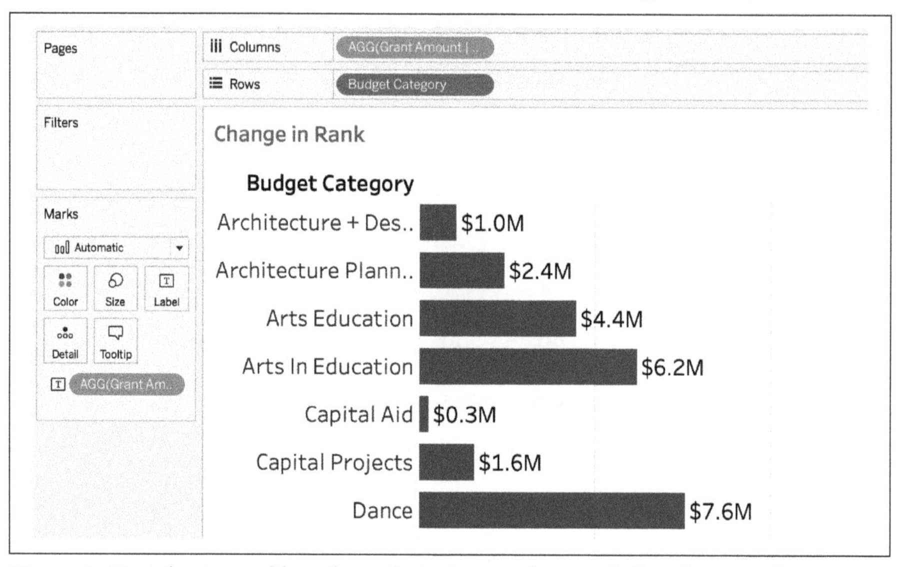

Figure 3-13. A horizontal bar chart depicting total grant dollars for 2019 by category

Several rank calculations are built into Tableau, including `RANK()`, `RANK_DENSE()`, `RANK_MODIFIED()`, `RANK_PERCENTILE()`, and `RANK_UNIQUE()`. Although these calculations would make sense to use, we are going to use the `INDEX()` function. Since it doesn't do any comparisons of the ranking number within the calculation, it tends to be more reliable and flexible.

2. Create the rank calculation as follows:

   a. Create a new calculation called **`[Index | Grant Amount | 2019]`**:

   ```
   // Index | Grant Amount | 2019
   INDEX()
   ```

   `INDEX()` creates a running count of values based on the way you specify your table calculation. We'll set up how [Index | Grant Amount | 2019] performs this running count by adjusting the table calculation on the view.

b. Add [Index | Grant Amount | 2019] to Label on the Marks card. You'll notice that the function just adds a running value from top to bottom by mark on the view (Figure 3-14).

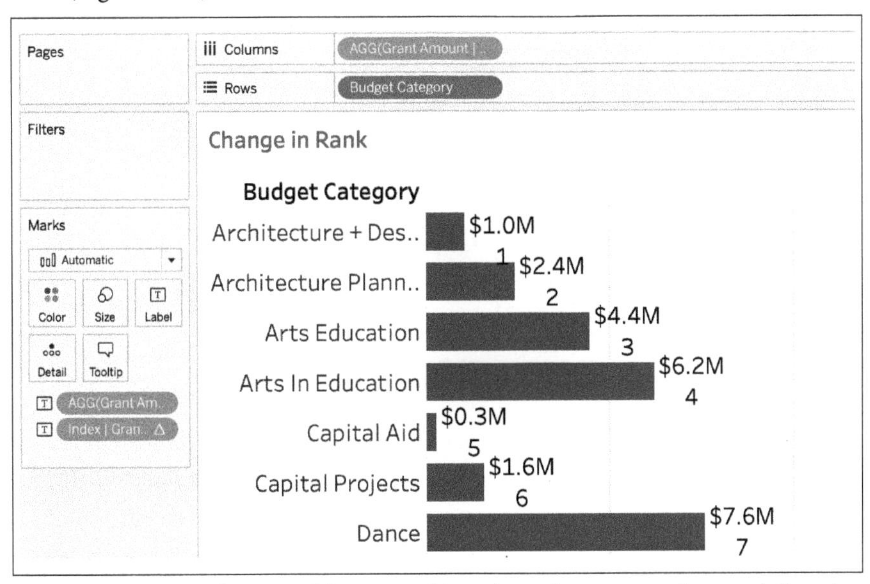

Figure 3-14. An unformatted horizontal bar chart showing total grant dollars in 2019 and an unsorted `INDEX()` value

c. To make the calculation work like a rank, edit the table calculation by right-clicking the value on the Marks card and editing the Table Calculation. Select Specific Dimensions and make sure Program is selected. Now you can edit the sort order. Choose Custom and Descending on [Grant Amount | 2019]. The function should now look like a unique rank (Figure 3-15)!

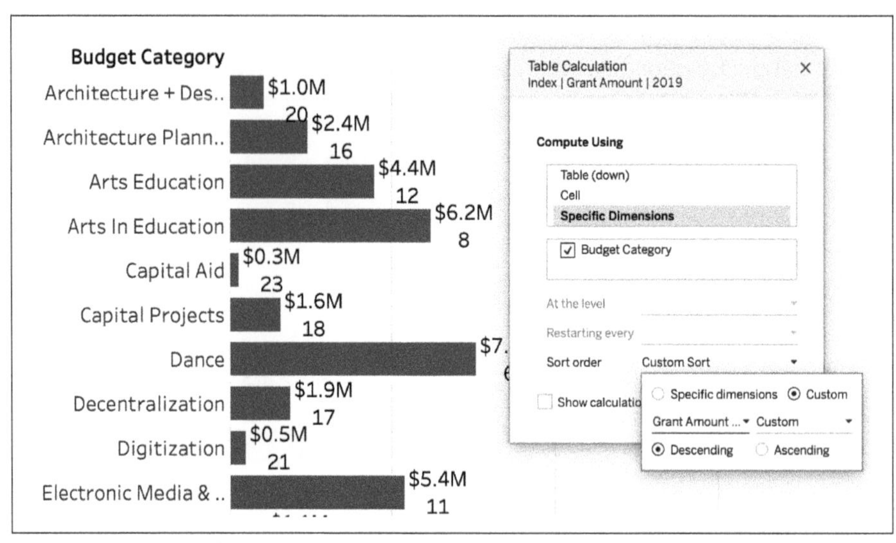

Figure 3-15. An unformatted horizontal bar chart showing total grant dollars in 2019 and a sorted INDEX() value based on the total grant amounts in 2019

3. Create the difference-in-rank calculation:

   a. Create a duplicate of [Index | Grant Amount | 2019] and call the new calculation **`[Index | Grant Amount | 2018]`**:

   ```
   // Index | Grant Amount | 2018
   INDEX()
   ```

   b. For your next calculation, you are going to take the difference between the two index calculations. Create a calculation called **`[Index Change | Grant Amount]`**:

   ```
   // Index Change | Grant Amount
   [Index | Grant Amount | 2018] - [Index | Grant Amount | 2019]
   ```

   c. Add this calculation to Text on the Marks card. Edit the table calculation. You will have the ability to customize both [Index | Grant Amount | 2018] and [Index | Grant Amount | 2019]. These table calculations are shown in Figure 3-16.

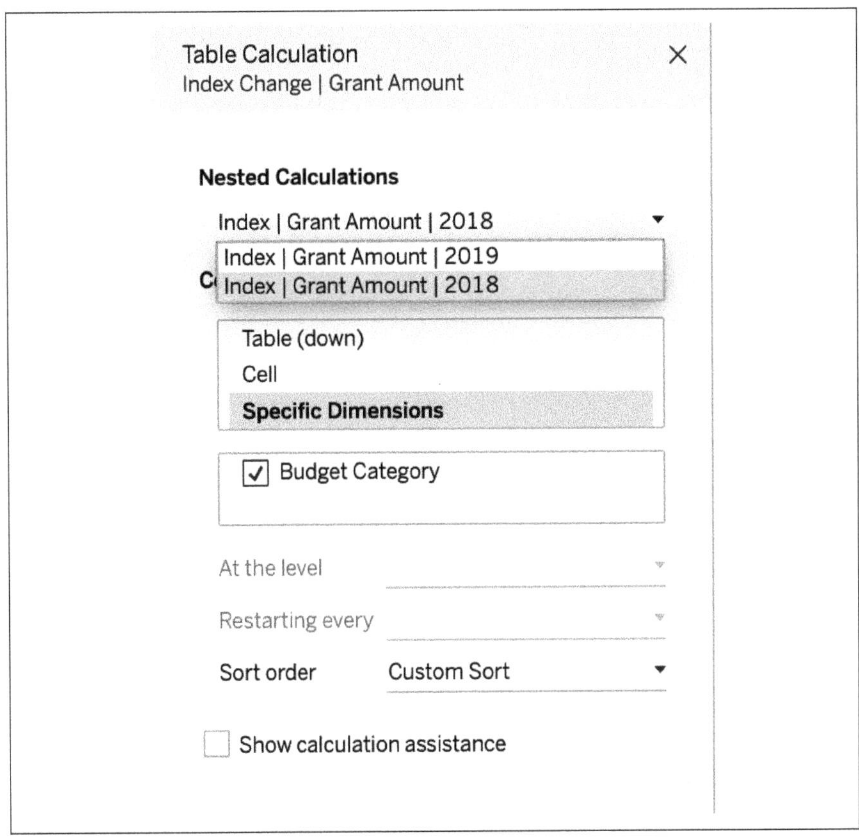

Figure 3-16. Be sure to edit the table calculations for both 2018 and 2019. Use the drop-down to select the table calculation you would like to update.

This is an important step because the next calculation relies on a correct sort for both [Index | Grant Amount | 2019] and [Index | Grant Amount | 2018].

d. Be sure to do a custom sort in descending order for [Index | Grant Amount | 2019] by using [Grant Amount | 2019] and a custom sort in descending order for [Index | Grant Amount | 2018] by using [Grant Amount | 2018]. This will give you a proper calculation showing the change in rank from 2018 to 2019.

4. Once the calculation is displaying results properly, right-click and edit the text formatting of [Index Change | Grant Amount]. Use custom formatting and write `(↑0);(↓0);""`, as shown in Figure 3-17. You can type the up arrow by holding down the Alt key while typing 24, and the down arrow by pressing Alt-25.

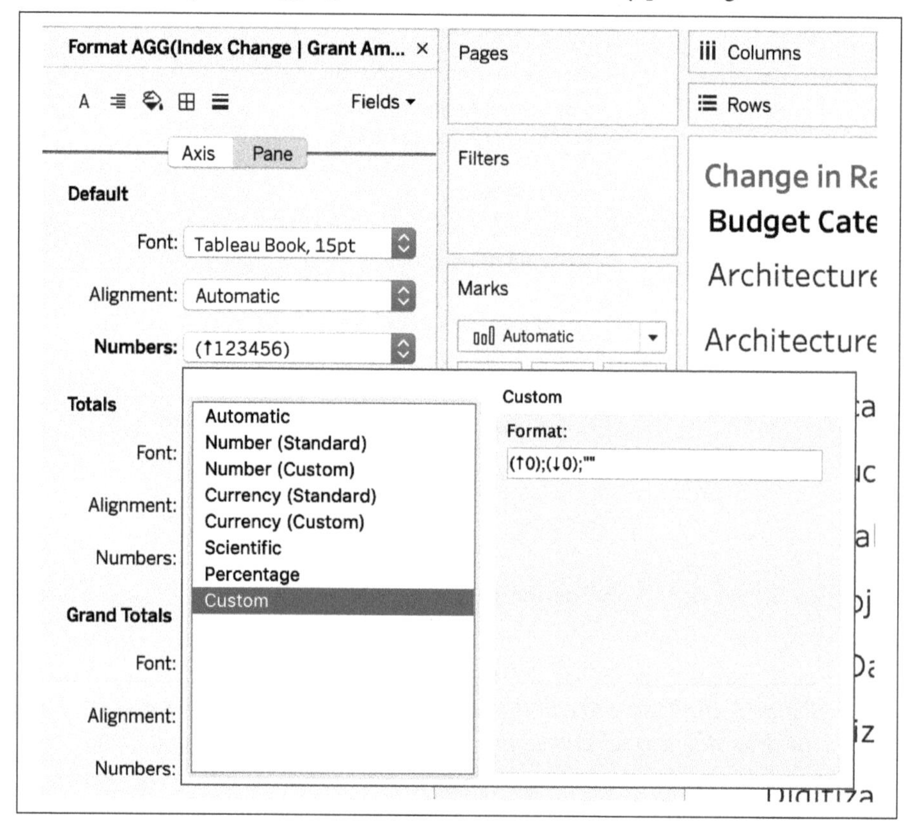

Figure 3-17. Use custom text formatting to add arrows to both positive and negative values

This custom formatting formats values one of three ways: when values are positive, when values are negative, or when values are zero. These are specified in an order of positive, negative, and zero and are separated by the semicolon. Based on these rules, the positive values will be in parentheses with an up arrow preceding the value, negative values will be in parentheses with a down arrow preceding the value, and if there is no change in rank, nothing will show up because we are using empty quotations.

Once you've formatted the display of your labels, the visualization will look something like Figure 3-18.

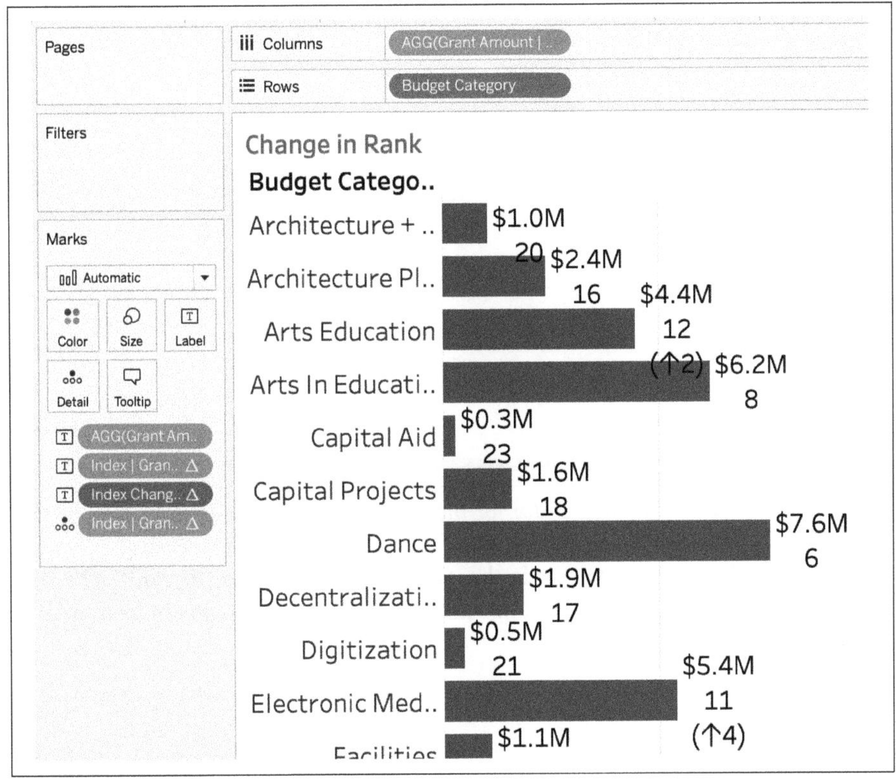

Figure 3-18. An unformatted horizontal bar chart showing total grant dollars in 2019, a rank of total grant dollars in 2019 using an unsorted `INDEX()` value, and a change in rank using nested `INDEX()` functions in a single calculation

5. Now that you have all the components on the dashboard, you need to finalize the view by formatting how the text is displayed on the visualization, formatting lines and dividers, and sorting the categories.

   Start by editing the text label. For this strategy, you're going to place all the measures on a single line and add two slashes to help separate the ranks from the values (Figure 3-19).

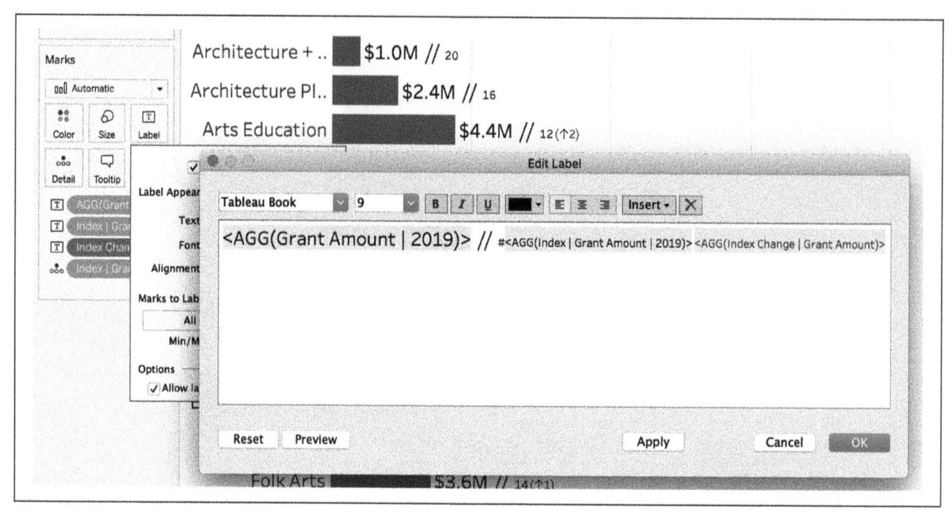

Figure 3-19. A view of the text editor for the labels of the change-in-rank bar chart

What's great about this strategy is the display of the change in rank. The element allows the audience to see how change affects the order of categories from one period to the next (Figure 3-20).

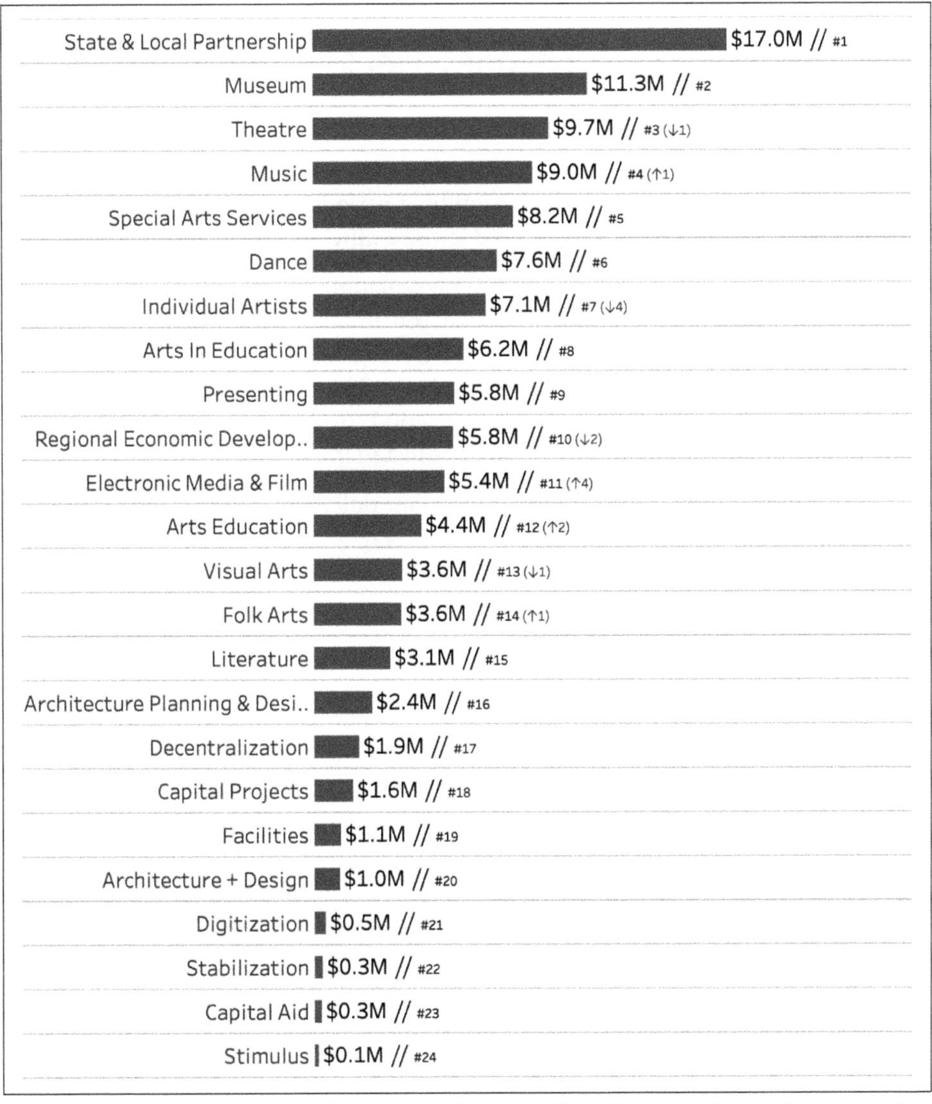

Figure 3-20. The final version showing total grant dollars for 2019, the rank compared across 24 categories, and the change in rank from 2018 to 2019

Bump Charts for Rank Changes over Multiple Periods: AP Case Study

The previous strategy focused on the change in rank over just two periods. What if you are trying to track this period over multiple periods? Certainly text with arrows for several bar charts won't work. We need to use a different chart type.

The board at AP was so impressed by the presentation on funding by industry segment for 2018 and 2019 that they've asked you for more information. They have requested an assessment on the same industry segments that indicate how much grant funding was distributed on an annual basis. This time, they would like to see this information represented for a five-year period, from 2015 to 2020. How would you create a visualization that is still clear to read and impactful?

First, let's look at a line chart showing total grant dollars by quarter and consider it from the perspective of your audience (Figure 3-21).

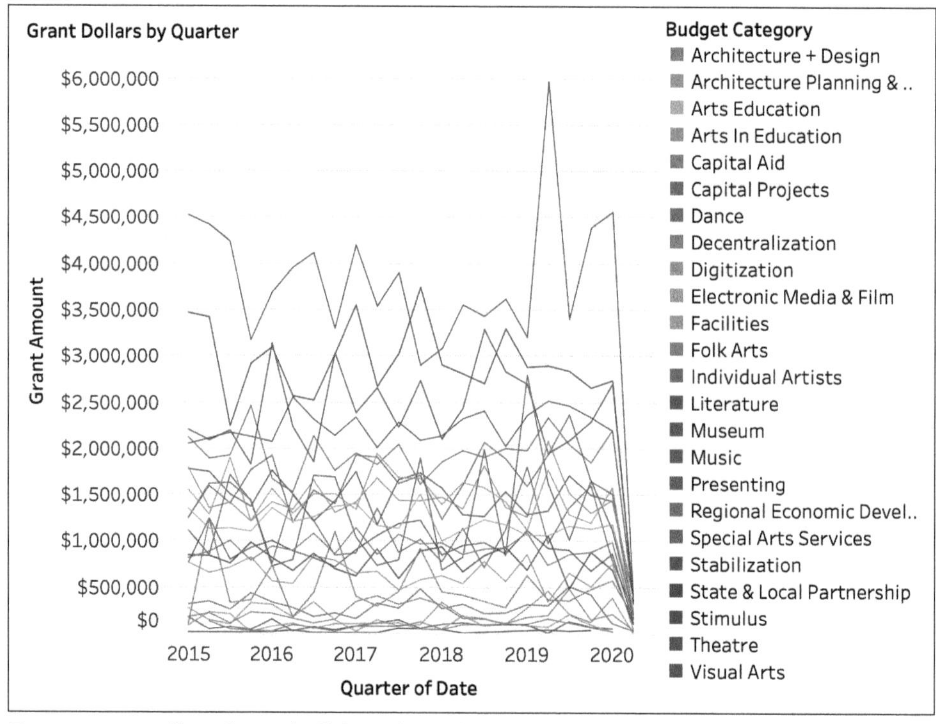

Figure 3-21. A (less-than-ideal) line chart showing total grant dollars by quarter for each category

This line chart has a lot of lines, but they are surprisingly not too difficult to track. Normally, you should avoid more than eight lines on a line chart because often too many overlap. But the audience is more interested in how any category compares to others, and this plot doesn't show that change over time. What's needed is a plot that describes this change in terms of rank.

One of the best chart types for change in rank over time is a *bump chart*. Using a bump chart allows you to see how changes in rank occur over time, without getting completely lost in the information. Instead of building the preceding chart, let's build a bump chart that allows us to select a single category (Figure 3-22).

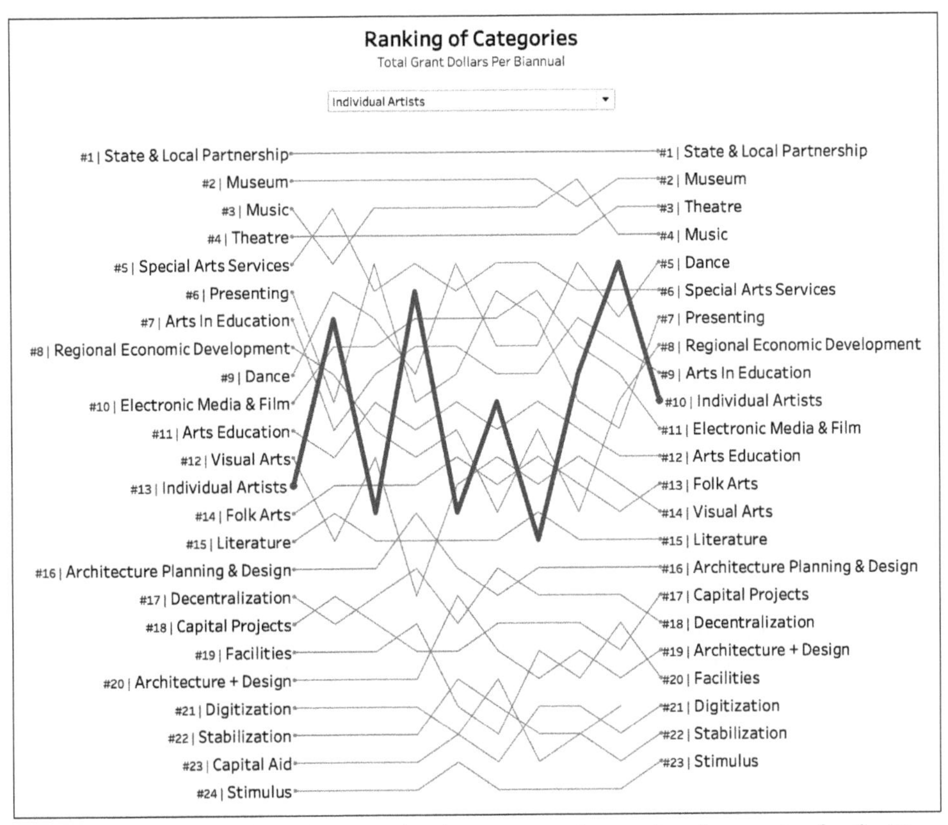

Figure 3-22. A bump chart showing the change in rank over time by quarter for the 24 categories

So how do you create a bump chart?

Strategy: Make a Bump Chart

A bump chart is fundamentally different from a line chart. This bump chart strategy shows how the ranks of categories change biannually from 2015 through 2019. The information does not care how close the underlying grant totals are; rather, it concerns only the overall rank. In this case, a bump chart shows the cyclical nature of grants for individual artists relative to other categories:

1. Create a line chart as follows:

   a. Add `SUM([Grant Amount])` to Rows and [Program] to Detail on the Marks card.

   b. Create a new calculated field called **`[Date | Biannual]`** that will return dates to the nearest biannual—either January 1 or July 1:

   ```
   // Date | Biannual
   IF MONTH([Date]) > 6
   THEN DATEADD('month', 6, DATETRUNC('year', [Date]))
   ELSE DATETRUNC('year', [Date])
   END
   ```

   Place this calculation on Columns and then right-click and change to Exact Date.

   c. Add a context filter using [Date | Biannual] and select all years but 2020.

   A *context filter* is slightly different from a standard dimension filter because it filters data from being included in a fixed LOD calculation. Standard dimension filters are applied to a view after LOD calculations are completed.

   d. Add a rank table calculation to `SUM([Grant Amount])` on Rows. Set the rank to be unique and descending (Figure 3-23). Compute the calculation on [Program].

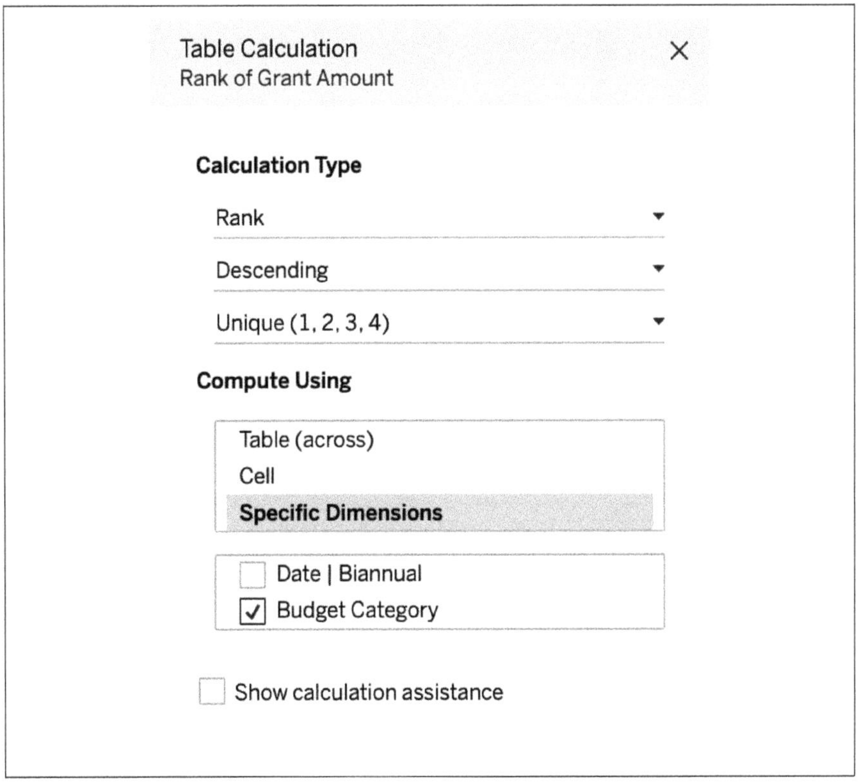

Table Calculation
Rank of Grant Amount

Calculation Type

Rank

Descending

Unique (1, 2, 3, 4)

Compute Using

Table (across)
Cell
Specific Dimensions

☐ Date | Biannual
☑ Budget Category

☐ Show calculation assistance

Figure 3-23. Creating a unique descending rank of total grant dollars for every date by Program

e. Drag a copy of the measure to the Tables area of the Data pane. This will create a new saved calculation. Edit the name of the calculation to **`[Grant Amount | Rank]`**.

   Once you've created the calculation, your visualization will look like Figure 3-24.

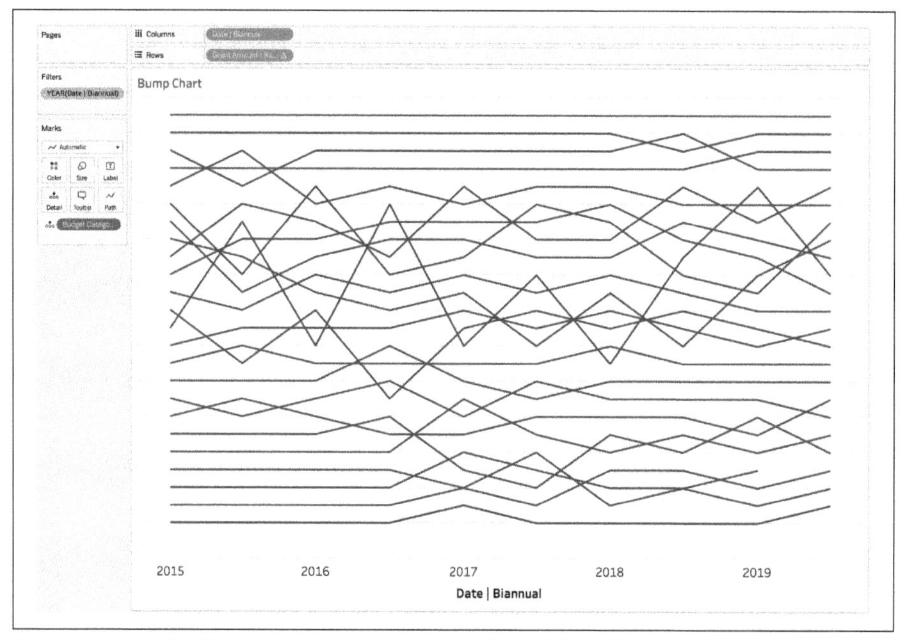

Figure 3-24. A look at the work-in-progress after adding the rank calculation

2. Prepare the labels as follows:

   a. Edit the rank axis and reverse the axis. Hide the rank axis.

   b. To add the labels, create two calculations. The first should be called **`[Bump Category Label]`**:

```
// Bump Category Label
IF [Date | Biannual] = {MIN([Date | Biannual])}
OR [Date | Biannual] = {MAX([Date | Biannual])}
THEN [Program]
END
```

   Call the second **`[Bump Rank Label]`**:

```
// Bump Rank Label
IF MAX([Date | Biannual]) = MIN({MAX([Date | Biannual])})
OR MIN([Date | Biannual]) = MAX({MIN([Date | Biannual])})
THEN [Grant Amount | Rank]

END
```

   We will discuss parts of these calculations in Chapter 4. Just know for now that both calculations will show labels on the starting and ending dates for each category and corresponding rank.

3. Build the labels as follows:

   a. Add [Bump Category Label] as an attribute to labels.

   b. Add [Bump Rank Label] to Label and change the table calculation to specific dimensions on [Program].

   c. Edit the axis of [Date | Biannual] to range from 1/1/2013 to 1/1/2022, and then hide the axis. We are setting the range of the axis to go beyond the values in the data to make space for labels.

   d. Right-click and format [Bump Rank Label] so that there is a **#** in front of the rank values (Figure 3-25).

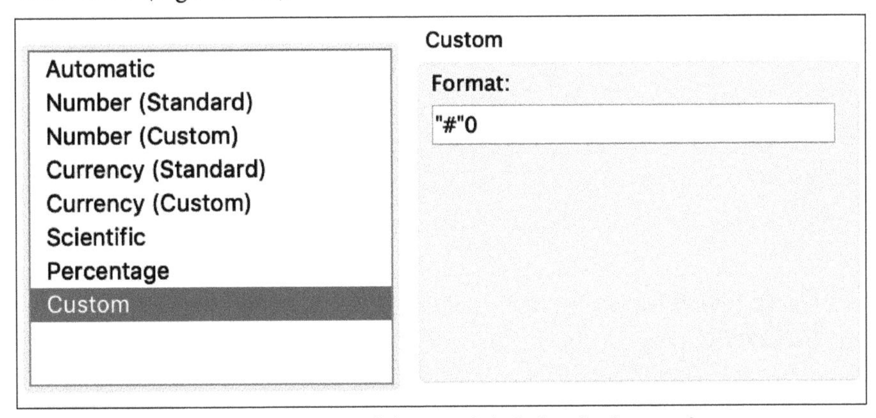

Figure 3-25. Custom formatting of the text labels for the bump chart

   e. After formatting the text, align the labels to the middle. Figure 3-26 shows the chart at this point.

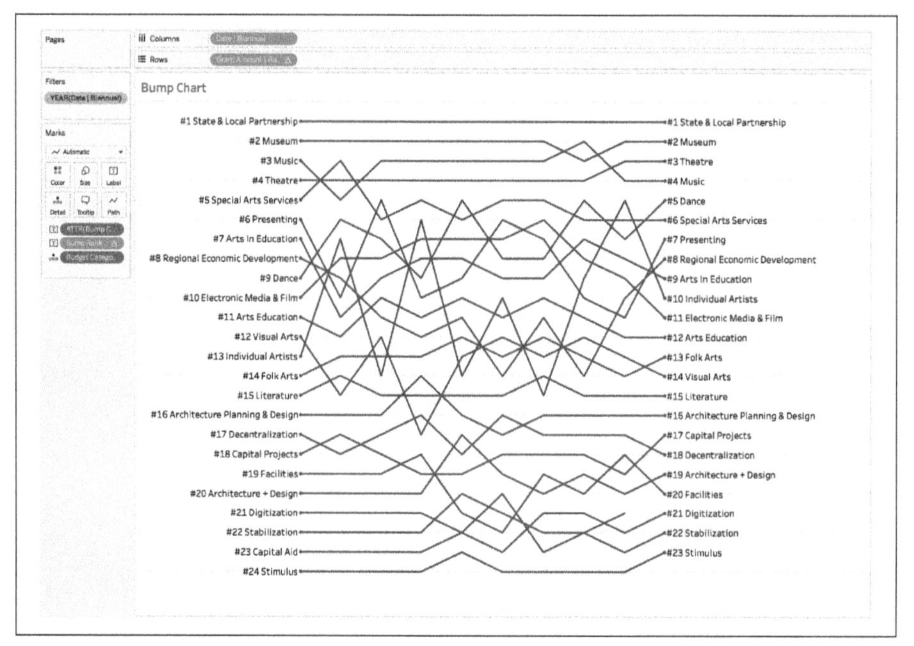

Figure 3-26. A look at the work in progress for the bump chart prior to adding highlighting

4. Highlight a single category:

   a. Create a parameter from [Program] called **`[Program Parameter]`**.

   b. Create a calculation from the dimension and parameter called **`[Program | TF]`**:

   ```
   // Program | TF
   [Program] = [Program Parameter]
   ```

   c. Place this new dimension on Size and Color on the Marks card.

   d. Edit the colors so False values are gray and True values are blue. Place the True values in front of the False values, and make the size of the True values larger than that of the False values.

   e. Edit the table calculation for both [Grant Amount | Rank] and [Bump Rank Label] to include both [Program] and [Program | TF] as the specific dimensions used to compute the table calculations (Figure 3-27).

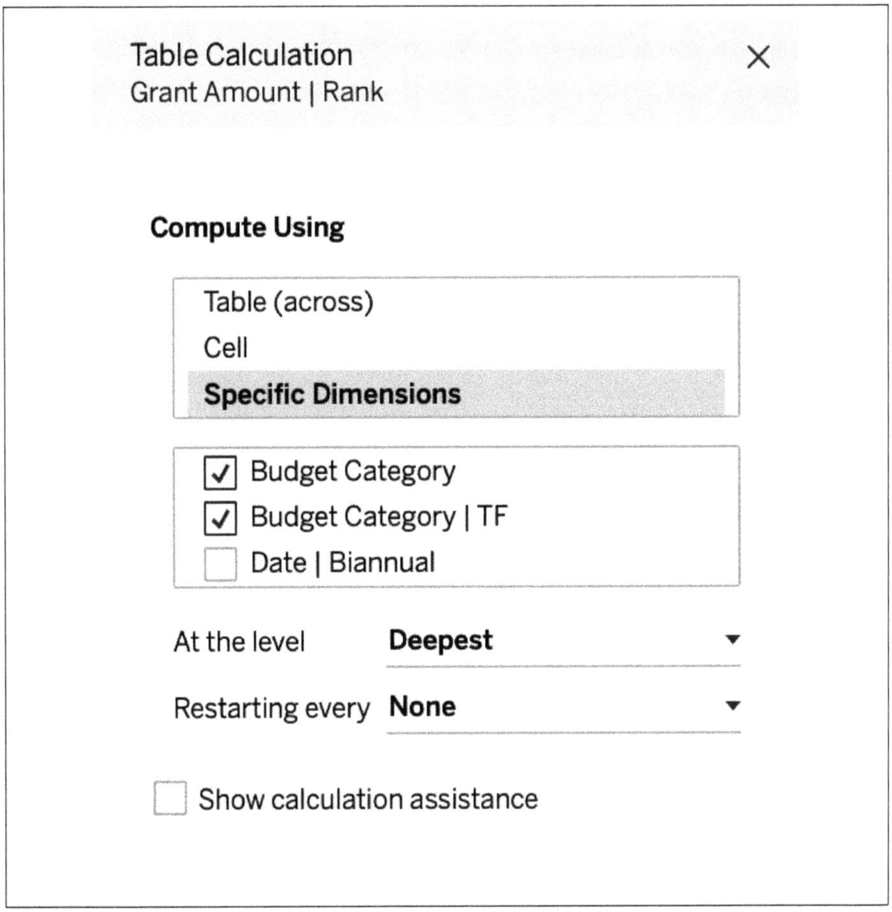

Table Calculation
Grant Amount | Rank

Compute Using

Table (across)
Cell
Specific Dimensions

☑ Budget Category
☑ Budget Category | TF
☐ Date | Biannual

At the level **Deepest**

Restarting every **None**

☐ Show calculation assistance

Figure 3-27. Adjust the table calculation of your rank calculation to include [Program | TF] as a selected dimension

After completing this step, your visualization will look like Figure 3-28.

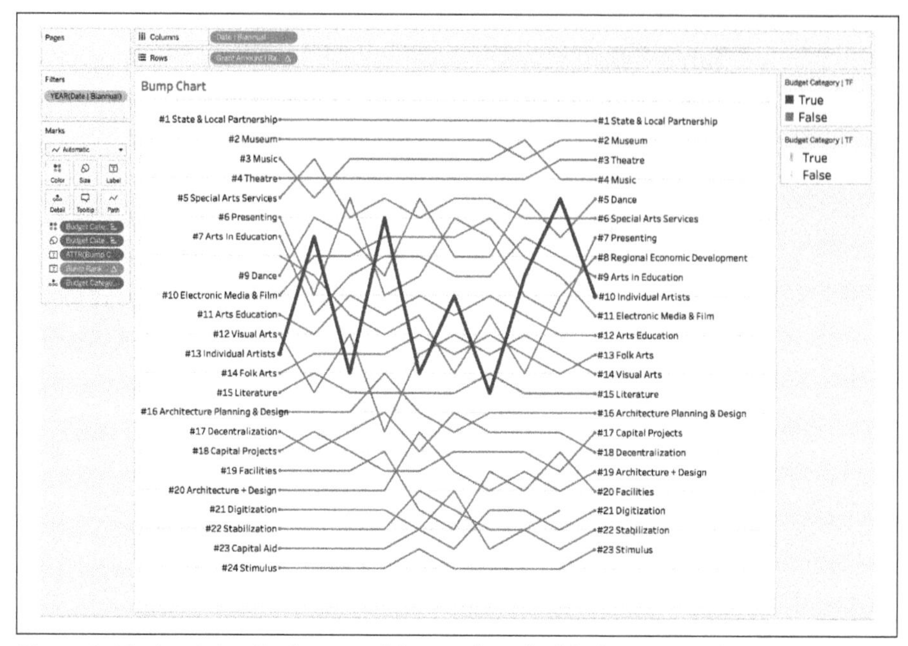

Figure 3-28. A minimally formatted bump chart highlighting a single category

5. Format the bump chart as follows:
   a. Remove all lines and dividers.
   b. It's also worth formatting your tooltips to include `SUM([Grant Amount])`. We prefer to go beyond the default formatting of tooltips.

   Figure 3-29 shows the resulting bump chart.

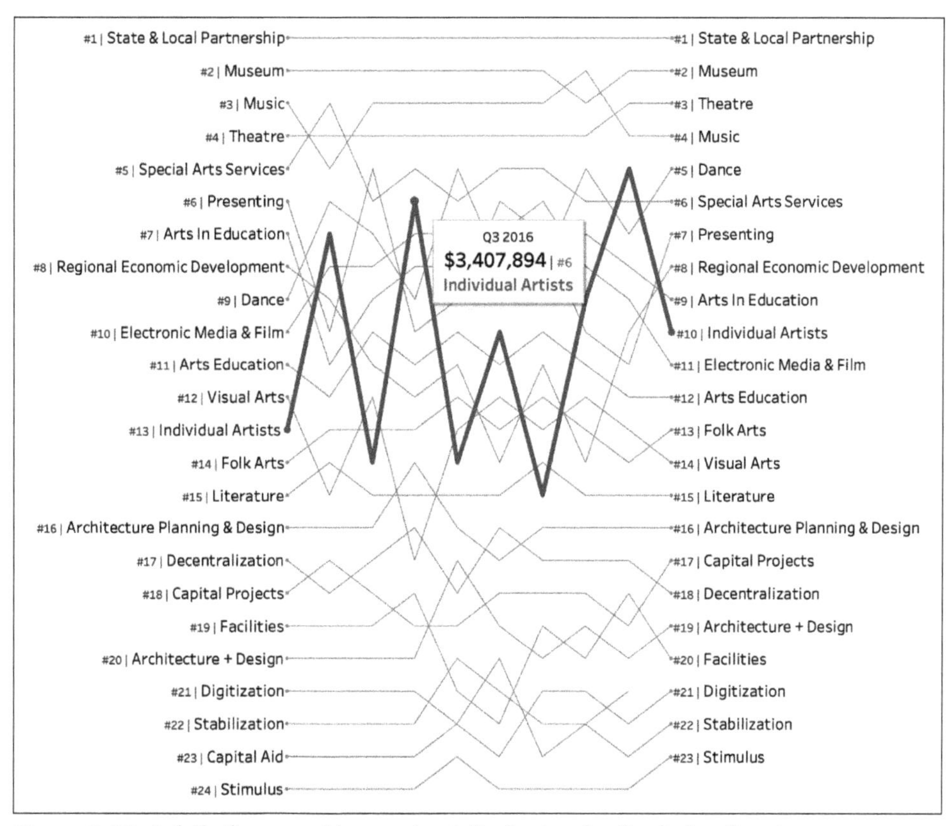

Figure 3-29. A fully formatted bump chart displaying the final visualization

Barbell Plots for Hierarchical Data: Office Essentials Case Study

Comparisons go beyond bar charts and rankings. Sometimes you have to compare data that is hierarchical in nature. One of the most common approaches for hierarchical comparisons is a barbell plot. (Surprise!)

The comparisons we will make for the rest of the chapter could be used in any industry. However, for this strategy—and the remaining strategies in the chapter—we're going to switch from community grant data and transition to using retail data.

The VP of Sales at OE has requested that the data team provide an analysis of the company's product line sub-categories that shows sales volume by region. He is looking to hire additional team members in the coming year, and needs to better understand sell-through opportunities. OE has 17 sub-categories and 4 sales regions.

What is the best way to clearly present the data, given that many data points will need to be represented? One way you might look at this data is as a single column of bar charts (Figure 3-30).

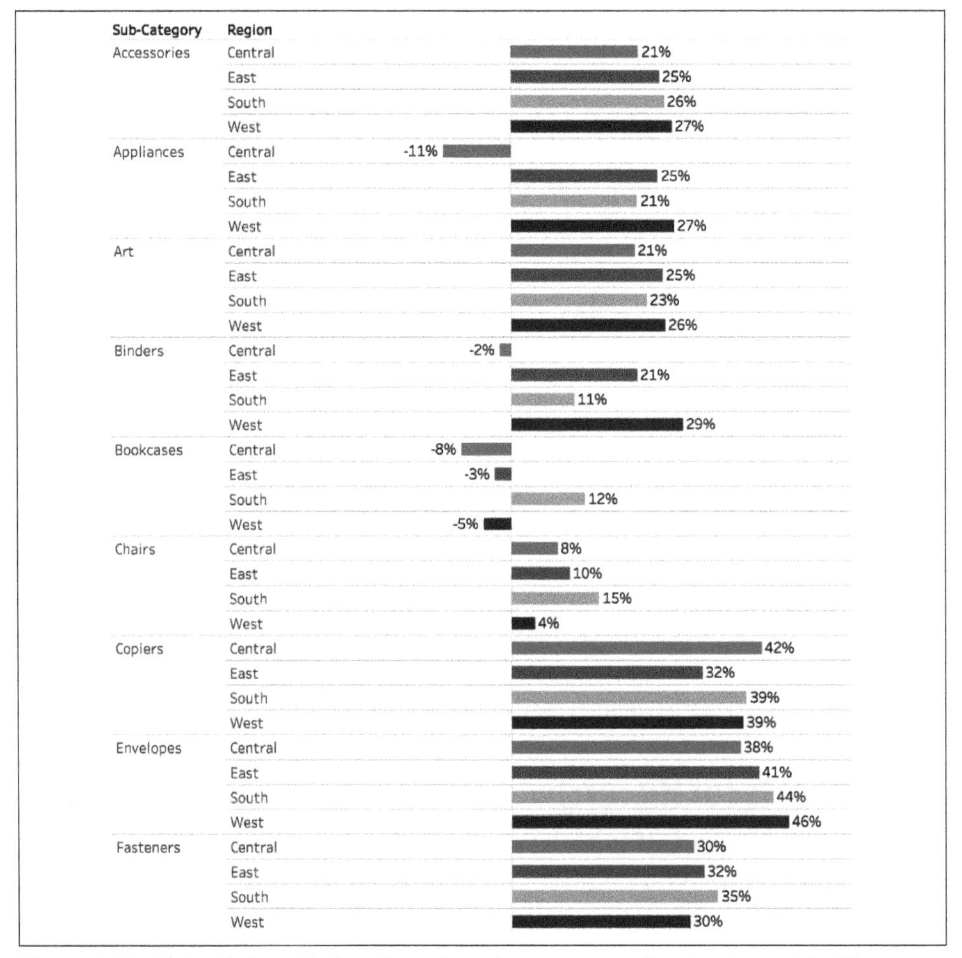

Figure 3-30. This side-by-side bar chart for sub-category and region is not ideal because of the space it consumes

We can't say this is a terrible solution: if our audience looks at any sub-category, they will be able to easily compare regions. The problem is that if they are interested in the Storage sub-category, for example, they have to go through dozens of data points until they get to the information they are interested in. When they do get to it, they have to reorient themselves to the chart.

An alternative option is to move [Region] to Columns, and [Percent of Margin] to Rows, to create vertical bar charts within each sub-category. This decreases the amount of cognitive load on the audience: they'll be able to more quickly process the information because each row is now one sub-category (not a combination of sub-category and region). Now, when they scan from left to right, they'll be able to quickly compare regions within a sub-category (Figure 3-31).

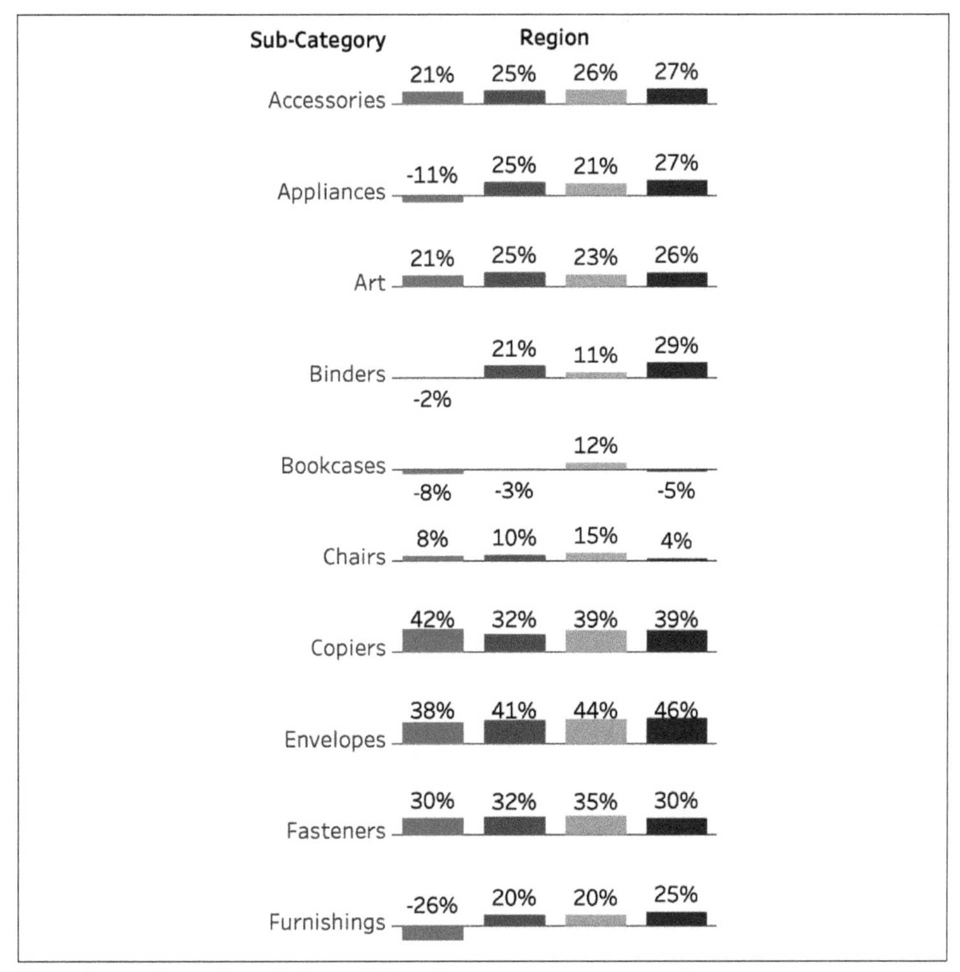

Figure 3-31. Even this side-by-side bar chart consumes too much space

While this chart provides clarity within each region, it's long and limits the reader's ability to make comparisons across sub-categories. An alternative to bar charts that might work well here is the *barbell plot*—a sibling to the dot plot.

The *barbell plot* gets its name from looking like a barbell, a straight bar with a weight on each side. In data visualization, this concept can be extended to include more dots, where the first and last dot (and also the length of the line) represent the minimum and maximum of the measure being plotted.

The barbell plot places dots—representing one dimension—along several lines. Each line represents the second dimension, and the measure of interest is the continuum of the line. Figure 3-32 shows an example.

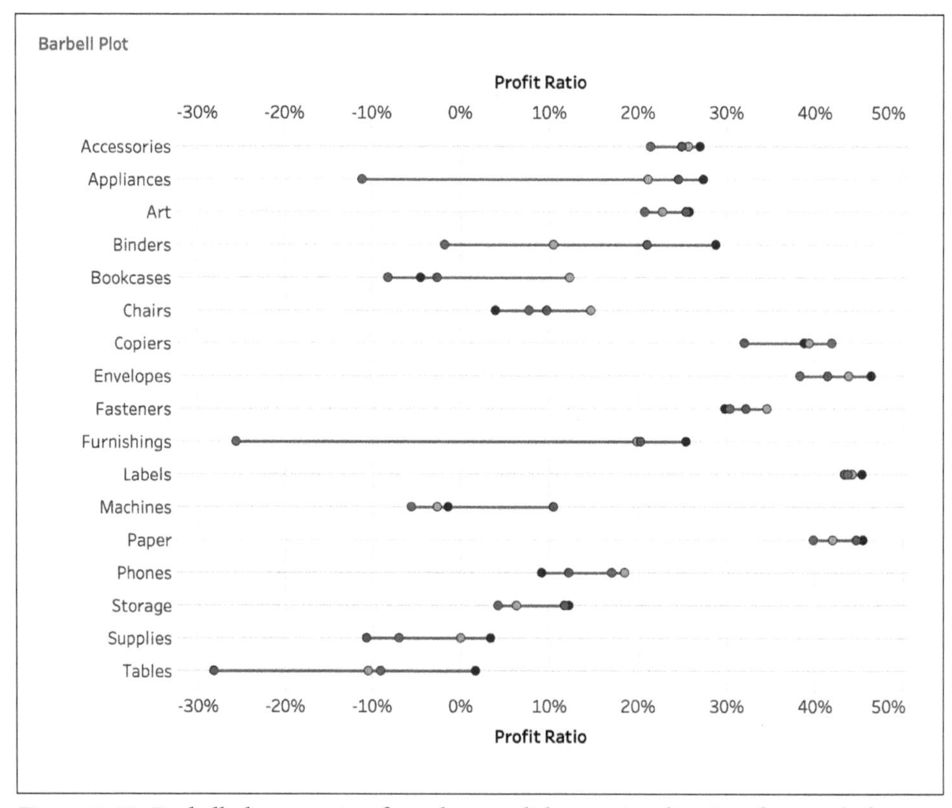

Figure 3-32. Barbell plots can significantly consolidate a visualization that might have otherwise been depicted in side-by-side bar charts

The result is a concise visual representation that allows your audience to quickly understand the performance within a measure while still grasping overall performance. The success of the barbell plot is not just in the plot design but also in the formatting. With the chart in Figure 3-32, your audience has a guideline that allows them to follow an individual sub-category from left to right or to quickly scan up or down to compare values.

So how do you build this visualization?

Strategy: Build a Barbell Plot

Follow these steps to build our barbell plot:

1. Create the base visualization:
   a. Connect to the Sample – Superstore dataset. Add [Sub-Category] to Rows.
   b. Add [Region] to Color. For this example, we've customized the colors.
   c. Create a calculation called **[% Margin]**:

      ```
      // % Margin
      SUM([Profit]) / SUM([Sales])
      ```

   d. Add the [% Margin] calculation twice to the Columns shelf (Figure 3-33).

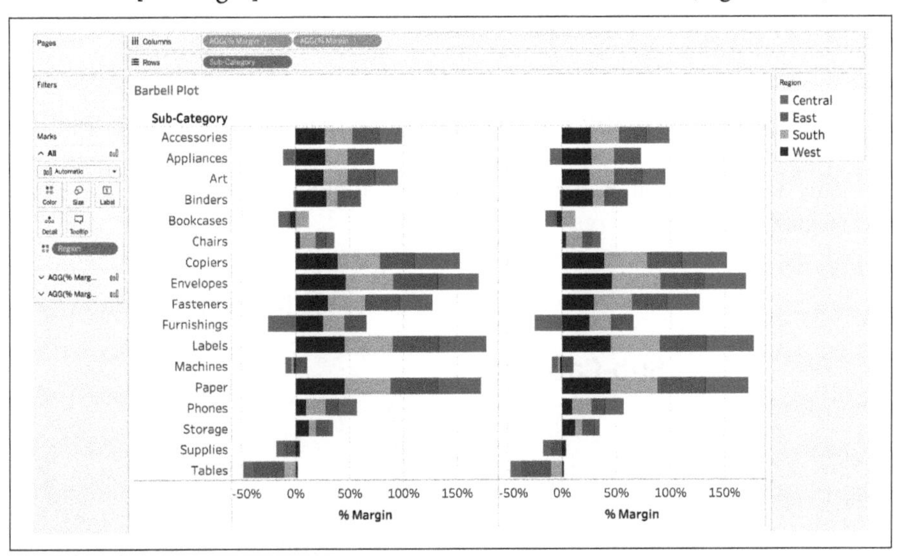

Figure 3-33. To create a barbell plot, start with stacked bar charts

2. Customize the Marks cards:
   a. For the leftmost [% Margin] Marks card, change the mark type from Automatic to Line. Move [Region] from Color to Path.
   b. For the rightmost [% Margin] Marks card, change the mark type from Automatic to Circle. Format the color by adding a border to the circle marks.
   c. Create a synchronized dual-axis chart.

3. Perfect the formatting:

   a. You are going to add an in-line calculation to the Rows shelf. Double-click to the right of [Sub-Category] on the Rows shelf and type **`MIN(0.0)`**, as shown in Figure 3-34. This will create the ruler that audiences will use to move from sub-category to values on the axis. Press Enter.

Using `MIN(0.0)`, `MIN(0)`, and `MIN(1)` are common conventions in Tableau development to create dummy continuous axes. They are often used to add a different metric to a label, or for added formatting.

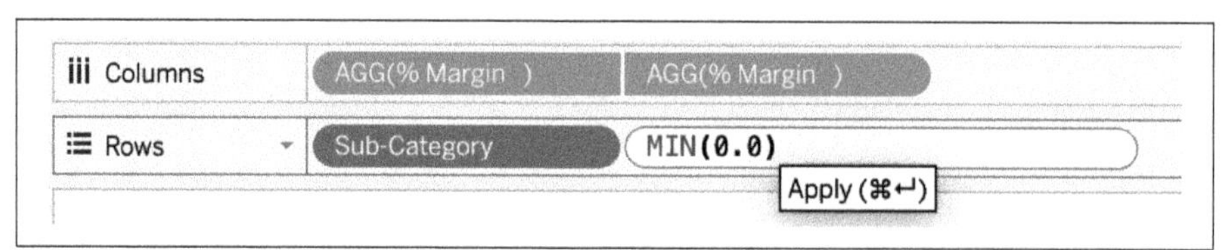

Figure 3-34. Adding `MIN(0.0)` as an ad hoc calculation

This will give you a new vertical axis. Right-click and hide the new axis (Figure 3-35).

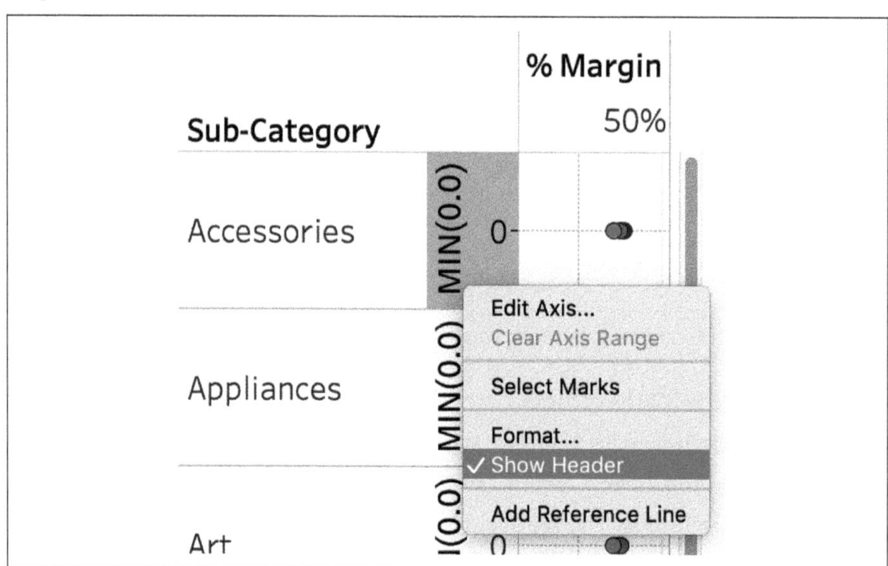

Figure 3-35. Uncheck Show Header to hide the axis

4. Resize the sub-categories so they belong to a concise view:
   a. Format and remove all border lines except for column grid lines and zero axis for rows.
   b. Format the line type and color of these to be identical.
   c. On the [Sub-Category] header, right-click and select Hide Field Labels for Rows (we'll do that a lot through the book).
   d. Right-click any sub-category and format the header so the members are right-aligned.
   e. Right-click the [MIN(0.0)] header and uncheck Show Header.
   f. Size the circles so they are noticeable but not distracting for your audience. Keep the line thinner, but don't use the smallest (to distinguish it from the guide).

   Figure 3-36 shows the final result.

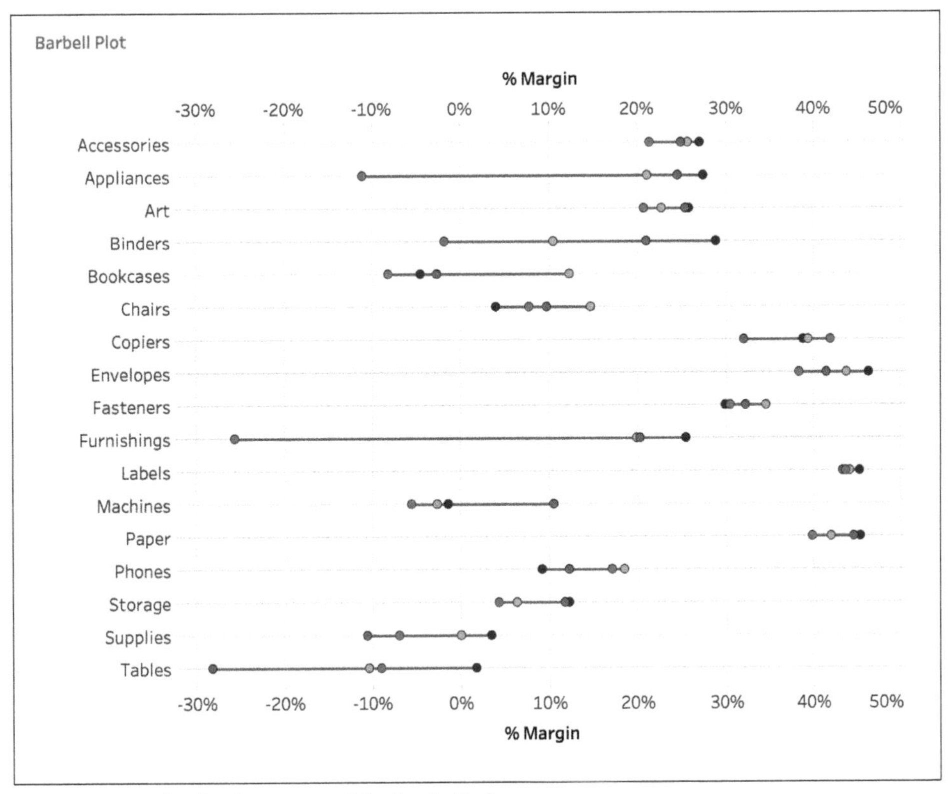

Figure 3-36. The final version of the barbell plot

Another alternative is to make the lines slightly larger than the circles (Figure 3-37). Whether you go with wide or narrow lines, you can use the border color of the circles to help distinguish the color and shape of the circles from the color of the lines.

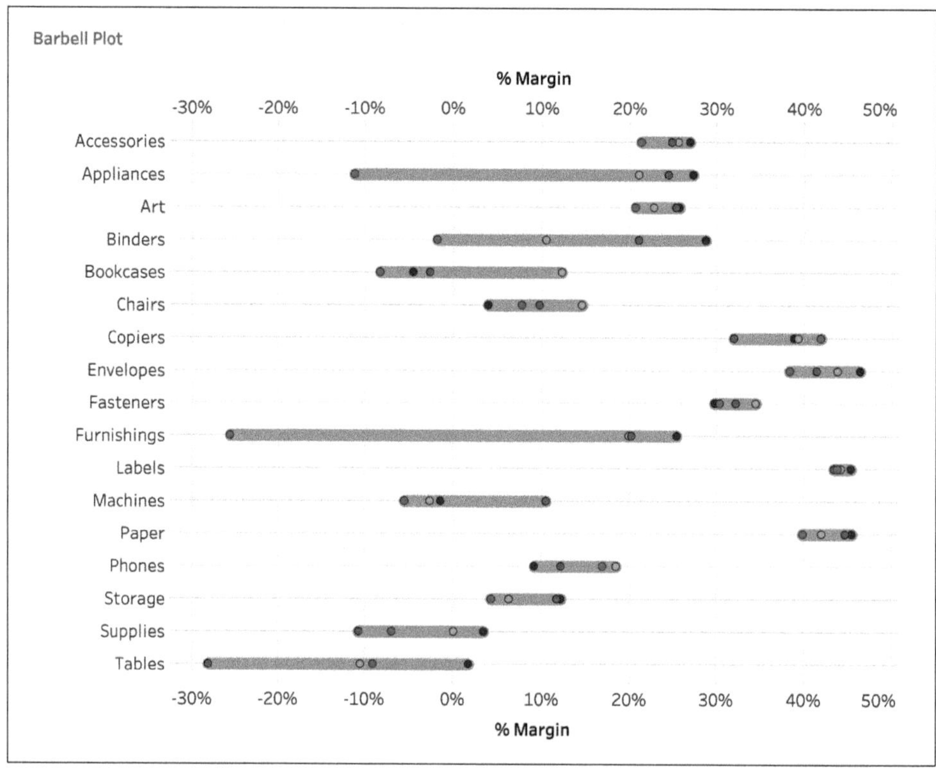

Figure 3-37. An alternative barbell plot in which the lines between the dots are much larger

The result of the barbell plot is a concise chart that your audience will be able to use to quickly compare values within a group—or even look across groups.

Trellis Charts/Small Multiples: Office Essentials Case Study

Consider the line chart in Figure 3-38 that shows total sales by quarter over four years and is divided into the 17 sub-categories.

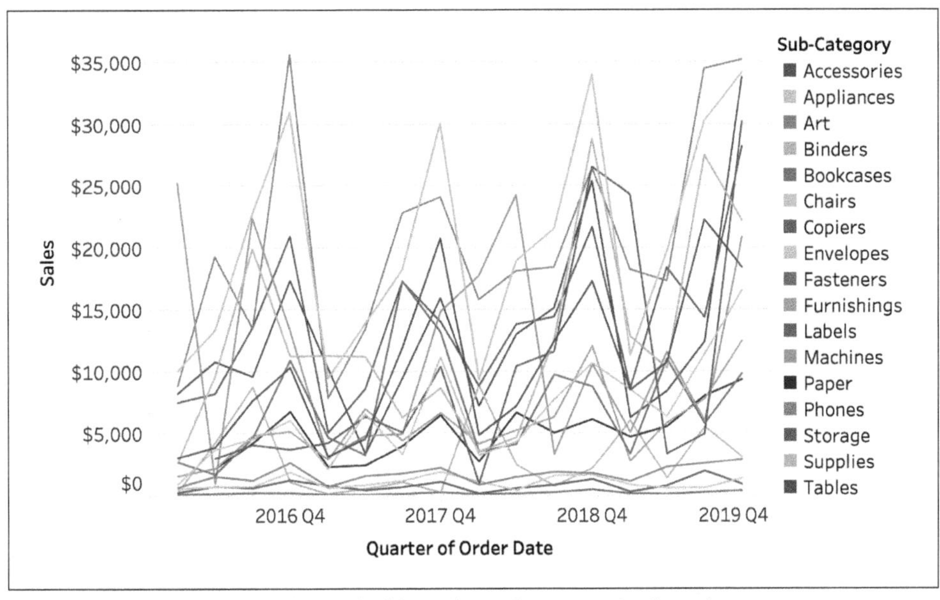

Figure 3-38. Another less-than-ideal line chart showing sales by sub-category

Pretend you are someone in your audience. Your response to this chart would be, "What the heck is this crap? It's too messy!" The reality is, almost no one is going to want to see a line plot with more than eight lines (seriously)! Your audience wants to trace a single line, and it's impossible to follow just one line in this chart—even with color encoding. You have to find a better way to communicate this information. One way we prefer to share this information is with a small multiple or trellis chart.

A *trellis chart* is actually a series of charts that use the exact same axes, most often ordered in a grid. Rather than showing a single line plot with 17 lines, we can choose to show 17 individual charts—for this example, as bars—to show the patterns in the data (Figure 3-39).

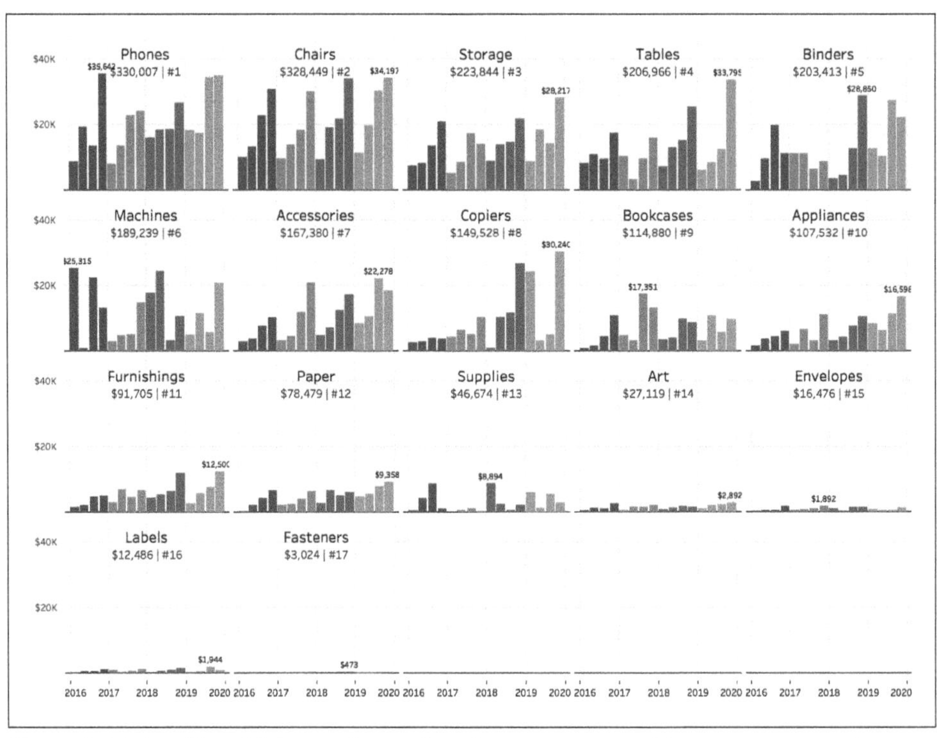

Figure 3-39. Small multiples can be used to show patterns within a group across time

Back at OE, the VP of Sales so appreciated the insights the data team provided that he has requested more information. When making hiring decisions, he'd like to see an even more granular view, showing sales by quarter. How can you clearly represent this information?

Figure 3-39 shows total sales by quarter for our 17 sub-categories. Each year has a distinct color to aid with analysis. We also added labels to the maximum value within each sub-category to provide additional context.

With small multiples, your audience can track patterns in the data. In this case, they will be able to see the seasonality of the data as well as the overall performance of sub-categories. When they look at this chart, they can see that Phones and Chairs are the top sellers. They can note the consistent seasonality in Chairs, Tables, Accessories, and Paper. And they can note that in Q2 and Q3 of 2019, copier sales dipped significantly before picking up to normal values in Q4. Now your audience can track individual values.

One detractor to this chart type is not knowing the actual values for any particular bar—but this is the struggle of almost every visualization. Additionally, comparing values across sub-categories can be difficult. But that's the battle with any chart; any chart type requires trade-offs. Your job is to weigh the options and select the best type.

When it comes to building a trellis chart in Tableau, difficulty varies based on the desired result. You could create a simple trellis by examining sales region and business segment, as shown in Figure 3-40.

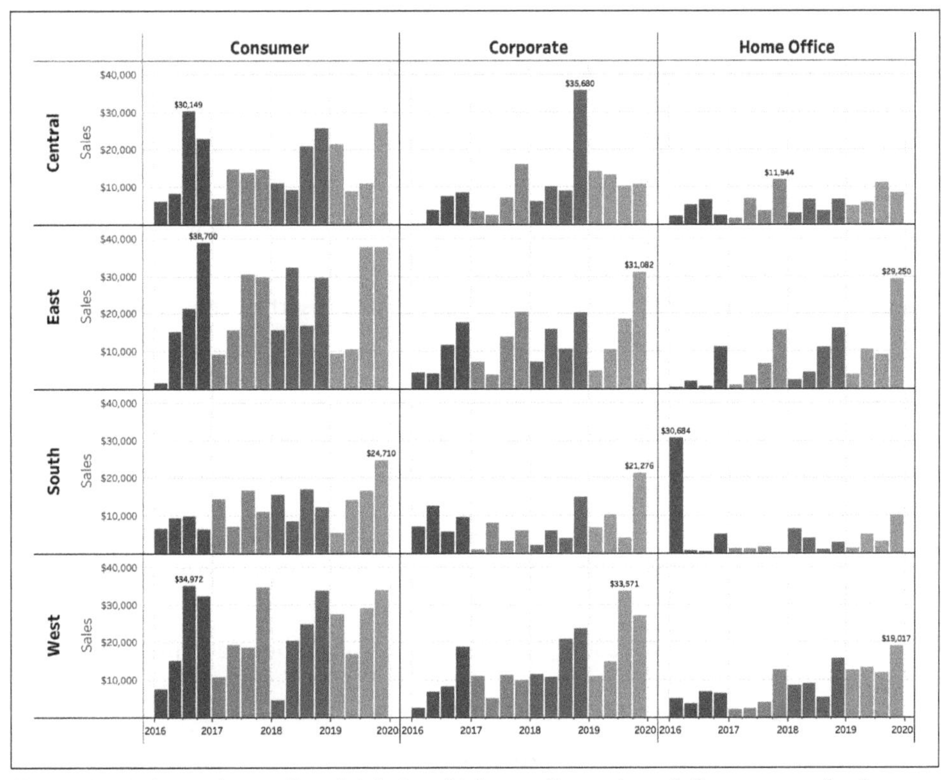

Figure 3-40. A simple small multiple in which two dimensions define rows and columns

This type of trellis is easy to complete in Tableau: it requires at least one dimension on Rows and another dimension on Columns (Figure 3-41).

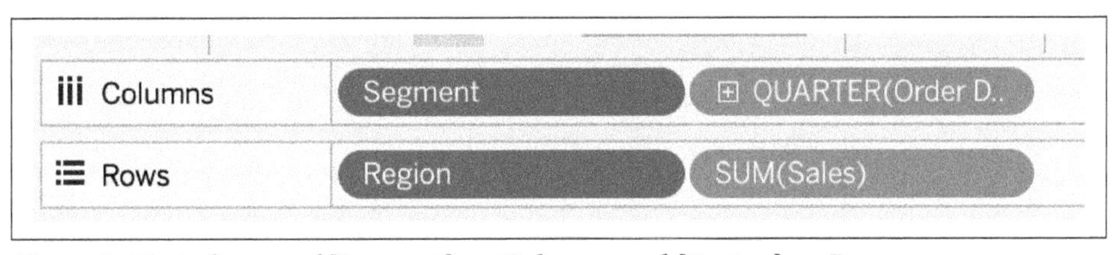

Figure 3-41. A closeup of [Segment] on Columns and [Region] on Rows

The real challenge is creating a grid of small multiples using a single dimension. In the following strategy, you will create a trellis with just a single dimension for the grid. This is not an out-of-the-box chart by any means. Creating this visualization requires you to think about how you are going to design your grid and add labels.

One last thing: because you are using table calculations, these charts can go awry if any data is missing for a particular member of a dimension for any part of your view.

Strategy: Create a Trellis Chart for a Single Dimension

Let's do this! Using the Sample – Superstore dataset, create a new sheet and follow these steps:

1. Create the grid. This requires using table calculations. You will write three calculations and create one parameter in this step:

   a. Create a new calculation called **[Index]** that will sort members displayed on the grid:

   ```
   // Index
   INDEX()
   ```

   You are going to use this calculation inside two other calculations that will form the columns and rows.

   b. Create a new integer parameter called **[Total Columns]**. This will provide a way to dynamically change the number of columns in our trellis chart. Set the value to 5.

   c. Create a calculation called **[Columns]** that will encode each member to the appropriate column:

   ```
   // Columns
   ([Index] - 1) % [Total Columns]
   ```

   This calculation takes the INDEX() function, a calculation that creates a running count of values from 1 to any number, and applies the modulo function.

   The modulo operator calculates the remainder of any division problem. Remember back in grade school when you calculated the remainder of a division problem and not the decimal? We bet you never thought you'd be using it as an adult!

   The output from this calculation provides you with members of a dimension divided into a number of groups, based on whatever number you specified in the Total Columns parameter: here, five groups. Figure 3-42 shows the mapping you'd expect to see.

Calculation	Result											
Index	1	2	3	4	5	6	7	8	9	10	11	12
Columns	0	1	2	3	4	0	1	2	3	4	0	1

Figure 3-42. Output of the Index and Columns calculations

d. Now that you have the [Columns] calculation, all you need is the [Rows] calculation:

```
// Rows
((([Index] - 1) - [Columns]) - 1) / [Total Columns]
```

Add [Sub-Category] and [Index] to Detail. Add [Columns] to the Columns shelf and [Rows] to the Rows shelf. Change [Columns] and [Rows] to discrete.

e. To build the grid on the sheet, edit the table calculations for [Column], [Rows], and [Index], and select Specific Dimensions and choose [Sub-Category]. Right-click [Sub-Category] on the Marks card and create a custom sort by using sum of sales, descending. For the [Rows] calculation, you have two table calculations to edit. Make sure the table calculations use the same Compute Using and Sort Order options (Figure 3-43).

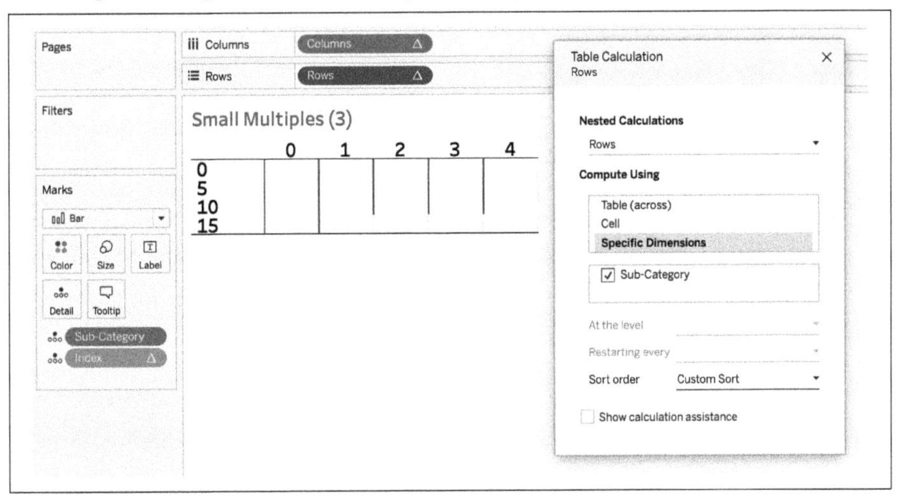

Figure 3-43. Use table calculations to build the grid for the small multiples

2. Now that you have established the grid, you can begin to build the visualization. You will look at sales based on order date:

   a. Click and drag [Order Date] onto Columns. Then right-click and select a continuous quarter date type (Figure 3-44).

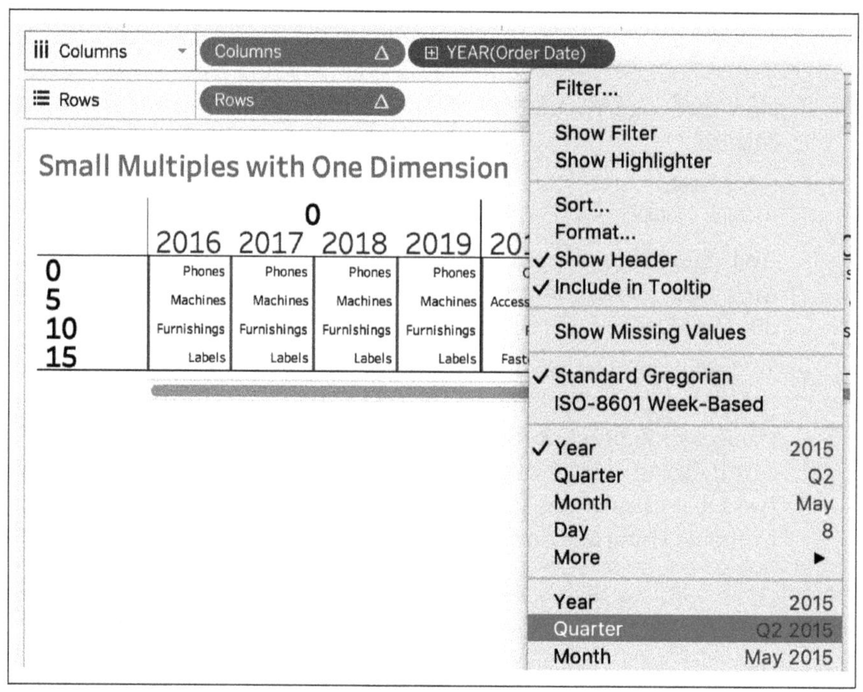

Figure 3-44. Change order date to a date value by quarter

   b. Add SUM([Sales]) to Rows. Add YEAR([Order Date]) to Color. You may need to change your mark type to Bar. This leaves you with the visualization in Figure 3-45.

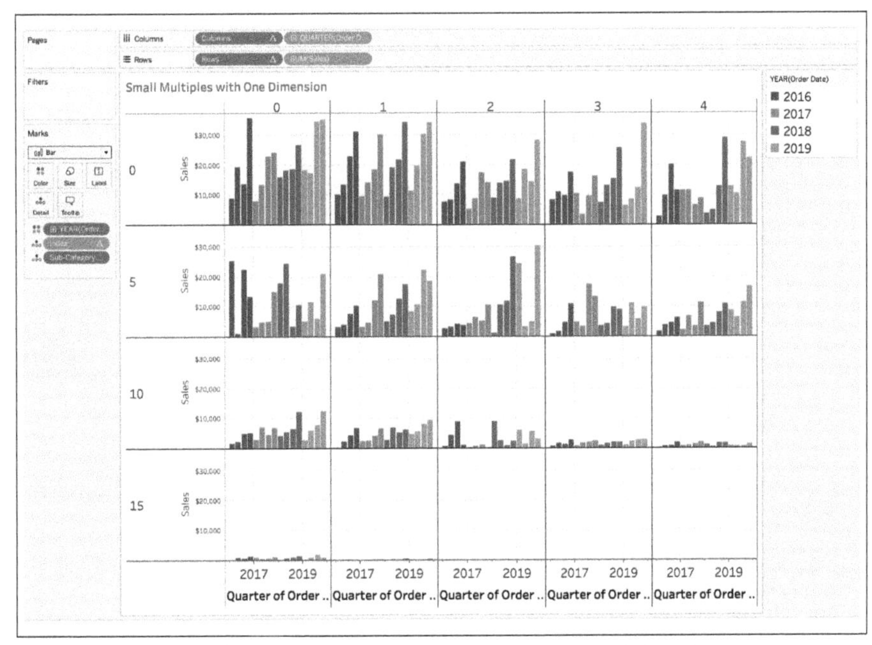

Figure 3-45. A small multiple chart without sub-category labels for each section of the grid

We've completed the base of our visualization. Now, labels for each box would be nice. To do this, you are going to create a dual-axis chart with two custom calculations.

3. To add a label that is centered above each multiple, create a custom date calculation that identifies the most middle date:

   a. We can do this by identifying the ends of our dates and then taking the average of the two. Write the following calculation, called **`[Order Date | SM Label]`**:

   ```
   // Order Date | SM Label
   {MIN([Order Date])} + (({MAX([Order Date])} - {MIN([Order Date])})/2)
   ```

   This finds the first and last order date and then finds the middle of the two dates. (You'll learn more about this in Chapter 4.)

   b. Add this calculation as an exact date to the Columns shelf. You'll notice a single stacked bar chart that corresponds to the new date calculation (Figure 3-46).

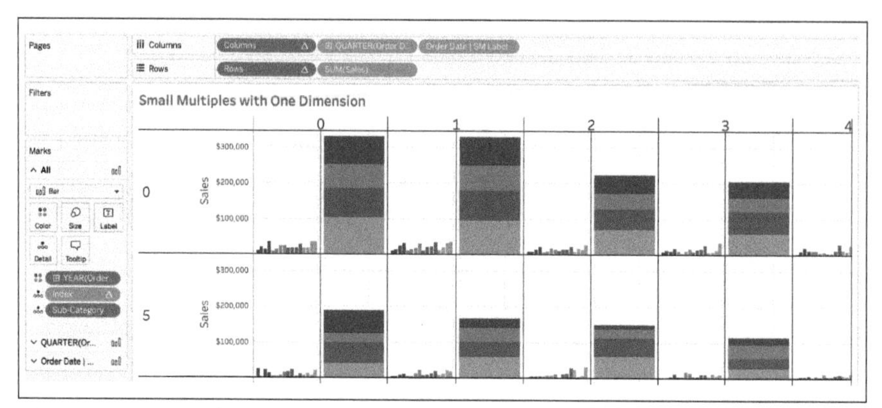

Figure 3-46. A work in progress for obtaining proper labels

c. Edit the Marks card for [Order Date | SM Label]. Change the mark type to Text and remove YEAR([Order Date]) from Color.

d. Create a synchronized dual axis. On this same Marks card, move [Sub-Category] and [Index] from Detail to Text.

   One issue you'll see is that the text doesn't align across each sub-category (Figure 3-47). This is because for the [Order Date | SM Label] axis, each label is positioned at the overall total for that sub-category. We need to find some magic (or a nifty way) to align these labels.

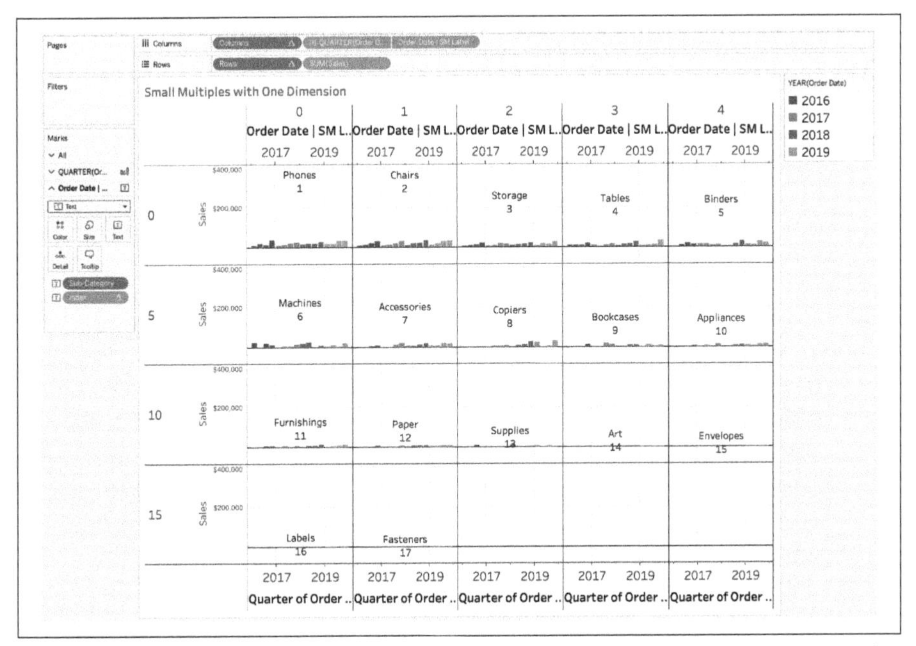

Figure 3-47. Labels are now showing for the small multiples, but the height varies by cell of the small-multiples grid

4. Labels can be aligned by finding the single largest quarterly total for sub-category and then just adjusting the label to be a multiple higher than that value. If the highest total sales for any category by quarter came, hypothetically, from Copiers in Q4 of 2018, and that value was $100, we'd want the labels for all the sub-categories to be slightly higher on the chart—perhaps centered at a 120.

   But here's the thing: you need text to be centered, and you need the bars to appear at the same height. And you can do this with only a single calculation! How do you accomplish it? By figuring out how many marks are occurring per calculation by using the `SIZE()` function!

`SIZE()` is another useful table calculation. It finds the total number of data points within a partition of data. As an example, if you had a simple text table with 10 rows of data, `SIZE()` would be 10.

a. Create a calculation called **`[Bars + Label]`**:

```
// Bars + Label
IF SIZE() > 1
THEN SUM([Sales])
ELSE 1.1 * AVG({
  MAX(
    {FIXED [Sub-Category], DATETRUNC("quarter", [Order Date]) :
     SUM([Sales])
  })
})
END
```

In this calculation, `SIZE()` first counts the marks based on how you set the table calculation. For the text marks, this will return a value of 1. For the bar charts, `SIZE()` will return the number of quarters displaying for each sub-category. Since this value is 16 for the bars, the calculation will return the total sales. For the text, it returns 1.1 times more than the highest total for a single sub-category by quarter.

b. Take this `[Bars + Label]` calculation and replace `SUM([Sales])` and [Rows] with the calculation. Edit the table calculation and select by Quarter of Order Date. Add `SUM([Sales])` to Label on the [Order | SM Label] Marks card. This leaves you with the visualization in Figure 3-48.

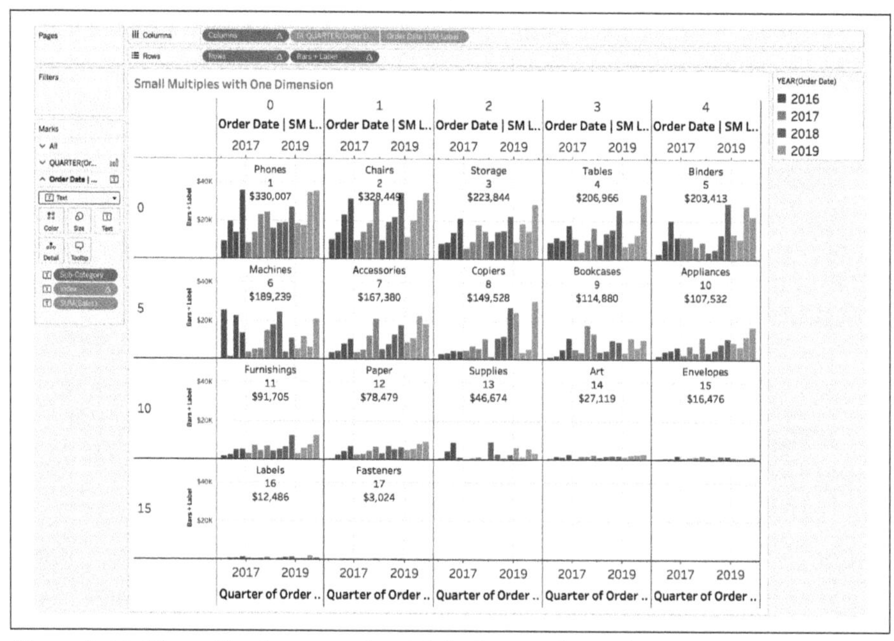

Figure 3-48. The trellis chart now has labels that are located in the same position, but just need formatting

5. To finalize the visualization, you'll want to format the text showing on [Order Date | SM Label] as follows:

    a. You can format this text as you prefer; we are putting Sub-Category on the first line, and the Index (which acts like a rank) and `SUM([Sales])` on the second line (Figure 3-49).

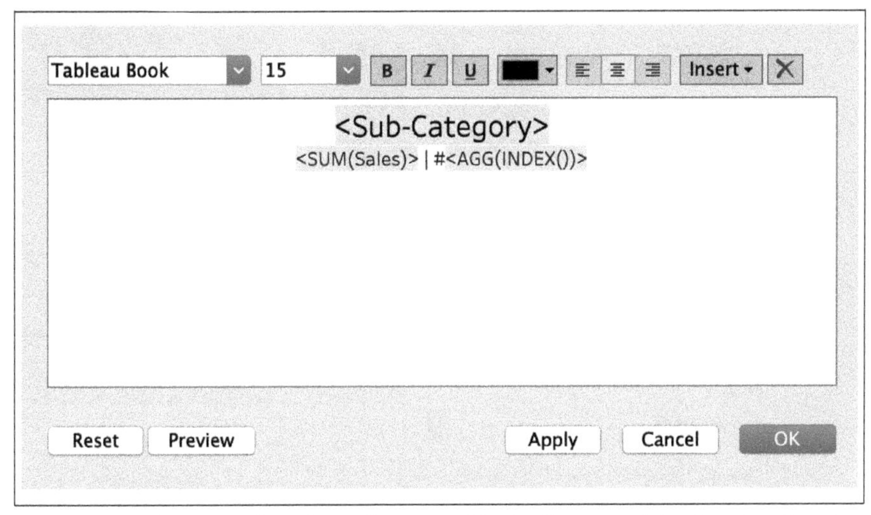

Figure 3-49. An example of how the text might be formatted for the visualization

    b. Add `SUM([Sales])` to Label on the [Quarter of Order Date] Marks card. We're setting the font to size 7, and then showing the top value per sub-category, which can be done by selecting Pane in the Scope section and then selecting "Label maximum value" (Figure 3-50).

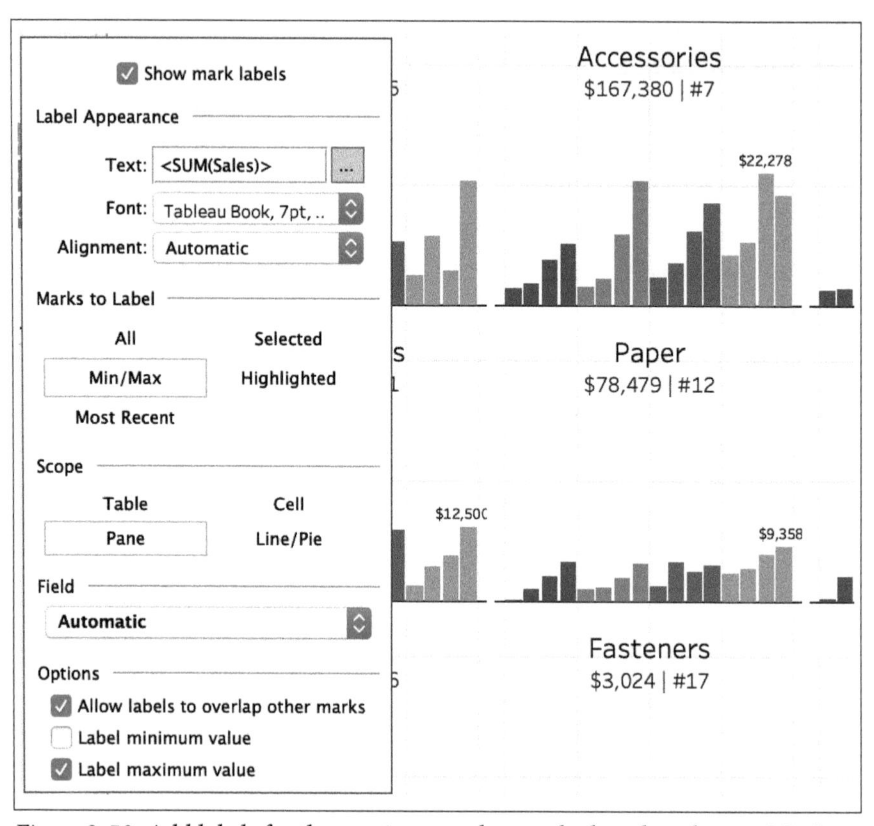

Figure 3-50. Add labels for the maximum value on the bars by selecting Min/ Max, then Pane, then checking "Label maximum value"

c. Hide headers for columns and rows. Adjust the lines and borders to your preference. For this example, we prefer to keep grid lines and include row and column dividers that are set to the background color. The final result is a 5 × 4 grid of small multiples (Figure 3-51).

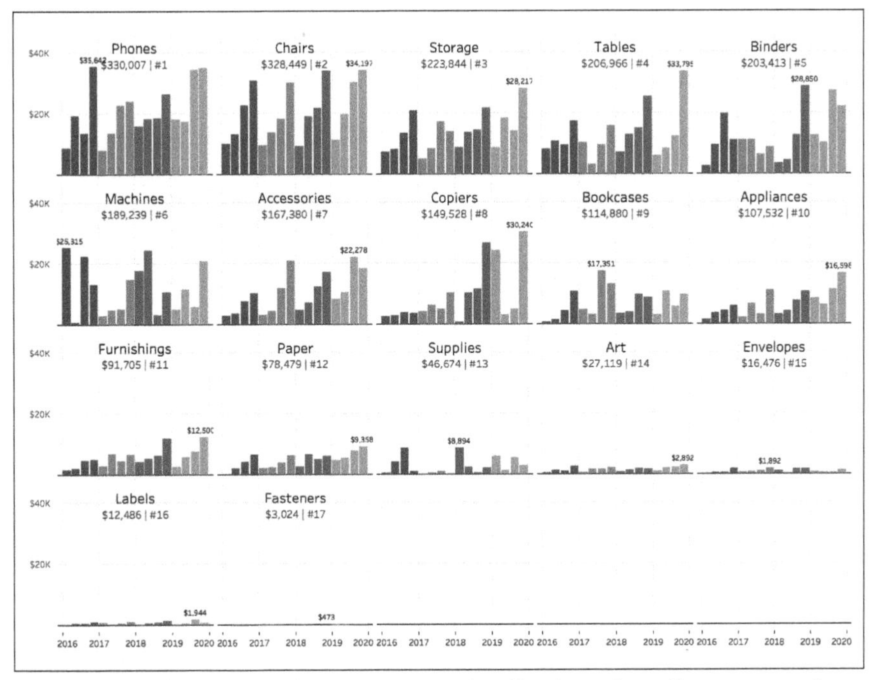

Figure 3-51. The final result is an organized trellis chart that allows your audience to examine patterns within sub-categories

d. If you are showing hundreds of multiples, you may want to filter to, perhaps, the top 20. You can do this by filtering using your Index calculation, and filtering to a range between 1 and 20. Just be sure your table calculation on the filter is set up correctly.

With a trellis, you can create multiple plots with the same base framework. By using a trellis, you are able to examine how subsets of your data look relative to each other. This allows for comparisons both within a subset and across subsets.

You can create a trellis chart in Tableau with multiple dimensions, or with a single dimension and a few handy table calculations. This particular example tackled a single dimension trellis with dates. This is perhaps the most difficult to create!

Parallel Coordinates Plots for Multiple Measures: OE Case Study

For our last strategy of the chapter, let's tackle a comparison that analysts are often forced to make: comparing multiple members of a single dimension across multiple measures. The most common way we see this done is with a bar chart (or table), as shown in Figure 3-52.

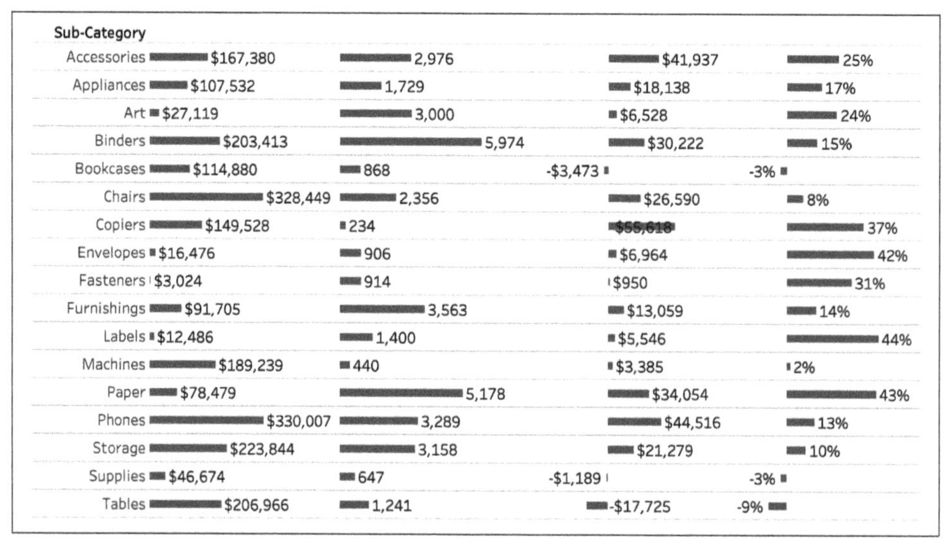

Figure 3-52. Multiple bar charts showing sales, units sold, margin, and percent of margin. This chart works, but there are alternatives.

The one downfall is that you could have too many members/rows to manage; you may prefer to summarize your data in a more concise format.

You can do this with a *parallel coordinates plot*. Parallel coordinates are used for plotting data that has multiple dimensions and comparing the relationships across them. With this chart type, each variable has its own axis. These axes are often on different units of measurement and are normalized for each measure. This allows the scales to remain uniform. A line is then used to connect each member of a dimension across the measures and their corresponding axes.

Because many members are often plotted at a single time, the plot can become cluttered very quickly. We recommend that you either highlight a single member relative to the entire group or minimize the number of members that are being compared. We prefer highlighting a single member of a dimension (Figure 3-53).

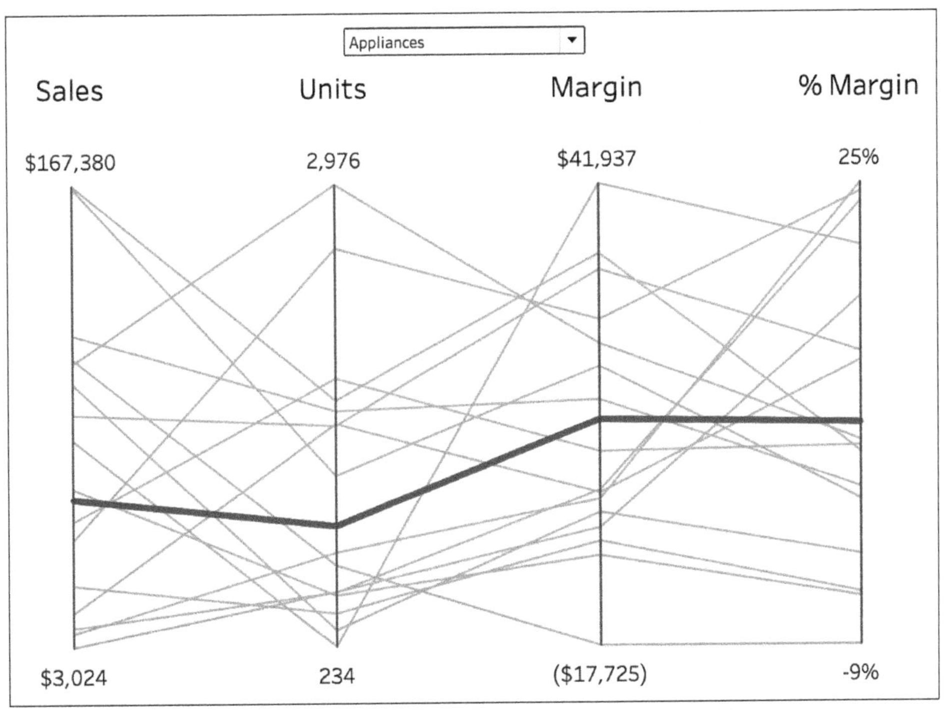

Figure 3-53. A parallel coordinates plot can allow audiences to quickly compare any member across multiple dimensions

A parallel coordinates plot can quickly compare any member of a dimension across multiple measures—providing context faster than a bar chart often can. Additionally, the plot isn't best for comparing the relationship between any two variables; this is better left for a scatter plot.

A parallel coordinates chart may be just what is needed at OE. Before making a hiring decision, the VP of Sales has asked for one more piece of information. The data team has been tasked with comparing the 17 sub-categories of a retailer across four metrics: total sales, total units solid, the margin or profit in dollars, and the margin as a percent of sales. The report should compare performance of each sub-category relative to the others.

How can you best represent this data while showing the relationships among sub-categories? The best way to do this is with a parallel coordinates chart.

Strategy: Build a Parallel Coordinates Chart

First, you need to build *normalized metrics*. This means taking any subset of metrics and transforming them from their existing scale to a scale that goes from 0 to 1. The format for this calculation is simple:

```
value - lowest value
------
highest value - lowest value
```

1. Since you are working with four metrics, you have to normalize four calculations. These calculations are **`[Sales | Normalized]`**, **`[Margin | Normalized]`**, **`[% Margin | Normalized]`**, and **`[Units | Normalized]`**:

```
// Sales | Normalized
(SUM([Sales]) - WINDOW_MIN(SUM([Sales])))
/
(WINDOW_MAX(SUM([Sales])) - WINDOW_MIN(SUM([Sales])))


// Units | Normalized
(SUM([Quantity]) - WINDOW_MIN(SUM([Quantity])))
/
(WINDOW_MAX(SUM([Quantity])) - WINDOW_MIN(SUM([Quantity])))


// Margin | Normalized
(SUM([Profit]) - WINDOW_MIN(SUM([Profit])))
/
(WINDOW_MAX(SUM([Profit])) - WINDOW_MIN(SUM([Profit])))


// % Margin | Normalized
((SUM([Profit])/SUM([Sales])) - WINDOW_MIN(SUM([Profit])/SUM([Sales])))
/
(WINDOW_MAX(SUM([Profit])/SUM([Sales])) - WINDOW_MIN(SUM([Profit])/SUM([Sales])))
```

You can also do this with nested LODs. Replace the `WINDOW_MAX()` and `WINDOW_MIN()` functions with fixed LOD expressions at the appropriate levels.

You'll notice each calculation has a similar structure.

2. Build the plot as follows:

    a. Place [Sub-Category] on Detail. Add [Measure Names] to Columns and [Measure Values] to Rows. Add [Sales | Normalized], [Units | Normalized], [Margin | Normalized], and [% Margin | Normalized] to the [Measure Values] Marks card.

    b. Edit each table calculation so that it is computed across sub-categories. Change the mark type to Line.

    Figure 3-54 shows the plot after this step.

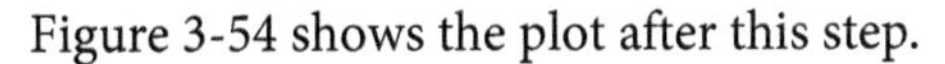

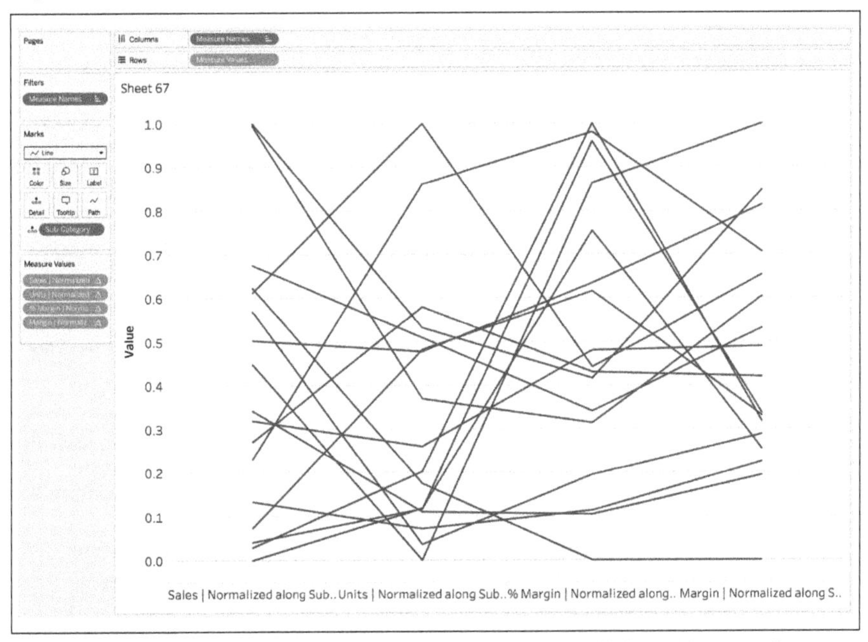

Figure 3-54. A sneak peek into the beginning of the parallel coordinates plot

    c. Add a second [Measure Values] to Rows. Edit the leftmost [Measure Values] Marks card and move [Sub-Category] from Detail to Path.

    d. Change the size of the lines to be narrower and change the color to be a medium gray.

    e. Create a synchronized dual-axis chart. Format and remove all lines and borders. Then hide the headers. This will create Figure 3-55.

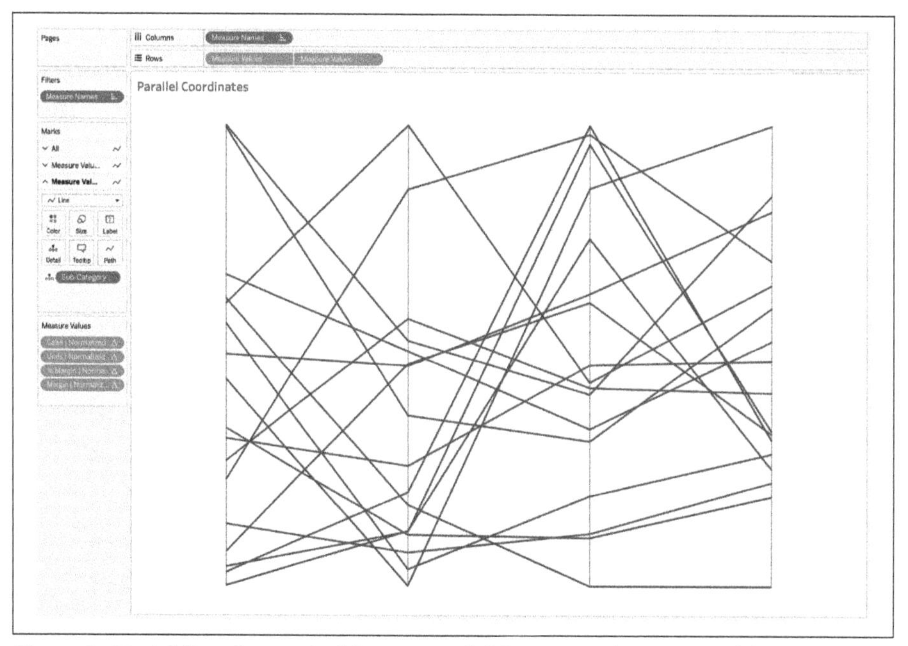

Figure 3-55. Adding the vertical lines is useful for our audience as it delineates where change occurs

3. Highlight an individual member of the [Sub-Category] dimension as follows:

   a. Create a parameter from [Sub-Category] and call it **`[Sub-Category Parameter]`**.

   b. Create a new calculation called **`[Sub-Category | TF]`**:

   ```
   // Sub-Category | TF
   [Sub-Category] = [Sub-Category Parameter]
   ```

   c. On the rightmost [Measure Values] Marks card, add [Sub-Category | TF] to Size and Color. On the leftmost [Measure Values] Marks card, add [Sub-Category | TF] to Detail.

   d. You need to edit your table calculations again to show the proper values with color and size. When you edit your table calculations in the Measure Values pane, be sure to compute using both [Sub-Category] and [Sub-Category | TF].

   e. Set a color for the True and False values. Set the True value to have greater luminosity relative to the False value. For this example, we've set the values to #7C00B2 for True and #CEB8D8 for False. Make sure the True value is layered on top of the False value. Make the True value slightly wider than the False value.

f. Add the actual values to the tooltips of the second Marks card and format the tooltip. Remove the tooltip from the first Marks card by deleting all the values inside the tooltip prompt.

g. Edit the axis so that the values run from exactly 0 to 1.

The core work required for this example is done (Figure 3-56).

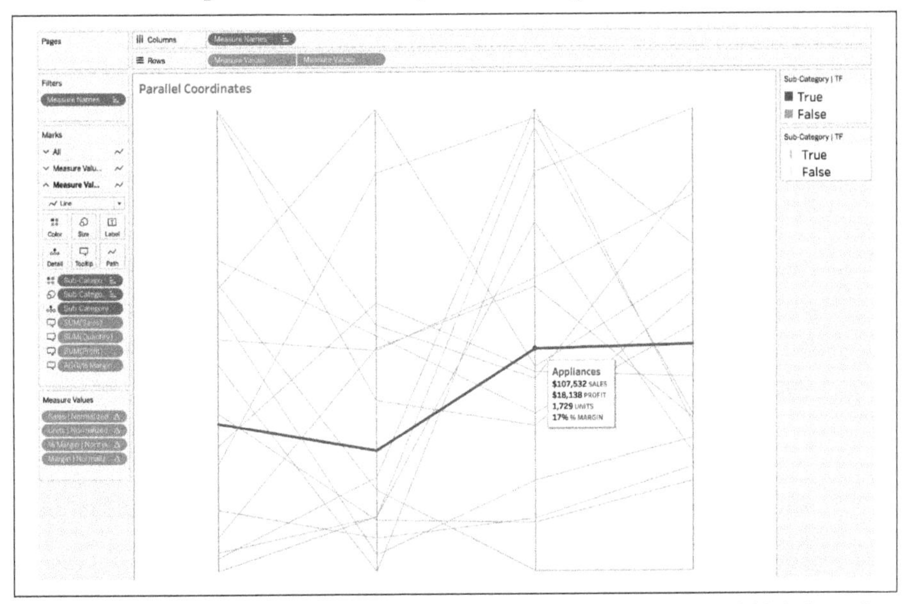

Figure 3-56. In a formatted parallel coordinates plot, you can't just add labels to the top and bottom of the chart because of limitations. Instead, you'll create separate sheets.

If you want simple labels for this chart, you could edit the leftmost [Measure Values] Marks card and add [Measure Names] to Label, and then show the values on the minimum and maximum values. However, we're going to show you a multi-sheet approach that will give more context to the data.

4. At the top of each axis, you want to display the maximum value. This will give context to your audience. To do this, you need to create a separate sheet:

   a. Start by creating four identical ad hoc calculations on Columns (Figure 3-57). Double-click and type **`MIN(0.0)`**. Do this again three more times. This will create four separate Marks cards on which we can place labels.

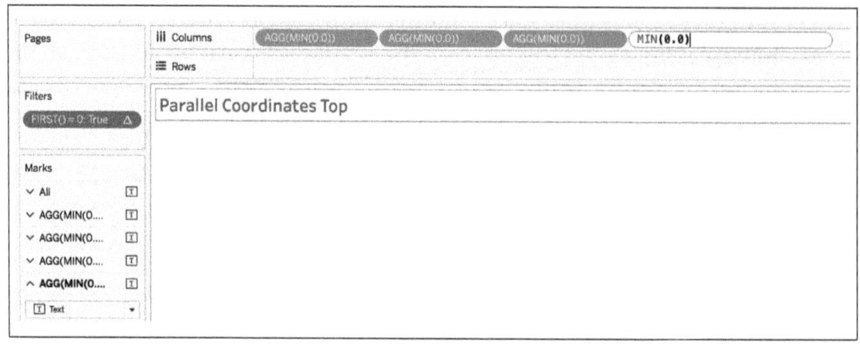

Figure 3-57. Add `MIN(0.0)` as many times as necessary to build the top labels of the parallel coordinates

b. Change all of the Marks cards to Text marks. On each Marks card, you are going to add a label for the measure type and the maximum value on the axis.

c. On the first Marks card, create a new calculation called **`[Sales | Window Max]`**:

```
// Sales | Window Max
WINDOW_MAX(SUM([Sales]))
```

Add this calculation to Text. Edit the text. Write **`Sales`** on the first line and place the calculation on the second line (Figure 3-58).

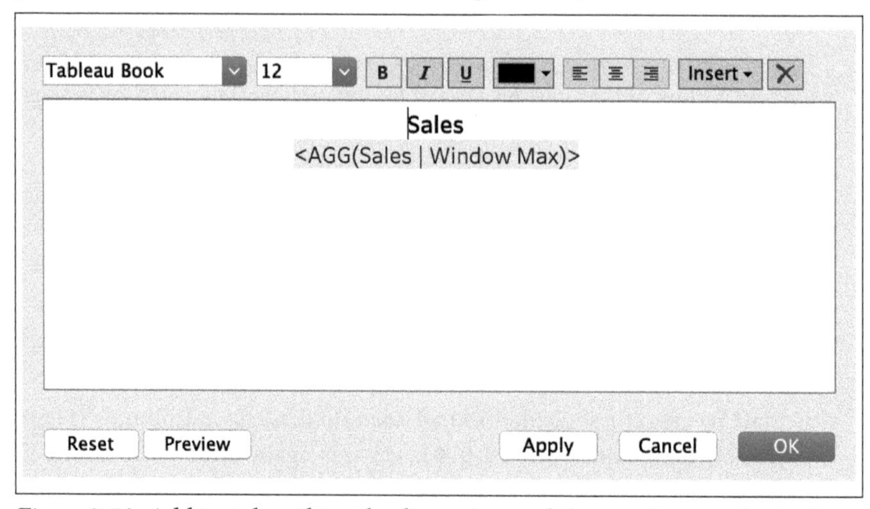

Figure 3-58. Add text describing the dimension and the maximum value in the window

d. Repeat this task on the second MIN(0.0) card for [Units | Window Max]. Repeat for [Margin | Window Max] and [% Margin | Window Max] on the third and fourth Marks cards, respectively:

```
// Units | Window Max
WINDOW_MAX(SUM([Quantity]))

// Margin | Window Max
WINDOW_MAX(SUM([Profit]))

// % Margin | Window Max
WINDOW_MAX(SUM([Profit])/SUM([Profit]))
```

e. Add [Sub-Category] to Detail on each of the Marks cards. Then remove all lines and borders. Finally, hide the headers. This will leave you with the messy visualization in Figure 3-59.

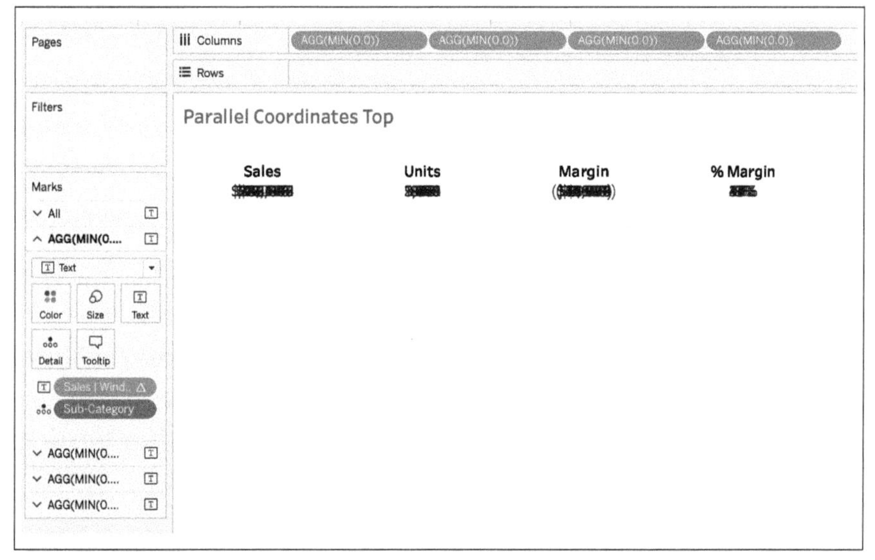

Figure 3-59. A sneak peek at the work in progress for the top labels before adding a filter to show a single value

f. To clean this up, we need to create and add a single calculation to our filters. Create a calculation called **`[First]`**:

```
// First
FIRST() = 0
```

Then place the calculation on Filters. Don't worry about selecting True or False right away. You'll need to edit your table calculation before anything works anyway.

g. Set your table calculation to compute across Sub-Category and then edit your filter for when values are True. Finally, turn off the tooltips.

Now you'll have a header that's worth sharing (Figure 3-60)!

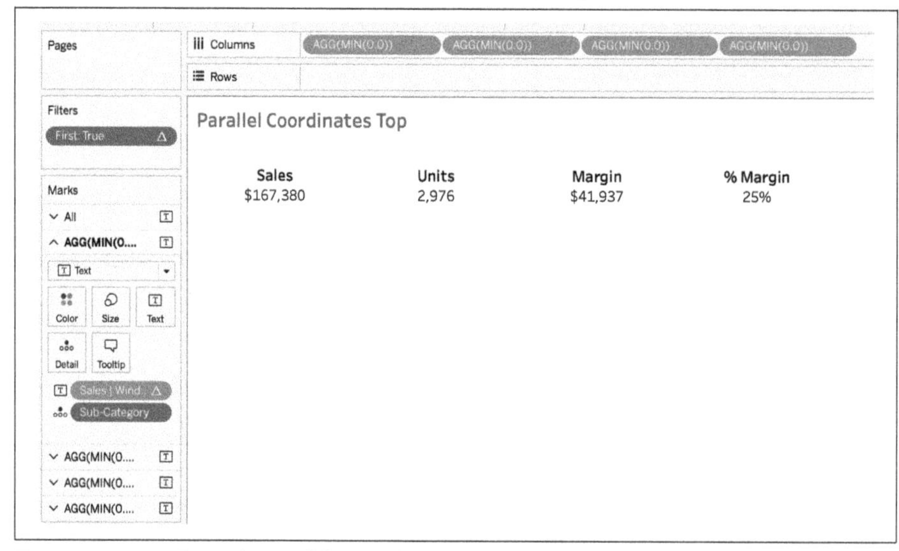

Figure 3-60. A sheet that will be used as a header for the parallel coordinates

5. The last sheet we need to create will display the lowest values for each of the measures and will look like it's labeling the bottom axis:

   a. Repeat the same process from the previous step, adding custom MIN(0.0) calculations to Columns.

   b. On the first Marks card, calculate the window minimum for sales, the window minimum for units, and then the minimum for margin and percent of margin. Each will be placed on Text:

```
// Sales | Window Min
WINDOW_MIN(SUM([Sales]))


// Units | Window Min
WINDOW_MIN(SUM([Quantity]))


// Margin | Window Min
WINDOW_MIN(SUM([Profit]))


// % Margin | Window Min
WINDOW_MIN(SUM([Profit])/SUM([Profit]))
```

c. Add [Sub-Category] to Detail for each. Then format your sheet.

d. Add [First] as a filter, repeating the steps we discussed in step 4f.

The result is four values representing the lowest values possible on each of our axes in our parallel coordinates (Figure 3-61).

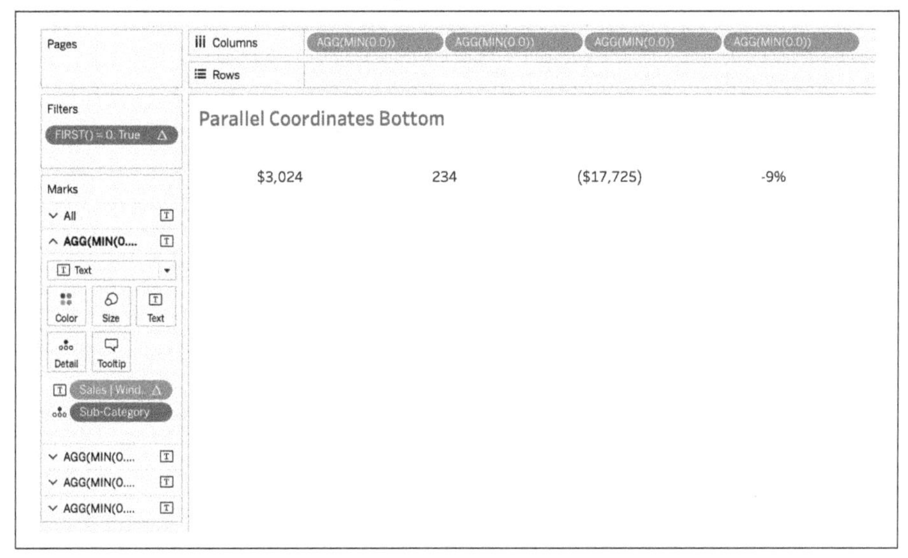

Figure 3-61. A sheet that will be used as a footer for the parallel coordinates

6. Now that you have three sheets with the key data needed, you need to stitch these together on a dashboard. The best way to do this is by using a vertical container on a dashboard.

   Ideally, this visualization would be paired with a table (yes, a table) or a series of bar charts that show the individual values. But for the sake of this strategy, we are going to have a dashboard with a single visualization:

   a. Create a new dashboard and set the size to 800 pixels × 500 pixels.

   b. Add a tiled vertical container.

We'll dive more into vertical containers in Chapter 12. For now, it's enough to know that a container allows you to place multiple sheets inside and precisely control the height of each sheet.

   c. Add the visuals: in the container, add the top of the parallel coordinates, then the parallel coordinates themselves, and then the footer. For each sheet, make sure it uses the entire view. You should also hide the titles.

d. Neither of the legends provides a ton of value, so remove them from the dashboard.

e. Adjust the spacing for the header and the footer. We recommend fixing the height for the header to 60 and the footer to 30.

f. Add the parameter for [Sub-Category] onto the dashboard.

The result is a very clean-looking parallel coordinates plot (Figure 3-62).

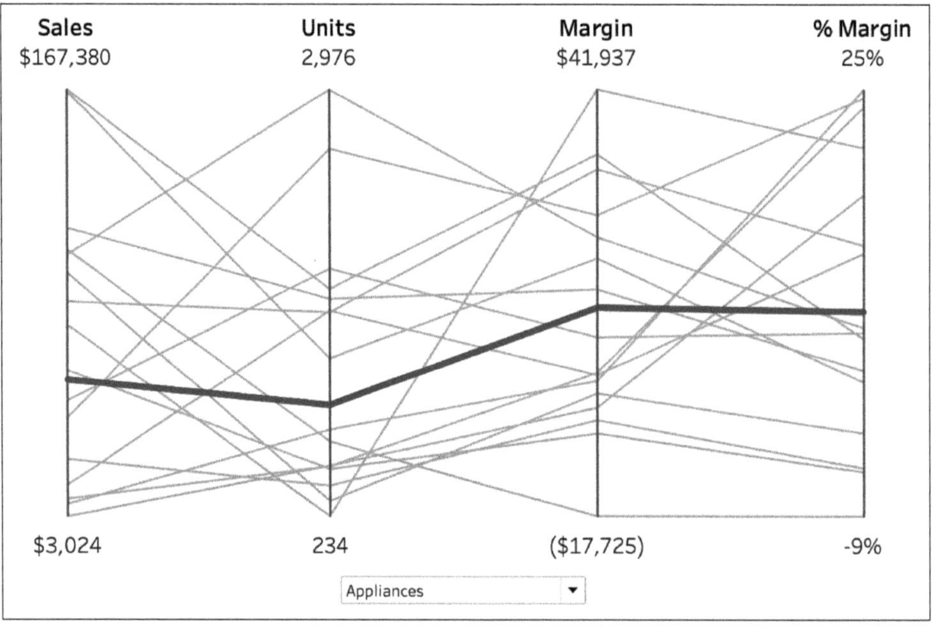

Figure 3-62. A single dashboard displaying the header, parallel coordinates, and footer

As mentioned earlier, a parallel coordinates plot is designed to summarize a specific group's performance across multiple measurements relative to other groups. Unlike a scatter plot, it does not provide specifics of the relationship between two or more metrics. It also does not describe the performance of other groups on those same metrics. The real purpose is to examine how one group performs across these multiple measurements.

In our example, if we looked at Appliances, we'd probably note how average of a category this is. It doesn't stand out on any particular metric! If we looked at a different sub-category, Copiers, we'd see a different pattern (Figure 3-63).

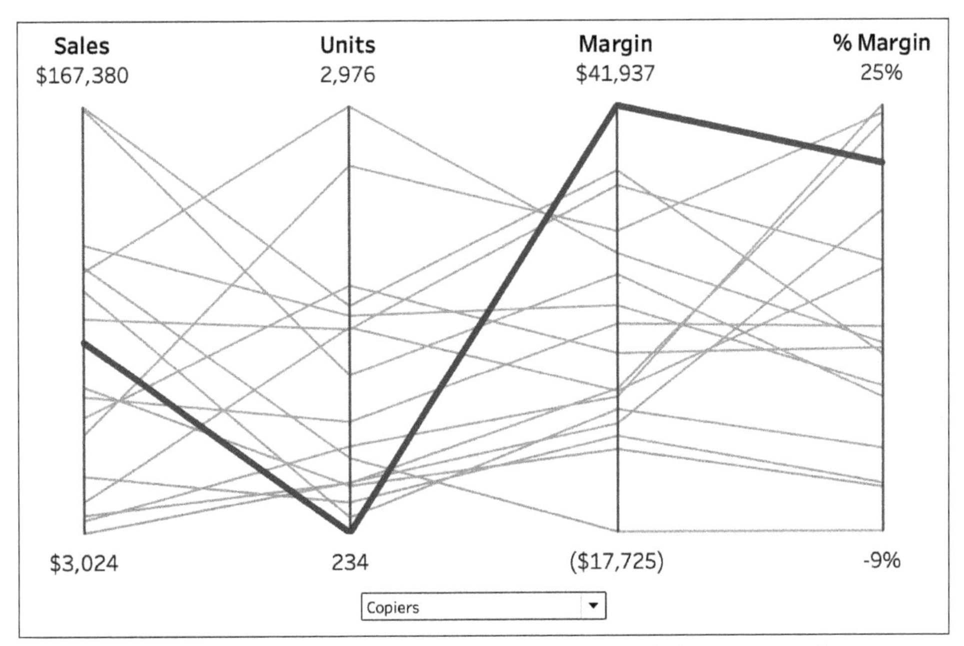

Figure 3-63. An updated view of the parallel coordinates with the Copiers sub-category selected

Copiers, like Appliances, have average sales, but that's where things begin to differ. Copiers are the lowest for total units, but they produce the highest total margin and a very high percent of margin. The result is a plot that concisely tells two different stories for the two sub-categories. While a lot of effort goes into creating a parallel coordinates plot, the payoff can be grand! And if you are unsure whether you want to go down the route of creating a parallel coordinates plot, you can always use a bar chart.

Conclusion

In this chapter, you explored various ways to compare data. Often we are comparing members of a single dimension across a single measure. For this type of comparison, we suggest you use a bar chart.

If your audience gets tired of bar charts, you can try other alternatives—which, while not technically bar charts, serve the same purpose. These alternatives include the lollipop chart and the Cleveland dot plot. The reality is that you don't always need to change the chart type to be something different; sometimes you just need to format your bar chart to look a little different.

You also explored how to calculate rankings without explicitly using the ranking calculation. By using `INDEX()` to sort, you have a more reliable and faster way to calculate unique rankings (particularly when rendering on a server). By using `INDEX()` for separate ranking calculations, you learned how to show the change in rank over time and display that change in rank, either up or down.

You also learned how to create a bump chart to illustrate the order of multiple members over a time series. With the bump chart, you had to reverse the axis and use an LOD calculation to identify the start and end of the lines to provide labels. You also focused on highlighting a single member from the bump chart to emphasize insight.

Next, you tackled scenarios requiring you to compare multiple dimensions or multiple measures. Some comparisons are straightforward. With the barbell plot, you learned that you can use a dual-axis chart and some formatting to really highlight differences across multiple measures.

You also learned about trellis or small-multiple charts. Sometimes creating these is easy when you have unique dimensions on rows and columns. Other times we're faced with the challenge of creating a grid by using just a single dimension. With this example, you saw that simple arithmetic and well-designed calculations can create a grid with ease. You also learned that adding a label to a grid is not always as easy as it seems. But once again, we can use the strength of Tableau to determine the number of marks on a part of the view and provide dynamic calculations based on Detail on the Marks card.

Finally, you tackled multidimensional data by creating a parallel coordinates plot. This required you to normalize our data. It also required a few extra sheets to create values that look like they are on the ends of each of the axes of the parallel coordinate systems. With this example, you conquered `WINDOW_MAX()`, `WINDOW_MIN()`, and `FIRST()`. The final result was a visualization that allowed for a concise comparison of one group versus others on multiple dimensions.

And remember our last piece of advice from our parallel coordinates strategy: bar charts work for all of the examples in this chapter. Bar charts are extremely useful and versatile. However, when you are looking to communicate information in a more concise form, consider some of the other charts discussed in this chapter.

In the next chapter, we'll certainly utilize some of the chart types from this chapter. However, our focus will be on showing you how to get the most out of dates and the calculations you can derive from dates. Chapter 4 also focuses on automating insights, rather than updating filters and parameters as data is updated.

CHAPTER 4

Working with Time

When we were debating the content of this book, no chapter contained more examples than this one. We had so many examples about time that we added them in practically all chapters—including earlier ones.

What makes time interesting is that we can have a single column in our data related to time, and it makes our data extremely flexible to analyze. This is because time data is naturally hierarchical. Whether seconds, minutes, hours, days, weeks, months, quarters, or years, a column including time data gives you flexibility no other column of data will have.

In Chapter 3, we discussed categorical data. Categorical data is ideal for making comparisons. One underlying assumption about all comparisons is consistency. When we, the developers, visualize two non-time-based values, our audience assumes that we are making comparisons that are appropriate. Your audience probably assumes they are the same time period.

Imagine you are doing call-center analytics and are comparing total calls. If today is May 17, 2020, you probably wouldn't want to compare total calls in 2019 to 2020 because they are two, much different lengths! Whether we are trying to compare year-to-date from the current year, or comparing one month's performance to the same month a year ago, our audience assumes consistency with time periods—it's practically an unspoken contract between the developer and the user.

The goal of this chapter is to discuss the natural hierarchy of dates and datetimes and to help you develop calculations that can standardize and automate your visualizations.

In this chapter, we'll walk you through the foundational calculations to make you an expert in date calculations. After we've built out this foundational understanding, we'll transition to strategies for specific, but regular challenges we've faced in working with dates data. And remember: in the remainder of the book, you will see plenty of examples that utilize dates too.

In This Chapter

For this chapter, you need two data sources: the Call Center dataset and the Sample – Superstore dataset. In this chapter, you will use a custom color palette. You can add the custom color palette by editing your *preferences.tps* file in your My Tableau Repository and adding this chunk of code:

```
<workbook>
  <preferences>
    <color-palette name="blackbody" type = "ordered-diverging">
      <color>#000000</color>
      <color>#8f0000</color>
      <color>#e63200</color>
      <color>#e6b900</color>
      <color>#eee154</color>
      <color>#f9f4bf</color>
      <color>#f0f6ff</color>
      <color>#c8dfff</color>
      <color>#a0c8ff</color>
    </color-palette>
  </preferences>
</workbook>
```

If you already have custom color palettes, you can add everything from the `<color-palette>` code sections.

In this chapter, you'll learn how to do the following:

- Navigate the hierarchies of date and time fields in Tableau, including using the `DATEPART()`, `DATENAME()`, and `DATETRUNC()` fields
- Plot points to the nearest hour and minute
- Plot points to the nearest 15 seconds and 15 minutes (or whatever time interval you choose)
- Build effective heatmaps
- Create continuous time calculations when the dates may be different
- Use table calculations to analyze month-over-month and year-over-year sales data
- Automatically filter data to the most recent 13 months by using date calculations
- Work with nonstandard fiscal calendars including mid-year fiscal starts, ISO-8601, and 4-5-4 retail calendars

Understanding Dates and Time

Tableau provides users with a very simple interface for working with time. By default, Tableau creates a hierarchy that allows users to easily navigate any datetime possibility. If you drag any date or datetime field onto the visualization, the field will automatically appear aggregated to the nearest year. This is the default of dates in Tableau.

In addition to Year as the default setting, Tableau places time values into a hierarchy. If you drill into the hierarchy, additional detail will be added to your view by quarter, month, week, day, hour, minute, and second.

This hierarchy will be available to your audience too. If you do not want that functionality, you will need to create custom date calculations—which we highly recommend.

While the default setting for a time field is a year, you can right-click and edit the year calculation and select the appropriate time.

If you do not want a year to be the default value, you can click and hold the Option key on a Mac, or right-click on a PC, while dragging the time field onto the view. From there, you can select the date part or value of interest (Figure 4-1).

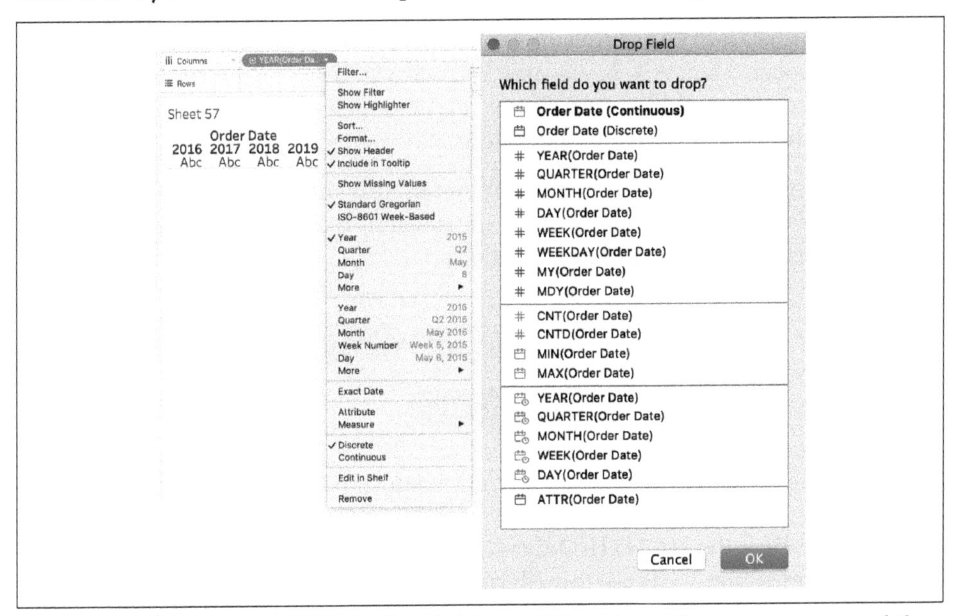

Figure 4-1. The view you will see after clicking and dragging a date onto a view and then right-clicking to change the date type (left); the view you will see after clicking and holding Ctrl (or Option on a Mac) while dragging a date onto the view (right)

Date Parts and Date Values

The way you can work with dates opens up many possibilities. When you look at Figure 4-1, you see many of the options for working with dates. You can choose date *parts* like Year (2015), Quarter (Q2), or Month (May), and you can choose date *values* like Quarter (Q2 2015), Month (May 2015), or Day (May 8, 2015). The difference between these is that you're either choosing a particular part of the date or truncating the date to the most recent date value.

Date Calculations

Whether it's figuring out the date part or the value to truncate to, Tableau offers the same flexibility in calculated fields through the `DATEPART()`, `DATENAME()`, and `DATETRUNC()` functions. `DATEPART()` returns a single numeric value for the part of the date of interest. For example, May 8, 2015 would return a value of 5 if set to return the month. `DATENAME()` returns a string for the part of the date of interest, so May 8, 2015 would return May if set to return the month. It's a subtle difference, but the way the information is displayed will be different. Unlike `DATEPART()` and `DATENAME()`, `DATETRUNC()` returns an actual date or datetime value. With `DATETRUNC()`, values are rounded or truncated to the most recent specified date value.

Now, if the preceding two paragraphs sound familiar, it's because we wanted it that way to show you this next point. While Tableau's fields on your view might say YEAR, QUARTER, or MONTH, they actually are using these calculations behind the scenes. Let's take a look at Figures 4-2, 4-3, and 4-4, and you'll see we have the date part of Month and date value of Year on columns. If you double-click either of these values, Tableau will show the underlying calculations.

First, you have a discrete month and continuous year on the Columns shelf. These fields look nice and clean—and easy to understand. But underneath each of these fields are actually more-complicated functions.

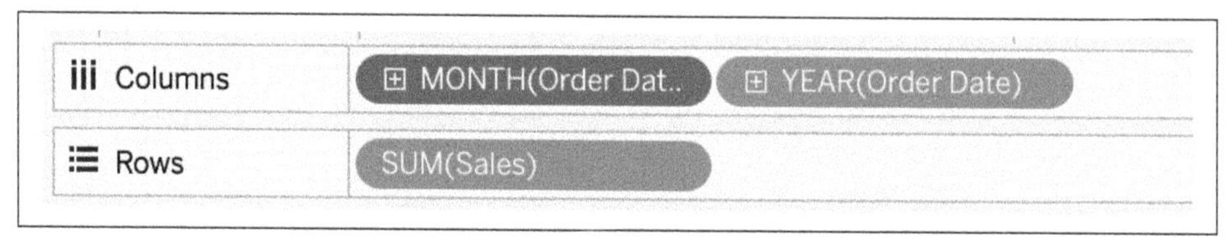

Figure 4-2. The Columns shelf showing discrete month and continuous year

If you double-click the MONTH(Order Date) discrete field, this will open up the ad hoc calculation editor. We'll use this editor quite a bit in this book. Figure 4-3 shows the underlying calculation for MONTH(Order Date). It turns out this calculation uses the `DATEPART()` function and specifies `month` as the first argument in the function. This helps return the month part of the order date.

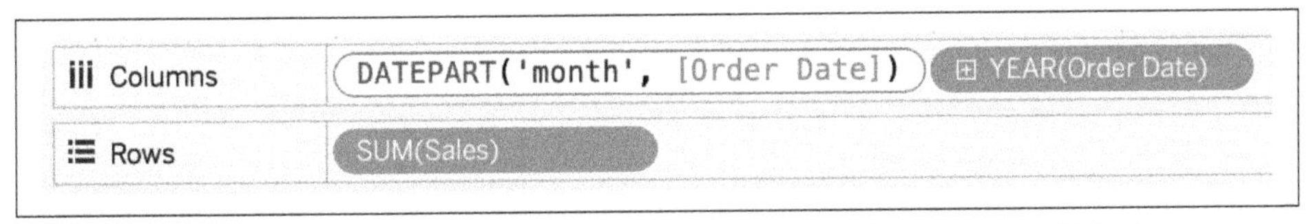

Figure 4-3. The `DATEPART()` function that makes up the infrastructure of the discrete MONTH(Order Date) field

And if you double-click the continuous field of YEAR(Order Date), the underlying function is `DATETRUNC()`, which as previously mentioned, rounds a field down to the specified level. This is shown in Figure 4-4. In this case, `year` will return January 1 of the year in [Order Date].

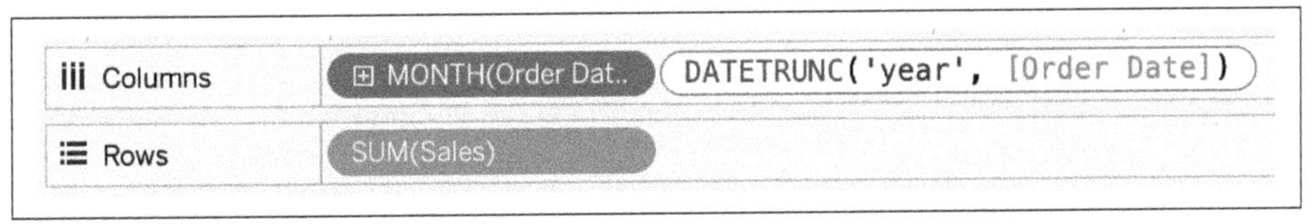

Figure 4-4. The `DATETRUNC()` field makes up the infrastructure of the continuous YEAR(Order Date) field

What's great about seeing these calculations is that it helps you learn date calculations inside Tableau while you're using the default settings of any time field. What's also great is that you can change these calculations quickly—even in meetings to share insights as conversation occurs.

Date Hierarchies and Custom Dates

You'll notice that all the dates in Figure 4-4 include hierarchies—this is that little + button to the left of the field names on your date fields on Columns. By default, a date field will automatically create these hierarchies. When you add a date field to your view, your audience will have the ability to interact with that date hierarchy. If you are looking to interact with a specific hierarchy of a date field, working with dates can be a challenge.

Luckily, there are ways to circumvent the automated date hierarchies. You can do this by creating a custom date: you right-click the original time field in the data source and choose Create → Custom date (Figure 4-5). From there, you must select the date part or date value of interest. Once you've created this field, place it somewhere in the view and you'll notice the date has no hierarchy associated with it.

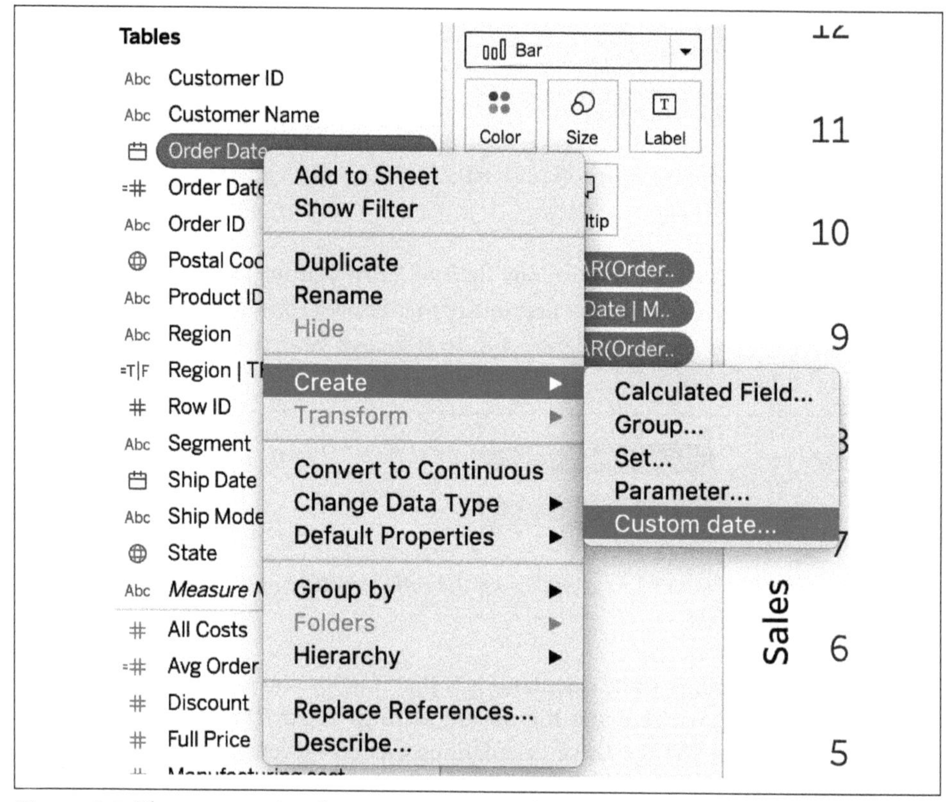

Figure 4-5. The menu action for creating custom dates

Discrete Versus Continuous Dates

To better understand how dates are visually represented in Tableau, let's look at four visualizations in which we vary two components on the horizontal axis. Let's look at the difference between date part and date value, and let's also look at how discrete and continuous axes vary for the two chart types. This gives you four options to explore (Figure 4-6).

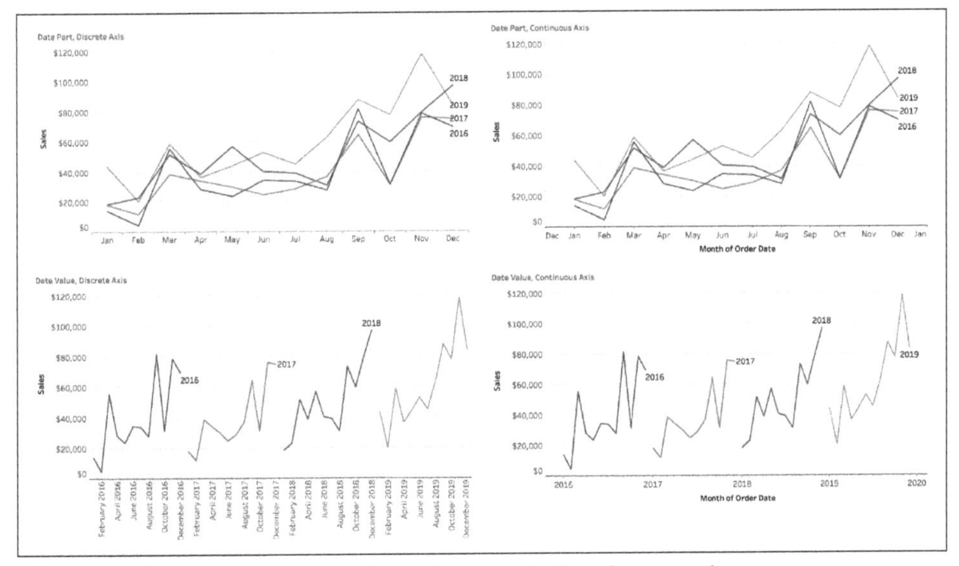

Figure 4-6. Visual output differs depending on whether dates are discrete or continuous and whether you're working with date parts or date values

You can see how changing just two options can yield four charts that operate quite differently. The two big differences that are worth reiterating: a date part selection returns only a single part of a date field, and the default option for a date part is typically Discrete. This creates *bins* that separate each date part.

You can still convert discrete date parts to continuous to make a continuous axis. The chart types will look similar, but the axes function differently. If you look at the continuous date part visualization at the top right, you will see that the axis ticks for each month are centered on the month name. With discrete date parts at the top left, the visualization has buckets for each month, and no ticks to align to the values for each month. These ticks make it easier for our audience to understand which months they are looking at, so for this reason we prefer working with continuous fields.

This extends to your date value options too. If you choose a date value of Month, Tableau rounds each value down to the start of each month by default. This gives you month-by-year views of the data. By default, date values are continuous fields, but you can convert them to discrete. You'll notice on the axis on the bottom left of Figure 4-6 that the axes are distinct buckets of label names. The continuous date values on the bottom right, however, show a single axis with ticks.

Remember that discrete dimensions will create headers only for data that exists. If you have gaps in your dates, you may want to use a continuous axis to preserve any unrepresented dates.

There is a place for discrete date axes. If you are planning on using a bar chart, you can use discrete date parts. If you are working with line charts, we recommend continuous axes.

Call Frequency: Chips and Bolts Call Center Case Study

The call center of a car part manufacturer, Chips and Bolts (CaB), is looking to improve its customer satisfaction scores. The executives need to better understand the basics before they can assess their performance. As a first step, they want to know how many calls they receive. They're looking to track this data at 15-minute increments over a 2.5-year window to better understand how call volume may relate to their customer satisfaction surveys. How can this information be represented?

Working with time in Tableau is not without challenges. But it's easy to place hours and minutes on a chart in Tableau: if you are working with a datetime field, Tableau automatically places it in the datetime hierarchy. Creating a plot on total calls, for instance, wouldn't be that difficult.

With the continuous date part example shown previously, it's fairly easy to convert a single date part into a continuous axis to plot multiple time periods. Using continuous date values truncates a date up to a certain date part, but sometimes you want to do the opposite. In this example, we want a continuous axis for hour, minute, and day. Take a look at Figure 4-7, showing total phone inbound calls received from the call center every 15 minutes over 2.5 years.

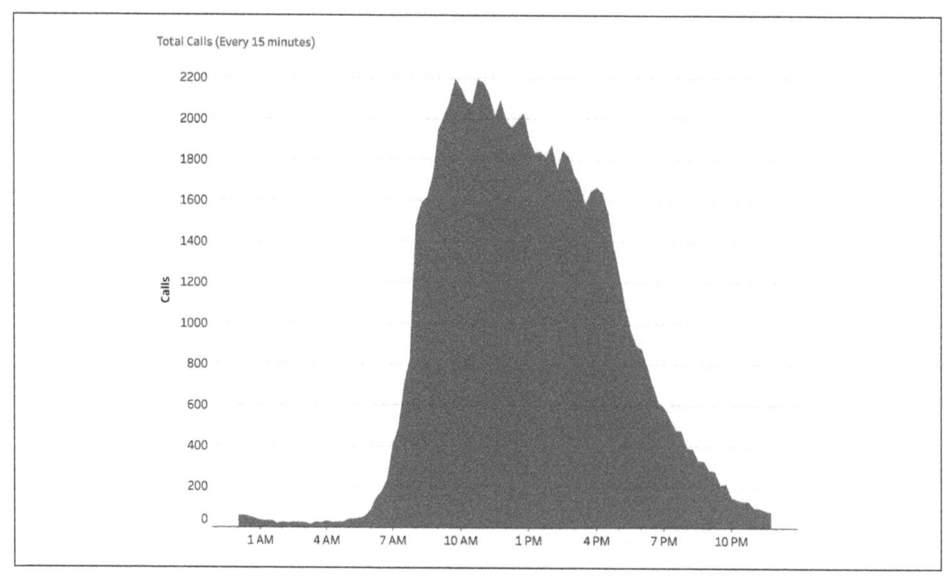

Figure 4-7. An area plot of total calls every 15 minutes

For the first four strategies, you will convert a datetime to various levels of aggregation. To complete the final analysis, call data aggregated on a continuous axis by every 15 minutes, you'll need to get every date to be the exact same—but retain the time of each row in the data.

Strategy: Determine Total Call Time by Hour

In this strategy, we will tackle the challenge of working with time—particularly on a continuous axis. The result of this strategy is shown in Figure 4-8.

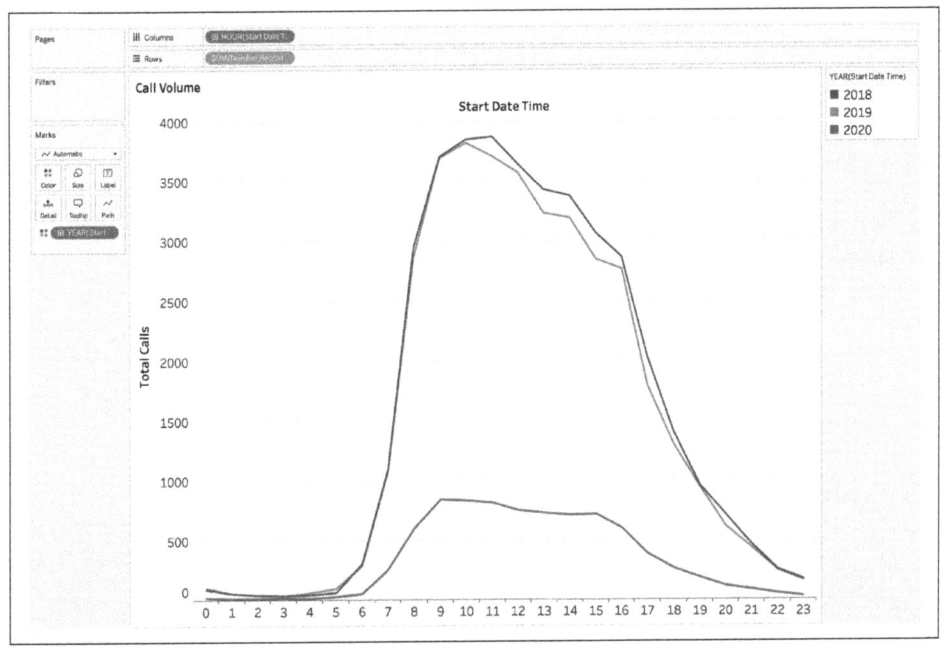

Figure 4-8. A line plot of sales by hour for each of the three years of data

In this strategy, we will plot the total calls by hour, broken down by year:

1. Create a new sheet and set it to fit the entire view.
2. Place [Start Date Time] on Columns. Then change the display to discrete hours.
3. If you are using Tableau Desktop 2020.1 or older, your data source already comes with the [Number of Records] field. If you are working with Tableau 2020.2 or newer, create a calculated field called **`[Number of Records]`**:

```
// Number of Records
1
```

4. Add [Number of Records] as a sum to Rows.
5. Add the discrete year of [Start Date Time] to Color.

Doing this allows you to see that—regardless of year—the calls begin to escalate around 7 a.m. but really pick up by 8 a.m. You also see that calls in 2020 are down across the board.

What's missing from this analysis is a deeper investigation. Do calls occur at the beginning of each hour, or at mid-hour? If you are creating a staffing plan, the exact times might be more useful.

Strategy: Create a Plot to Measure Total Call Time by Minute

We will expand on the previous strategy by creating a visualization that plots total calls by minute of the day:

1. Sort the data by minute by clicking the + on the HOUR hierarchy. Figure 4-9 shows the resulting visualization.

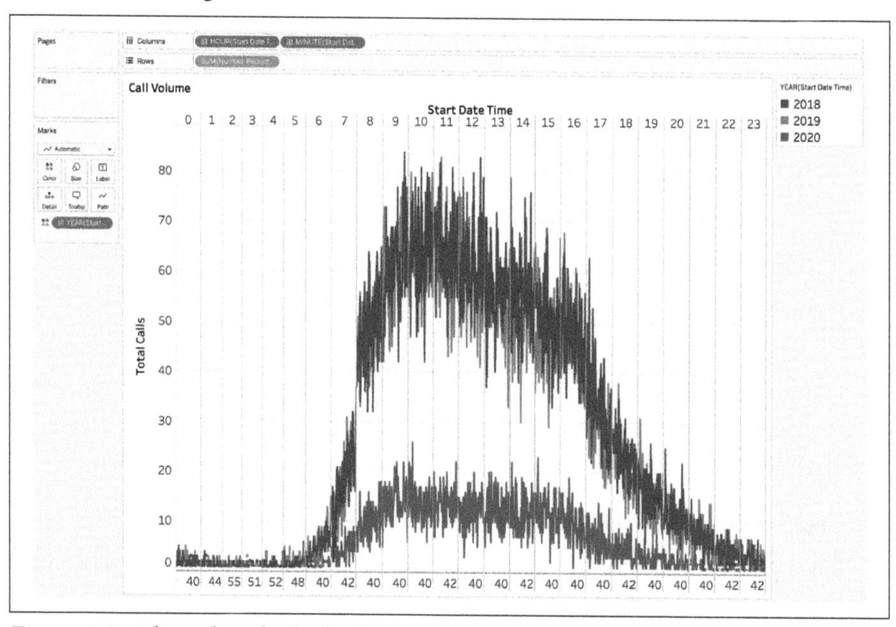

Figure 4-9. A line plot of sales by hour and minute for each of the three years of data

The result shows calls by minute in the day, but now we have two discrete axes: the top axis partitioned by hour, and a second axis for minutes, also creating up to 60 individual partitions within each hour. This creates partitions only where data exists. So if there is no data for hour 0, minute 53 (and there isn't), the partition doesn't exist.

2. Add any missing partitions. If you want to include missing partitions of any specified date part, right-click the date part and select Show Missing Values (Figure 4-10).

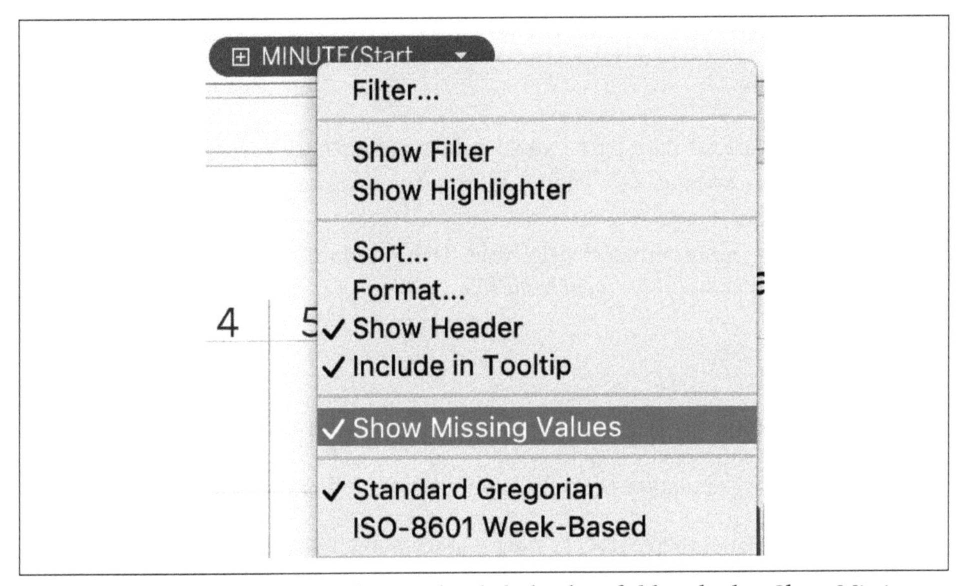

Figure 4-10. To show missing values, right-click the date field and select Show Missing Values

If you take a look at Figure 4-9, you'll see that your visualization has two sets of discrete bins: one for hours and one for minutes. Using discrete values for hours and minutes is difficult because you end up with 1,440 partitions (24 hours × 60 minutes).

So how can you make a single axis? For this calculation, we are going to rely on two commonly used calculations: `DATEADD()` and `DATEDIFF()`.

`DATEADD()` adds or subtracts dates or time and requires three inputs:

- A specified date part, written inside quotation marks and in lowercase. This shows the units we are adding to a date, whether it's seconds or hours or years.
- Any integer indicating the amount of time we want to add or subtract. If a negative number is specified, time will be subtracted from the value.
- The initial datetime field.

This extremely versatile calculation is one we use all the time.

`DATEDIFF()` calculates the difference between two dates based on the date part of interest and requires three inputs:

- A specified date part
- A starting date
- An ending date

You can specify various date parts: `year`, `quarter`, `month`, `dayofyear`, `day`, `weekday`, `week`, `hour`, `minute`, `second`, `iso-year`, `iso-quarter`, `iso-week`, and `iso-weekday`.

For more technical help, read Tableau's documentation on date functions (*https://oreil.ly/blPRT*).

Strategy: Create a Continuous Datetime Axis by the Second

You'll now create a calculation that allows you to have a single axis for time:

1. Create a calculated field called **`[time]`** and write the following calculation:

```
// time
DATEADD(
    "day",
    DATEDIFF(
        "day",
        [Start Date Time],
        {MAX(DATETRUNC("day", [Start Date Time]))}
    ),
    [Start Date Time]
)
```

   This calculation will change all dates in your dataset to be equal to the maximum date in your dataset. The time (hour, minutes, and seconds) will remain the same.

2. Create a continuous axis as follows:

   a. Remove all calculations on the columns.

   b. Add the [time] field as an exact date to the columns.

   This will produce the visualization in Figure 4-11.

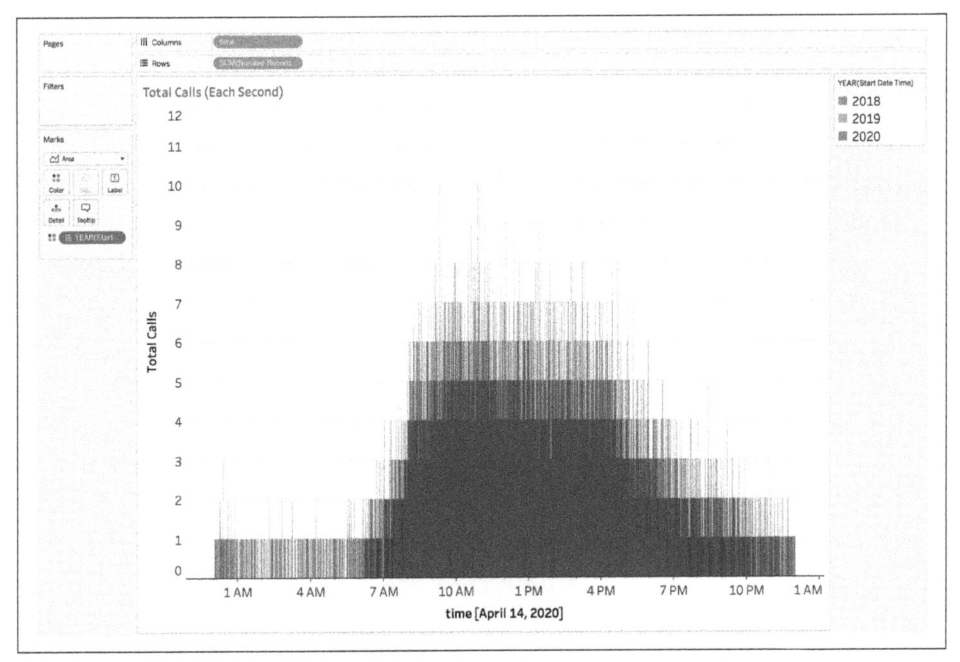

Figure 4-11. Total calls per second of the day using a continuous axis

You now have a continuous axis. However, the analysis is to the second, which isn't extremely helpful or insightful. Instead of a per-second analysis, maybe you want to capture data every 15 seconds.

Strategy: Create a Continuous Datetime Axis for 15-Second Intervals

You will continue exploring datetime by creating a custom calculation that aggregates calls based on every 15 seconds of the day:

1. Use the visualization from our preceding strategy.
2. Create a new calculation and call it **`[time / 15 sec]`**.
3. Write the following:

   ```
   // time / 15 sec
   DATEADD(
     "second",
     -(DATEPART("second",[time]) % 15),
     [time]
   )
   ```

   Here you're first calculating the seconds for the time field. You are then using the modulo operator (`%`) to calculate the total seconds every 15 seconds. Therefore, rather than counting to 60, you are counting to 14; then, instead of continuing to 15, you restart at 0.

The result of this calculation is a datetime truncated to the most recent 15 seconds.

4. Click and drag [time / 15 sec] to replace the time field as a continuous axis. This produces the visualization in Figure 4-12.

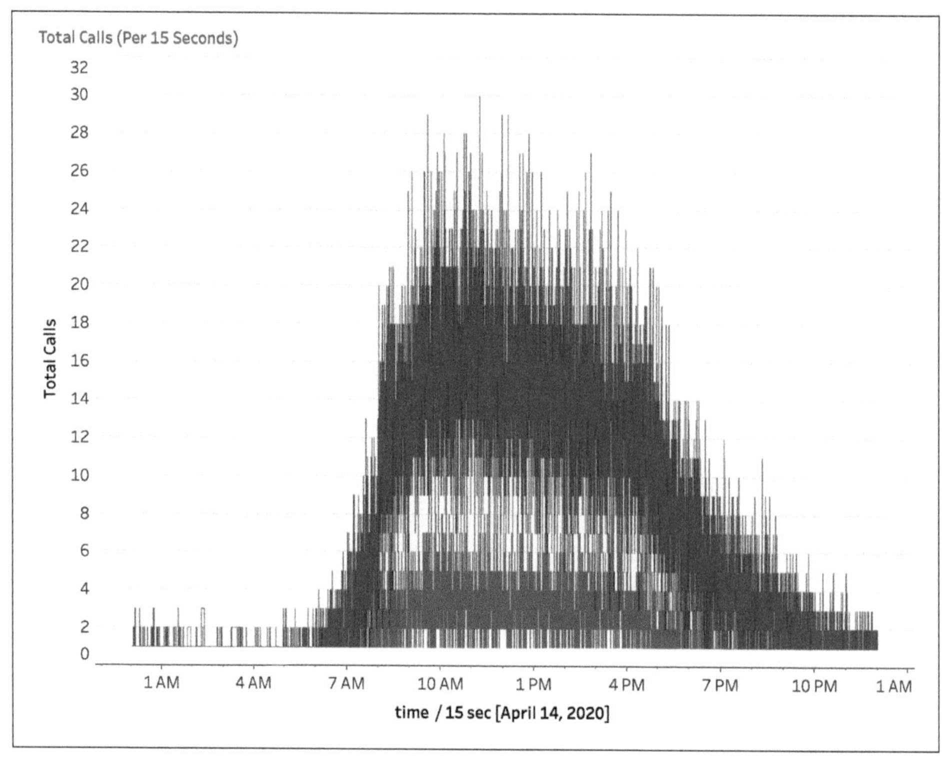

Figure 4-12. Total calls every 15 seconds of the day using a continuous axis

You're starting to see patterns like those with the hourly plot, but this view is still too granular. Instead of every 15 seconds, what if you looked at every 15 minutes?

Strategy: Create a Continuous Datetime Axis for 15-Minute Intervals

When we started working with this data, we saw that information at the hour LOD was interesting, but we needed to see more information to get more specific. From there, we looked at the plots by every minute and every 15 seconds. Those plots were too detailed. In this strategy, we create a calculation that truncates time to every 15 minutes. The result will be a plot that is far more actionable than the previous three.

1. Create a new calculation called **`[time / 15 min]`**.
2. Write the following:

```
// time / 15 min
DATEADD(
  "minute",
  -(DATEPART("minute", [time]) % 15),
  DATETRUNC("minute", [time])
)
```

   The format for [time / 15 min] looks almost the same, except we've replaced `second` with `minute`, and our third argument is now `DATETRUNC("minute", [time])` instead of `time`. This is because our analysis with [time / 15 sec] was already at the lowest level in Tableau (seconds). Since we are working at a higher level of data, we need to roll up all the values to the nearest minute.

3. Roll up to the nearest minute:
   a. Click and drag to replace [time / 15 sec] with [time / 15 min].
   b. On the Marks card, click Path and change the line type to Step. We like using a step path instead of a straight line because we know that the line represents all values across this 15-minute increment.

   Finally, in Figure 4-13, we have a single axis where we can see patterns in the data at 15-minute intervals. This plot gives us lots of great information about how quickly calls are scaling up each morning, with much greater precision than to the nearest hour (but not too precise).

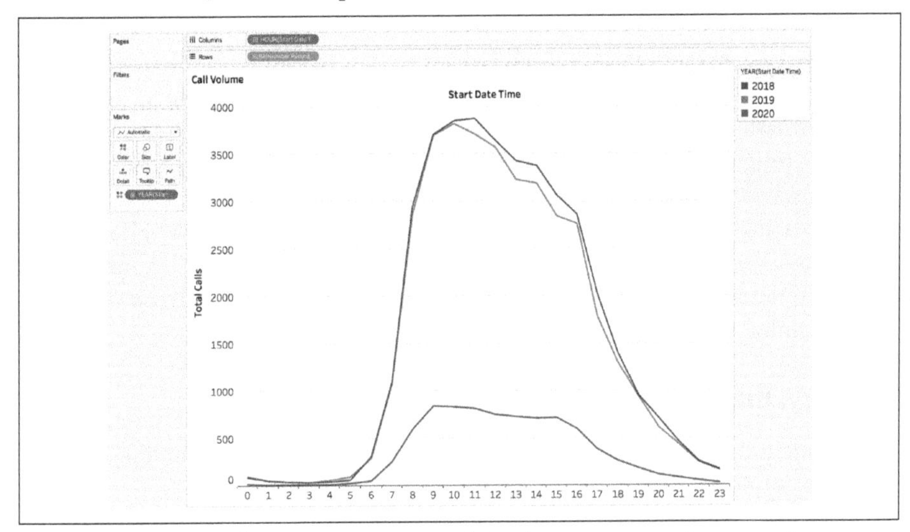

Figure 4-13. Total calls every 15 minutes of the day using a continuous axis

You still have the red line—representing 2020—much lower than all other values. This might be because we've collected data only up until April 14, 2020. And because business might be seasonal, comparing time periods that are alike might be worthwhile. But before we do that, we want to take a deeper look at calls per 15 minutes.

4. Let's look at calls per 15 minutes, by day of the week, by adjusting the dimension on Color. Right-click YEAR(Start Date Time) and change the date type to a discrete date part of `weekday` (this is located under the More section of the date part). Feel free to edit the colors afterward. Figure 4-14 shows the resulting visualization.

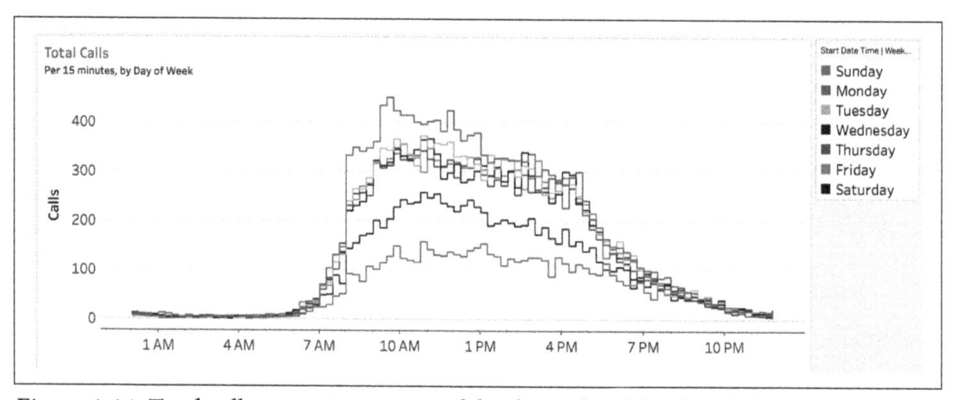

Figure 4-14. Total calls every 15 minutes of the day, colored by day of the week and using a continuous axis

Heatmaps (Highlight Tables)

The information in Figure 4-14 is extremely useful, but there are just too many lines to read through the insights. When we're working with line charts that have eight or more lines, we immediately consider other chart types. Our go-to chart type for this scenario is the *heatmap*—though Tableau calls it a *highlight table*. Heatmaps allow audiences to see change via color, intensity, or hue rather than through direction. This heatmap is displayed as a matrix so that anyone can easily track changes for a single member in a dimension.

Strategy: Build an Essential Heatmap

Let's create a heatmap that reimagines the same analysis from the line chart in our preceding strategy:

1. Create a new sheet.
2. Change the mark type to Square.

3. Create a custom date for the date part of weekday by using [Start Date Time] and place that on Columns.
4. Place [time / 15 min] on Rows but choose the Hour date part.
5. Place [Number of Records] on Color.
6. Choose a color palette that works for you. We're choosing a custom color palette (we'll talk about that more in Chapter 12).

   The result is the heatmap in Figure 4-15.

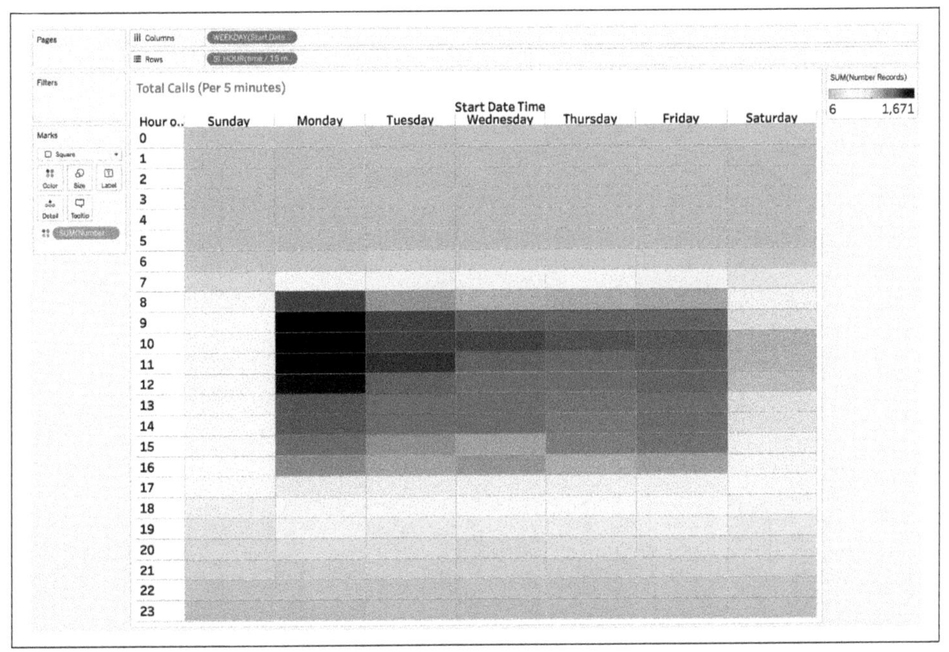

Figure 4-15. Total calls every hour of the day and day of the week using a heatmap

Once again, you see that calls begin to pick up at 8 a.m., through change in color. But you are also able to spot that Mondays, particularly in the morning, are extremely busy. Weekends are quieter, and Sunday is exceptionally slow. We also see that calls begin falling off after 5 p.m. (depicted as 17 in the chart).

Strategy: Create a More Detailed Heatmap

A heatmap can be extremely helpful even when the data is fairly granular. Heatmaps' value only increases as more complexities and detail are added to a visualization. Follow these steps to add more detail to your heatmap:

1. Duplicate your visualization from the preceding strategy.

2. Create a custom date for the date part of Month by using [Start Date Time].
3. Place it on Columns and to the left of your weekday calculation on Columns.
4. Click the + on HOUR(time / 15 min) on Rows. This will show time rounded to the nearest 15 minutes.
5. You'll notice some whitespace where there are no values on your dashboard. If you want to add marks for those locations, you'll need to use a lookup calculation: `ZN(LOOKUP(SUM([Number of Records]),0))`. (We'll talk about this in more detail in Chapter 6.) Add this calculation to Color.
6. Format your column and row headers to be more readable. Figure 4-16 shows the resulting visualization.

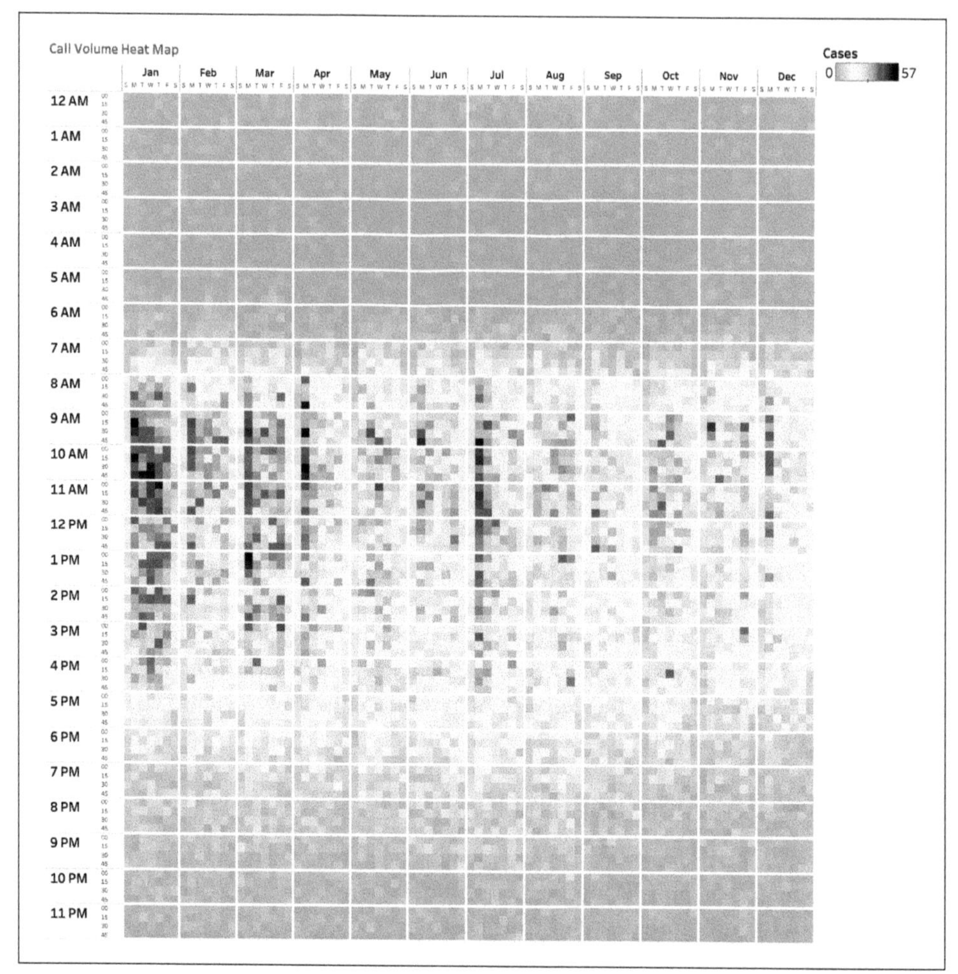

Figure 4-16. Total calls every 15 minutes of the day by month and day of the week using a heatmap

We've chosen to format dividers at the Hour and Month level. This allows your audience to quickly navigate to segments of analysis.

This chart shows call volumes by 15-minute increments by month and day of the week. So what insights can you glean from this data?

- Calls pick up at 8 a.m. on weekdays almost every month of the year.
- More calls tend to occur in the later evening during the summer months.
- Regardless of day of the week, a lot of calls typically occur in January.
- In April, July, and December, more calls happen on Mondays. This might be because a heater or air conditioner that broke over the weekend results in calls for service on Monday.

Identifying call patterns from this visualization could be extremely helpful in staffing. Mondays are always busy, but it might make sense to bulk up on those days in April, July, and December. With call volumes higher later into the evening in the summer, staffing might be needed until 6 p.m. instead of 5 p.m. This could be offset with staff working fewer hours on non-Mondays during November and December.

We love heatmaps. They are underrated tools for representing time and are particularly useful when data can be represented with many members of a dimension.

Comparing Values Year-to-Date: CaB Call Center Case Study

Now that the call center better understands the number of calls coming in, the staff would like to understand how this has changed over time. It's currently April, and they want to see the current year represented in the analysis. How can you create a like-to-like comparison against prior years?

For this case study, we will continue our analysis of call center data. In this final analysis, we are concerned with making appropriate comparisons across years. To do so, let's go back to our plot of total calls per 15 minutes by year (Figure 4-17).

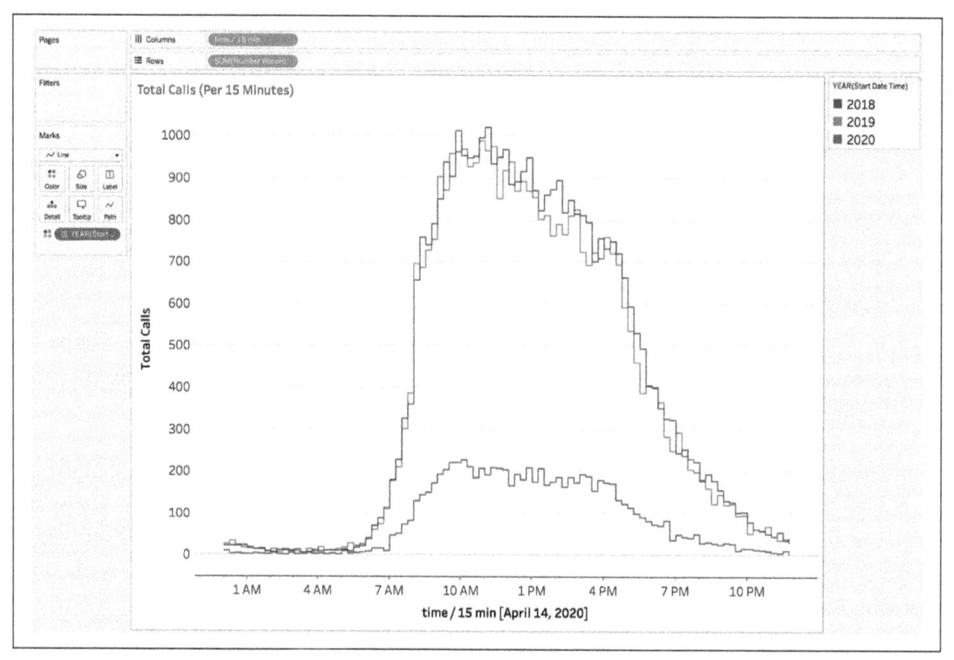

Figure 4-17. Total calls every 15 minutes of the day by year

We see that calls are down, but that's because we're at a different point in time in 2020 than the other years. This is partly because our data is only through April 14, 2020. What might be fairer is to compare calls for 2018, 2019, and 2020 through April 14 instead of their overall totals.

This is a challenge we face quite often, regardless of data type: being able to compare similar time periods. So how can you solve the problem? With a well-designed date calculation.

The goal of the next two strategies is to create a calculation to compare the most recent date for the most recent year to the same date in prior years. For the next strategy, you will re-create a bar chart that shows progress to the total. This will allow your audience to keep an eye on overall values while simultaneously displaying comparable year-to-date values. Then, we'll apply a filter to our line chart, allowing for a proper year-to-date comparison, rather than the visualization shown in Figure 4-18.

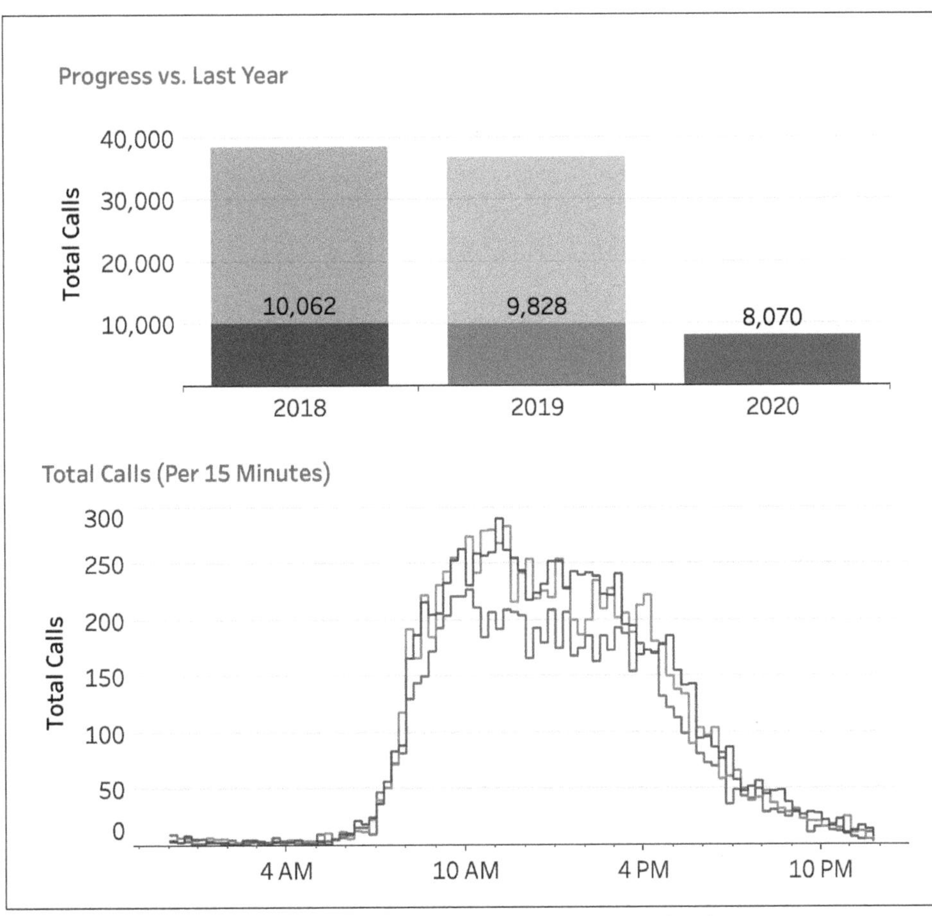

Figure 4-18. Visualizations showing total calls versus the total for the two previous years (top), and total calls filtered to the same day of the year for three years (bottom)

Strategy: Show Progress to the Total by Using Two Bar Charts

In this strategy, you'll use a bar chart to show progress to the total:

1. Build the calculations as follows:

   a. Normalize the dates to the same year by creating a calculation called **`[Start Date Time | Same Year]`**:

   ```
   // Start Date Time | Same Year
   DATEADD(
     "year",
     DATEDIFF("year", {MAX([Start Date Time])}, [Start Date Time]),
     DATETRUNC("day", [Start Date Time])
   )
   ```

b. Create a second calculation called **`[Start Date Time | Same Year | TF]`**. This is a Boolean that detects whether a date is less than or equal to the day of the year for the most recent year:

```
// Start Date Time | Same Year | TF
DATEPART("dayofyear", [Start Date Time | Same Year])
  <= DATEPART("dayofyear", {MAX([Start Date Time])})
```

c. Create a third calculation called **`[Total Calls | YTD]`**:

```
// Total Calls | YTD
SUM(
  IF [Start Date Time| Same Year | TF]
  THEN [Number Records]
  END
)
```

This will return year-to-date values for each year.

2. Build the visualization:

   a. Add [Number of Records] to Rows.

   b. Add [Total Calls | YTD] to the right of [Number of Records] on the Rows shelf.

   c. Create a synchronized dual-axis chart.

   d. Add [Start Date Time] as a discrete year name to Columns and to Color.

   e. Set the opacity on the `SUM([Number of Records])` Marks card to 40%.

   f. Format your visualization by removing the column and row dividers, adding a darker column axis ruler and tick marks, styling your grid lines, showing just a left axis, and renaming the axis.

   g. Show labels on the AGG(Total Calls | YTD) Marks card.

   This results in the bar-on-bar chart in Figure 4-19.

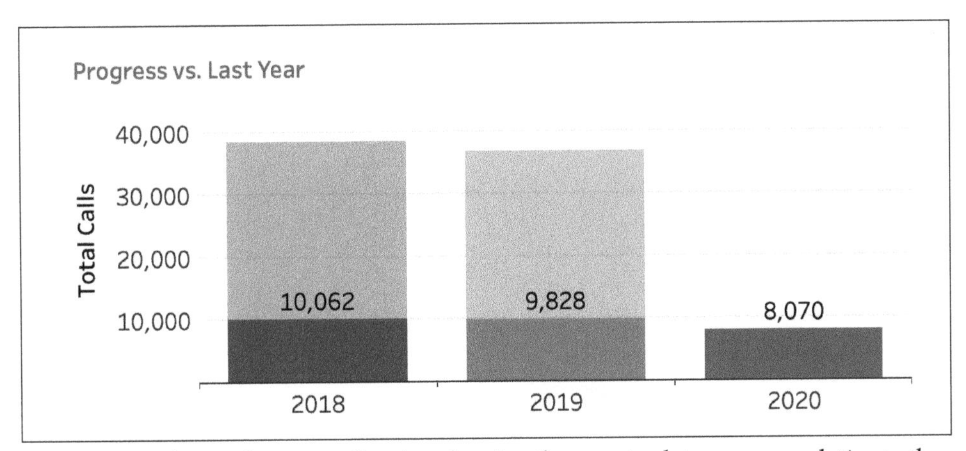

Figure 4-19. The resulting visualization showing the year-to-date progress relative to the current date for the most recent year

In this visualization, we embedded an `IF` statement inside the aggregation to do the filtering inside the calculation. This is something you should regularly do to allow a visualization to be dynamic. As the year continues, we'll see the 2020 bar increase. The 100% opaque bars in 2018 and 2019 will also continue to grow. These bars will eventually cap at the values shown in the 40% opaque bars shown on the same axis.

Take a look back at Figure 4-17. That visualization shows total calls for all dates in 2018 and 2019. It also shows just dates through April 14 in 2020. It's not a like-for-like comparison. It would be great if we could compare these values.

Strategy: Compare Similar Periods on a Line Chart

Unlike the preceding strategy, where we used an `IF` statement to filter our data, we are going to explicitly place a calculation created on the Filters shelf. Our goal with this strategy is to use a visualization created earlier in the chapter and add a year-to-date filter:

1. Duplicate your final visualization from the earlier continuous data strategy shown in Figure 4-13.
2. Ensure that you have completed step 1a and step 1b from the preceding strategy.
3. Edit Weekday of [Start Date Time] currently on Color by changing the date type to discrete year.
4. Add [Start Time Date | Same Year | TF] to the Filters shelf, select True, and click OK.
5. Edit the axis and remove the axis title. Figure 4-20 shows the result.

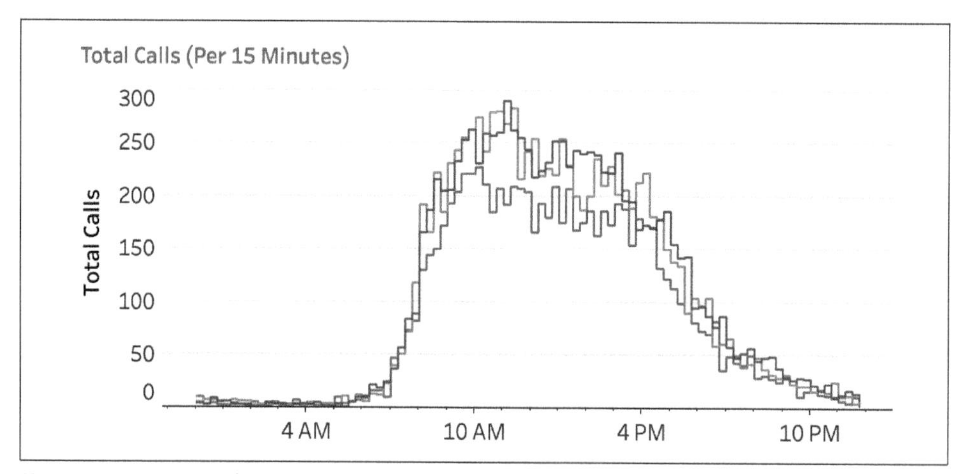

Figure 4-20. A visualization showing total calls every 15 minutes of the day filtered to the same day of the year for 2018, 2019, and 2020

When you place the results from our preceding two strategies next to each other, the end result is a miniature year-to-date dashboard that allows your audience to track total calls year to date as well as when the calls occurred (Figure 4-21).

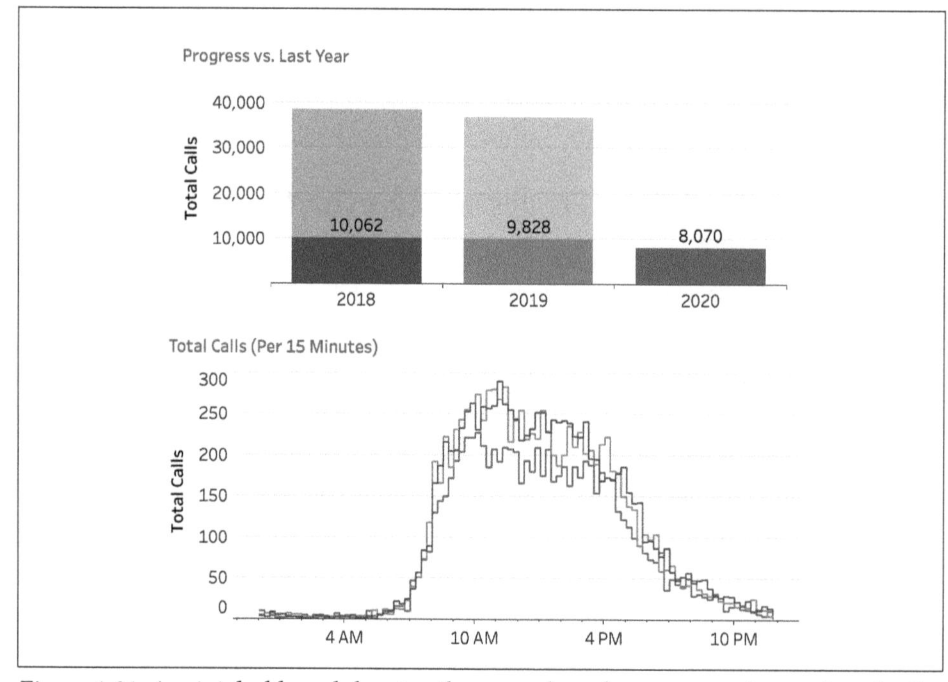

Figure 4-21. A mini dashboard showing the same-day-of-year comparisons of total calls and every 15 minutes of the day for 2018, 2019, and 2020

Automated Reports

In this section, we are going to look at automating reports by using custom calculations. Whether it's call center data, financial reports, or student enrollment numbers, we spend a lot of time developing tables that automatically update at the end of a month. While implementing these tables takes a little bit of time, the effort goes a long way in saving time. One of the most common actions we see from novice users of Tableau is manually updating data, then going to a dashboard and editing a filter to include updated data. The goal of the next strategy is to show how to automatically update a dashboard based on the data that is on the dashboard.

Automating Reports for Month-over-Month and Year-over-Year Change: CaB Call Center Case Study

Now that the CaB call center is starting to understand overall call volume by year, the employees want to take a closer look at how manufacturing cycles and large orders impact their satisfaction scores. They have requested a month-over-month view in addition to the year-over-year reporting already provided. This view will show more-granular data. How would you build a report that is still easy to understand? What steps would you take to automate this report?

Take a look at the table in Figure 4-22. It shows the average calls per day as well as the percentage of change in calls month-over-month and year-over-year. The table shows data from March 2019 through March 2020. We decided to not report anything for April because our data reports only through April 14. When we have data reported for the final day of April, the report will automatically update so that the table shows metrics from April 2019 through April 2020. Additionally, this table shows a breakdown of calls by call reason and aggregates to totals.

Call Reason		Mar-19	Apr-19	May-19	Jun-19	Jul-19	Aug-19	Sep-19	Oct-19	Nov-19	Dec-19	Jan-20	Feb-20	Mar-20
Total	Calls/Day	30.0	102.3	105.2	113.1	116.3	111.9	105.1	99.7	97.1	85.3	82.6	77.8	75.3
	MoM	-67%	241%	3%	8%	3%	-4%	-6%	-5%	-3%	-12%	-3%	-6%	-3%
	YoY	-64%	65%	2%	6%	3%	-4%	-7%	-5%	-3%	-12%	-3%	-5%	-8%
3	Calls/Day	12.0	48.2	51.5	52.2	55.2	51.2	48.4	43.3	44.0	37.6	37.8	38.1	36.8
	MoM	-73%	301%	7%	1%	6%	-7%	-5%	-11%	2%	-14%	0%	1%	-4%
	YoY	-75%	74%	6%	1%	6%	-9%	-6%	-11%	2%	-14%	0%	1%	-11%
1	Calls/Day	13.0	36.9	38.6	40.7	41.5	39.6	36.6	37.5	34.7	30.7	29.6	27.1	27.8
	MoM	-61%	184%	5%	5%	2%	-5%	-8%	2%	-7%	-11%	-4%	-8%	3%
	YoY	-57%	58%	4%	5%	2%	-5%	-8%	2%	-9%	-12%	-3%	-7%	5%
2	Calls/Day	4.0	10.9	10.1	13.7	14.5	14.8	14.7	13.6	13.6	12.9	11.2	8.6	7.0
	MoM	-58%	173%	-7%	36%	6%	2%	0%	-8%	0%	-5%	-13%	-24%	-18%
	YoY	-55%	58%	-6%	17%	6%	2%	-1%	-9%	0%	-6%	-16%	-28%	-39%
Null	Calls/Day	1.0	6.4	4.9	6.4	5.0	6.3	5.5	5.6	4.9	4.3	4.1	4.1	4.3
	MoM	-80%	537%	-23%	31%	-22%	26%	-13%	2%	-13%	-13%	-4%	-1%	5%
	YoY	-48%	56%	-16%	15%	-17%	19%	-15%	2%	-14%	-12%	-3%	-1%	20%

Figure 4-22. Average calls per day and percentage of change in calls, month-over-month and year-over-year, March 2019 through March 2020

Strategy: Automated Rolling Table

In this strategy, you will re-create the table shown in Figure 4-22. This will always show the last 13 fully completed months based on the date with the last entries:

1. Build the base table for this visualization. Create the metric for our table, **`[Calls/Day]`**:

   ```
   // Calls/Day
   SUM([Number Records])/COUNTD(DATETRUNC("day", [Start Date Time]))
   ```

   Start by adding [Calls/Day] to Text. Add [Call Reason] to Columns and sort the dimension in descending order by calls per day. Create a new custom date called **`[Start Date Time | Month]`** that returns monthly date values. Place this as a discrete value on Columns. Add column totals and place them at the top, as shown in Figure 4-23.

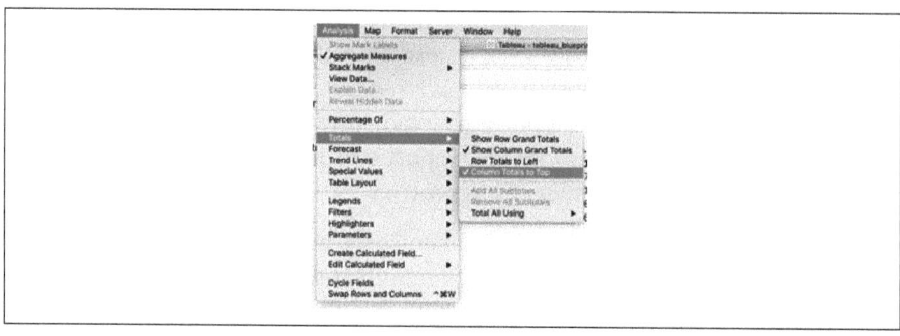

Figure 4-23. To add totals to the top of the columns, choose Analysis → Totals and then select Show Column Grand Totals and Column Totals to Top

2. Format your table so that only row dividers exist, as shown in Figure 4-24. Add band color to your totals only. This will serve as the base for your month-over-month calculation, your year-over-year calculation, and the automations you will create.

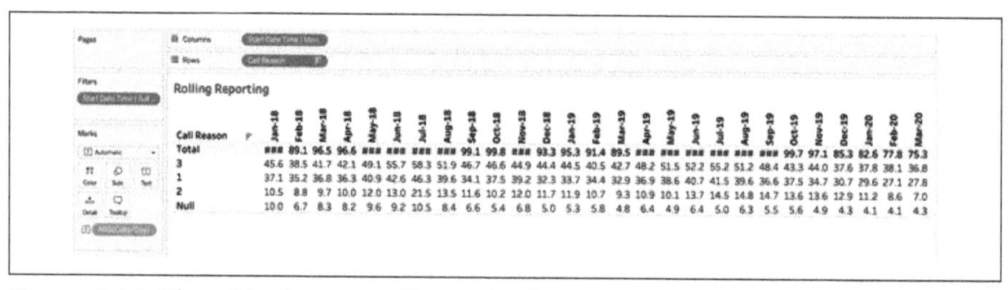

Call Reason	Jan-18	Feb-18	Mar-18	Apr-18	May-18	Jun-18	Jul-18	Aug-18	Sep-18	Oct-18	Nov-18	Dec-18	Jan-19	Feb-19
Total	###	89.1	96.5	96.6	###	###	###	###	99.1	99.8	###	93.3	95.3	91.4
3	45.6	38.5	41.7	42.1	49.1	55.7	58.3	51.9	46.7	46.6	44.9	44.4	44.5	40.5
1	37.1	35.2	36.8	36.3	40.9	42.6	46.3	39.6	34.1	37.5	39.2	32.3	33.7	34.4
2	10.5	8.8	9.7	10.0	12.0	13.0	21.5	13.5	11.6	10.2	12.0	11.7	11.9	10.7
Null	10.0	6.7	8.3	8.2	9.6	9.2	10.5	8.4	6.6	5.4	6.8	5.0	5.3	5.8

Call Reason	Mar-19	Apr-19	May-19	Jun-19	Jul-19	Aug-19	Sep-19	Oct-19	Nov-19	Dec-19	Jan-20	Feb-20	Mar-20
Total	89.5	###	###	###	###	###	###	99.7	97.1	85.3	82.6	77.8	75.3
3	42.7	48.2	51.5	52.2	55.2	51.2	48.4	43.3	44.0	37.6	37.8	38.1	36.8
1	32.9	36.9	38.6	40.7	41.5	39.6	36.6	37.5	34.7	30.7	29.6	27.1	27.8
2	9.3	10.9	10.1	13.7	14.5	14.8	14.7	13.6	13.6	12.9	11.2	8.6	7.0
Null	4.8	6.4	4.9	6.4	5.0	6.3	5.5	5.6	4.9	4.3	4.1	4.1	4.3

Figure 4-24. The table showing calls per day by call reason and month

3. Create the month-over-month calculation. This is done with a table calculation. Call this new calculation **`[Calls/Day | % Change 1]`**:

   ```
   // Calls/Day | % Change 1
   (ZN([Calls/Day]) - LOOKUP(ZN([Calls/Day]), -1))
   /
   ABS(LOOKUP(ZN([Calls/Day]), -1))
   ```

   This calculation creates a percent change based on the previous value, in this case the previous month. Because a business may be seasonal (and many are), it's often better to compare values to the previous year. To do this, create a new calculation called **`[Calls/Day | % Change 12]`**:

   ```
   // Calls/Day | % Change 12
   (ZN([Calls/Day]) - LOOKUP(ZN([Calls/Day]), -12))
   /
   ABS(LOOKUP(ZN([Calls/Day]), -12))
   ```

   You'll notice that the calculation is just slightly different: –1 has been changed to –12. Double-click [Calls/Day | % Change 1] and [Calls/Day | % Change 12]. This converts the table to include [Measure Names], and the text is now [Measure Values]. The table calculations in [Calls/Day | % Change 1] and [Calls/Day | % Change 12] need updating, but you can wait until we have all the components of the visualization on the view.

4. Let's now move on to showing only full months. Start by calculating the maximum date of [Start Date Time] by writing a calculation called **`[Start Date Time | Max Date]`**:

   ```
   // Start Date Time | Max Date
   {MAX([Start Date Time])}
   ```

   Here we use an LOD calculation to calculate the maximum date in our dataset. We'll monitor this date so we know when to update our table. This will allow you to calculate relevant time periods dynamically. You just need to find the start and end points of your dynamic table. Let's first calculate the last day of the last full month and call the calculation **`[Last Day of Last Full Month]`**:

   ```
   // Last Day of Last Full Month
   DATETRUNC("month", [Start Date Time | Max Date] + 1) - 1
   ```

   Use this calculation to create a Boolean to filter data in the current month—which is not yet complete. Create a calculation called **`[Start Date Time | Full Months]`**:

   ```
   // Start Date Time | Full Months
   [Start Date Time] <= [Last Day of Last Full Month]
   ```

   Add this calculation to Filters and select True.

5. Filter this visualization down to the most recent 13 complete months. Create a new calculation called **`[Start Date Time | Last 13 Months]`**:

```
//Start Date Time | Last 13 Months
[Start Date Time] > DATEADD("month", -12,
  [Last Day of Last Full Month] + 1) - 1
```

Place this calculation to the left of [Start Date Time | Month]. Right-click the False header and select Hide. Then deselect Show Header from the same menu (Figure 4-25).

Figure 4-25. Hiding the header by deselecting Show Header

6. Finalize the table calculations by editing the [Calls/Day | % Change 1] and [Calls/Day | % Change 12] table calculations in the [Measure Values] Marks card so that only [Call Reason] is deselected, as shown in Figure 4-26.

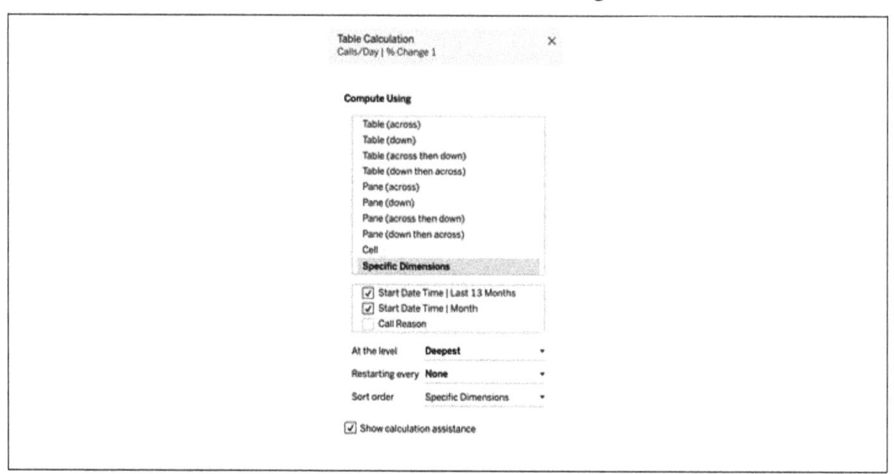

Figure 4-26. The table calculation settings for [Calls/Day | % Change 1] and [Calls/Day | % Change 12]

7. Be sure to format both month-over-month and year-over-year calculations as percentages. Finally, right-click and edit the alias (Figure 4-27). [Change Calls/Day | % Change 1] to MoM (month-over-month) and change [Calls/Day | % Change 12] to YoY (year-over-year).

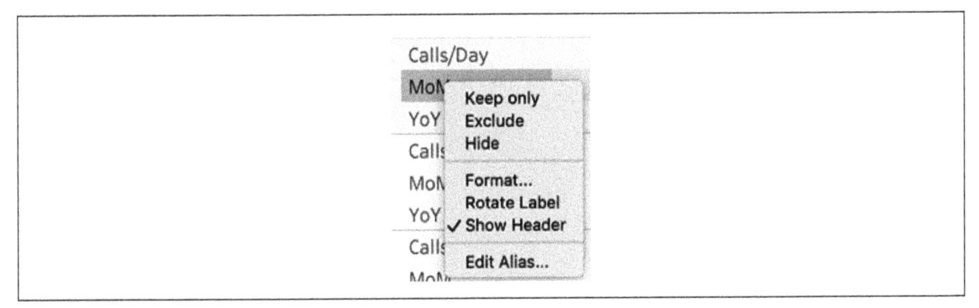

Figure 4-27. Right-click [Measure Names] and select Edit Alias

The result of all this work (shown previously in Figure 4-22) is a humble table that provides a lot of great insights and is automatically updated each month.

For the most part, we as developers spend very little time thinking about what a year, month, or week even means. We just assume that a year goes from January 1 to December 31. But when it comes to organizations, a fiscal year is defined in many ways. This next section provides a brief overview of working with fiscal dates in Tableau.

Nonstandard Calendars

A *fiscal year* can start at any time; it can be January 1, June 5, or even the fifth Monday of the standard calendar year. It's all relative. Tableau provides some flexibility.

If your calendar year starts at the beginning of a month, you can standardize this by right-clicking and then navigating to Default Properties → Fiscal Year Start → Month of Fiscal Year (Figure 4-28). This simplifies the hierarchy associated with that particular date measure.

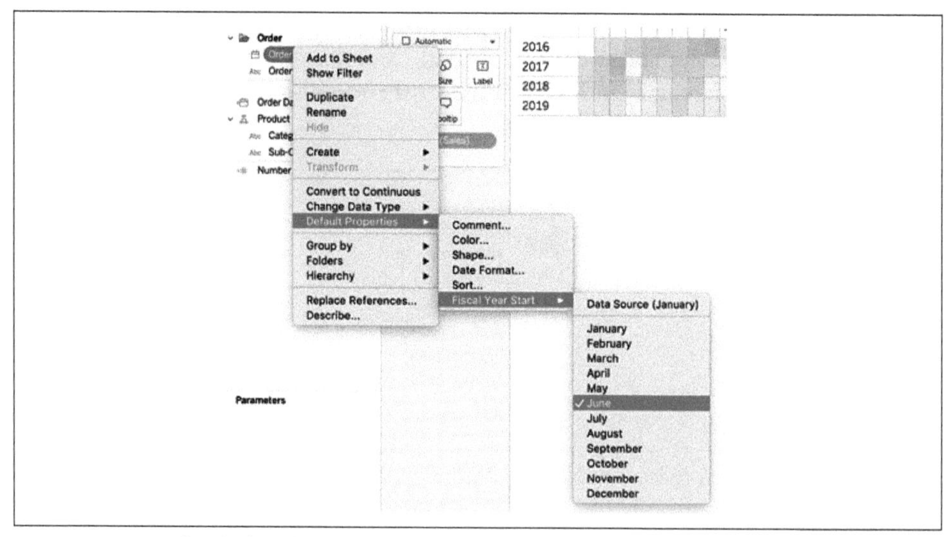

Figure 4-28. Right-click a date to change the default properties of the fiscal year start (in this example, the fiscal year start is set to June)

Some organizations work with the standard Gregorian calendar as their fiscal year: January 1 thru December 31. Other organizations, however, start the fiscal year or month on the first day of the week that month starts. So if January 1 is on a Tuesday, the fiscal year would start on December 30. This calendar type is called *ISO-8601*. While the name is funky, just know that the calendar is week-based. You can specify the calendar type by right-clicking the date value on your view and selecting ISO-8601 Week-Based. (We'll just call it an *ISO calendar* in this section.)

In Figures 4-29 and 4-30, you can see how data from a standard calendar can differ ever so slightly from an ISO calendar.

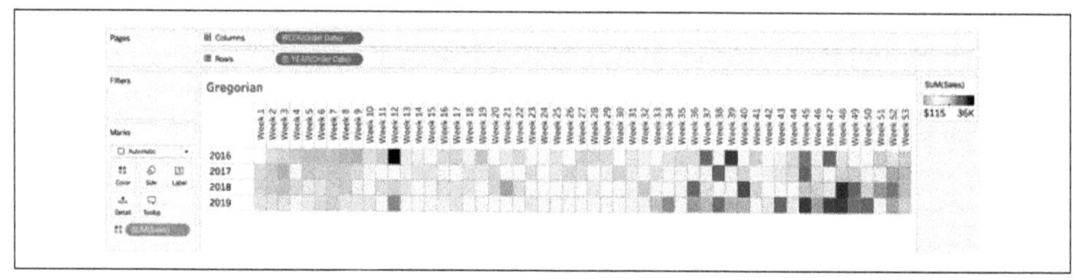

Figure 4-29. Data in a standard calendar

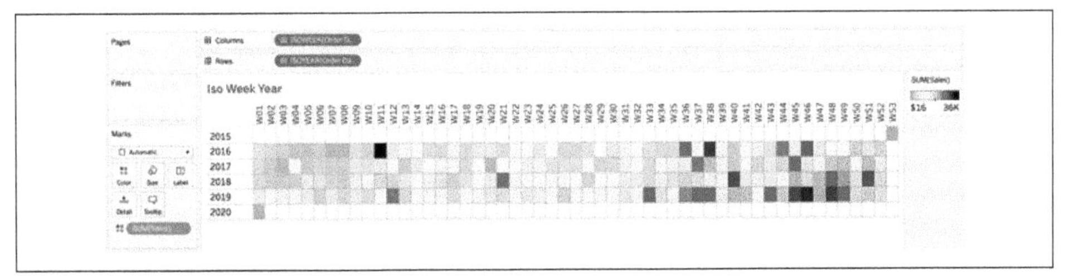

Figure 4-30. Data in an ISO calendar

Strategy: Build a Monthly Bar Chart with a June 1 Fiscal Year Start

Let's take a second to build a visualization with a June fiscal start. To keep it simple, imagine you are building a bar chart that shows total sales by month and fiscal year. You will replicate Figure 4-31:

1. Connect to the Sample – Superstore dataset.
2. Duplicate the [Order Date] field and call it **`[Order Date | June]`**.
3. Right-click [Order Date | June] and change the default fiscal year start to June.
4. Add [Order Date | June] as a continuous data value by month to Columns.
5. Add `SUM([Sales])` to Rows.
6. Add YEAR([Order Date | June]) to Color.
7. Change the mark type to Bar.
8. Right-click the September 2016 bar and add an annotation to the mark, displaying the date and the total sales.

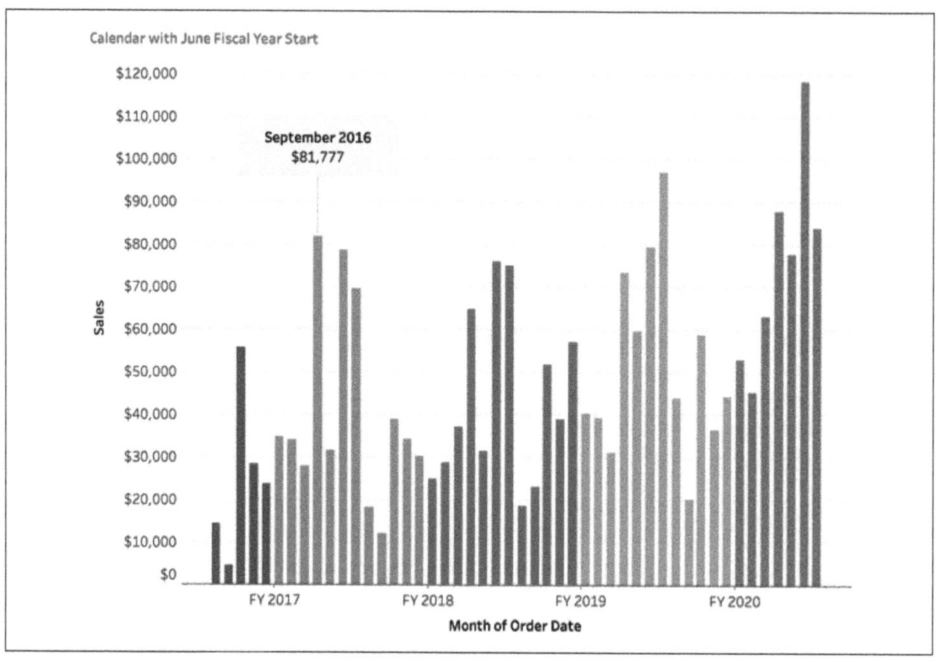

Figure 4-31. Sales by month, colored by fiscal year, using a June start to the fiscal year

Visualizing the 4-5-4 Calendar: Office Essentials Case Study

Our large retail store, OE, has relied on data metrics tied to a calendar year. The company would like to redesign some of its standard reports to now follow a 4-5-4 calendar. How would you complete this task?

Retailers often use the *4-5-4 calendar*. This calendar allows them to compare sales by dividing the year into months based on a repeating four weeks, five weeks, and four weeks. Retailers use this calendar because holidays tend to line up, and because the same number of Saturdays and Sundays are displayed in comparable months. The 4-5-4 sales calendar is not perfect: because the calendar is based on 52 weeks, or 364 days, this leaves an extra day each year to be accounted for. To adjust for this, a week is added to the fiscal calendar every five to six years. These occurred in 2012 and 2017, and will happen again in 2023.

The 4-5-4 calendar year varies from year to year. When February 1 occurs on Thursday, Friday, or Saturday, the calendar year starts the Sunday after February 1. If February 1 occurs on Sunday, Monday, Tuesday, or Wednesday, the calendar year starts the Sunday of the week of February 1.

The next strategy is focused on building date components for the 4-5-4 retail calendar. These include week of the year, month of the year, quarter of the year, and week of the quarter. After you build the components, you will build a visualization highlighting some of those calculations.

Strategy: Build a Bar Chart Using the 4-5-4 Retail Calendar

1. Create a calculation that calculates February 1. Call the calculation **`[Feb 1]`**:

```
// Feb 1
DATEADD("month", 1, DATETRUNC("year", [Order Date]))
```

2. Calculate the start of the calendar year based on whether February 1 is after Wednesday in the week. Name the calculation **`[454 Year Start]`**:

```
// 454 Year Start
IF DATEPART('weekday', [Feb 1]) > 4
THEN DATETRUNC('week', DATEADD('week', 1, [Feb 1]))
ELSE DATETRUNC('week', [Feb 1])
END
```

3. Determine the start of the 4-5-4 calendar year for the prior year by calculating February 1 for the prior year. Name the calculation **`[Feb 1 | PY]`**:

```
// Feb 1 | PY
DATEADD('year', -1, DATEADD("month", 1, DATETRUNC("year", [Order Date])))
```

   Calculate the start of the previous calendar year. We will use this calculation with the current year values to determine the week number of the calendar year. Label the calculation **`[454 Prior Year Start]`**:

```
// 454 Prior Year Start
IF DATEPART('weekday', [Feb 1 | PY]) > 4
THEN DATETRUNC('week', DATEADD('week', 1, [Feb 1 | PY]))
ELSE DATETRUNC('week', [Feb 1 | PY])
END
```

4. Parse the retail weeks of the year:

```
// Retail Week
IF [454 Year Start] <= [Order Date]
THEN DATEDIFF('week', [454 Year Start], [Order Date]) + 1
ELSE ({FIXED [Feb 1] : MAX(DATEDIFF('week',
  [454 Prior Year Start], DATETRUNC('year',[454 Year Start])))}
+
DATEPART('week', [Order Date])
     )
END
```

5. Now that you have the week, you can create components like Retail Quarter, Retail Month of Quarter, Retail Week of Quarter, Retail Month, and Retail Week of Month. Build out each of these calculations:

```
//Retail Quarter
FLOOR(([Retail Week]-1)/13)+1


//Retail Week of Quarter
(([Retail Week] - 1) % 13) + 1


// Retail Month of Quarter
IF [Retail Week of Quarter] <= 4
THEN 1
ELSEIF [Retail Week of Quarter] > 4
AND [Retail Week of Quarter] <= 9
THEN 2
ELSEIF [Retail Week of Quarter] > 9
AND [Retail Week of Quarter] <= 13
THEN 3
END


//Retail Month
IF [Retail Week] <= 4
THEN "February"
ELSEIF [Retail Week] > 4 AND [Retail Week] <= 9
THEN "March"
ELSEIF [Retail Week] > 9 AND [Retail Week] <= 13
THEN "April"
ELSEIF [Retail Week] > 13 AND [Retail Week] <= 17
THEN "May"
ELSEIF [Retail Week] > 17 AND [Retail Week] <= 22
THEN "June"
ELSEIF [Retail Week] > 22 AND [Retail Week] <= 26
THEN "July"
ELSEIF [Retail Week] > 26 AND [Retail Week] <= 30
THEN "August"
ELSEIF [Retail Week] > 30 AND [Retail Week] <= 35
THEN "September"
ELSEIF [Retail Week] > 35 AND [Retail Week] <= 39
THEN "October"
ELSEIF [Retail Week] > 39 AND [Retail Week] <= 43
THEN "November"
ELSEIF [Retail Week] > 43 AND [Retail Week] <= 48
THEN "December"
ELSEIF [Retail Week] > 48 AND [Retail Week] <= 52
THEN "January"
END
```

```
// Retail Week of Month
IF [Retail Week of Quarter] <= 4
THEN [Retail Week of Quarter]
ELSEIF [Retail Week of Quarter] > 4
AND [Retail Week of Quarter] <= 9
THEN [Retail Week of Quarter] - 4
ELSEIF [Retail Week of Quarter] > 9
AND [Retail Week of Quarter] <= 13
THEN [Retail Week of Quarter] - 9
END
```

Having all of these calculations is extremely useful for any visualization using a retail calendar. Let's create a visualization that we like to use to showcase the retail calendar.

6. Add a discrete dimension of [Retail Month of Quarter] to Columns. Add a discrete dimension of [Retail Quarter] to Rows. Add [Retail Week of Month] as a continuous dimension to Rows. Add 0.0 as an ad hoc continuous dimension to the right of [Retail Week of Month]. Create a synchronized dual axis.
7. Add details as follows:
    a. Set the Marks card of [Retail Week of Month] to a bar chart. Add `[Profit Ratio] * (SUM([Profit])/SUM([Sales]))` to Color. Add `SUM([Sales])` and the continuous dimension of [Retail Week] to Text. Format the text so it reads as shown in Figure 4-32.

    <SUM(Sales)> | Week <Retail Week>

    Figure 4-32. Text labels for the 4-5-4 bar chart

    b. On the Marks card of the 0.0 value, set the mark type to Text. Add [Retail Month] to Text.
8. For the last part, we need to place a value on Columns that will control both axes. Create a calculation called **`[bar]`** and add it to Columns:

```
// bar
IF COUNTD([Retail Week]) = 1
THEN SUM([Sales])/WINDOW_MAX(SUM([Sales]))
ELSE .9
END
```

This calculation will show sales as a percentage of the maximum sales for a retail week for the bars and will place a label of each state at 0.9. Change the table calculation and select all values, as shown in Figure 4-33.

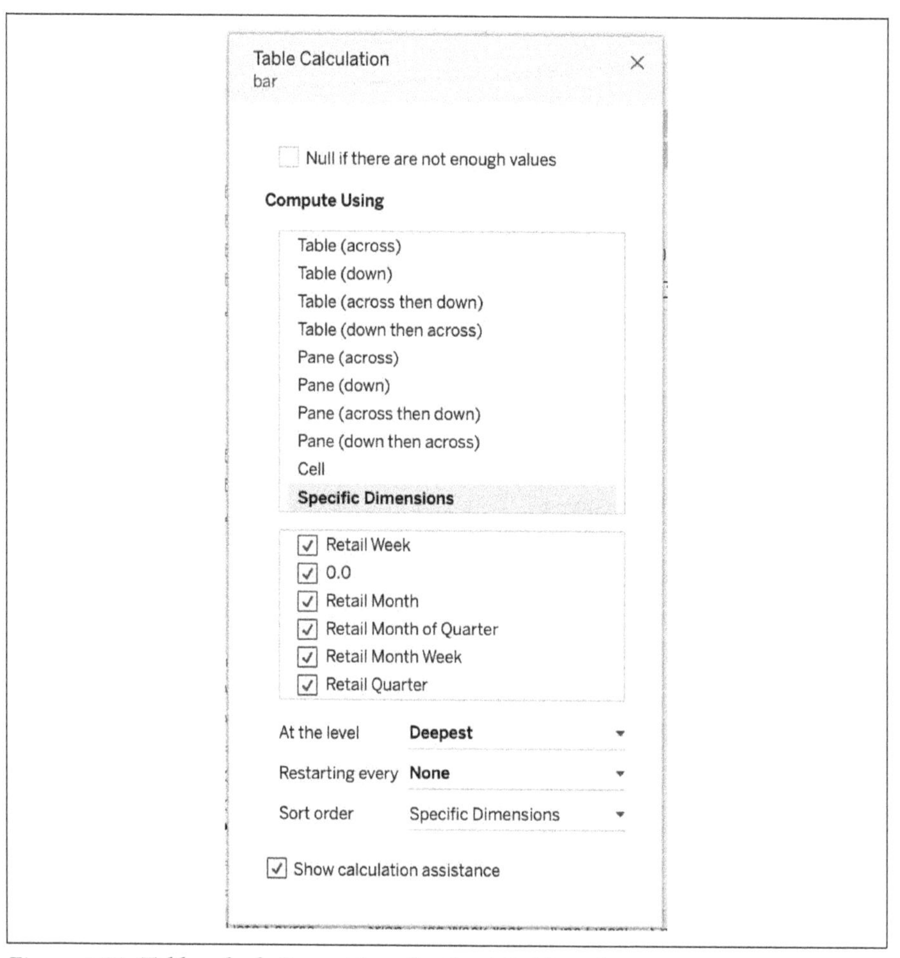

Figure 4-33. Table calculation settings for the 4-5-4 bar chart

Set the bar axis range between 0 and 1.8, and then hide the axis. Reverse the axis of the [Retail Week of Month] field and then hide the axis.

9. Finally, format the chart according to your design standards.

   The resulting visualization is shown in Figure 4-34.

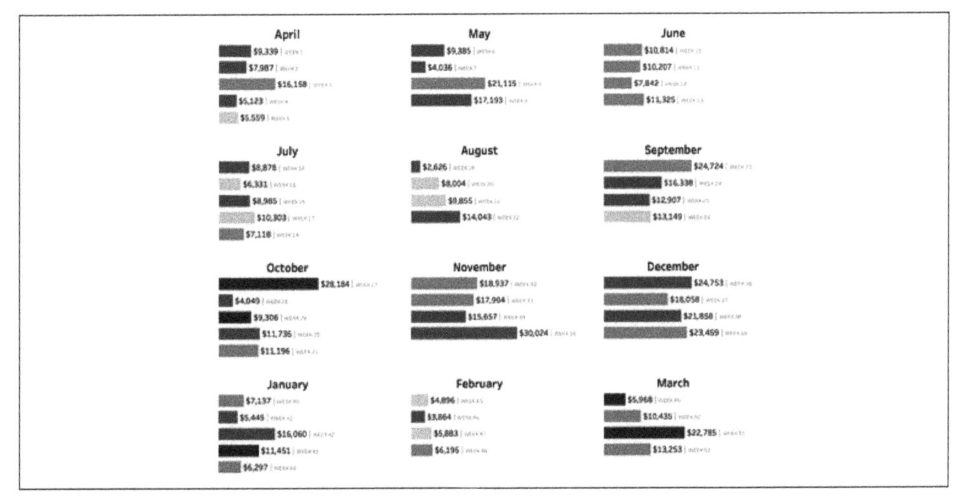

Figure 4-34. Bar chart showing sales and profit by 4-5-4 retail month and week

Conclusion

In this chapter, we scratched the tip of the dates-and-times iceberg. Because dates and times are naturally hierarchical, and because dates and times can be treated as either dimensions or measures, a single date field has a nearly unlimited number of combination options.

In the chapter opening, we discussed the importance of understanding the difference between a date *part* and a date *value*. With date parts, Tableau returns a single component of a date or time field. The underlying calculation for date parts is the `DATEPART()` function.

In the first two strategies, we showcased how to plot parts of a date. Combining multiple date parts (in this case, hour and minute) can lead to too many discrete values being shown.

With date values, Tableau returns a date rounded down to the specified part of the date. This means a date part of Month will return the month name or number, while a date value of Month will return the year and month combination. The underlying calculation for a date value is the `DATETRUNC()` function.

Tableau defaults date parts to dimensions, and date values to continuous values. While those are the default values, we can convert either to discrete or continuous values. Our selection of discrete or continuous values affects what our axes might look like and therefore the type of chart we are most likely to select.

The final fundamental we discussed was Tableau making a date hierarchy available to your audience by default. If you are looking to limit the availability of this hierarchy, you must use a custom date.

For most audiences, working with a continuous axis is more intuitive because that's how people think about time! This lends itself to the selection of line charts. One challenge of working with dates is creating continuous plots of a single date part. We tackled this challenge of creating continuous axes. In each of these challenges, we moved from showing data rounded to the nearest second of the day, to the nearest 15 seconds, and to the nearest 15 minutes.

While line plots—and occasionally bar charts—are two common ways in which date and time values are visualized, we like to use heatmaps (which only Tableau calls *highlight tables*). Line charts provide a limited ability to describe more than five lines—any more, and it becomes difficult, even for the most data-savvy audience, to interpret patterns. One simple alternative to line charts is heatmaps. We highlighted the flexibility of this chart type when working with date and time fields.

Another regular challenge when working with dates is comparing previous years to the current year to date. We showed you a calculation you can use to calculate the year-to-date values of any date. This can be used inside aggregate calculations and can be used to create a chart that shows progress to a total. We extended this same year-to-date calculation to adjust line charts to make fair comparisons.

We also looked at how to create automated reports by using both LOD and table calculations. By using discrete date values and the `LAST()` function, you can easily show the last *N* months. We also showcased how you can use a table calculation to complete month-over-month and year-over-year calculations.

We wrapped up the chapter talking about non-Gregorian calendars. Gregorian calendars are the standard, but some organizations have fiscal years that start at different months other than January. We even went beyond the traditional calendar and discussed the 4-5-4 calendar, which is common with retail and consumer-packaged goods companies.

The next chapter covers key performance indicators (KPIs). KPIs are standalone metrics that your audience can use to drive day-to-day operations or guide strategic decisions. While KPIs are focused on the metric itself, they are often defined by a specific time period. If you become fluent with date fields in Tableau, creating dynamic, automated KPIs will come naturally.

CHAPTER 5

Key Performance Indicators

Let's talk about one more controversial topic in data visualization: *key performance indicators*, or *KPIs*. The controversy around KPIs arises because you're presenting only a single number and, in fact, you're not using any particular visualization at all—you're just showing a number or value as text. Still, there is an art to this text, and in this chapter, we will discuss the art of creating a KPI.

This type of analysis aids in answering common business questions such as these:

- What is our revenue this month?
- Did we meet our goal for fundraising dollars last month?
- How does time to ship compare to the same time last year?
- What is the lift of our 20% off promotion?
- How does profit margin change over time (where time is the dimension)?

KPIs are utilized across virtually every organization that uses data. This is because goals established within an organization are often measured using a single metric. The definition of this metric is typically established by leaders and regularly reviewed as a snapshot in time. As the developer of a dashboard, your job is to make this metric readily available. KPIs are usually measured across a single point in time, whether it's a week, month, or year. This makes it extremely important for any dashboard developer to know the particular time encompassed within a KPI and exactly how to measure that metric.

Developing useful KPIs means providing more than just a single number; it means providing additional context to that metric. This context could be in the form of a percent change month-over-month or year-over-year, or as part of a *sparkline* (a very small line chart, typically drawn without axes or coordinates) or other microcharts without axes or labels.

Now, not every data point collected—nor every dashboard created—will need a KPI, but whether your audience is an executive, a salesperson, or a machinist working on the shop floor, they likely all work with their own KPIs. Each of these individuals will recognize that their KPIs are just one element of measurement, and the KPIs often don't articulate the drivers of the actual outcome.

For instance, the executive might be interested in growth in profitability. Even a simple growth in profitability has several drivers: current period sales, prior period sales, current period costs, and prior period costs. A machinist, who might measure productivity in number of units created, has to consider that number in context. Units created in a particular month might vary by the amount of vacation taken, machine downtime due to repairs, or complexity of the part.

Just remember that KPIs are examined at different frequencies depending on the metric and the audience. A machinist might watch a gauge every 15 minutes to make sure a machine doesn't break down and reduce output, because the machinist's job is to keep that machine running in the moment. An exec might examine plant output once a month because their job is to steer the long-term future of the organization and requires keeping an eye on dozens of indicators.

Everyone uses KPIs, but the KPI as a standalone value often doesn't provide enough context. Is the current value an anomalous single spike in the data, or is this consistent for its performance? As the developer, you have an obligation not only to share a KPI, but also to enhance the KPI with contextual features that aid decision making. Ask yourself the following:

- In what direction is this value moving, compared to the previous measure of the KPI?
- What is the one-year trend of your KPI?
- Does this KPI cross a threshold and require an alert?

Use indicators to show directionality, show sparklines, and add color; don't let the KPI stand alone. Remember, you are the developer and you need to support the proper use of the KPIs!

In This Chapter

In this chapter, you'll use the Sample – Superstore dataset. You'll need to download a 1 × 1 transparent image from *https://www.tessellationtech.io/wp-content/uploads/2020/05/NULL.png*. In this chapter, you'll learn how to do the following:

- Create automated KPIs that show the most recent month (even while the month is partially complete) and how to show these as a single value
- Create percentages that can show change week over week, month over month, or quarter over quarter
- Enhance the charts by using microcharting techniques like sparklines, spark bar charts, and percent-to-target indicators

Displaying KPIs

Before diving too deep into how to create KPIs, let's start with a scenario that we most typically see when working with beginner Tableau developers: a series of large numbers for KPIs that provide no context. Additionally, these values need to be manually updated each month.

When developing the Tableau dashboard, try to automate as many calculations as possible. Let's take a look at Figure 5-1, which shows four KPIs: Sales, Profit, Profit Ratio, and Quantity.

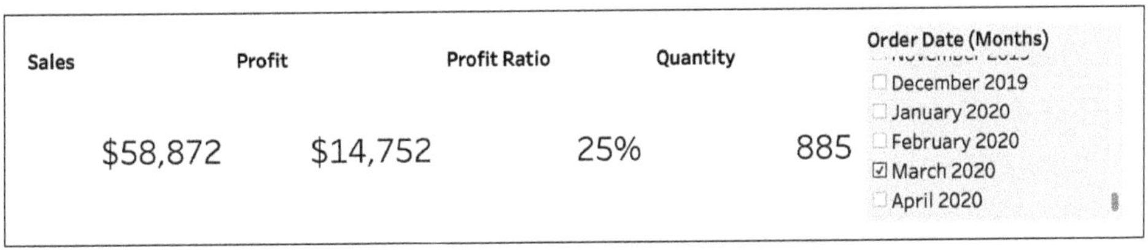

Sales	Profit	Profit Ratio	Quantity
$58,872	$14,752	25%	885

Order Date (Months)
December 2019
January 2020
February 2020
March 2020
April 2020

Figure 5-1. Novice KPIs providing no context to the values shown, and being filtered manually

When we are working with novice developers, we often see exactly the same solution shown in this figure: individual sheets for each KPI, with numbers that aren't necessarily aligned with the title of the sheet and values that need to be updated manually.

We've been around too many organizations not to know that *automation* hardly ever means calculations that automatically update values at the end of any month, week, or year. More often, it means developers updating a series of hidden filters that might be just off the dashboard or a few clicks away after the data has been updated for the dashboard. As a developer, you shouldn't have to open a dashboard and update a filter every day, month, or week. You should design calculations that automatically update with the dates selected.

Otherwise, if you've been reporting KPIs in March 2020, but April has just finished and executives are expecting to see new values, you have to go into the reports and update the filters to represent April 2020 before leaders see the report. Not only is this not the most efficient way to solve the problem, but such analysis can be automated with well-designed calculations.

Displaying KPIs: Office Essentials Case Study

In an effort to reduce duplicated effort and hours spent on reporting, the executive team at OE is looking to improve its KPI tracking. In particular, executives have asked the data team to automate as many functions as possible. How can you format the new reports so that KPIs are accurate?

For the entirety of this chapter, we'll use the Sample – Superstore dataset. We'll build KPIs for sales, profit, profit as a percent of sales, and total units. We'll also look at each of these metrics at the monthly level. To add appropriate complexity, in all of these examples we will set the date to March 15, 2020. Executives rarely react to KPIs at the daily or weekly level. Sometimes they just want to look at the values on a monthly cadence.

To set the date to March 15, 2020, you'll need to edit the Sample – Superstore data source:

1. Click Extract in the Connection section at the top right.
2. Then select Add in the filters selection.
3. Use the Order Date field and set the range as 11/1/2017 through 3/15/2020.
4. Click OK and OK again.
5. Then go to a sheet to execute the extraction.

This will set the range of dates in your dataset as 1/1/2017 through 3/15/2020 and will simulate your visualizations as if the current date is March 15!

In this case study, you will learn how to format your KPIs correctly, automate your calculations so you don't have to update filters each month or week, and add context to help your end users understand the KPIs. Your KPIs will be reported at the monthly level because the executive team's reoccurring schedule has 30 minutes to discuss these KPIs once a month.

We will use the KPIs shown previously in Figure 5-1 as a starting point. Each KPI is developed on a separate sheet and uses the default settings of Tableau. To the novice developer, there is nothing wrong with these visuals. But so much more can be done.

If you wanted to build novice KPIs, you'd just have to do the following:

1. Connect to the Sample – Superstore dataset.
2. Create four new sheets titled Sales, Profit, Profit Ratio, and Quantity.
3. On the Sales sheet, add SUM(Sales) to Text.
4. On the Profit sheet, add SUM(Profit) to Text.
5. On the Profit Ratio sheet, add [Profit Ratio] to Text.
6. On the Quantity Sheet, add SUM(Quantity) to Text.
7. On each of the four sheets, add a filter for 2020 by using the year of Order Date. Set the sheet to take up the entire view.
8. Create a new dashboard 1,000 pixels wide and 500 pixels tall. Add a horizontal container and place the worksheets inside it.

If you normally make KPIs like this, that's OK. We're going to give you some great tips to improve your KPIs!

Strategy: Designing Clear KPIs

When creating your KPIs, make sure their titles align with the values (Figure 5-2). This will increase the readability of the document.

Sales	Profit	Profit Ratio	Quantity
$58,872	$14,752	25%	885

Figure 5-2. Matching the title and value alignment of the KPIs

Notice that we changed the alignment of the sheet titles to match with the KPIs and that readability has been improved? This is a great improvement, but we can go much further. Next, instead of showing sheet titles of a dashboard, we'll add the titles on the sheet itself (Figure 5-3).

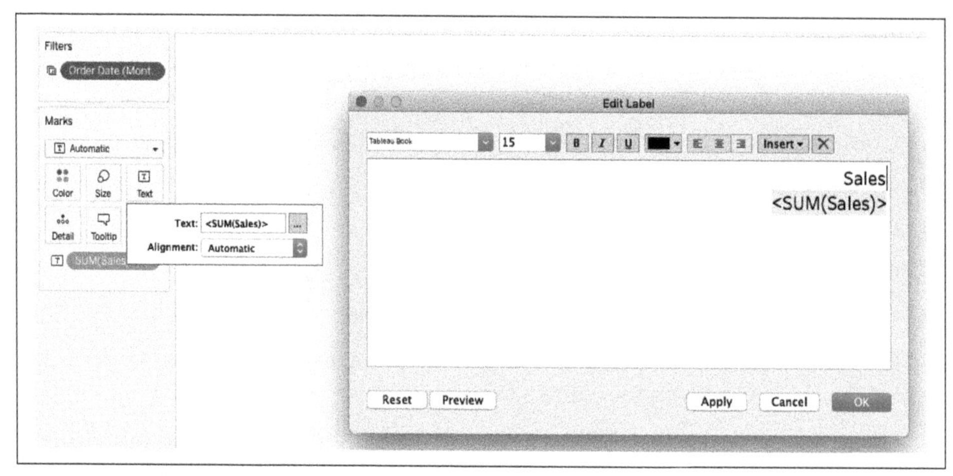

Figure 5-3. Editing the text of a KPI

Here are the steps:

1. Click the sheet, click the Text property on the Marks card, and then click the ellipsis next to the Text box.
2. This opens the Edit Label dialog. Add the title from your sheet names as text above the measure and click OK.
3. Hide the sheet title.

   The result is significantly less space between the title and the KPI (Figure 5-4).

Sales	Profit	Profit Ratio	Quantity
$58,872	$14,752	25%	885

Figure 5-4. KPIs with titles on Text of a Marks card

One final introductory-level tip for KPIs: make the title smaller than the value—but be sure it remains legible. In Figure 5-5, the KPI title uses 10-point Tableau Book font, and the KPI value uses 20-point Tableau Semibold font. We usually keep the color the same, but make the titles lighter than the values. The final result is a set of no-frills KPIs that look extremely professional.

SALES	PROFIT	PROFIT RATIO	QUANTITY
$58,872	**$14,752**	**25%**	**885**

Figure 5-5. Formatted no-frills KPIs

Strategy: Create Single-Sheet KPIs

One of the challenges of creating KPIs on separate sheets is having to toggle between multiple sheets and adjust each one incrementally. Another challenge is performance. Even though the KPIs are extremely straightforward, you can optimize the performance of your dashboard by placing all of your KPIs on a single sheet. This can be done using the `MIN(0.0)` technique discussed in Chapter 3:

1. Create an ad hoc calculation on columns by double-clicking the Columns shelf. This opens up a prompt to write a calculation.
2. Type **`MIN(0.0)`**.
3. Repeat three more times. This will create five Marks cards, one for all Marks cards and one for each of the `MIN(0.0)` calculations. The resulting visualization is shown in Figure 5-6.

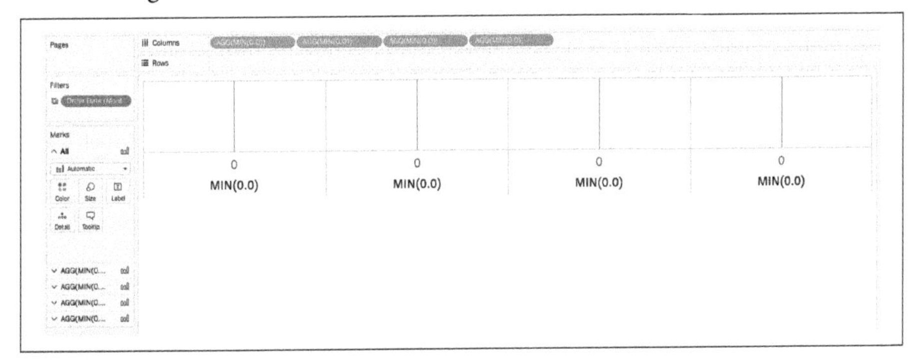

Figure 5-6. The visualization with four `MIN(0.0)` measures

4. Edit each axis and set the values to range from –1 to 0.1. Then hide the axes.
5. Add SUM(Sales) to the first MIN(0.0) Marks card. Repeat for SUM(Profit), Profit Ratio, and SUM(Quantity). Hide the axes when finished.
6. Align all text to the Left Middle for all Marks cards.
7. Change the mark type to Shape.
8. Outside of Tableau, download a 1 × 1 transparent image from *https://tessellationtech.io/wp-content/uploads/2020/05/NULL.png*.
9. On your desktop, go to the Shapes folder in your My Tableau Repository, and add the transparent image to the KPI folder.
10. Back in Tableau Desktop, edit the Shape on the All Marks card, and then select More Shapes. Click the Reload Shapes button, and then from the Select Shape Palette drop-down, select KPI. Next, select the Transparent image. Click OK. If you attempt to hover over the custom shape, it will not highlight because it's a fully transparent shape.

11. On the sheet formatting, remove all the lines and borders.

    This produces a single sheet that includes all of the relevant KPIs (Figure 5-7).

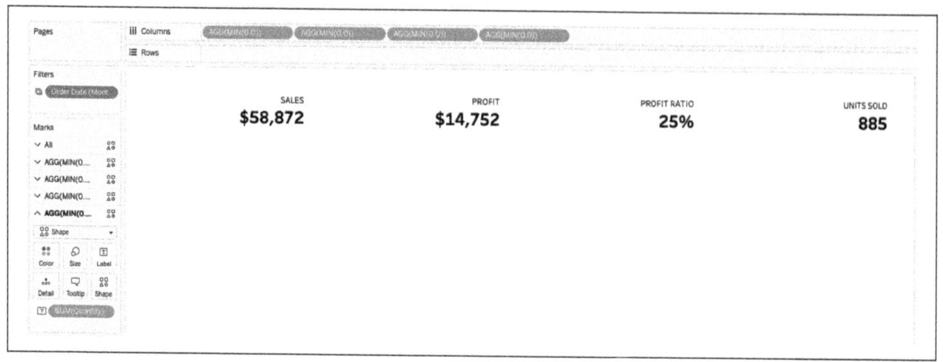

Figure 5-7. KPIs on a single sheet

Adding all KPIs to a single sheet certainly requires more steps than having one sheet per KPI. But spending the time now will result in faster-performing dashboards that will be easier to adjust in the future. You will continue to update this sheet throughout the chapter as you enhance our KPIs.

You've already addressed several problems with the initial visualization in Figure 5-1, but your filters are hidden out of sight from your dashboard audience. Instead of using a filter on your dashboard, you need to build a custom calculation that filters data for you.

For our example data, the Sample – Superstore dataset, we have access to line items (individual products) on order data that is provided daily. However, for sharing KPIs, only values for fully completed months should be shared. For your next strategy, you are going to build a series of calculations that will automatically update a KPI when the time is right. Specifically, you will create automated KPIs that show the most recent value for the last full month of data.

This means that if it's the middle of May 2020 (namely, May 17, 2020), data will show for only April 2020. If the date is May 30, 2020, the data will still show April. But as soon as data is available for May 31, the last day of the month, the KPIs will update to show values for the now-completed month of May 2020.

Strategy: Automate Your KPIs

In this strategy, you will create KPIs that automatically return the most recent full month of data. This will be done on a single sheet rather than multiple sheets:

1. Edit your data source to change the end date from your data filter on your extraction from March 15, 2020 to May 17, 2020.

2. Create a calculation called **[Max Date]**. This calculation will also calculate the maximum date in the dataset:

```
// Max Date
{MAX([Order Date])}
```

This is the single most important LOD expression to remember when working with time. You're finding the maximum date in your dataset, which is extremely useful and provides flexibility with other calculations.

3. Calculate the last day of the last fully completed month. Call the calculation **[Last Day of Last Full Month]**:

```
// Last Day of Last Full Month
DATETRUNC("month", [Max Date] + 1) - 1
```

4. Calculate the start of the last fully complete month. Call the calculation **[First Day of Last Full Month]**:

```
// First Day of Last Full Month
DATETRUNC("month", [Last Day of Last Full Month])
```

5. Now create another calculation called **[Most Recent Month]**:

```
// Most Recent Month
[Order Date] >= [First Day of Last Full Month]
AND
[Order Date] <= [Last Day of Last Full Month]
```

When we work with KPIs, we use Boolean calculations as much as possible. They are super-fast to work with and process, and very easy to debug.

6. From here, you can start building out a dynamic calculation for sales for the last full month called **[Sales | LM]**:

```
// Sales | LM
SUM(
  IF [Most Recent Month]
  THEN [Sales]
  END
)
```

Repeat for **[Profit | LM]**, **[Profit Ratio | LM]**, and **[Quantity | LM]**:

```
// Profit | LM
SUM(
  IF [Most Recent Month]
  THEN [Profit]
  END
)

// Profit Ratio | LM
[Profit | LM]/[Sales | LM]

// Quantity | LM
SUM(
  IF [Most Recent Month]
  THEN [Quantity]
  END
)
```

7. Right-align your text and set all lines and borders to None.
8. Place the calculations onto their respective Marks cards on the single sheet that displays the KPIs. The final result, shown in Figure 5-8, is a sheet with no measures or dimensions on Filters.

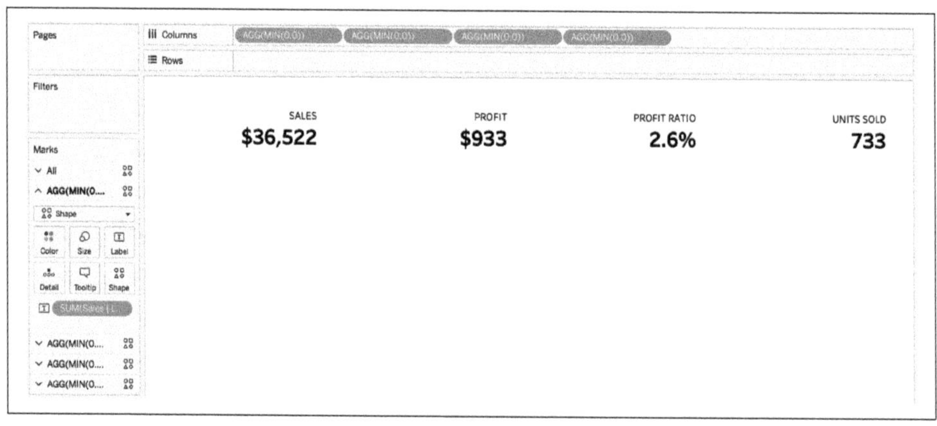

Figure 5-8. Single-sheet KPIs with automation

With this strategy, you've begun automating your KPIs. The next challenge is providing context for these KPIs.

Right now, the only takeaway is that the amount of sales last month was $36,522. Here are the next questions the executives are going to ask:

- How is this metric trending?
- How does that compare to last month?
- How does it compare to the same period last year?
- How much more growth occurred this period versus last period?
- How close is OE to hitting its targets?

It is your job to anticipate these questions. It's not necessary to provide all modes of context for a KPI, but providing *at least one* additional data point will be key. For our next strategy, we'll walk through how to create dynamic year-over-year calculations.

Strategy: Create Year-over-Year Calculations

Here you will build calculations that calculate the year-over-year change. These will provide valuable context to your KPIs:

1. Connect to Sample – Superstore.
2. Set an extract filter to include all dates up to and including May 17, 2020 from the [Order Date] field.
3. Follow steps 1–6 in "Strategy: Automate Your KPIs" on page 184.
4. Calculate the last day of the last full month, but from the prior year:

   ```
   // Last Day of Last Full Month Prior Year
   DATEADD("year", -1, [Last Day of Last Full Month] + 1) - 1
   ```

5. Calculate the first day of the last full month, but from the prior year:

   ```
   // First Day of Last Full Month Prior Year
   DATETRUNC("month", [Last Day of Last Full Month Prior Year])
   ```

6. Create a Boolean calculation that identifies whether a date was during the last full month of data, but for the prior year:

   ```
   // Most Recent Month Prior Year | TF
   [Order Date] <= [Last Day of Last Full Month Prior Year]
   AND
   [Order Date] >= [First Day of Last Full Month Prior Year]
   ```

7. Calculate the sales for the last full month, but for the previous year:

```
// Sales | LMLY
SUM(
  IF [Most Recent Month Prior Year | TF]
  THEN [Sales]
  END
)
```

8. Calculate the difference between the sales for the last full month versus the same time last year:

```
// Sales | Delta
[Sales | LM] - [Sales | LMLY]
```

9. Calculate the percent change between the two time periods:

```
// Sales | % Delta
[Sales | Delta] / [Sales | LMLY]
```

10. Create an arrow that will point upward when the change is positive:

```
// Sales | Delta | Arrow Up
IF [Sales | Delta] > 0
THEN "▲"
END
```

11. Create an arrow that will point downward when the change is negative:

```
// Sales | Delta | Arrow Down
IF [Sales | Delta] < 0
THEN "▼"
END
```

12. On the existing Sales KPI, add [Sales | % Delta], [Sales | Delta | Arrow Up], and [Sales | Delta | Arrow Down] to Text on the Sales MIN(0.0) Marks card of your sheet with the KPIs. ([Sales | LM] should already be on Text.)

13. Format your KPIs. When we format our KPIs, we prefer to keep the contextual information much smaller (in terms of font) than the metric itself. We also often choose a lighter-color font. With the use of the separate arrow calculations, we are also able to have two colors for the arrows. This way, we can use color as a visual indicator for our audiences to note the changes in the metrics. The result is shown in Figure 5-9.

Figure 5-9. A snapshot of Tableau Desktop

14. Repeat steps 5–9 for each KPI you intend on creating.

The final result is a KPI that provides context (Figure 5-10). In this case, the $36,522 in sales is actually down 6% from last year. This is why you always need to provide context with KPIs. Your audience has no idea what a static value means. Even if you look at the same information from week to week or month to month, it's hard to keep track of the context. That's where you, the analytics developer, need to remember that adding context is critical!

SALES	PROFIT	PROFIT RATIO	UNITS SOLD
$36,522	$933	2.6%	733
LY: ▼6%	LY: ▼69%	LY: ▼67%	LY: ▲15%

Figure 5-10. A full screenshot of the full visualization

In this example, you gave your KPIs context with a single value, allowing your audience to start to grasp exactly what each KPI means.

Sparklines

If you feel that your audience will want more than just a single value to justify the direction of your KPIs, we highly suggest using *sparklines*. These are very small line charts drawn without axes or coordinates, emphasizing the shape and trend of the data; they were first introduced by statistician and professor Edward Tufte and popularized by Microsoft Excel. They provide a simple and condensed method for displaying data. They work well in small multiples but are also excellent when paired with a KPI.

Sparklines have been popular for a long time. Even today they can be found on front pages of some of the largest digital media sites. For instance, on the *New York Times* website, sparklines describe the trend of COVID-19 cases in the United States (Figure 5-11). These are paired next to KPIs for new cases and new deaths.

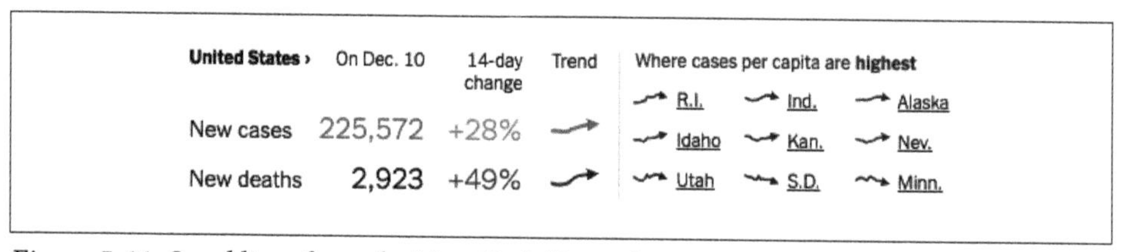

Figure 5-11. Sparklines from the New York Times front page on December 11, 2020

Unlike single-sheet KPIs, these sparklines will have to be created on separate sheets.

Sparklines: Office Essentials Case Study

As part of your job, you've been given feedback to provide additional context to KPIs. After a few rounds of feedback, it's decided that you need to show sparklines that display the last 13 full months of data.

Strategy: Create Sparklines

Follow these steps to create sparklines:

1. Connect to Sample – Superstore.
2. Set an extract filter to include all dates up to and including May 17, 2020 from the [Order Date] field. Create a custom date value at the Month level of detail.
3. Create a custom date by right-clicking [Order Date] (Figure 5-12). Add this new calculation, called [Order Date (Months)] to Columns.

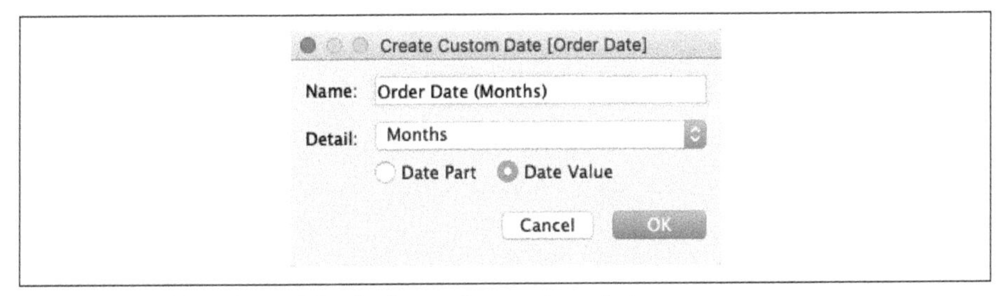

Figure 5-12. Creating a Month date value custom date

4. Create a new calculation called **[Full Months]**. This calculation will be used as a filter to help us display months for which data is complete:

```
// Full Months
[Order Date] <= DATETRUNC("month", {MAX([Order Date])} + 1) -1
```

5. Place [Full Months] on the Filters shelf and set it to True.
6. Create a new calculation called **[Last 13]**. This calculation will help us show the last 13 months of data:

```
// Last 13
LAST() <= 12
```

7. Add [Last 13] to the Filters shelf and set it to True. You should now be showing the last 13 months—April to April.
8. From there, you need to add your key metric to the chart. In this case, it is [Sales].
9. Edit the [Sales] axis and deselect "Include zero" (Figure 5-13). We'll talk about why we do this at the end of this strategy.

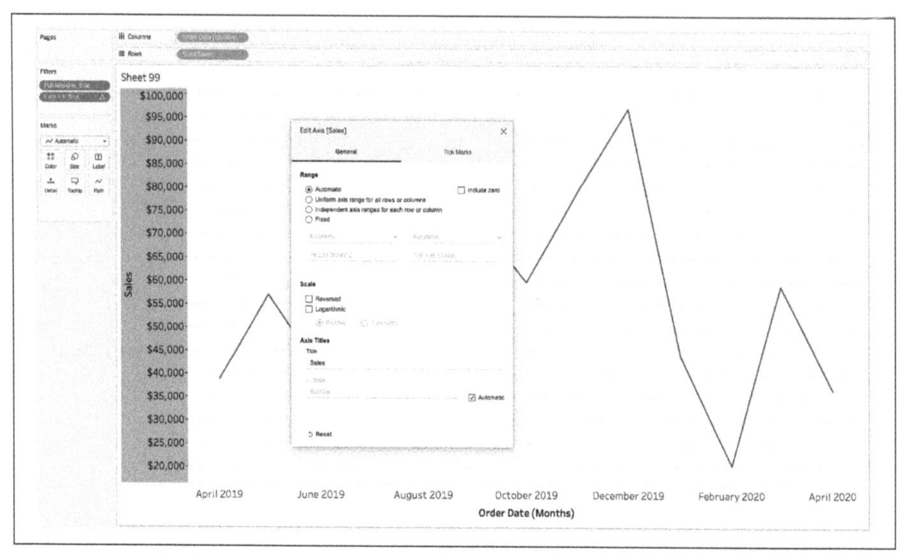

Figure 5-13. The line chart that will eventually be turned into a sparkline

10. Format the chart by changing the Size slider on the Marks card to the lowest value. Next hide the axes. Finally, hide all lines and borders on the sheet.
11. Repeat for each of your KPIs.
12. Create a new dashboard and do the following in order:
    a. Add a vertical container to your dashboard.
    b. Inside that container, add your KPIs.
    c. Add a horizontal container below your KPIs, inside the vertical container.
    d. Add each of your sparkline sheets to the horizontal container.
    e. Hide the title of all the sheets and make sure they take up the entire view.
    f. Adjust the left and right padding so the sparkline is aligned with your KPIs. This will be a manual process that depends on the width and height of your containers.

For this example, the container is 900 pixels wide and 140 pixels high. We have set our left inner padding to 100 px and our right inner padding to 20 px for our sparklines. We also fixed the horizontal container height to 25 px. The final visualization includes sparklines that accent the KPIs with essential information (Figure 5-14).

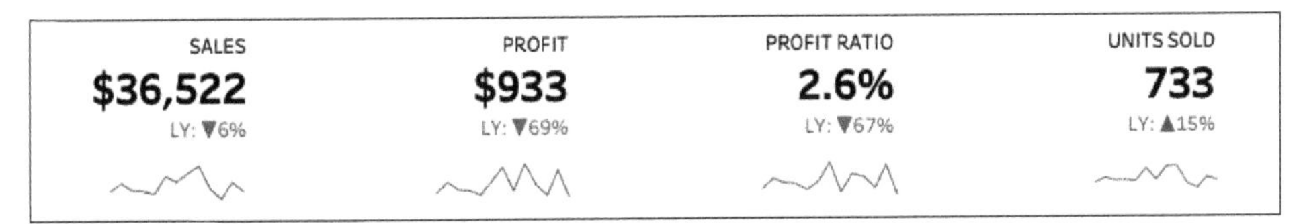

Figure 5-14. The final visualization after creating each of the sparklines and adding to the dashboard

What we love about sparklines is their ability to add context to our data products without taking up too much space. Novice developers often create sparklines as full-blown charts that occupy significant real estate on a dashboard; that isn't necessary if we can provide context in simpler, easier-to-understand ways.

Other methods for creating sparklines can be useful. In Figure 5-14, we are showing the last 13 months of data, but sometimes showing an alternative is worthwhile. For instance, you might want to show how a year-to-date sparkline would compare to a prior-year sparkline. There are two options for doing this. The first is displaying two lines: one for the most recent 12 months and one for the prior 12 months. The other is to display a single line plotted by month that shows the difference between the selected month and the same month from the previous year (we'll call this a *Delta spark bar* because it's showing the difference between the two months and will be a bar chart).

Strategy: Create Year-to-Date Sparklines

Year-to-date sparklines allow your end users to compare how this year is trending against last year. The order of the two lines makes it easy to interpret because left to right becomes January through December, rather than just showing a rolling 12 months:

1. Connect to Sample – Superstore.
2. Set an extract filter to include all dates up to and including May 17, 2020 from the [Order Date] field. Create a custom date value at the Month level of detail.
3. Add [Order Date] as continuous monthly date part to Columns.
4. Add [Order Date] as a discrete year to Color.
5. Add the [Full Months] calculation (found in the previous strategy) to Filters and select True.
6. Create a calculation called **`[Last Two Years]`**. Add it to Filters and select True:

```
// Last Two Years
DATETRUNC("month", [Order Date]) >= DATEADD("year", -1,
  DATETRUNC("year", {MAX([Order Date])}))
```

7. Add [Sales] to Rows.

8. Format the colors. Typically, we use a darker, more prominent color for the current year and a lighter gray for the comparison period.
9. Repeat for each sparkline.

   The product, shown in Figure 5-15, is a set of clear sparklines that allow audiences to track progress throughout the year.

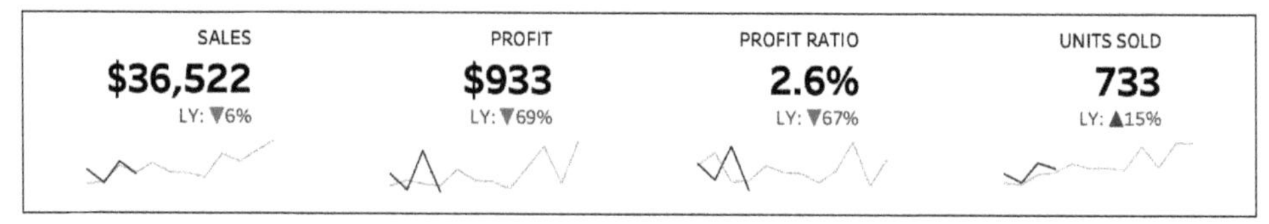

Figure 5-15. Year-over-year comparison sparklines within a KPI visualization

Strategy: Create Rolling 24-Month Sparklines

Sometimes an individual sparkline isn't enough. Sometimes you need two sparklines per metric. This strategy shows the rolling 12-month sparkline with a second sparkline indicating the prior year, to compare the sparkline to its previous year's value:

1. Connect to Sample – Superstore.
2. Set an extract filter to include all dates up to and including May 17, 2020 from the [Order Date] field. Create a custom date value at the Month level of detail.
3. Create a custom date value at the Month level of detail. Add this calculation to Detail on the Marks card.
4. Create an ad hoc calculation for Columns:

   ```
   (INDEX()-1) % 12
   ```

   Edit the table calculation and select Order Date (Months).
5. Add an ad hoc calculation to the Marks card and then set it to the color:

   ```
   [Order Date (Months)] < DATEADD("year",
     -1, DATETRUNC("month", {MAX([Order Date])}+1)-1)
   ```

6. Add the [Full Months] calculation to Filters and select True.
7. Create a calculation called **`[Last 24]`**. Add it to Filters. Edit the table calculation and select Order Date (Months) and select True:

   ```
   // Last 24
   LAST() < 24
   ```

8. Add [Sales] to Rows.
9. Format the lines so they match the design in Figure 5-16.
10. Repeat for each sparkline, and add each to a horizontal divider.

The final result is sparklines that have quite a bit of historical context (Figure 5-16). When you think of what our job as dashboard developers is, it is to give context to numbers. With this more historically oriented set of sparklines, we are able to provide that context.

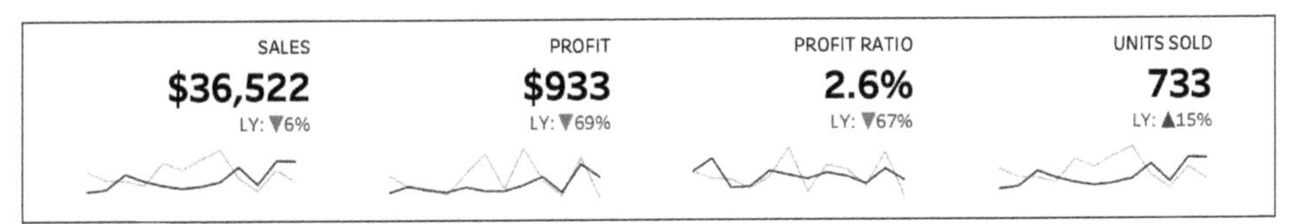

Figure 5-16. The final visualization with rolling 24-month sparklines

Strategy: Build Delta Spark Bars

When we work with executives, we often find that they are most interested in seeing a spark chart of the change metric that is the most valuable to them. They specifically want to know whether the change is consistent. One of the best ways to do this is with a *spark bar*, a feature similar to sparklines, but represented as a bar chart instead. This allows you to show whether the value is above or below zero, presenting a more nuanced data visualization for this scenario:

1. Connect to Sample – Superstore.
2. Set an extract filter to include all dates up to and including May 17, 2020 from the Order Date field. Create a custom date value at the Month level of detail.
3. Create a custom date value at the Month level of detail. Add this custom date to Columns.
4. Create a calculation called **`[Last 12]`**. Add it to Filters. Edit the corresponding table calculation and select Order Date (Months). Change the filter to select values that are True:

```
// Last 12
LAST() < 12
```

5. Add the [Full Months] calculation to Filters and select True.
6. Create a calculation called **`[Sales | Delta 12]`** that calculates the percent change from 12 periods ago and add the calculation to Rows:

```
// Sales | Delta 12
(ZN(SUM([Sales])) - LOOKUP(ZN(SUM([Sales])), -12))
/
ABS(LOOKUP(ZN(SUM([Sales])), -12))
```

7. Format so they match the bars in Figure 5-17.
8. Repeat for each sparkline, and add each to a horizontal divider.

The result is a series of bar charts that represent the last 12 months of change over time (Figure 5-17).

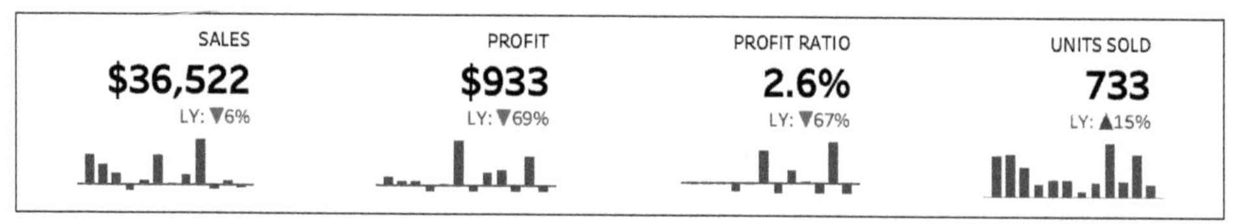

Figure 5-17. Spark bars showing the year-over-year change by month for each KPI

Our executives can easily scan this visual to see whether a bar is above or below the zero line. In many cases, the values represented by spark bars become the top-line metrics.

For instance, let's revisit our COVID-19 visualization from the *New York Times* in Figure 5-11. While the KPI might be new cases (225,572), the more important value is the change metric over the past 14 days (+28%). Very often executives aren't interested in the raw value, but rather the change.

Think about that. Executives know the raw value means so little that they are more—and correctly—focused on the change metric as the true KPI. They know that +28% provides far more understanding about the current state than 225,572. And in the context of COVID-19, both are bad together!

Strategy: Progress to Target

Sometimes you'll be asked to track progress of KPIs—specifically, whether they've met a threshold (a preset goal or a historical value). In this strategy, we'll build out a visual beneath the KPIs to show that progress (Figure 5-18).

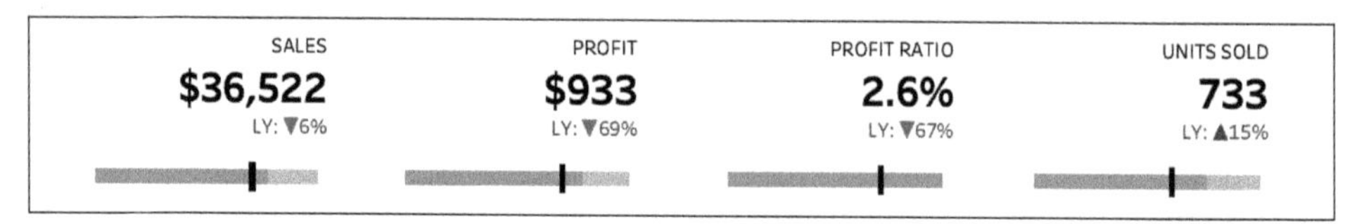

Figure 5-18. Percent-to-progress charts shown with KPIs

Here are the steps:

1. Create an ad hoc calculation of **`MIN(1.0)`** on the Columns shelf.
2. Set the MIN(1.0) axis from –0.1 to 1.1.
3. Set the color of this bar to light gray.
4. Create a float parameter called **`[Sales Target]`**. Set the value to 1.1. This value will be used to set a target for sales to be 110% higher than the previous year.

5. Create a Boolean calculation called **[Current Year Full Months]** that calculates full months for the current year:

```
// Current Year Full Months
[Order Date] < DATETRUNC("month", {MAX([Order Date])})
AND
[Order Date] >= DATETRUNC("year", {MAX([Order Date])})
```

6. Create a Boolean calculation called **[Last Year Full Months]** that calculates the same full months for the previous year as the current year:

```
// Last Year Full Months
[Order Date] < DATEADD('year', -1, DATETRUNC("month", {MAX([Order Date])}))
AND
[Order Date] >= DATEADD('year', -1, DATETRUNC("year", {MAX([Order Date])}))
```

7. Calculate the percent of sales to date for the prior year versus the total for the prior year. Call the calculation **[Sales | % Expected]**:

```
// Sales | % Expected
SUM(
    IF [Last Year Full Months]
    THEN [Sales]
    END
)/
SUM(
    IF YEAR([Order Date]) = YEAR({MAX([Order Date])})
    THEN [Sales]
    END
)
```

8. Calculate the percentage of sales compared to the expected total for the year. Call the calculations **[Sales | % Target]**:

```
//Sales | % Target
SUM(
    IF [Current Year Full Months]
    THEN [Sales]
    END
)/
SUM(
    IF YEAR([Order Date]) = YEAR({MAX([Order Date])})
    THEN [Sales] * [Sales Target]
    END
)
```

9. Place a cap on the overall total in case it goes over the target by creating a calculation called **`[Sales | % Target | Capped]`**:

```
// Sales | % Target | Capped
IF [Sales | % Target] < 1
THEN [Sales | % Target]
ELSE 1
END
```

10. Add [Sales | % Target | Capped] to the right of MIN(1.0) on Columns. Create a synchronized dual axis. Remove [Measure Names] from Color on both Marks cards.

11. Create a calculation that compares the percent expected versus the percent to target. Call the calculation **`[Sales | % Expected vs % Target]`**:

```
// Sales | % Expected vs % Target
[Sales | % Expected] < [Sales | % Target]
```

Place this on Color. Set True colors to teal and False colors to red.

12. Add [Sales | % Expected] to Detail on the [Sales | % Target | Capped] Marks card.

13. Right-click [Sales | % Target | Capped] and add a reference line, using [Sales | % Expected]. Set the line width to the thickest option and set the color to black.

14. Repeat steps 1–13 for [Profit], [Profit Ratio], and [Total Customers]. The result of this step is shown in Figure 5-19.

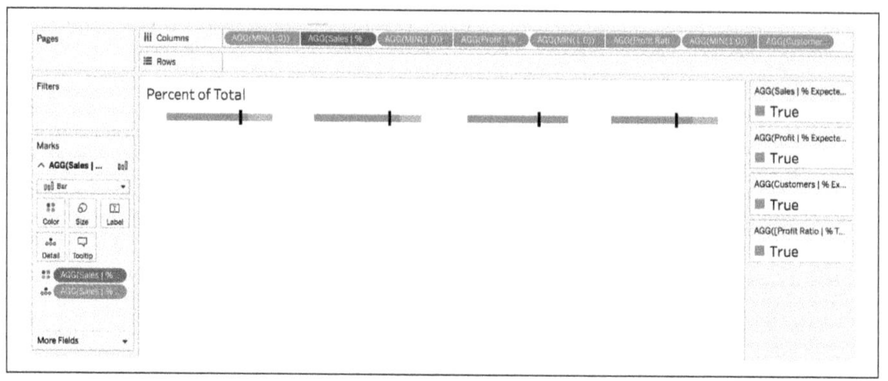

Figure 5-19. Percent-to-progress charts shown after step 14

15. Place your KPI sheet and your percent-to-progress sheet next to each other on the same dashboard and format them to reduce space between. The result will be Figure 5-20.

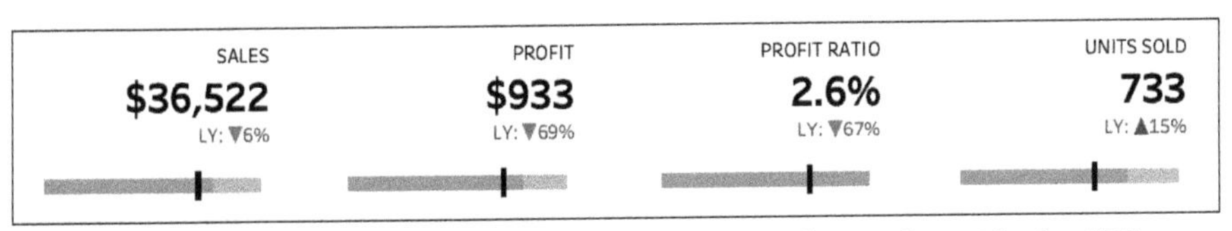

Figure 5-20. The final result of the percent-to-progress charts alongside the KPIs

Multiple KPIs as a Scorecard

Sometimes a KPI as a standalone metric isn't good enough. Sometimes leaders use a series of metrics, which we'll call a *scorecard*—all of which must be in line—to measure success. These sets of KPIs are typically an all-or-nothing measurement of success.

Aggregate KPIs, or scorecards, are not a new concept. Success itself is rarely measured as a binary value. In our jobs, bonuses are often calculated based on whether we hit a series of targets. In retail, you are measured not just on total sales, but the number of orders, and the total units sold. If you hit all your targets, you get a bonus. If you are one short, it's a miss and you don't get the bonus!

This concept applies for a unit in any organization where you need to track multiple KPIs simultaneously.

KPI Scorecard: Office Essentials Case Study

In this case study, you are tasked with tracking four KPIs for each sub-category in the Sample - Superstore dataset. You need to see how sales, total orders, and total units perform toward a target. Then you need to calculate an aggregate KPI scorecard to highlight the percent of those KPIs that are hitting their targets. If any KPI is not hitting 100% of its targets, you'll want to call that out with an indicator (in this case, a red dot).

In this scenario, it's less important to know the value of the KPI and more important to know that each KPI is hitting its target. The result of the next strategy is shown in Figure 5-21.

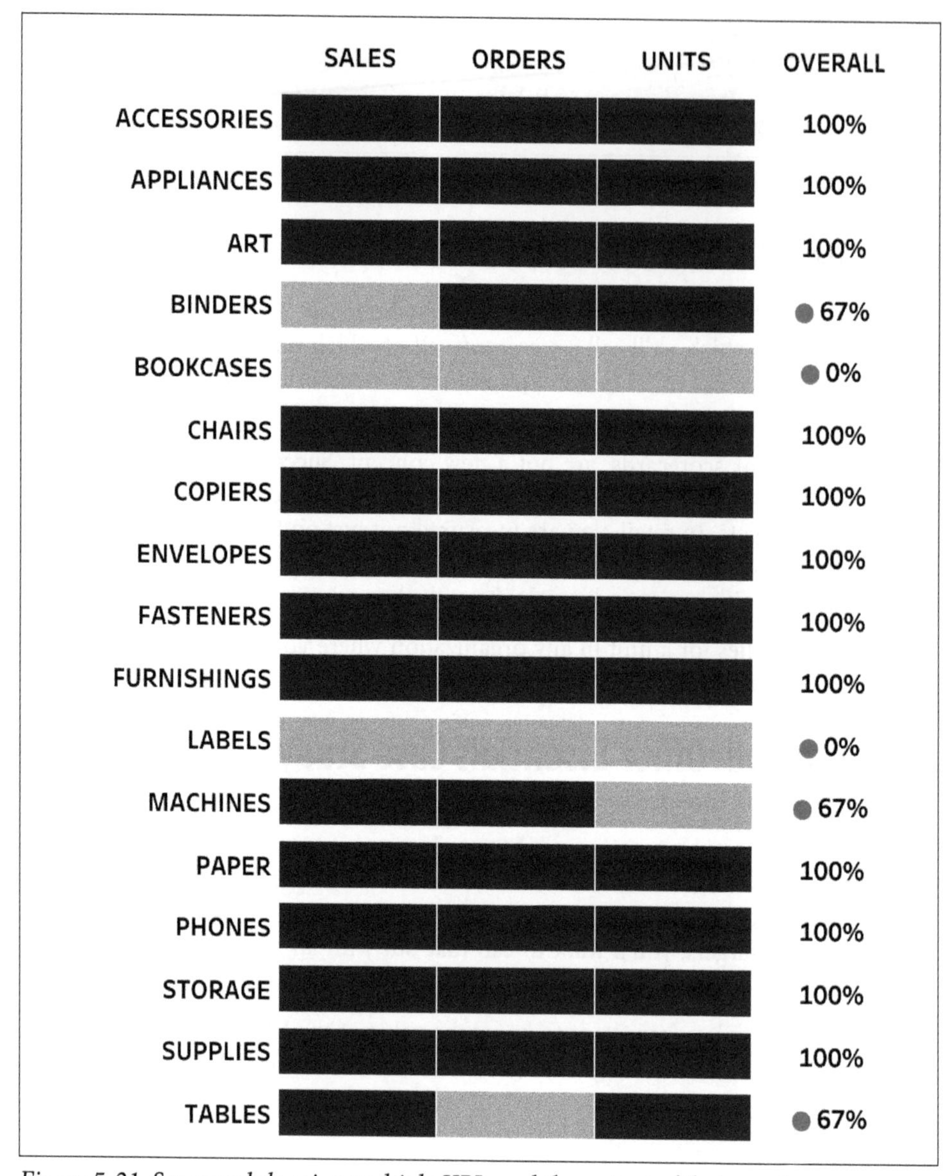

Figure 5-21. Scorecard showing multiple KPIs and the output of their result as an aggregate or overall score

Strategy: Create Aggregate KPIs

In this strategy, you will build the visualization shown in 5-21:

1. Add [Sub-Category] to Rows.
2. Create an ad hoc calculation on Columns and type **`MIN(1)`**. Duplicate the measure by holding Ctrl while clicking and dragging the measure to the right of itself. Repeat this three more times.
3. Edit each of the axes to range from 0 to 1.
4. On the first MIN(1) Marks card, add the [Sales | % Delta] calculation from earlier in the chapter to Color. Edit the color: choose a Custom Diverging palette. Select a gray for the left value and a color for the right value. Used stepped color with two steps. Set the center to 0. (See Figure 5-22.)

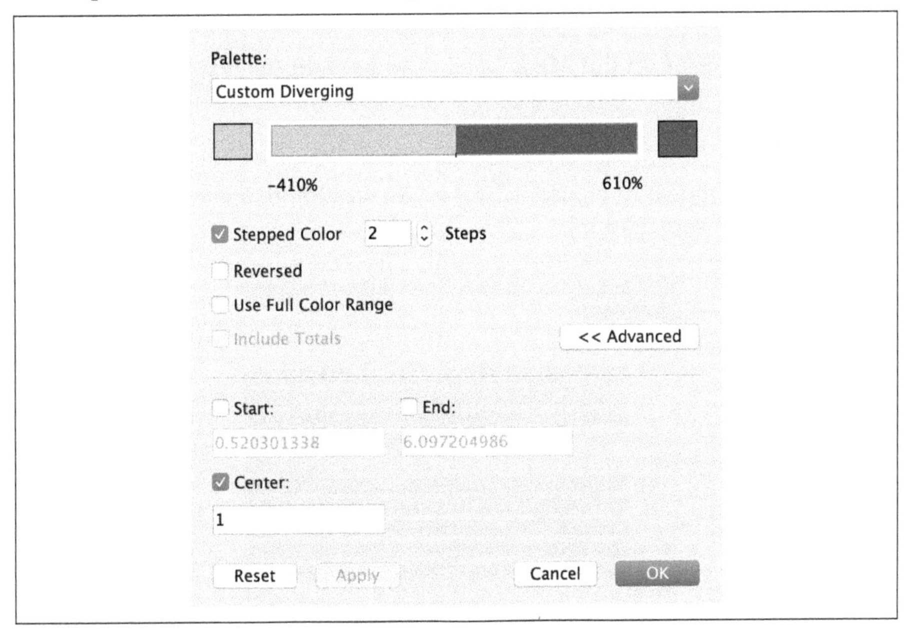

Figure 5-22. Editing the color palette for each KPI

5. Repeat for the second MIN(1) Marks card, using the [Profit | % Delta] calculation created earlier in the chapter. Repeat the color process from step 4.
6. Repeat for the third MIN(1) Marks card, using the [Quantity | % Delta] calculation created earlier in the chapter. Repeat the color process from step 4.

7. Create a calculation called **`[Total Percentage]`** and add this to Text on the fourth MIN(1) Marks card:

```
// Total Percentage
(
  IIF([Sales | % Delta] > 0, 1, 0) +
  IIF([Profit | % Delta] > 0, 1, 0) +
  IIF([Quantity | % Delta] > 0, 1, 0)
)/3
```

8. Create a calculation called **`[Alert Dot]`**. Also add this calculation to the fourth MIN(1) Marks card on Detail:

```
// Alert Dot
IF [Total Percentage] < 1
THEN '●'
ELSE ''
END
```

9. Change the color on this Marks card to white. Edit the text. Place [Alert Dot] to the left of [Total Percentage] on the same line. Change the [Alert Dot] color to red.

The final result, shown in Figure 5-23, is a visualization that allows us to see the performance of each sub-category.

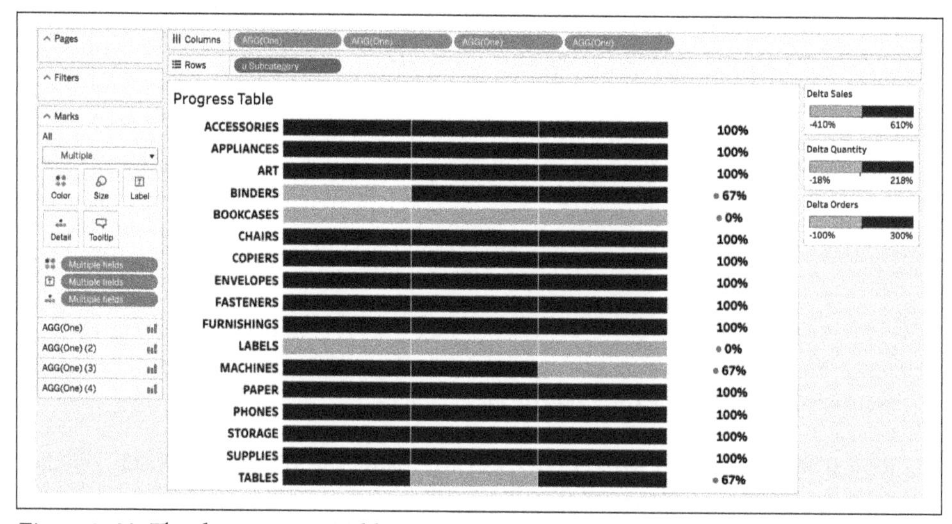

Figure 5-23. The sheet view in Tableau Desktop showing how the aggregate KPI was created

Tracking Daily Changes: Office Essentials Case Study

OE has seen a steady rise in its sales. In addition to monthly reports, executives now want to track cumulative daily values so they can see how the business is performing each morning. How can you represent these daily values in context?

For our last strategy of this chapter, we will build KPIs that can track daily changes in cumulative values versus a previous time period. In this visualization, we'll display a KPI along with a line and an area chart. This system of visualizations is built for tracking daily, rather than for having systematic discussions about a single value over several days. They're great for executives to look at on a more frequent basis, or when you think there's a need to have more visual real estate for context.

These charts show progress toward the prior-year value compared to progress for the current year. Common use cases may be tracking sales data, reporting on unit production, and understanding the volume of customer engagements. All of these scenarios rely on tracking performance against the prior year to better understand how the business is trending against expectations.

Adding these month-to-date lines give context to the current performance this month versus last year. When a value is better than the same time of the month versus last year, the line is green. When it is below last year, it is red. While these charts look extremely simple in Tableau, multiple steps are required to build these lines.

This next case study will be divided into three strategies but will be combined into a single dashboard. The output, shown in Figure 5-24, highlights sales performance by month.

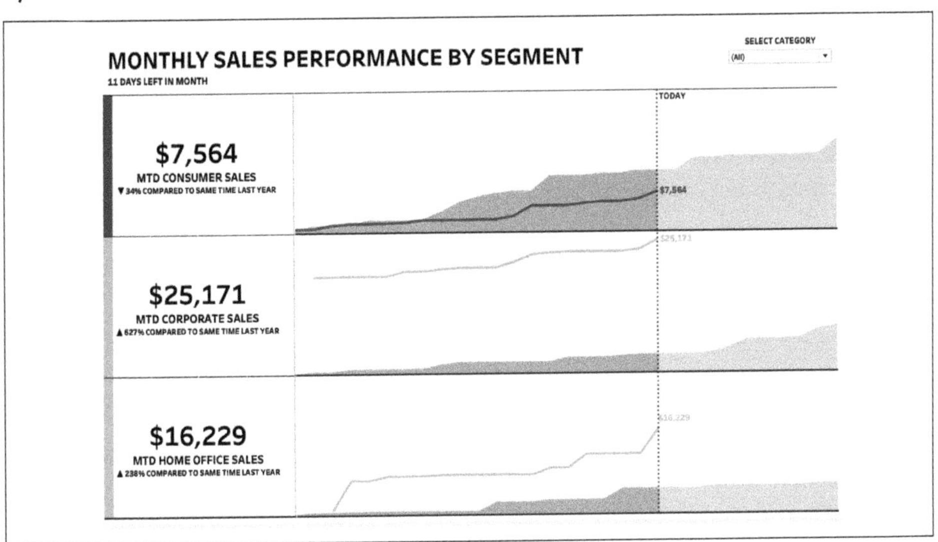

Figure 5-24. The output for the next three strategies

Strategy: Create Month-to-Date Sparklines

This strategy creates the line and area chart shown in Figure 5-24:

1. Connect to Sample – Superstore.
2. Set an extract filter to include all dates up to and including September 17, 2020 from the [Order Date] field. Create a custom date value at the Month level of detail.
3. Add [Day of Order Date] to Columns as a continuous date part.

   For the next part, you are going to build a series of calculations that will help you create the visualization.
4. Create a calculation called **`[Today]`**. Normally, you would place the function `TODAY()` in the calculation, but for the sake of the example, use the max date in our dataset—which, after the extract filter is applied, is May 17:

   ```
   // Today
   {MAX([Order Date])}
   ```
5. Create a Boolean calculation called **`[CMCY | TF]`** that represents True or False if the data is in the same month as [Today] (CMCY = current month, current year):

   ```
   // CMCY | TF
   DATETRUNC('month' ,[Order Date]) = DATETRUNC('month', [Today])
   AND DATEPART('day', [Order Date]) <= DATEPART('day', [Today])
   ```
6. Create a Boolean called **`[SDLY | TF]`** that represents True or False if the data is the same day of the month as [Today], but last year (SDLY = same day, last year):

   ```
   // SDLY | TF
   DATETRUNC('month', [Order Date]) = DATETRUNC('month', DATEADD('year', -1,
     [Today]))
   AND DATEPART('day', [Order Date]) <= DATEPART('day',[Today])
   ```
7. Create a Boolean called **`[SMLY | TF]`** that indicates whether the month is the same month as [Today], but a year prior (SMLY = same month, last year):

   ```
       // SMLY | TF
   DATETRUNC('month', [Order Date]) = DATETRUNC('month', DATEADD('year', -1,
     [Today]))
   ```
8. Create a calculation for sales for the current month of [Today], called **`[Sales | CMCY]`**:

   ```
   // Sales | CMCY
   IF [CMCY | TF]
   THEN [Sales]
   END
   ```

9. Create a calculation called **`[Sales | SMLY]`** that calculates sales if they occurred the same month as [Today] but for the last year:

   ```
   // Sales | SMLY
   IF [SMLY | TF]
   THEN [Sales]
   END
   ```

10. Create a calculation called **`[Sales | SDLY]`** that calculates sales that occurred the same day and month of last year as [Today]:

    ```
    // Sales | SDLY
    IF [SDLY | TF]
    THEN [Sales]
    END
    ```

    Now that you've built the base calculations for this visualization, we can start putting it all together.

11. Create a calculation called **`[Sales | CMCY | Hidden]`**:

    ```
    // Sales | CMCY | Hidden
    IF MAX(DAY([Order Date])) > MAX(DAY([Today]))
    THEN NULL
    ELSE RUNNING_SUM(SUM([Sales | CMCY]))
    END
    ```

12. Add [Sales | CMCY | Hidden] to Rows. Set the mark type to Line. Edit the table calculation. Choose Specific Dimensions and select only Day of Order Date.

13. Create a calculation called **`[Sales | SDLY | Hidden]`**:

    ```
    // Sales | SDLY | Hidden
    IF MAX(DAY([Order Date])) > MAX(DAY([Today]))
    THEN NULL
    ELSE RUNNING_SUM(SUM([Sales | SDLY]))
    END
    ```

14. Add [Measure Values] to Rows. In the new [Measure Values] Marks card, make sure that only SUM(Sales | SMLY) and [Sales | SDLY | Hidden] are included. Right-click SUM(Sales | SMLY) on the [Measure Values] Marks card and add a running total table calculation (Figure 5-25).

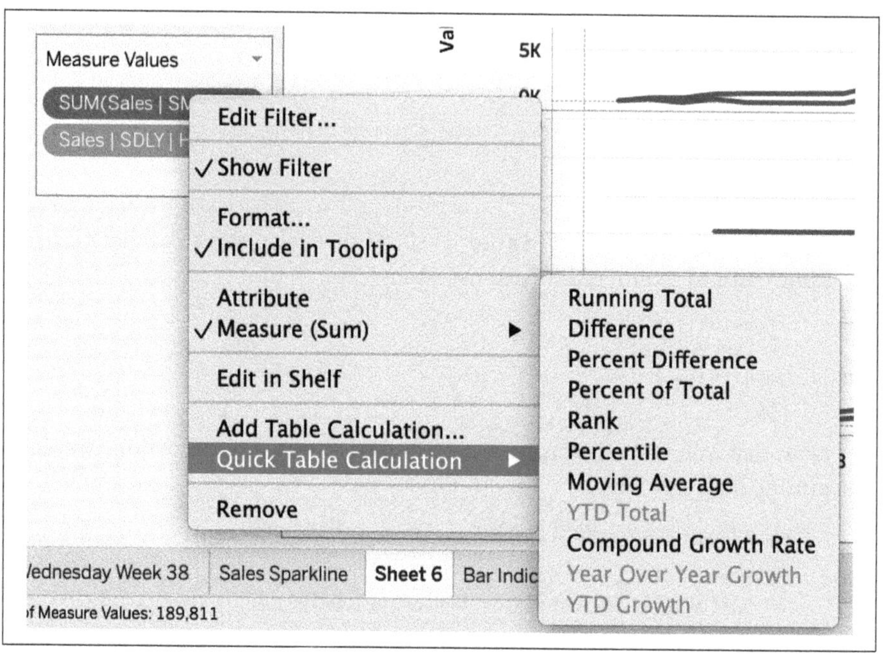

Figure 5-25. Right-click a measure on the [Measure Values] Marks card to add a quick table calculation of running total

15. Change the [Measure Values] mark type to Area. Add [Measure Names] to Color, set SUM(Sales | SMLY) to a light gray, and [Sales | SDLY | Hidden] to a medium gray.

16. You'll notice our two area charts are stacked. You don't want this. From the top menu, choose Analysis → Stack Marks → Off (Figure 5-26).

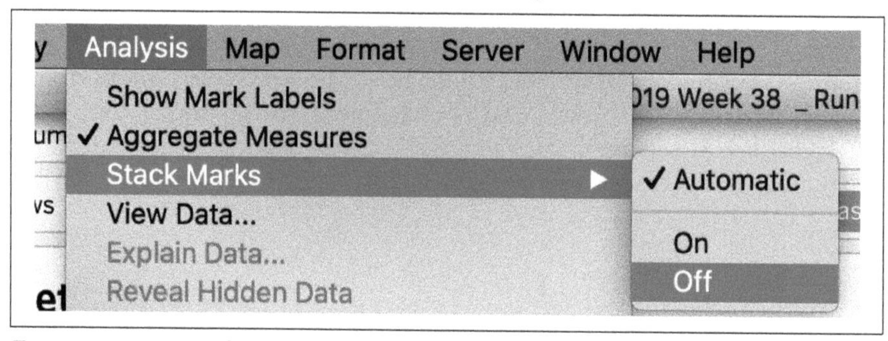

Figure 5-26. Turn off stack marks from the top menu

17. Create a synchronized dual axis on your Rows shelf between [Sales | CMCY | Hidden] and [Measure Values]. After that, right-click the [Sales | CMCY | Hidden] axis and select "Move marks to front."

18. Create a calculation called **[Today Reference Line]**:

```
// Today Reference Line
DATEPART('day', [Today])
```

Click and drag this calculation to Detail of the [Sales | CMCY | Hidden] Marks card. Change the aggregation of the calculation to maximum. Right-click the [Day of Order Date] axis and add a reference line.

For the entire table, use the maximum value of [Today Reference Line]. Show no label or tooltip. Change the line type to Short Dashes. Uncheck "Show recalculated line for highlighted or selected data points." Figure 5-27 shows the chart at this point.

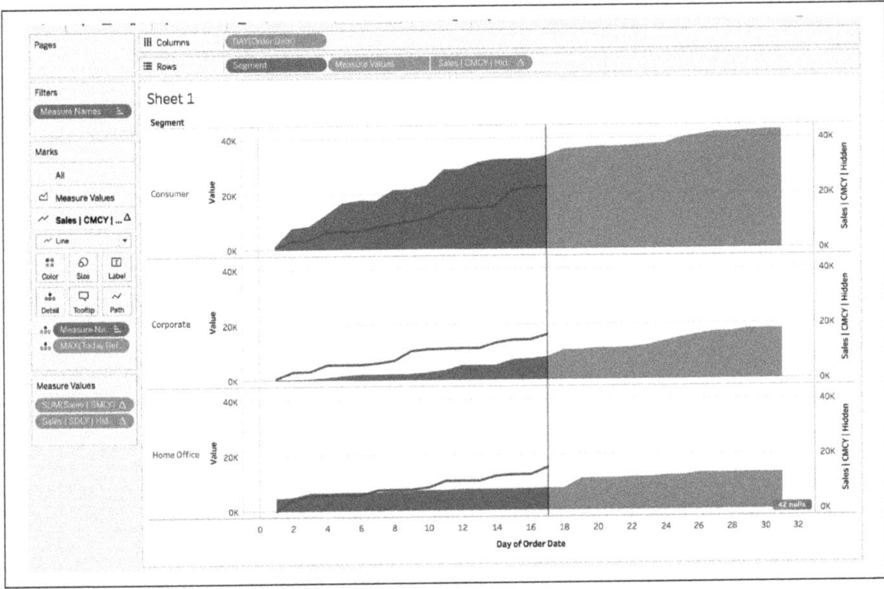

Figure 5-27. Progress on the percent-to-target area and line chart

19. Create a calculation called **[Sales | MTD | % Delta]**. This will calculate the change in sales as a percentage from the day and month to date compared to the year prior:

```
// Sales | MTD | % Delta
(SUM([Sales | CMCY]) - SUM([Sales | SDLY]))
/
SUM([Sales | SDLY])
```

20. Create a calculation called **[Color Line]**:

```
// Color Line
{EXCLUDE [Order Date]: SIGN([Sales | MTD | % Delta])}
```

Add [Color Line] to Color on the [Sales | CMCY | Hidden] Marks card. Change the mark type from Continuous to Discrete. Edit the color so –1 is red and 1 is green.

21. On the [Sales | CMCY | Hidden] Marks card, select the "Show mark labels" checkbox. For the Marks to Label option, select Line End and select only Label Ends. Finally, under the font drop-down, select Match Mark Color. Figure 5-28 shows these options.

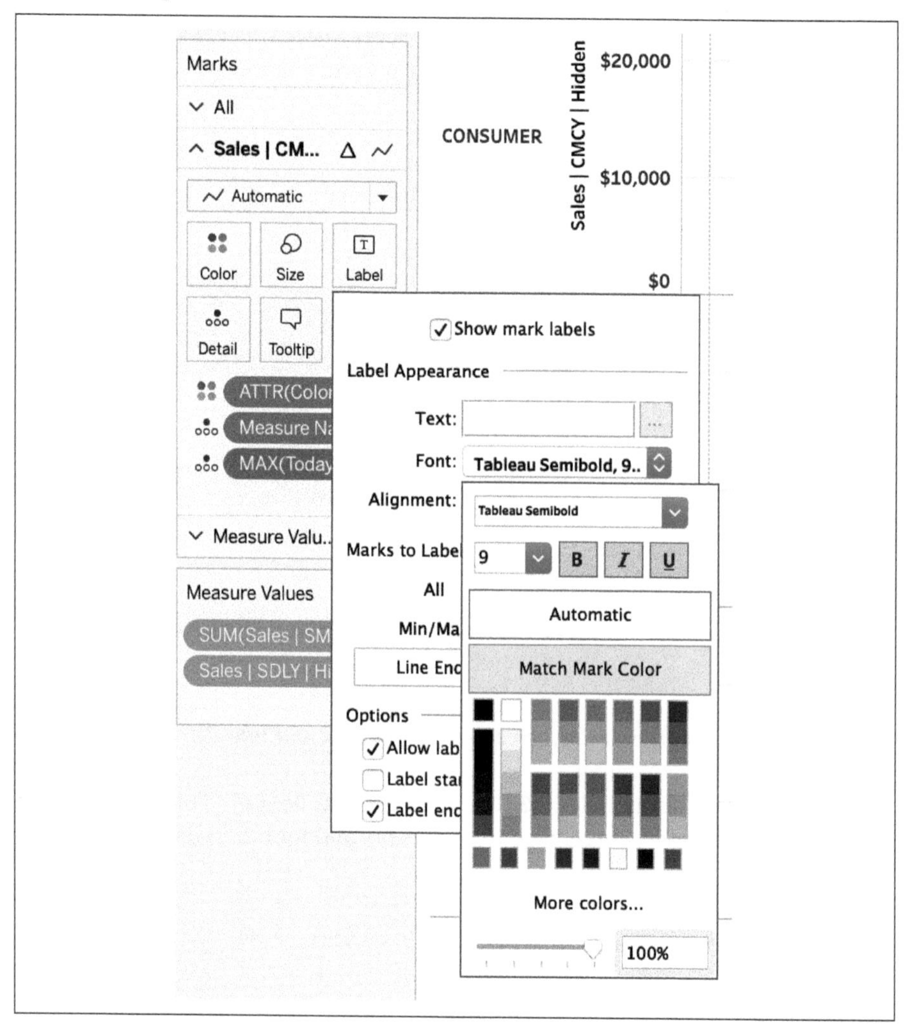

Figure 5-28. Match the mark color by using the Match Mark Color option from Label on your Marks card

22. Finally, you need to format your visualization. Hide the Value axis, the [Sales | CMCY | Hidden] axis, and the [Day of Order Date] axis. Then remove the

dashed zero line and column dividers. In the top-left corner of the visualization, right-click [Segment] and select Hide Field Labels for Rows.

The result, shown in Figure 5-29, is beautiful lines that compare this year versus an area chart of last year. Of course, the chart is useful, but you will want to combine these with KPIs for the month as well.

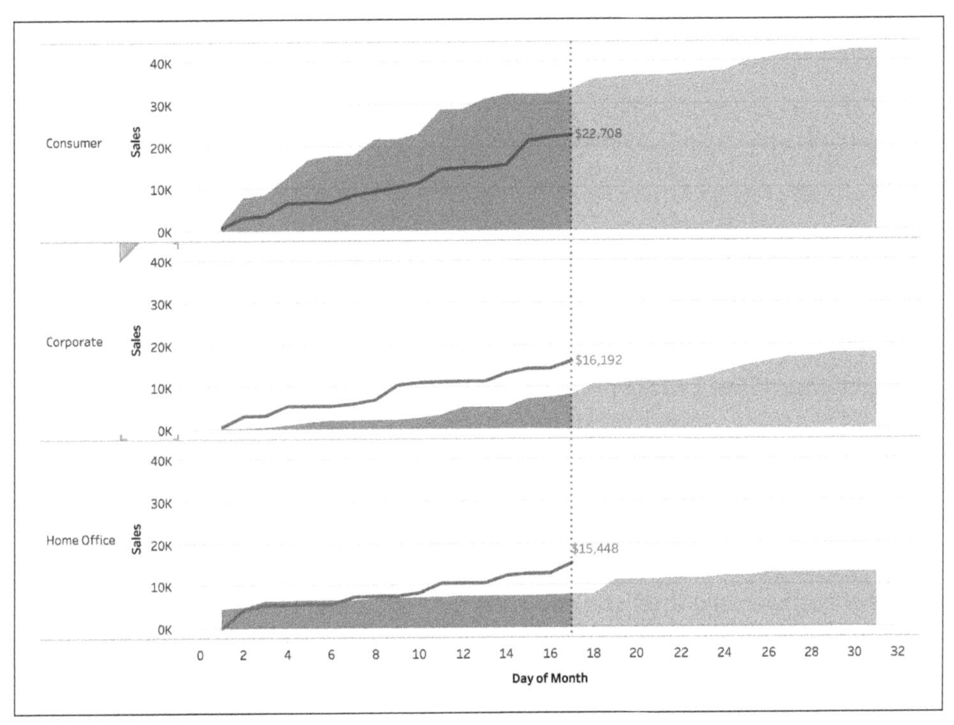

Figure 5-29. The final product

Strategy: Create Month-to-Date KPIs

In this strategy, we'll walk through the process of combining our previous KPIs with those for the month. Luckily, we've built many of the calculations we need when we built out the lines, so this build should be a little more straightforward:

1. Add [Segment] to Rows.
2. Type **`MIN(10)`** as an ad hoc calculation on Columns. Set the mark type to Text. To format, set the grid lines, zero lines, axis rulers, and axis ticks to None. Edit the axis and fix the range from 0 to 20.
3. Add SUM(Sales | CMCY), [Sales | MTD | % Delta], and [Segment] to Label. Edit the text as shown in Figure 5-30.

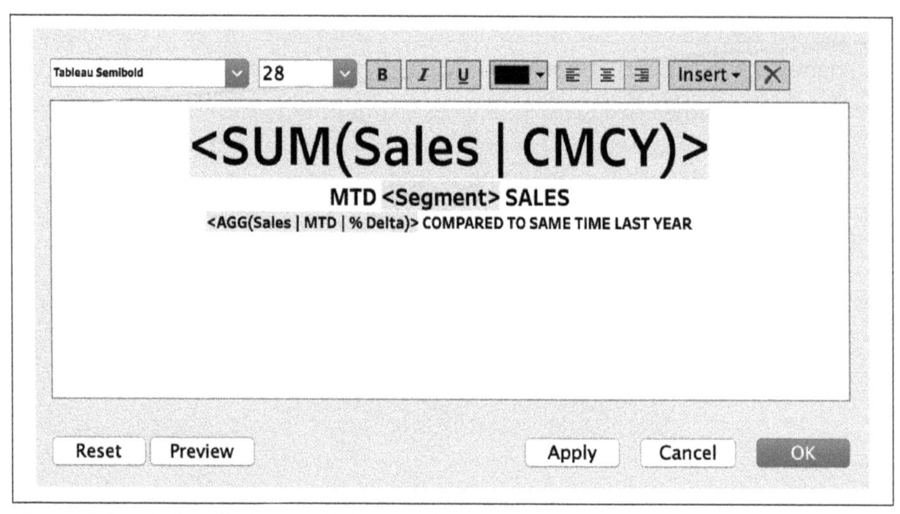

Figure 5-30. A detailed view of the text formatting

4. Right-click [Sales | MTD | %Delta] on the Marks card and select Format. Under font formatting, select Pane (if it's not already selected). Click the Numbers drop-down. Under the Default section, near the top of the formatting, choose Custom and then type **`"▲ 0%;▼ 0%;▬"`**. This will place an up arrow before positive percentages, a down arrow before negative percentages, and a large dash for zero values.
5. Type **`MIN(1)`** as an ad hoc calculation on Columns, to the right of MIN(10). Create a synchronized dual axis and then hide the axes. Also remove the [Measure Names] values that were automatically added to Color on both Marks cards.

   Change the mark type to Bar. Set the bar type to the maximum width. On Color, set the border to None.

   Remove the grid lines and column dividers. Match the formatting of the row dividers on the sheet with the running line and area charts.
6. Add the [Color Lines] calculation to the color of the MIN(1) Marks card. This color should match the color of your lines from "Strategy: Create Month-to-Date Sparklines" on page 204.

   Once you have completed this step, you've created your KPIs (Figure 5-31).

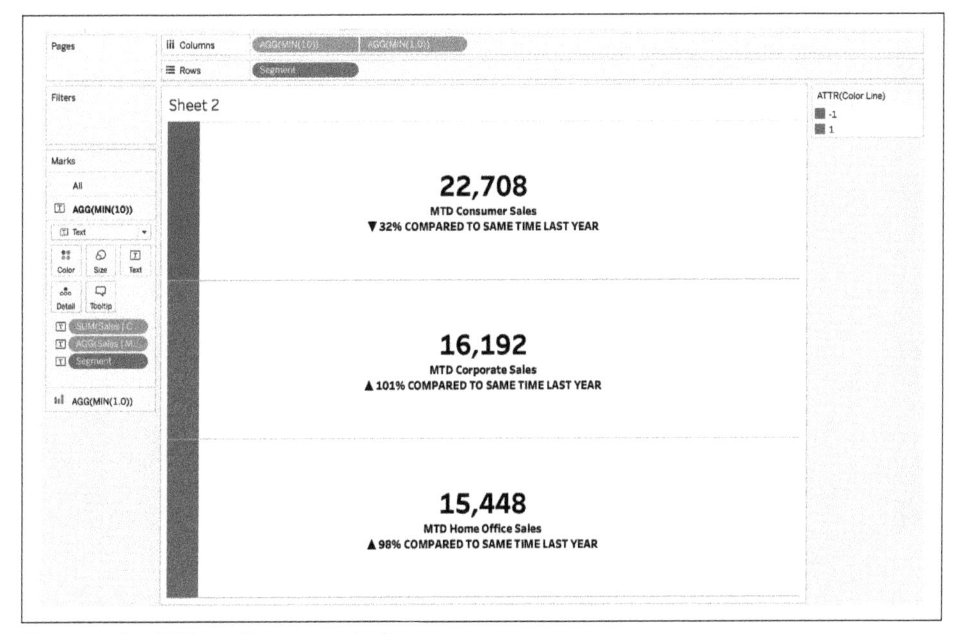

Figure 5-31. KPIs and a success indicator

Strategy: Combine KPI Visualizations in a Dashboard

Now that you have completed the KPIs and microcharting in the preceding two strategies, it is time to bring the visualizations together on a dashboard:

1. Add a horizontal container to your dashboard.
2. Add the KPIs to the horizontal container. Set all outer or inner padding to zero. Hide the title. Select the option that fits KPIs to the entire view.
3. Add the microcharting to the right of the KPIs inside the horizontal container. Hide the Segment header on the right. Set all outer or inner padding to zero and hide the title. Select the KPIs to fit to the entire view.
4. Remove the container with the legends.
5. Adjust the size of your KPI sheet to the desired width and to accommodate space for your line chart.

   The result is two visualizations that appear as a single component on a dashboard (Figure 5-32).

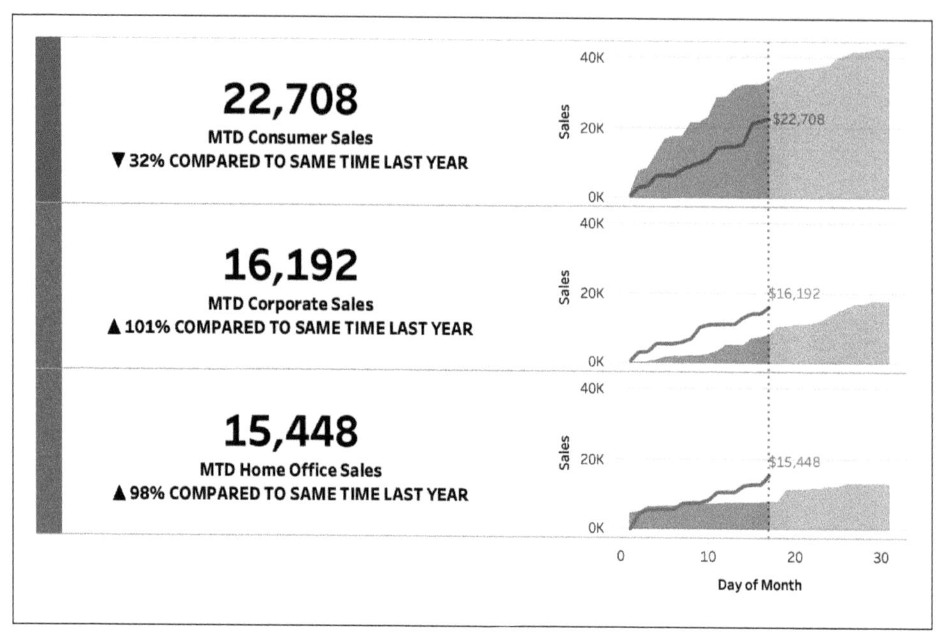

Figure 5-32. The final result of the strategy

Conclusion

Now that you have had a chance to work with and expand your KPIs, let's recap.

We started the chapter by discussing the importance of formatting your KPIs so they are easily readable for your audience. Simple features like text size, text alignment, and font selection matter. The overall goal is to increase readability so your audience can glance at a value and gain the necessary information extremely quickly.

We showed how to use the calculation MIN(0.0) as a placeholder to build multiple KPIs on a single sheet. Developing your KPIs on a single sheet offers you two major benefits. First, the KPIs are all in one location, which simplifies and speeds up development. Instead of hopping from sheet to sheet, you now can quickly swap to different Marks cards to edit. It also simplifies the formatting process. The other, much larger benefit is the performance increase due to working from a single sheet. If you can find a way to reduce the number of sheets you are using, you should take that chance. Fewer sheets often means better performance.

In much of the chapter, we displayed examples of adding context to KPIs. Calculating year over year (or month over month) is critical for most business situations. Imagine trying to tell a story just by using the odometer in your car.

Noting only your speed at any given point in time gives little context:

- What if you could say if that speed was above or below a target, like the speed limit?
- What if you could say how the speed at that moment compared to previous trips at the same point along the way?
- What if you could show snapshots at fixed periods—say, every 15 minutes?

This context would improve understanding. You might be able to tell whether you are speeding, are making the trip faster than usual, or know where along the way you've lost time.

All of this translates to the strategies in this chapter. We showed you how to calculate year-over-year comparisons by using LOD calculations. This allowed you to understand the direction of your metric.

We also addressed automating your KPIs. This is because audiences crave up-to-date information and think there's no added lift in the maintenance of their data products. Automated KPIs (and other calculations) are life-saving. And we've witnessed that! We can personally attest that this strategy has returned people's weekends, and as a result, improved a company's employee-retention rate! How crazy is that?! With well-designed calculations, you can automate your KPIs (and the rest of your dashboards) not only so your end users have the most up-to-date information, but also so analysts get a huge chunk of their lives back.

Throughout the strategies presented in this chapter, we showcased how to display automated sparklines depicting the last 13 months, overlaid current year-to-date progress with the previous year, and combined ideas to show rolling 24-month sparklines. Each sparkline option enhances the KPI and allows your audience to see the progress to the path without overanalyzing the data.

We also showcased a spark bar chart that showed the year-over-year difference of a metric as the value driving the microchart. Just as important, this strategy showed an alternative method—using table calculations—for calculating year-over-year metrics.

If you are tracking toward a cumulative target (which is extremely common with sales), a progress-to-target bar chart is extremely useful. While we used 110% of prior year sales for our targets, targets could be set using other methods. The important part is showing that we can track progress!

While many of the strategies presented in this chapter focused on monthly metrics, they can also be reported anywhere from the second (water volume per second) to the lifetime (customer lifetime value). Toward the end of the chapter, we changed the focus from monthly to daily. The results are running daily totals of sales and a snapshot of daily values being above or below a specified target.

KPIs themselves are fairly straightforward. Each indicates a single value that is driving conversation for your audience. And while we've mentioned it throughout the chapter, we want to end by saying it's your job as the developer to add context to that value so that members of your audience are aligned.

In the next chapter, we stay focused on the essentials and discuss table design and analysis. While most data visualization experts avoid talking about tables, we want to provide this guidance because every developer receives a request for a table. Like KPIs, a well-designed table with appropriate context can be extremely useful for your audience.

CHAPTER 6
Building Impactful Tables

Quick—without thinking too much, answer the following questions:

- What chart type is used the most when displaying data?
- What chart type is the hardest to design?
- What chart type is the least customizable?
- What chart type do data practitioners hate most?

The answer to all these questions is the humble data table. This chapter is all about building tables that are full of information, without sacrificing design.

Like the visual in Figure 6-1, most information is still shared in tables. But when there's a special point we want to communicate about the data, it's likely to get lost amid a sea of information.

Analysts and designers complain about tables, but the reality is that people use tables in their decision making. Heck, it's probably your go-to visualization when trying to validate data. It's our job as analytics developers and data visualization practitioners to give our users the tools that allow them to do their job. If your audience asks for tables, this means building tables!

Product Name	Revenue	Revenue/ Item	Profit	Profit/Item	Total Orders	Items Sold	% Margin
Canon imageCLASS 2200 Advanced Copier	$61,600	$3,079.99	$25,200	$1,260.00	5	20	69%
Fellowes PB500 Electric Punch Plastic Comb Binding Machine with Manual ..	$27,453	$885.59	$7,753	$250.10	10	31	39%
Cisco TelePresence System EX90 Videoconferencing Unit	$22,638	$3,773.08	-$1,811	($301.85)	1	6	-7%
HON 5400 Series Task Chairs for Big and Tall	$21,871	$560.78	$0	$0.00	8	39	0%
GBC DocuBind TL300 Electric Binding System	$19,823	$535.77	$2,234	$60.37	11	37	13%
GBC Ibimaster 500 Manual ProClick Binding System	$19,025	$396.34	$761	$15.85	9	48	4%
Hewlett Packard LaserJet 3310 Copier	$18,840	$495.78	$6,984	$183.79	8	38	59%
HP Designjet T520 Inkjet Large Format Printer - 24" Color	$18,375	$1,531.24	$4,095	$341.25	3	12	29%
GBC DocuBind P400 Electric Binding System	$17,965	$665.37	-$1,878	($69.56)	6	27	-9%
High Speed Automatic Electric Letter Opener	$17,030	$1,548.21	-$262	($23.82)	3	11	-2%

Figure 6-1. A table with some color (but not all color) to highlight some key values

In This Chapter

You'll need the Sample – Superstore dataset. In this chapter, you'll learn how to do the following:

- Build, format, and customize tables
- Visually combine worksheets on a dashboard to create visual systems that look like tables
- Create dynamic tables that highlight key insights while retaining the format of a table

Building Great Tables

Before you start building tables, let's look at some tips that can help make designing tables more useful. People think tables are simple, but they are complex systems of information that must be formatted with great care to maximize understanding.

Let's start with a table we are all guilty of making at some point in our careers: the small-font, information-dense table; see Figure 6-2.

State	Sales	Profit	Profit Ratio	Discount
California	$457,688	$76,381	17%	7%
New York	$310,876	$74,039	24%	6%
Texas	$170,188	-$25,729	-15%	37%
Washington	$138,641	$33,403	24%	6%
Pennsylvania	$116,512	-$15,560	-13%	33%
Florida	$89,474	-$3,399	-4%	30%
Illinois	$80,166	-$12,608	-16%	39%
Ohio	$78,258	-$16,971	-22%	32%
Michigan	$76,270	$24,463	32%	1%
Virginia	$70,637	$18,598	26%	0%
North Carolina	$55,603	-$7,491	-13%	28%
Indiana	$53,555	$18,383	34%	0%
Georgia	$49,096	$16,250	33%	0%
Kentucky	$36,592	$11,200	31%	0%
New Jersey	$35,764	$9,773	27%	0%
Arizona	$35,282	-$3,428	-10%	30%
Wisconsin	$32,115	$8,402	26%	0%
Colorado	$32,108	-$6,528	-20%	32%
Tennessee	$30,662	-$5,342	-17%	29%
Minnesota	$29,863	$10,823	36%	0%
Massachusetts	$28,634	$6,786	24%	2%
Delaware	$27,451	$9,977	36%	1%
Maryland	$23,706	$7,031	30%	1%
Rhode Island	$22,628	$7,286	32%	2%
Missouri	$22,205	$6,436	29%	0%
Oklahoma	$19,683	$4,854	25%	0%
Alabama	$19,511	$5,787	30%	0%
Oregon	$17,431	-$1,190	-7%	29%
Nevada	$16,729	$3,317	20%	6%
Connecticut	$13,384	$3,511	26%	1%
Arkansas	$11,678	$4,009	34%	0%
Utah	$11,220	$2,547	23%	6%
Mississippi	$10,771	$3,173	29%	0%
Louisiana	$9,217	$2,196	24%	0%
Vermont	$8,929	$2,245	25%	0%
South Carolina	$8,482	$1,769	21%	0%
Nebraska	$7,465	$2,037	27%	0%
New Hampshire	$7,293	$1,707	23%	1%
Montana	$5,589	$1,833	33%	7%
New Mexico	$4,784	$1,157	24%	6%
Iowa	$4,580	$1,184	26%	0%
Idaho	$4,382	$827	19%	9%
Kansas	$2,914	$836	29%	0%
District of Columbia	$2,865	$1,060	37%	0%
Wyoming	$1,603	$100	6%	20%
South Dakota	$1,316	$395	30%	0%
Maine	$1,271	$454	36%	0%
West Virginia	$1,210	$186	15%	8%
North Dakota	$920	$230	25%	0%

Figure 6-2. A table with a small font and no spacing

When you make this table, you're not thinking at all about the end user or how you could design the table to improve decision making. Your only thought is, how can I cram as much information as possible into this table so I don't have to think about it ever again? But tables can be useful. And we shouldn't care about space because we basically have as much space as we want; it's digital, not paper, so don't be afraid to use pixels for good.

So what are some easy things you can do to maximize the readability of your tables?

Ensure a Clear Purpose for Your Table

It's easy to provide a table of data for any audience. It's quite another thing to figure out the exact needs of the audience, tailor the output of the table to highlight key insights for that audience, and use the additional data to provide context. Maybe the key value you want to communicate is the number of total customers. Maybe it's a change in customer numbers from the previous month, or the change in total customers from the previous month compared to the same change a year ago (this is known as the *delta of a delta*). Each of these three objectives needs a different kind of table.

Format Your Table to Maximize Readability

There are lots of ways to do this, but we recommend the following techniques:

Make font sizes readable
: You might be tempted to try to cram too much into a table by minimizing font size. Don't do it! A font size between 10 and 14 is usually appropriate.

Distinguish column headers from the body
: Use bold fonts or changes in the color of the header text to distinguish the header from the body.

Add spacing and padding to rows in the table
: Row padding should be between 0.5 to 1.0 times the font size both above and below the text.

Choose table-friendly fonts
: Be thoughtful about your font selections. Luke personally avoids serif fonts and chooses either sans serif or mono fonts; Ann likes to stick with the Tableau family of fonts since they've all been maximized for readability and data presentation at all font sizes. Overall, it's important to choose `a font in which numeric values are monotonically spaced`. This will allow digits to align across rows in a column.

Limit your use of color
: Color is useful for table design, but too much can be a distraction. With some customization, you can use color judiciously to highlight specific columns of a table.

Our first two strategies introduce the basic principles for creating tables that are ready for any audience. In the remainder of the chapter, you'll work with strategies that will build your Tableau table development skills by applying these principles.

Using Color in Tables: Office Essentials Case Study

Let's go back to OE, the office supply store you've seen in previous chapters. In this example, the data team has been asked to create a table that shows a breakdown of metrics by customer. This table will have lots of rows, which will make gathering insights difficult. How can the team use color in a way that gets the point across?

Color is a great way to highlight what you need to communicate, but too much color can be overwhelming. How can our OE data team highlight some values to communicate insights?

We'll look at a few strategies in this section. First, we'll walk you through creating a table that uses just a bit of color to make its point. Then we'll try a variation that uses dots to encode color.

Strategy: Create a Table Using Limited Color

You'll start by building a base table with the Sample - Superstore dataset:

1. Create the base table, shown in Figure 6-3, as follows:
    a. Add [Customer Name] to the Rows shelf.
    b. Add [Measure Names] to Columns and [Measure Values] to Text.
    c. In the Filters shelf, you'll see [Measure Names]. Edit this by selecting [Sales], [Orders], [Profit], and [Profit Ratio].

       If you don't have the Profit Ratio calculation, create it:

       ```
       // Profit Ratio
       SUM([Profit])/SUM([Sales]))
       ```

    d. Reorder the values from top to bottom: SUM(Sales), AGG(Orders), SUM(Profit), and AGG(Profit Ratio).

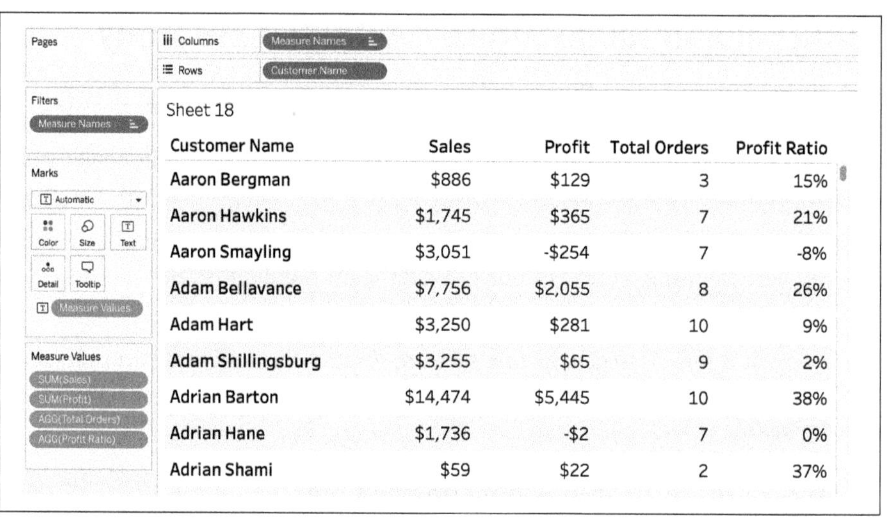

Customer Name	Sales	Profit	Total Orders	Profit Ratio
Aaron Bergman	$886	$129	3	15%
Aaron Hawkins	$1,745	$365	7	21%
Aaron Smayling	$3,051	-$254	7	-8%
Adam Bellavance	$7,756	$2,055	8	26%
Adam Hart	$3,250	$281	10	9%
Adam Shillingsburg	$3,255	$65	9	2%
Adrian Barton	$14,474	$5,445	10	38%
Adrian Hane	$1,736	-$2	7	0%
Adrian Shami	$59	$22	2	37%

Figure 6-3. A simple base table in Tableau

2. Apply readability principles to the table by formatting its text to be the same font style, size, and color as the row headers. You can do this by clicking the Text button and setting the font to Tableau Book, size 15, and selecting the second-to-last shade from the first column of black shades (Figure 6-4).

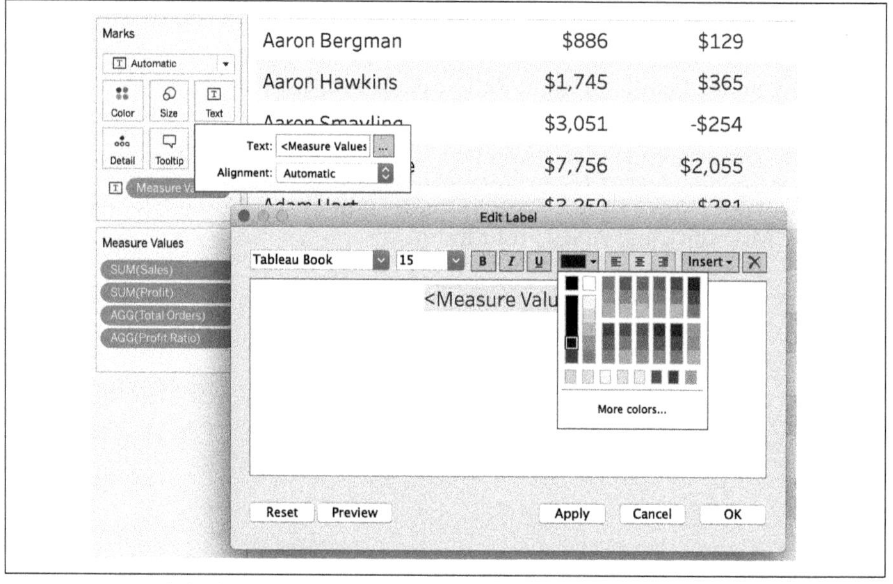

Figure 6-4. By default, Tableau does not use best practices with text color and size in tables. Be sure to update the text and the font.

Match this styling with the row header by right-clicking a customer name and selecting Format. Figure 6-5 shows the resulting options.

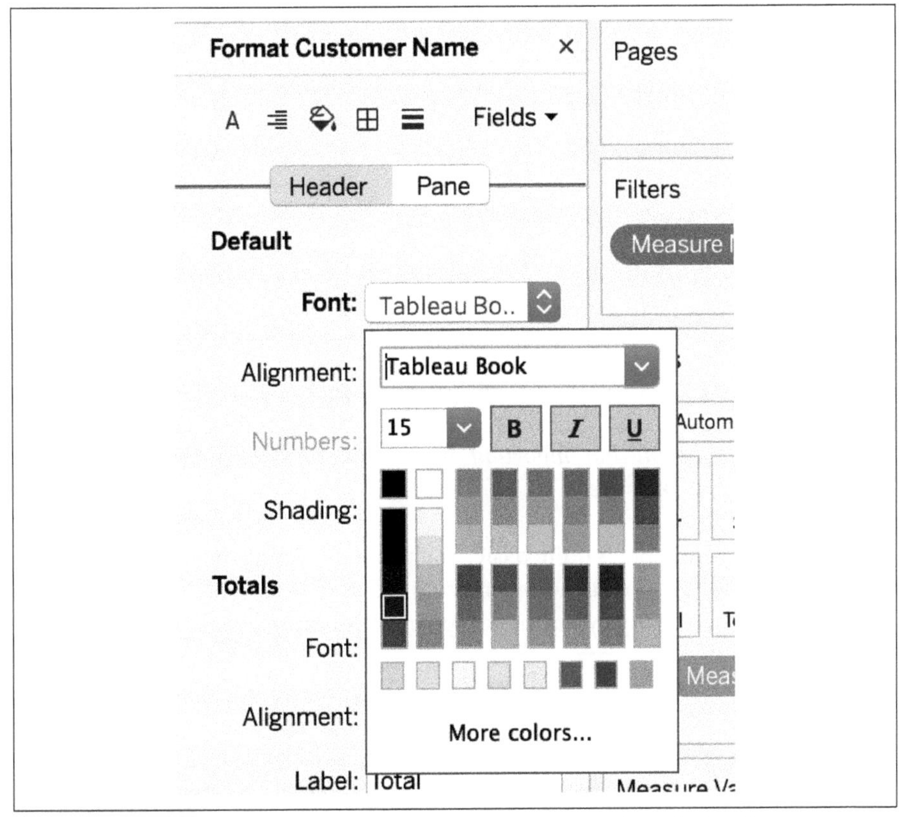

Figure 6-5. In addition to updating the text, be sure to update the header

Make sure the top header is a bold font (it should be). Finally, remove the row banding, but add row dividers for every row. Format the borders and set the row dividers to the lowest level of detail by moving the slider to the right. This will leave you with the table in Figure 6-6.

Customer Name	Sales	Profit	Total Orders	Profit Ratio
Aaron Bergman	$886	$129	3	15%
Aaron Hawkins	$1,745	$365	7	21%
Aaron Smayling	$3,051	-$254	7	-8%
Adam Bellavance	$7,756	$2,055	8	26%
Adam Hart	$3,250	$281	10	9%
Adam Shillingsburg	$3,255	$65	9	2%
Adrian Barton	$14,474	$5,445	10	38%
Adrian Hane	$1,736	-$2	7	0%
Adrian Shami	$59	$22	2	37%
Aimee Bixby	$967	$314	5	32%

Figure 6-6. The subtle but important difference in this table is that each row has the exact same font and style

3. Right now, your table is quite basic. You're simply listing data, not showcasing the power of Tableau. So let's add some color (Figure 6-7). Drag [Measure Values] to Color and then change your mark type from Automatic to Square.

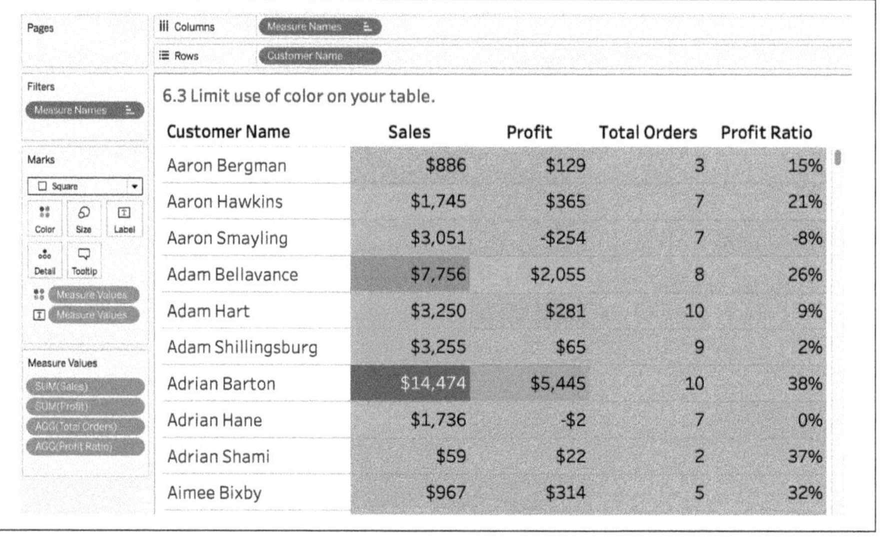

Customer Name	Sales	Profit	Total Orders	Profit Ratio
Aaron Bergman	$886	$129	3	15%
Aaron Hawkins	$1,745	$365	7	21%
Aaron Smayling	$3,051	-$254	7	-8%
Adam Bellavance	$7,756	$2,055	8	26%
Adam Hart	$3,250	$281	10	9%
Adam Shillingsburg	$3,255	$65	9	2%
Adrian Barton	$14,474	$5,445	10	38%
Adrian Hane	$1,736	-$2	7	0%
Adrian Shami	$59	$22	2	37%
Aimee Bixby	$967	$314	5	32%

Figure 6-7. Color by using [Measure Values] in a table

This is how you often see color used on tables in Tableau: one color using [Measure Values] for all measures. But you have four measures, all extremely different. The [Sales] measure consists of only positive values that go from zero to the tens of thousands. [Profit] can be negative or positive, and either can go into the thousands. [Total Orders] are represented as integers, but don't even reach 20. And [Profit Ratio] is a percentage, with the majority of values between –10% and 25%. A single-color encoding won't work for all four.

You can fix this by right-clicking [Measure Values] on Color and selecting Use Separate Legends (Figure 6-8).

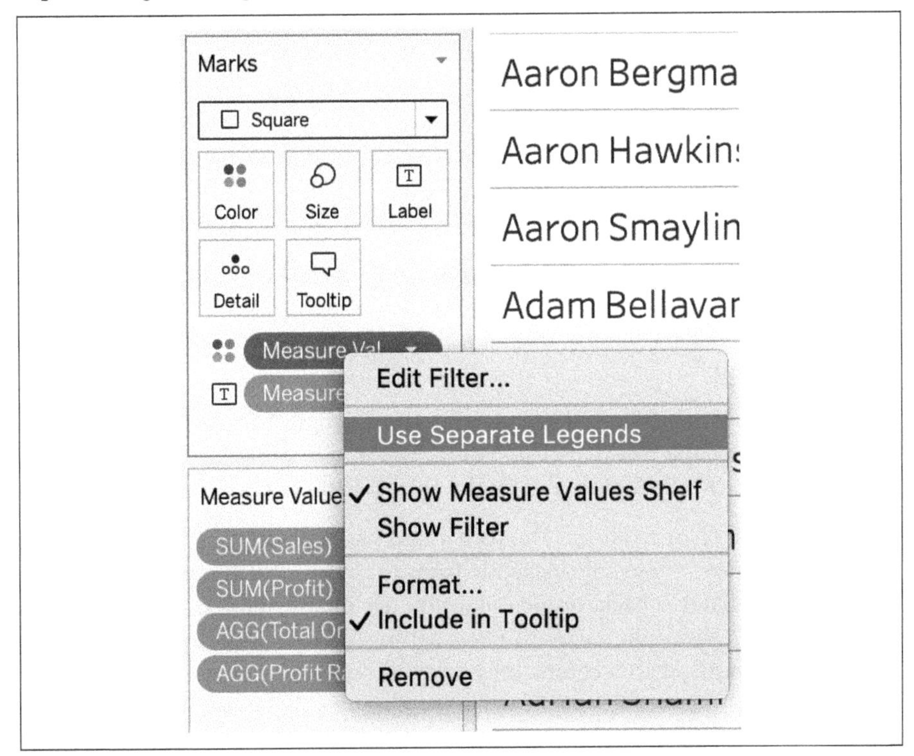

Figure 6-8. If you want different color for each measure in [Measure Names], be sure to select Use Separate Legends after right-clicking [Measure Values] on Color

This gives your table a different-color legend for each of the four measures (Figure 6-9).

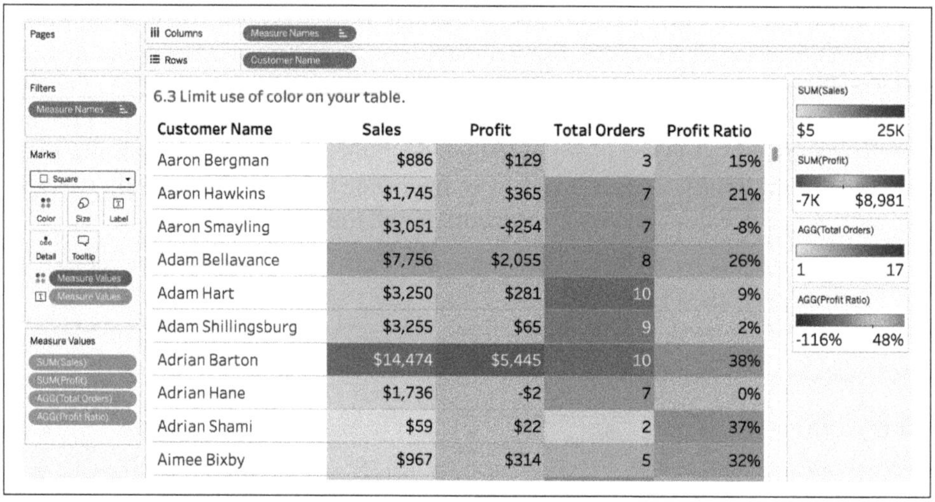

6.3 Limit use of color on your table.

Customer Name	Sales	Profit	Total Orders	Profit Ratio
Aaron Bergman	$886	$129	3	15%
Aaron Hawkins	$1,745	$365	7	21%
Aaron Smayling	$3,051	-$254	7	-8%
Adam Bellavance	$7,756	$2,055	8	26%
Adam Hart	$3,250	$281	10	9%
Adam Shillingsburg	$3,255	$65	9	2%
Adrian Barton	$14,474	$5,445	10	38%
Adrian Hane	$1,736	-$2	7	0%
Adrian Shami	$59	$22	2	37%
Aimee Bixby	$967	$314	5	32%

Figure 6-9. Default colors after selecting Use Separate Legends

This is amazing because you now have the ability to individually encode each measure. Remember, though, that limiting color is important as it enables your audience to more easily gather insights from the table. In this case, at most two measures should have color.

You'll design this table so that color exists on only [SUM(Sales)] and [AGG(Profit Ratio)]. First, edit the [SUM([Sales)] color palette and set the palette to Green-Gold. Next, edit [AGG(Profit Ratio)] and set the color palette to Gold-Purple Diverging.

Now, to remove color from [SUM(Profit)] and [AGG(Total Orders)]. Unfortunately, you cannot remove these color legends from the view. Edit the palettes so that they are white (or whatever background color you use in your tables). You can do this by creating custom diverging color palettes, setting the Stepped Color to 2 steps, and then changing both custom colors to white (#FFFFFF), as shown in Figure 6-10.

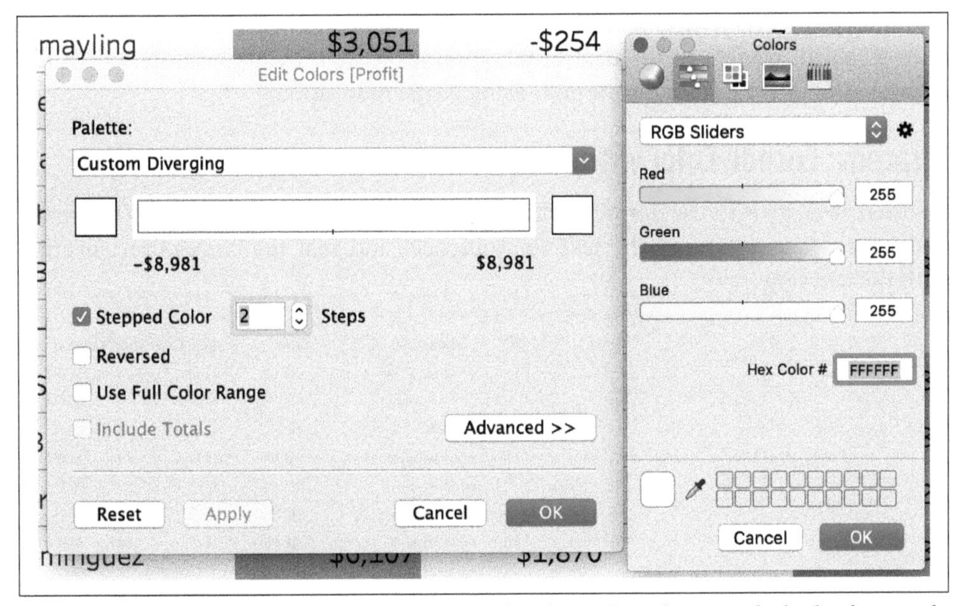

Figure 6-10. If you don't want to show a legend, select colors that match the background color of your dashboard

This leaves the table looking like Figure 6-11.

Pages

Columns: Measure Names

Rows: Customer Name

Filters: Measure Names

Marks: Square (Color, Size, Label, Detail, Tooltip; Measure Values, Measure Values)

Measure Values: SUM(Sales), SUM(Profit), AGG(Total Orders), AGG(Profit Ratio)

6.3 Limit use of color on your table.

Customer Name	Sales	Profit	Total Orders	Profit Ratio
Aaron Bergman	$886	$129	3	15%
Aaron Hawkins	$1,745	$365	7	21%
Aaron Smayling	$3,051	-$254	7	-8%
Adam Bellavance	$7,756	$2,055	8	26%
Adam Hart	$3,250	$281	10	9%
Adam Shillingsburg	$3,255	$65	9	2%
Adrian Barton	$14,474	$5,445	10	38%
Adrian Hane	$1,736	-$2	7	0%
Adrian Shami	$59	$22	2	37%
Aimee Bixby	$967	$314	5	32%

SUM(Sales): $5 – 25K

SUM(Profit): -9K – $8,981

AGG(Total Orders): 1.00 – 17.00

AGG(Profit Ratio): -116% – 48%

Figure 6-11. A table highlighting two columns with color using separate legends. Note that two other legends are set to all white.

The final product is a table that uses color in just two columns. This allows our audience to reach insights in the data faster, specifically identifying customers with the highest sales or the highest profit ratio using visual indicators.

Strategy: Encode Color with Dots

Another strategy is to use dots for color encoding. As you follow along, you'll want to make sure your dots keep the text right-aligned, and that the circles don't overlap with the text (Figure 6-12).

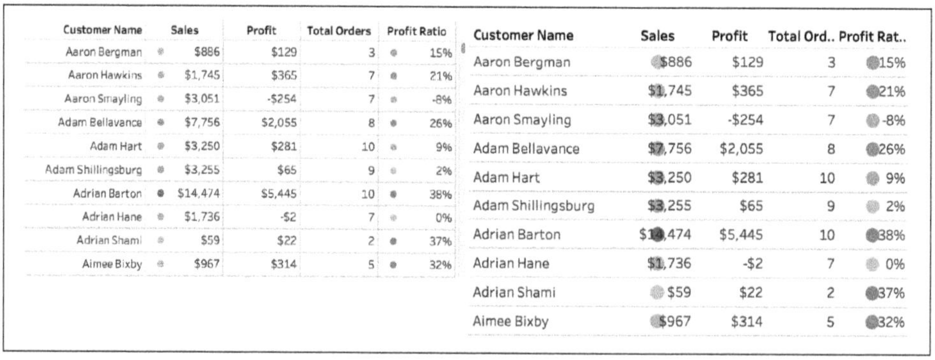

Customer Name	Sales	Profit	Total Orders	Profit Ratio
Aaron Bergman	$886	$129	3	15%
Aaron Hawkins	$1,745	$365	7	21%
Aaron Smayling	$3,051	-$254	7	-8%
Adam Bellavance	$7,756	$2,055	8	26%
Adam Hart	$3,250	$281	10	9%
Adam Shillingsburg	$3,255	$65	9	2%
Adrian Barton	$14,474	$5,445	10	38%
Adrian Hane	$1,736	-$2	7	0%
Adrian Shami	$59	$22	2	37%
Aimee Bixby	$967	$314	5	32%

Customer Name	Sales	Profit	Total Ord..	Profit Rat..
Aaron Bergman	$886	$129	3	15%
Aaron Hawkins	$1,745	$365	7	21%
Aaron Smayling	$3,051	-$254	7	-8%
Adam Bellavance	$7,756	$2,055	8	26%
Adam Hart	$3,250	$281	10	9%
Adam Shillingsburg	$3,255	$65	9	2%
Adrian Barton	$14,474	$5,445	10	38%
Adrian Hane	$1,736	-$2	7	0%
Adrian Shami	$59	$22	2	37%
Aimee Bixby	$967	$314	5	32%

Figure 6-12. When using dots to highlight values, make sure the dots don't overlap or distract from the values in the table (left), or you'll end up with a cluttered visualization (right)

If you just changed the mark type from Square to Circle, technically you would have the same chart, but you'd have text occasionally overlapping with circles, making it more difficult to read. Let's do it the correct way:

1. Change your mark type from Square to Circle.
2. To align the circles and text appropriately, you need to create custom measures that will center the values where you want in the table. Start by creating an ad hoc calculation. On the Columns shelf, double-click the whitespace to the right of [Measure Names]. This opens a prompt that will allow you to write a custom calculation. Type `MIN(0.2)`. Right now, this looks like it centers the circles and text on 0.2, but when it's all said and done, it will center only the dots on this value. This creates the table in Figure 6-13.

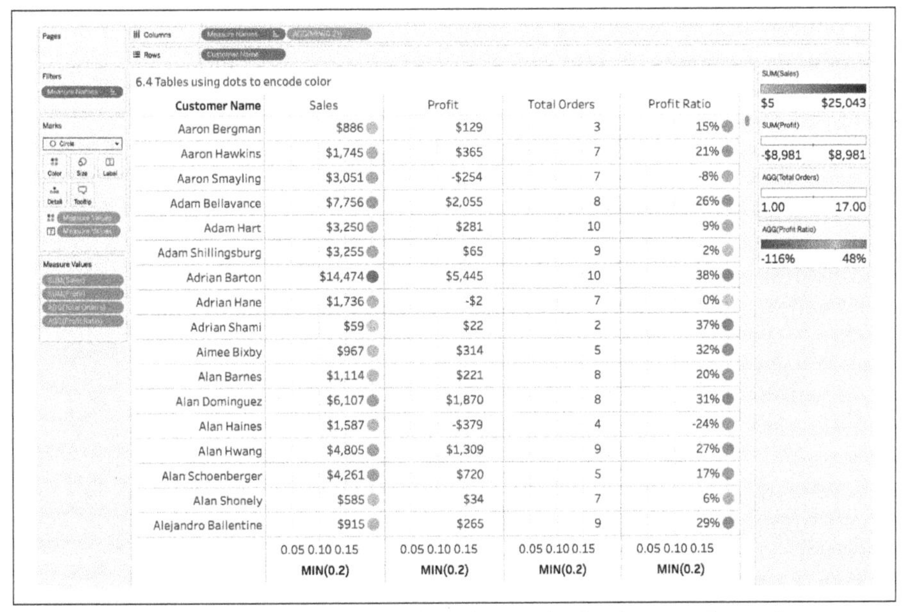

Figure 6-13. Our work in progress, adding dots and color for visual encoding

When you use `MIN(0.2)`, you're using a fixed measure to create a column in your data source. Using the minimum function ensures that the value gets placed directly at whatever value you specify. If you used `SUM()`, the value would be equal to the total records for that member of a dimension, multiplied by the value you placed in the `MIN()` function.

You will use this `MIN()` technique when you want to create several columns in your dataset or layer marks over each other.

This technique helps derive additional columns in your data. In this example, you are using `MIN(0.2)`. In almost all other use cases, however, it's better to use `MIN(0.0)` or `MIN(1.0)`; these are more natural values to use as placeholders.

If you wanted to keep the circles right-aligned to the text, you could just hide the header and remove grid lines, and that would be it. But since you are trying to put the circles to the left of the measures while also keeping the text right-aligned, there are a few more steps.

3. Double-click to the right of AGG(MIN(0.2)) and type **`MIN(1.0)`**. This doubles the number of marks on our view and creates a second Marks card. The result is Figure 6-14.

6.4 Tables using dots to encode color

Customer Name	Sales		Profit		Total Orders		Profit Ratio	
Aaron Bergman	$886	$886	$129	$129	3	3	15%	15%
Aaron Hawkins	$1,745	$1,745	$365	$365	7	7	21%	21%
Aaron Smayling	$3,051	$3,051	-$254	-$254	7	7	-8%	-8%
Adam Bellavance	$7,756	$7,756	$2,055	$2,055	8	8	26%	26%
Adam Hart	$3,250	$3,250	$281	$281	10	10	9%	9%
Adam Shillingsburg	$3,255	$3,255	$65	$65	9	9	2%	2%
Adrian Barton	$14,474	$14,474	$5,445	$5,445	10	10	38%	38%
Adrian Hane	$1,736	$1,736	-$2	-$2	7	7	0%	0%
Adrian Shami	$59	$59	$22	$22	2	2	37%	37%
Aimee Bixby	$967	$967	$314	$314	5	5	32%	32%
Alan Barnes	$1,114	$1,114	$221	$221	8	8	20%	20%

Figure 6-14. Adding another dummy data column, MIN(1.0), will double the marks, but we'll update the visualization so the marks aren't shown twice

4. Take the AGG(MIN(0.2)) and AGG(MIN(0.2)) measures and create a synchronized dual axis between the two. This will leave you with the chart in Figure 6-15.

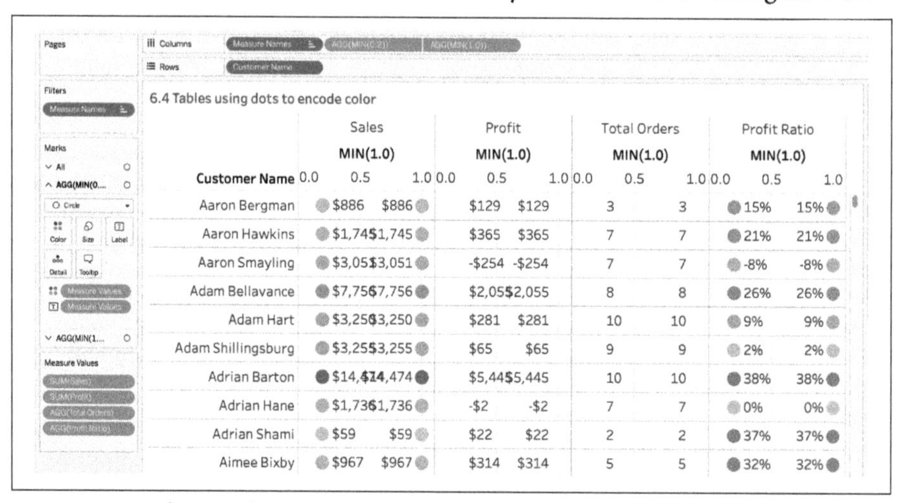

6.4 Tables using dots to encode color

Customer Name	Sales		Profit		Total Orders		Profit Ratio	
	MIN(1.0)		MIN(1.0)		MIN(1.0)		MIN(1.0)	
Aaron Bergman	$886	$886	$129	$129	3	3	15%	15%
Aaron Hawkins	$1,745	$1,745	$365	$365	7	7	21%	21%
Aaron Smayling	$3,051	$3,051	-$254	-$254	7	7	-8%	-8%
Adam Bellavance	$7,756	$7,756	$2,055	$2,055	8	8	26%	26%
Adam Hart	$3,250	$3,250	$281	$281	10	10	9%	9%
Adam Shillingsburg	$3,255	$3,255	$65	$65	9	9	2%	2%
Adrian Barton	$14,474	$14,474	$5,445	$5,445	10	10	38%	38%
Adrian Hane	$1,736	$1,736	-$2	-$2	7	7	0%	0%
Adrian Shami	$59	$59	$22	$22	2	2	37%	37%
Aimee Bixby	$967	$967	$314	$314	5	5	32%	32%

Figure 6-15. After synchronizing the axes, you'll notice a circle that's to the left of the text in one column and to the right of the text in the other

5. On the AGG(MIN(0.2)) Marks card, remove [Measure Values] from Text.

6. On the AGG(MIN(1.0)) Marks card, change the mark type to Gantt Bar, remove [Measure Values] from Color, set the color to white, and to fully ensure they won't ever overlap or conflict with the text, change the opacity to 0%. Click Label on the Marks card and change the horizontal alignment from Automatic to Left. Figure 6-16 shows the visualization at this stage.

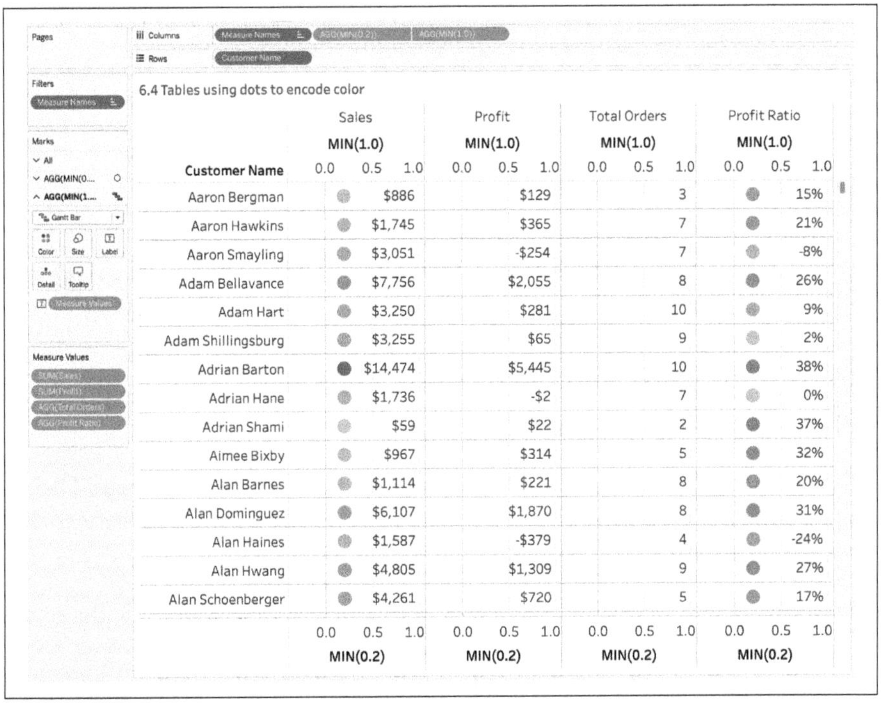

6.4 Tables using dots to encode color

Customer Name	Sales	Profit	Total Orders	Profit Ratio
Aaron Bergman	$886	$129	3	15%
Aaron Hawkins	$1,745	$365	7	21%
Aaron Smayling	$3,051	-$254	7	-8%
Adam Bellavance	$7,756	$2,055	8	26%
Adam Hart	$3,250	$281	10	9%
Adam Shillingsburg	$3,255	$65	9	2%
Adrian Barton	$14,474	$5,445	10	38%
Adrian Hane	$1,736	-$2	7	0%
Adrian Shami	$59	$22	2	37%
Aimee Bixby	$967	$314	5	32%
Alan Barnes	$1,114	$221	8	20%
Alan Dominguez	$6,107	$1,870	8	31%
Alan Haines	$1,587	-$379	4	-24%
Alan Hwang	$4,805	$1,309	9	27%
Alan Schoenberger	$4,261	$720	5	17%

Figure 6-16. After adjusting text on MIN(0.2) and MIN(1.0), the visual indicators start to take shape

7. You're almost there. To set the axis range, right-click the MIN(0.2) axis and choose Edit Axis. In the Edit Axis pop-up, set the range to be fixed from 0 to 1. Then exit the pop-up and hide both axes.

8. Right-click the chart and select Format. Under Format Lines, remove all grid lines and zero lines. This leaves you with the final chart (Figure 6-17), where the text is still right-aligned but you also have color-encoded circles.

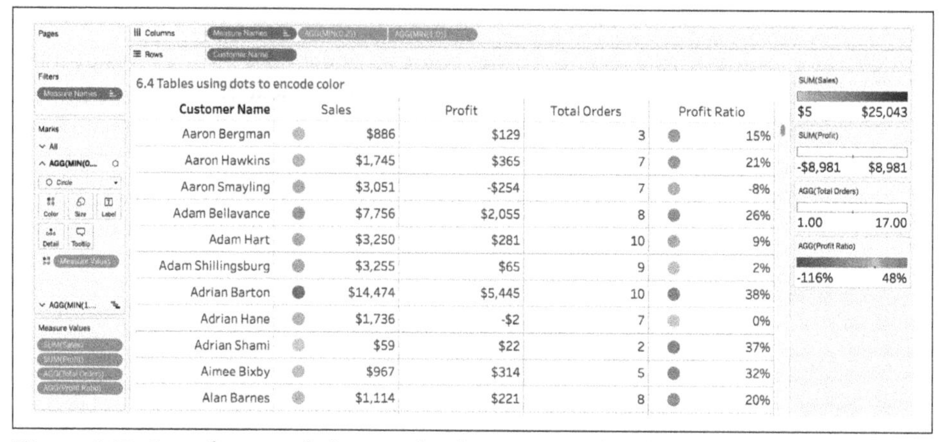

Customer Name	Sales	Profit	Total Orders	Profit Ratio
Aaron Bergman	$886	$129	3	15%
Aaron Hawkins	$1,745	$365	7	21%
Aaron Smayling	$3,051	-$254	7	-8%
Adam Bellavance	$7,756	$2,055	8	26%
Adam Hart	$3,250	$281	10	9%
Adam Shillingsburg	$3,255	$65	9	2%
Adrian Barton	$14,474	$5,445	10	38%
Adrian Hane	$1,736	-$2	7	0%
Adrian Shami	$59	$22	2	37%
Aimee Bixby	$967	$314	5	32%
Alan Barnes	$1,114	$221	8	20%

Figure 6-17. Once formatted, the visual indicators provide subtle hints to the direction of the data

Moving Beyond Measure Names and Values: Office Essentials Case Study

Our friends on the data team at Office Essentials have another challenge. Their manager, Ali, likes the table from Figure 6-17, but she thinks they can do better—and she has a wish list! Here's what she's asked for:

- A way to see the percent increase of sales year over year alongside the total sales metric
- A table like Figure 6-17, but that highlights the most and least profitable items
- A sales table that will automatically update when there is a full week of data for a particular week, and shows whether total orders have increased or decreased since the previous week

Sometimes you want to build tables, but the calculations you're trying to use require something other than measure names and measure values. All of Ali's requests involve providing more context to the team's tables. There are lots of ways to do this, such as adding dimensions to cells, or encoding your columns data with different mark types.

Take a look at Figure 6-18, which shows a "table" created in Tableau. Why the scare quotes? Because a true table in Tableau is made up of a single sheet, and this example is not.

These visualizations include, from left to right:

- A column that has additional context provided by a second metric (in this case, percentage change from month to month)
- A column that has specific thresholds for color set below $0 and above $1,000
- A third column that has color encoding changes based on whether the value is at or less than zero, or above zero. The third column also includes a different mark type.

This visual shows three mark types: Gantt, Bar, and Shape.

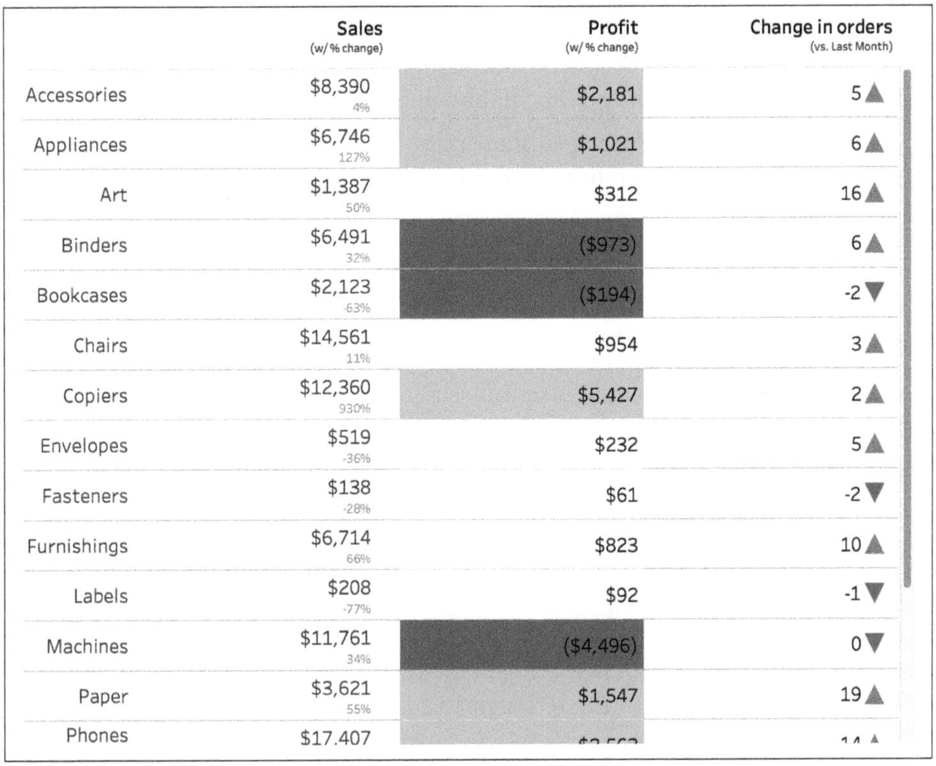

	Sales (w/ % change)	Profit (w/ % change)	Change in orders (vs. Last Month)
Accessories	$8,390 4%	$2,181	5▲
Appliances	$6,746 127%	$1,021	6▲
Art	$1,387 50%	$312	16▲
Binders	$6,491 32%	($973)	6▲
Bookcases	$2,123 -63%	($194)	-2▼
Chairs	$14,561 11%	$954	3▲
Copiers	$12,360 930%	$5,427	2▲
Envelopes	$519 -36%	$232	5▲
Fasteners	$138 -28%	$61	-2▼
Furnishings	$6,714 66%	$823	10▲
Labels	$208 -77%	$92	-1▼
Machines	$11,761 34%	($4,496)	0▼
Paper	$3,621 55%	$1,547	19▲
Phones	$17,407	[illegible]	[illegible]

Figure 6-18. Three ways to use placeholders to display values: adding context, adding color to a single column (without hacking it!), and adding a custom shape

This example highlights some fundamentals that we (the authors) would put into an individual visualization—but not all the features we'd include. In the next strategy, you should focus on how to create the components rather than the chart as a whole.

For this example, we will use `MIN(1.0)` to create placeholder columns. With this method, we create a new Marks card for each dimension we add to Rows. This allows you to mimic measure names and measure values, but gives significantly more flexibility to customize your tables based on individual metrics.

Strategy: Create a Table Body Without Measure Names or Measure Values

In this strategy, you will create the body of the table shown in Figure 6-18. This table includes three mark types:

1. Start by adding [Sub-Category] to the Rows shelf.
2. When you create a table without measure names or measure values, you need to create a column placeholder by double-clicking the Columns shelf and typing **`MIN(1.0)`**. This creates a column with bars that are 1 unit tall. Change the mark type to Gantt Bar. Set the opacity of the color to 0%.

   In this column, you're going to show sales for the most recent year. You're also going to show the difference in sales from the most recent year of sales to the prior year.
3. Write the following calculation, called **`[Sales | Current Year]`**, to show the most recent year of sales:

   ```
   // Sales | Current Year
   IF DATEPART("year", {MAX([Order Date])}) = DATEPART("year", [Order Date])
   THEN [Sales]
   END
   ```
4. Write **`[Sales | Prior Year]`** to calculate sales for the prior year:

   ```
   // Sales | Prior Year
   IF (DATEPART("year", {MAX([Order Date])}) - 1)
     = DATEPART("year", [Order Date])
   THEN [Sales]
   END
   ```

   Note that the only difference between the two calculations is the `-1` to the left of the Boolean argument in the `IF` statement. This will return sales for one year less than the maximum year.
5. Now you can calculate the difference between the two calculations—the change in sales—by creating a calculation called **`[Sales | CY vs PY]`**:

   ```
   // Sales | CY vs PY
   SUM([Sales | Current Year]) - SUM([Sales | Prior Year])
   ```

   Once you've created all three calculations, be sure to set the default formatting of the numbers to currency with no decimals.

6. Add the [Sales | Current Year] and [Sales | CY vs PY] calculations to Text on the Marks card. Format the text to be left-aligned (Figure 6-19).

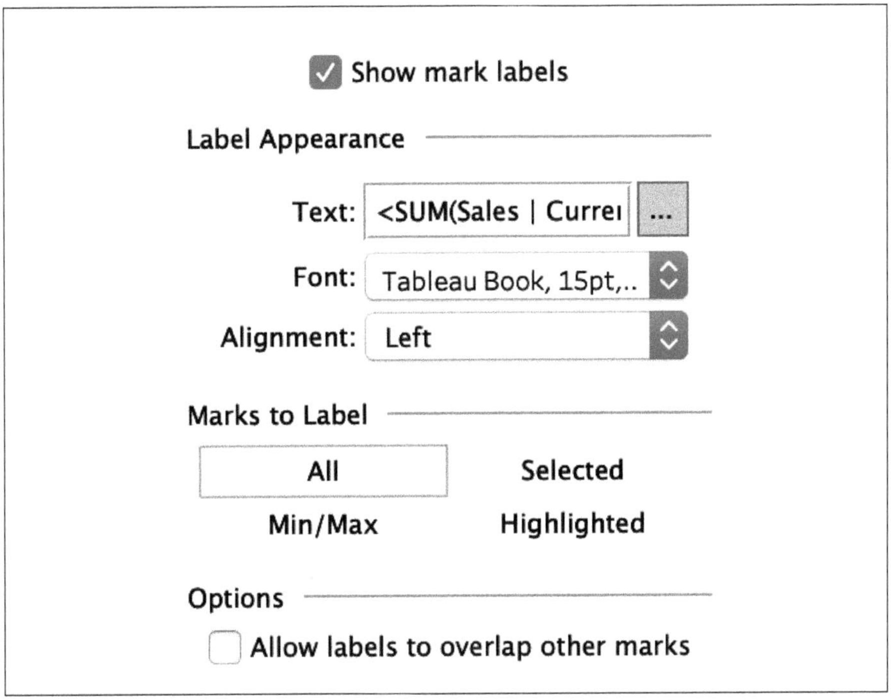

Figure 6-19. Be sure text is left-aligned on the Text button

7. Format the mark type as follows:
    a. Edit the text by clicking the ellipsis.
    b. Place [Sales | Current Year] on the first line and [Sales | CY vs PY] on the second line.
    c. Change the size of [Sales | CY vs PY] to be about 33% of the font size as [Sales | Current Year]. In Figure 6-20, the font size of [Sales | Current Year] is 15 pixels, while [Sales | CY vs PY] is 10 pixels in height.
    d. You should also set the color of the second line to be a slightly lighter tint. Set this font to the right in the text editor.

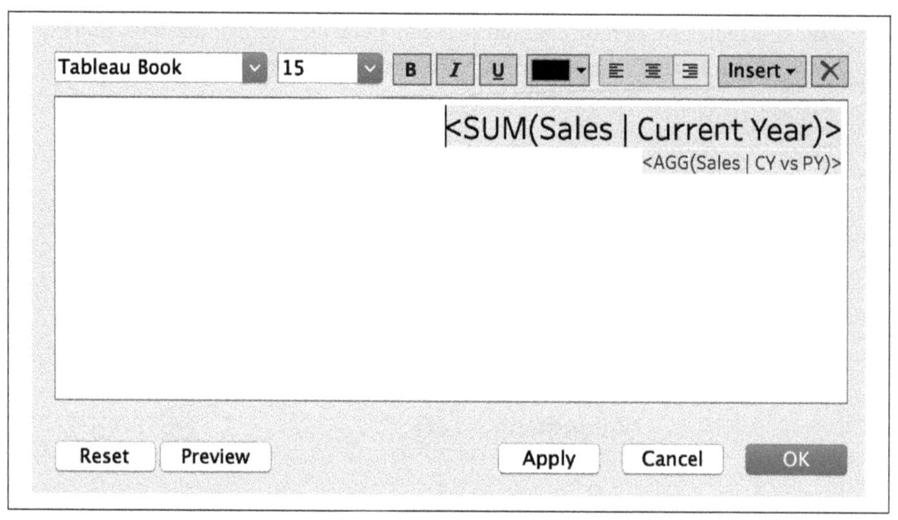

Figure 6-20. Right-align and format text in the text editor

8. Right-click and format the lines and remove grid lines, zero lines, axis rulers, and axis ticks. This will leave you with the visualization in Figure 6-21, which satisfies the first item on Ali's wish list.

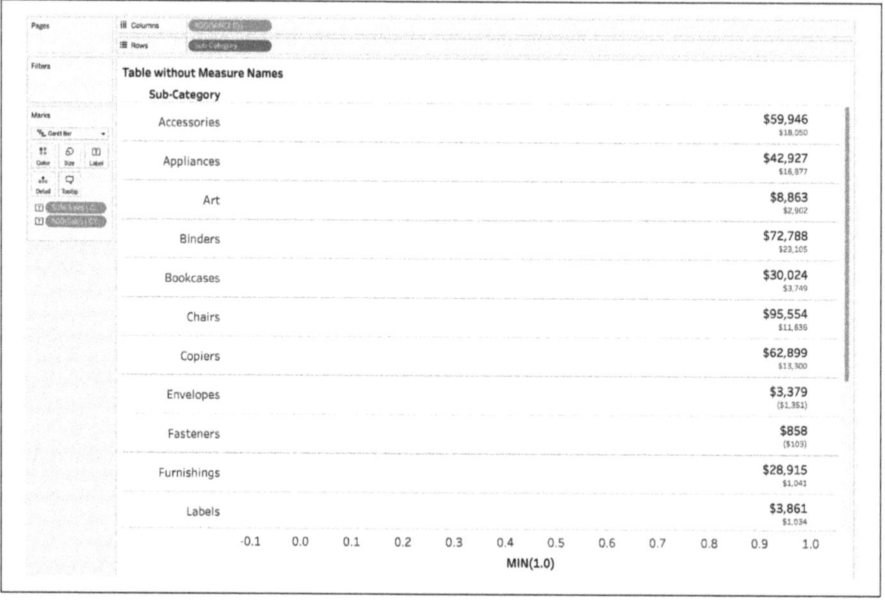

Figure 6-21. The result is a column that gives context to the data

Let's add another column. This will let you highlight the most profitable and least profitable items, the second thing on Ali's wish list.

9. Add a column that looks like a highlight table:

   a. Add another column. Double-click to the right of AGG(MIN(1.0)) and type **MIN(-1.0)**.

   b. Change the mark type for MIN(-1.0) to Bar. Edit the axis to a range from –1 to 0.

   c. Change the color back to 100% opacity. Change the size to just under the maximum value.

   d. Create a calculation of total profit for the most recent year called **[Profit | Current Year]**:

      ```
      // Profit | Current Year
      IF DATEPART("year", [Order Date]) = DATEPART("year", {MAX([Order Date])})
      THEN [Profit]
      END
      ```

   Next, format the number for this calculation, and then place the measure on Text for the AGG(MIN(-1.0)) Marks card.

10. By using this technique to create columns, you can set discrete colors with a custom calculation. For example, if profit is negative, create a category called Bad; if the value is greater than $10,000, add the members to a category called Good. All other members should be categorized as Neutral:

    ```
    // Profit | Current Year | Color
    IF SUM([Profit | Current Year]) < 0
    THEN "Bad"
    ELSEIF SUM([Profit | Current Year]) > 10000
    THEN "Good"
    ELSE "Neutral"
    END
    ```

    Add [Profit | Current Year | Color] to Color on the AGG(MIN(-1.0)) Marks card. Edit the colors. Set the colors so Bad is red (#D81159), Good is green (#75CF4A), and Neutral is white (#FFFFFF). Set the opacity to 70%.

    Your table should now look like Figure 6-22.

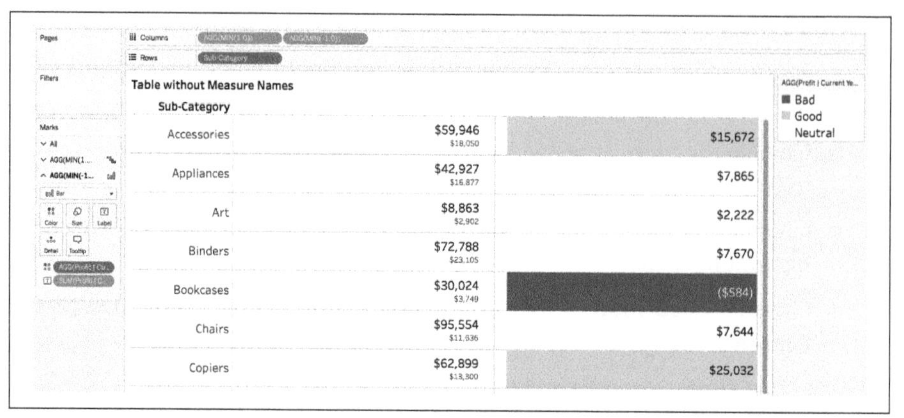

Figure 6-22. You can use custom color in a bar chart to make a column look like a table

11. Add a column that will show changes in total orders. You'll show the difference in total orders from the most recent year compared to the year before that:

    a. To the right of AGG(MIN(-1.0)), type **`MIN(1.0)`**. This creates another Marks card.

    b. Change the mark type to Shape and the opacity of the color to 100%.

    c. Create calculations for total orders this year:

    ```
    // Total Orders | Current Year
    COUNTD(
      IF DATEPART("year", {MAX([Order Date])})
        = DATEPART("year", [Order Date])
      THEN [Order ID]
      END
    )
    ```

    And create **`[Total Orders | Prior Year]`** to calculate orders for the prior year:

    ```
    // Total Orders | Prior Year
    COUNTD(
      IF (DATEPART("year", {MAX([Order Date])}) - 1) = DATEPART("year", [Order Date])
      THEN [Order ID]
      END
    )
    ```

    Then calculate the difference between the two totals, calling the metric **`[Total Orders | CY vs PY]`**:

    ```
    // Total Orders | CY vs PY
    [Total Orders | Current Year] - [Total Orders | Prior Year]
    ```

Add [Total Orders | CY vs PY] to Text on the new Marks card. Left-align the text.

d. Now create a calculation for shape and color:

```
// Total Orders | CY vs PY > 0
[Total Orders | CY vs PY] > 0
```

Add this calculation to Shape and Color. Set True on Shape to a filled up arrow, and False to a filled down arrow. Set True on Color to the same green as in step 10, and False on Color to the same as the Bad category (red color) before.

Hide the headers and remove the column dividers. Right-click [Sub-Category] on the header and select Hide Field Labels for Rows. The result is shown in Figure 6-23.

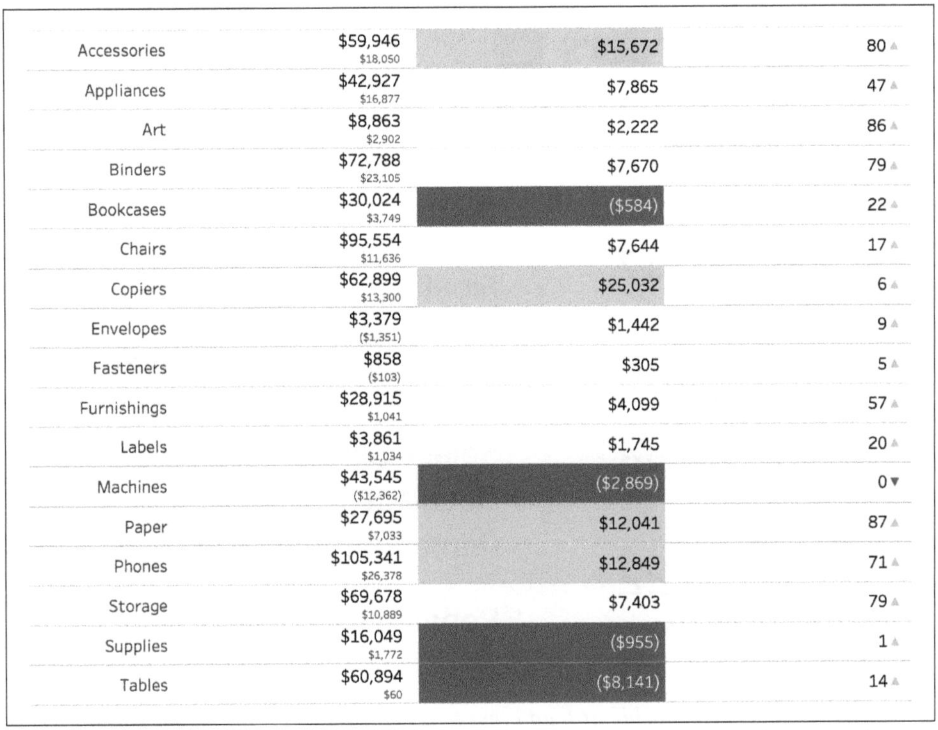

Accessories	$59,946	$18,050	$15,672	80 ▲
Appliances	$42,927	$16,877	$7,865	47 ▲
Art	$8,863	$2,902	$2,222	86 ▲
Binders	$72,788	$23,105	$7,670	79 ▲
Bookcases	$30,024	$3,749	($584)	22 ▲
Chairs	$95,554	$11,636	$7,644	17 ▲
Copiers	$62,899	$13,300	$25,032	6 ▲
Envelopes	$3,379	($1,351)	$1,442	9 ▲
Fasteners	$858	($103)	$305	5 ▲
Furnishings	$28,915	$1,041	$4,099	57 ▲
Labels	$3,861	$1,034	$1,745	20 ▲
Machines	$43,545	($12,362)	($2,869)	0 ▼
Paper	$27,695	$7,033	$12,041	87 ▲
Phones	$105,341	$26,378	$12,849	71 ▲
Storage	$69,678	$10,889	$7,403	79 ▲
Supplies	$16,049	$1,772	($955)	1 ▲
Tables	$60,894	$60	($8,141)	14 ▲

Figure 6-23. A third column displaying a custom shape alongside two columns with completely different designs

One downfall is that this technique builds a table with no headers, meaning you won't be able to rely on Tableau's built-in column sorting. We can create headers, either with a text box or with a sheet. If you don't need dynamic sorting, you can use text boxes; otherwise, you can use sheets or parameters to create sorting options.

Strategy: Create Table Headers Without Measure Names or Measure Values

In addition to the table body, you can combine the visualization with a table header. Let's try that here:

1. On the dashboard, add a vertical container. Add the table we just created into that container. Click and drag a horizontal container into the vertical container, but place it above the table.
2. Inside the horizontal container, add three text boxes, with the text **`Sales`**, **`Profit`**, and **`Total Orders`**, respectively. Select the horizontal container, choose Edit Height (Figure 6-24), and set the height to 50 pixels.

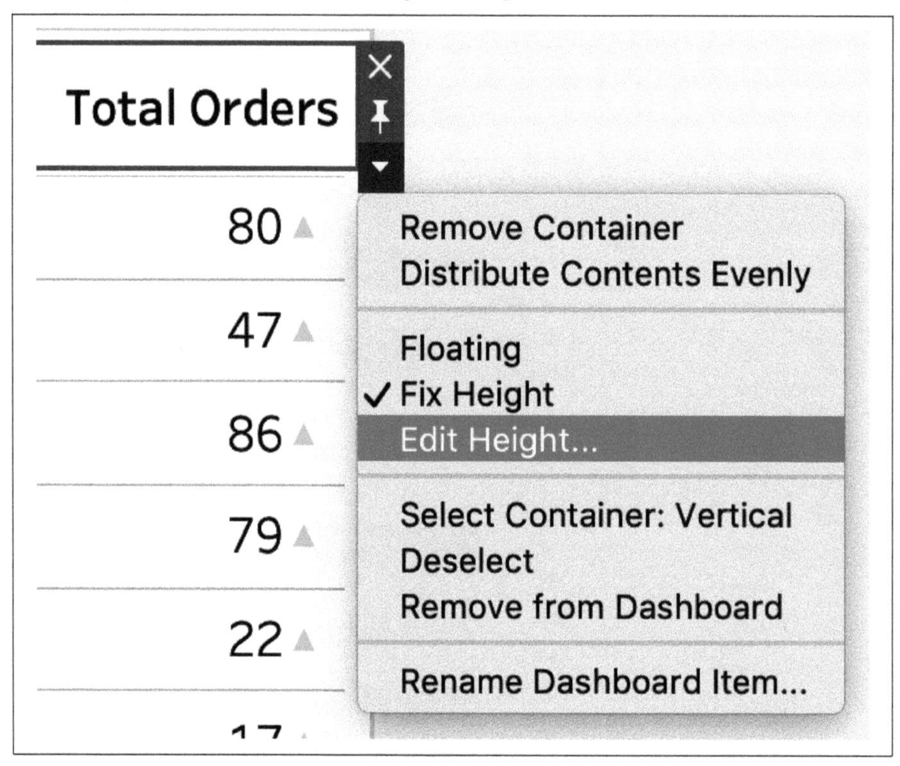

Figure 6-24. On the dashboard, add text to a horizontal container and edit the height of the container

3. Adjust the left padding of the horizontal container so that the text boxes align with the columns. Tableau's guides automatically show up inside the container, making this easier (Figure 6-25).

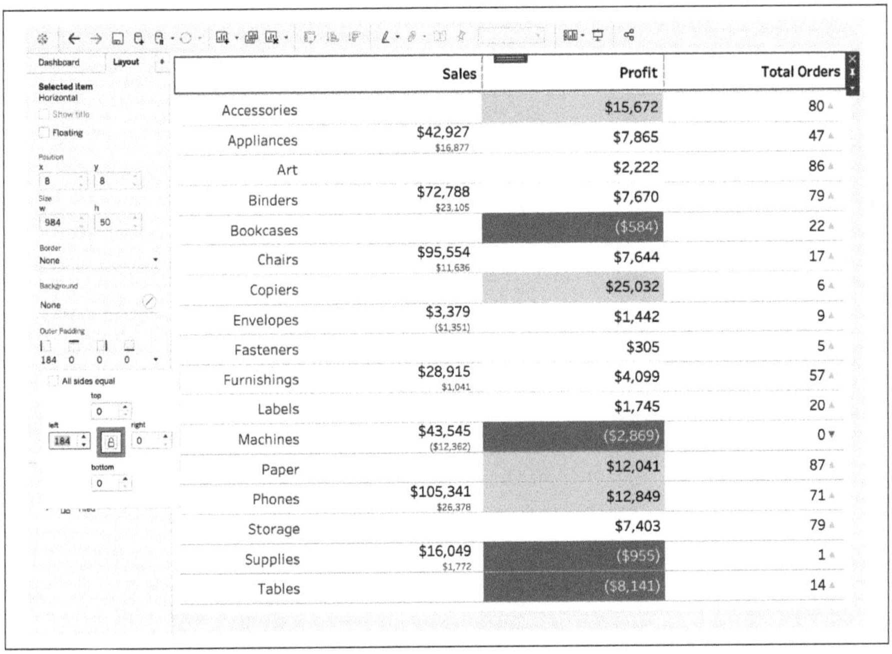

	Sales	Profit	Total Orders
Accessories		$15,672	80
Appliances	$42,927 $16,877	$7,865	47
Art		$2,222	86
Binders	$72,788 $23,105	$7,670	79
Bookcases		($584)	22
Chairs	$95,554 $11,636	$7,644	17
Copiers		$25,032	6
Envelopes	$3,379 ($1,351)	$1,442	9
Fasteners		$305	5
Furnishings	$28,915 $1,041	$4,099	57
Labels		$1,745	20
Machines	$43,545 ($12,362)	($2,869)	0
Paper		$12,041	87
Phones	$105,341 $26,378	$12,849	71
Storage		$7,403	79
Supplies	$16,049 $1,772	($955)	1
Tables		($8,141)	14

Figure 6-25. If you properly format the sheet and the text boxes, the result will be a visualization resembling a table

4. Finally, adjust the right padding from 4 to 14 for each of the text boxes. This will align the right side of the header text with the text in the table.

	Sales	Profit	Total Orders
Accessories	$59,946 $18,050	$15,672	80▲
Appliances	$42,927 $16,877	$7,865	47▲
Art	$8,863 $2,902	$2,222	86▲
Binders	$72,788 $23,105	$7,670	79▲
Bookcases	$30,024 $3,749	($584)	22▲
Chairs	$95,554 $11,636	$7,644	17▲
Copiers	$62,899 $13,300	$25,032	6▲
Envelopes	$3,379 ($1,351)	$1,442	9▲
Fasteners	$858 ($103)	$305	5▲
Furnishings	$28,915 $1,041	$4,099	57▲
Labels	$3,861 $1,034	$1,745	20▲
Machines	$43,545 ($12,362)	($2,869)	0▼
Paper	$27,695 $7,033	$12,041	87▲
Phones	$105,341 $26,378	$12,849	71▲
Storage	$69,678 $10,889	$7,403	79▲
Supplies	$16,049 $1,772	($955)	1▲
Tables	$60,894 $60	($8,141)	14▲

Figure 6-26. The final version of this "table"

The final product, shown in Figure 6-26, is a "table" that does not use measure names or measure values and that allows you to customize each column with a different mark type, multiple measures on text, and color. This is Luke's preferred method for building tables; Ann, less so. However, we both rely on this technique when required.

Strategy: Use Measures as Headers

Sometimes instead of placing an aggregate measure on Text within the Marks card, you want to include the value in the header as a dimension (even though it's an aggregate value). This strategy is perfect when you want to share an overall value for a set of dimensions or use data-driven indicators to call out a set of values.

Our next strategy addresses the third item on Ali's wish list: a table that will automatically update when there is a full week of data for a particular week (Figure 6-27). In it, you'll create a series of measures (both as LOD calculations and as table calculations) that you'll use as headers.

You will continue to use the Superstore – Sample dataset to write two LOD calculations: one that calculates the overall sales per week, and a second that produces an arrow to highlight the week with the highest level of overall sales. You will also create a calculation that will rank the last 13 weeks of sales.

You'll then split sales by category, show the rank of sales with each category by week, and highlight the top and bottom weeks.

			Furniture	Office Supplies	Technology
September 29, 2019	(#9)	$20,090	(#9) $5,083	(#9) $6,049	(#4) $8,959
October 6, 2019	(#12)	$12,229	(#8) $5,346	(#13) $2,358	(#10) $4,524
October 13, 2019	(#10)	$15,940	(#10) $5,011	(#1) $9,643	(#12) $1,286
October 20, 2019	(#4)	$25,911	(#11) $4,872	(#12) $3,561	(#1) $17,479
October 27, 2019	(#11)	$15,307	(#6) $5,951	(#11) $4,902	(#11) $4,454
▶ November 3, 2019	(#1)	$29,398	(#4) $7,906	(#10) $5,700	(#3) $15,792
November 10, 2019	(#7)	$21,517	(#7) $5,793	(#3) $7,898	(#6) $7,826
November 17, 2019	(#3)	$28,975	(#12) $4,438	(#2) $8,732	(#2) $15,805
November 24, 2019	(#2)	$29,342	(#1) $15,095	(#7) $6,860	(#7) $7,387
December 1, 2019	(#6)	$23,706	(#3) $8,058	(#6) $7,771	(#5) $7,877
December 8, 2019	(#5)	$24,290	(#2) $11,188	(#8) $6,365	(#8) $6,736
December 15, 2019	(#13)	$11,254	(#13) $2,362	(#5) $7,853	(#13) $1,039
December 22, 2019	(#8)	$20,950	(#5) $7,146	(#4) $7,865	(#9) $5,939

Figure 6-27. Sometimes you need to use measures as row headers—as in this example, where rank and total sales are measures-turned-dimensions

Here are the steps:

1. Create a new calculation called **`[Full Weeks | TF]`** that will determine whether the most recent week is complete. This calculation creates a Boolean that checks order dates to see whether they are before the start of the week for the maximum date plus one day:

   ```
   // Full Weeks | TF
   [Order Date] < DATETRUNC("week", {MAX([Order Date])} + 1)
   ```

   Add this new calculation to the Filters shelf and select True. Once the filter has been applied, right-click [Full Weeks | TF] on the Filters shelf and turn it into a context filter.

2. Since Ali wants this visualization to show only the most recent 13 weeks, let's create another filter to do that.

   Create a new calculation called **`[Last 13 Weeks]`**. It will look similar to the previous calculation, except this time you'll change the sign from less-than to greater-than-or-equal-to. You'll also nest the righthand side of the argument inside the `DATEADD()` function so you can go back:

   ```
   // Last 13 Week
   [Order Date] >= DATEADD("week", -13, DATETRUNC("week",
     {MAX([Order Date])}))
   ```

   Take this calculation and place it on Filters. Select True. Then make the filter into a context filter.

   Now that your filters are complete, you can start building the visualization.

3. Go to [Order Date] on the Data pane. Right-click and choose Create → Custom Date. In the Create Custom Date dialog box, set the Name to **`Order Date | Week`**. From the Detail list, select "Week numbers." Select Date Value and then click OK.

   Once the field is created, right-click it on the Data pane and select Convert to Discrete. This will make Tableau treat the calculation as a discrete value when it's on the visualization (which will save steps in the future).

   Now simply duplicate the [Order Date | Week] field. You can rename it **`[Order Date | Week 2]`**.

Why do you need two identical fields? Because you are going to hide one of the dimensions in the header. If you have the same calculation twice, on Rows or Columns, it will hide all the headers where the field is the same! Try it: add [Order Date | Week 2] twice to Columns and hide one of the headers —they both disappear.

Now you are going to add to the existing visualization. If you've tried it out, make sure you reselect Show Headers. You'll also want to remove both [Order Date | Week 2] fields from your view.

Take [Order Date | Week] and add the dimension to the Rows shelf. Do the same thing for [Order Date | Week 2], but place it to the left of [Order Date | Week] on Rows. You should have two identical-looking column headers (shown in Figure 6-28).

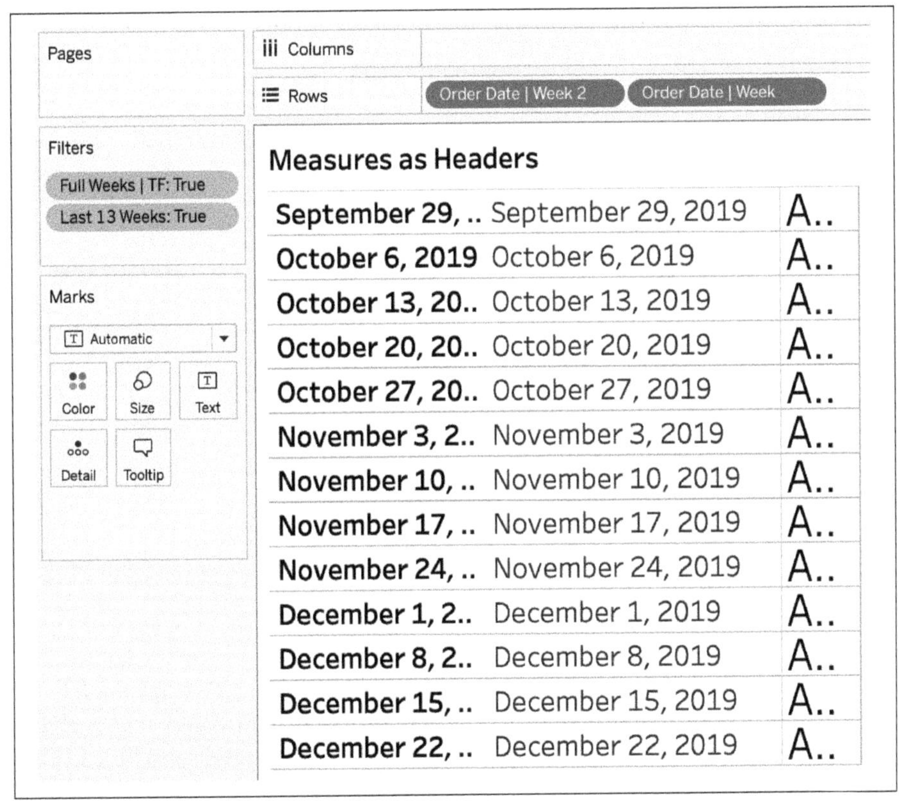

Figure 6-28. The duplicate fields, both as visible headers

4. From here, you can add dimensions and measures to your view. Add SUM(Sales) to Text on the Marks card, and [Category] to the Columns shelf. Make sure your text is right-aligned for the Sales values.

5. You are going to build three calculations, as shown in Figure 6-29. The first is the total sales for a particular week. The second is an arrow that will point to the week with the highest overall sales. The third ranks overall sales by week.

		Furniture	Office Supplies	Technology
September 29, ..	September 29, 2019	$5,083	$6,049	$8,959
October 6, 2019	October 6, 2019	$5,346	$2,358	$4,524
October 13, 20..	October 13, 2019	$5,011	$9,643	$1,286
October 20, 20..	October 20, 2019	$4,872	$3,561	$17,479
October 27, 20..	October 27, 2019	$5,951	$4,902	$4,454
November 3, 2..	November 3, 2019	$7,906	$5,700	$15,792
November 10, ..	November 10, 2019	$5,793	$7,898	$7,826
November 17, ..	November 17, 2019	$4,438	$8,732	$15,805
November 24, ..	November 24, 2019	$15,095	$6,860	$7,387
December 1, 2..	December 1, 2019	$8,058	$7,771	$7,877
December 8, 2..	December 8, 2019	$11,188	$6,365	$6,736
December 15, ..	December 15, 2019	$2,362	$7,853	$1,039
December 22, ..	December 22, 2019	$7,146	$7,865	$5,939

Figure 6-29. Building the base table

6. Create a calculation called **[Total Sales]** by using an LOD. Set the fixed value by your [Order Date | Week] calculation and place SUM(Sales) on the righthand side of the argument. This tells Tableau to calculate the total sales for each week:

```
// Total Sales
{FIXED [Order Date | Week] : SUM([Sales])}
```

7. Create a calculated field called **[Best Week]** by using the [Total Sales] calculation you just wrote. This will calculate the week with the best overall sales:

```
// Best Week
IF [Total Sales] = {MAX([Total Sales])}
THEN "►"
ELSE ""
END
```

This calculation nests an LOD calculation from [Total Sales] inside another LOD calculation to get the top overall value, aka the maximum. If a week is equal to that top value, you want it to return a right-pointing arrow by using an ASCII code.

ASCII encoding uses numeric codes to stand in for characters like English-language letters, numbers, and punctuation marks. (For example, the ASCII code for an uppercase *S* is 83.) Other encoding systems, like Unicode, offer more options. You typically won't type out ASCII codes by hand; you can copy and paste them from a site like Alt-Codes (*https://www.alt-codes.net*).

8. For the rank label for each week, we won't use the rank functions (surprise!). Instead, we'll use `INDEX()`. Create a calculation called **`[Rank Label]`** and write the following:

   ```
   // Rank Label
   "(#" + STR(INDEX()) + ")"
   ```

   This will return an index value plus formatting, so that any value is inside parentheses.

Have you noticed we avoided using the rank functions? We won't overload you with the technical details here, but it's important to know that sometimes the ordering of the rank values renders differently on Tableau Desktop versus Tableau Server. To avoid this issue, we stick to the `INDEX()` function. This more-direct function simply adds a numerical index value to rows in your partition.

9. Now that you've written the calculations, you can place them on your view and convert them into headers:

   a. Start with [Total Sales]. Click and drag this value from the Data pane out onto the Rows shelf. After the value shows up on the Rows shelf, right-click and select Discrete. Right-click again and choose Format on the Format window. Change the formatting of the value to a currency with no decimals. (Make sure you are formatting the header and not the pane.) Your current view should look like Figure 6-30.

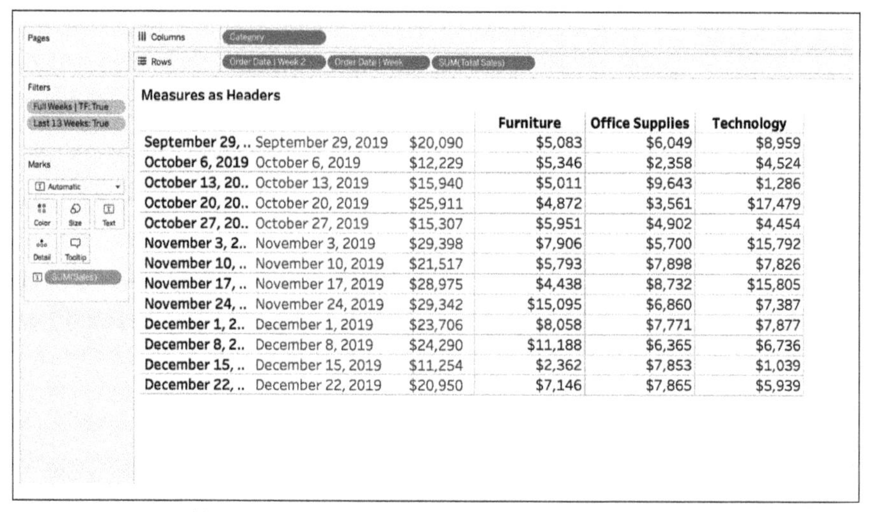
Measures as Headers

			Furniture	Office Supplies	Technology
September 29, ..	September 29, 2019	$20,090	$5,083	$6,049	$8,959
October 6, 2019	October 6, 2019	$12,229	$5,346	$2,358	$4,524
October 13, 20..	October 13, 2019	$15,940	$5,011	$9,643	$1,286
October 20, 20..	October 20, 2019	$25,911	$4,872	$3,561	$17,479
October 27, 20..	October 27, 2019	$15,307	$5,951	$4,902	$4,454
November 3, 2..	November 3, 2019	$29,398	$7,906	$5,700	$15,792
November 10, ..	November 10, 2019	$21,517	$5,793	$7,898	$7,826
November 17, ..	November 17, 2019	$28,975	$4,438	$8,732	$15,805
November 24, ..	November 24, 2019	$29,342	$15,095	$6,860	$7,387
December 1, 2..	December 1, 2019	$23,706	$8,058	$7,771	$7,877
December 8, 2..	December 8, 2019	$24,290	$11,188	$6,365	$6,736
December 15, ..	December 15, 2019	$11,254	$2,362	$7,853	$1,039
December 22, ..	December 22, 2019	$20,950	$7,146	$7,865	$5,939

Figure 6-30. A table, now including total sales as a measure-turned-dimension

b. Click and drag [Best Week] onto Rows between [Order Date | Week 2] and [Order Date | Week]. You should see a single right-arrow on your view.

c. Repeat for [Rank Label], placing the value between [Order Date | Week] and [Total Sales]. A warning: the visual is probably not going to look the way you think it should. This is because it's a table calculation, and you'll need to edit your aggregation.

d. Right-click to edit the table calculation for [Rank Label] on your Rows shelf. Select Specific Dimensions and select all values *except* [Category]. (That's because we want to complete this calculation across each category separately.)

   Then edit the sort order within the table calculation window by using a custom sort. Set the sort to be descending by the sum of sales. Your visualization should look like Figure 6-31.

Measures as Headers

				Furniture	Office Supplies	Technology
September 29, ..	September 29, 2019	(#9)	$20,090	$5,083	$6,049	$8,959
October 6, 2019	October 6, 2019	(#12)	$12,229	$5,346	$2,358	$4,524
October 13, 20..	October 13, 2019	(#10)	$15,940	$5,011	$9,643	$1,286
October 20, 20..	October 20, 2019	(#4)	$25,911	$4,872	$3,561	$17,479
October 27, 20..	October 27, 2019	(#11)	$15,307	$5,951	$4,902	$4,454
November 3, 2.. ▶	November 3, 2019	(#1)	$29,398	$7,906	$5,700	$15,792
November 10, ..	November 10, 2019	(#7)	$21,517	$5,793	$7,898	$7,826
November 17, ..	November 17, 2019	(#3)	$28,975	$4,438	$8,732	$15,805
November 24, ..	November 24, 2019	(#2)	$29,342	$15,095	$6,860	$7,387
December 1, 2..	December 1, 2019	(#6)	$23,706	$8,058	$7,771	$7,877
December 8, 2..	December 8, 2019	(#5)	$24,290	$11,188	$6,365	$6,736
December 15, ..	December 15, 2019	(#13)	$11,254	$2,362	$7,853	$1,039
December 22, ..	December 22, 2019	(#8)	$20,950	$7,146	$7,865	$5,939

Figure 6-31. Adding rank as a measure-turned-dimension

10. Now you can just right-click [Order Date | Week 2] and uncheck Show Header. That will leave you with the majority of the visualization complete (Figure 6-32).

				Furniture	Office Supplies	Technology
	September 29, 2019	(#9)	$20,090	$5,083	$6,049	$8,959
	October 6, 2019	(#12)	$12,229	$5,346	$2,358	$4,524
	October 13, 2019	(#10)	$15,940	$5,011	$9,643	$1,286
	October 20, 2019	(#4)	$25,911	$4,872	$3,561	$17,479
	October 27, 2019	(#11)	$15,307	$5,951	$4,902	$4,454
▶	November 3, 2019	(#1)	$29,398	$7,906	$5,700	$15,792
	November 10, 2019	(#7)	$21,517	$5,793	$7,898	$7,826
	November 17, 2019	(#3)	$28,975	$4,438	$8,732	$15,805
	November 24, 2019	(#2)	$29,342	$15,095	$6,860	$7,387
	December 1, 2019	(#6)	$23,706	$8,058	$7,771	$7,877
	December 8, 2019	(#5)	$24,290	$11,188	$6,365	$6,736
	December 15, 2019	(#13)	$11,254	$2,362	$7,853	$1,039
	December 22, 2019	(#8)	$20,950	$7,146	$7,865	$5,939

Figure 6-32. Adding the arrow to indicate the top-ranked week

11. This visualization is fine, but it's hard for Ali and her team to pick out the rankings visually. It would be easier if the table had some color. Let's try it.

    a. Double-click Columns and type **`MIN(-1.0)`**. Set the mark type to Bar, and then set the axis for the measure to go from –1 to 0. Set the size of the mark to the maximum amount. Click the Label property on the Marks card and change the alignment to left-aligned. (The label will actually go to the right side, since the origin of the bars is a negative value.) Hide the axis.

    b. To make the ranks stand out, you can use color to highlight the top value, the next two top values, the bottom value, and the next two bottom values. You can do that by using the `INDEX()` function and sorting on the view. Create the following calculation called **`[Color]`**:

    ```
    // Color
    IF INDEX() =  1
    THEN "TOP"
    ELSEIF INDEX() <= 3
    THEN "TOP 3"
    ELSEIF INDEX() = SIZE()
    THEN "BOTTOM"
    ELSEIF INDEX() >= SIZE() -2
    THEN "BOTTOM 3"
    ELSE "MIDDLE"
    END
    ```

    The output of the calculation is a label for each type of member. As you've seen in this chapter, you can use `INDEX()` to create a more reliable rank calculation in Tableau.

This calculation pairs `INDEX()` with `SIZE()`. The `SIZE()` function is great for automatically determining the number of rows in the view (your table)—in this case, the number of weeks.

c. After you've created this calculation, add the Color value to the Color property on the Marks card. When you add the value, you'll also need to right-click and edit the table calculation. Click and drag [Category] to the top of the list of the four dimensions. Select Specific Dimensions. In the "At the level" list, make sure Deepest is selected, and from the "Restarting every" list, select Category (see Figure 6-33).

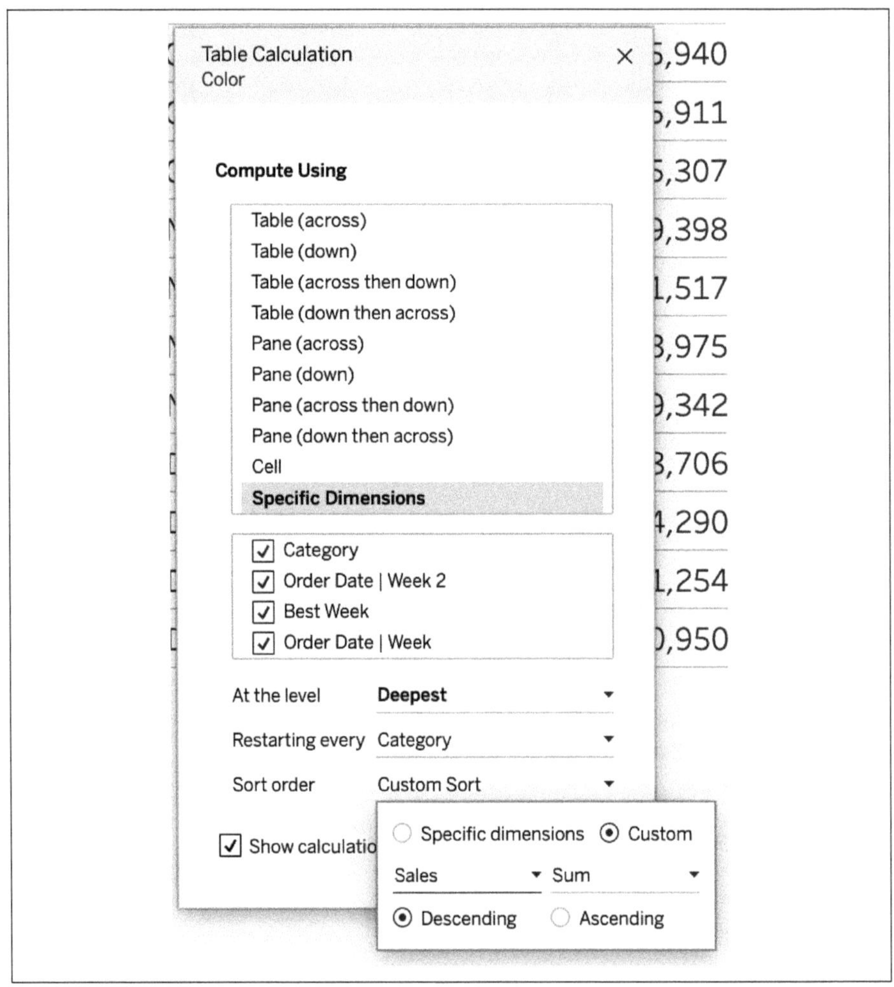

Figure 6-33. Editing the table calculation to ensure color is properly calculated

This will change the colors so that for each Category, there is one week in the TOP, one week in the BOTTOM, two more weeks in the TOP 3, and two more weeks in the BOTTOM 3. All others will be encoded to MIDDLE.

12. Now that you've adjusted the table calculation, you can assign the colors. Table 6-1 provides hexadecimal codes for the exact colors we've used in this strategy.

Table 6-1. Colors and labels

Label	Color
TOP	#989CA3
TOP3	#D3D3D3
MIDDLE	#FFFFFF
BOTTOM 3	#FCD8E5
BOTTOM	#F690B5

This will make your table look like the one in Figure 6-34.

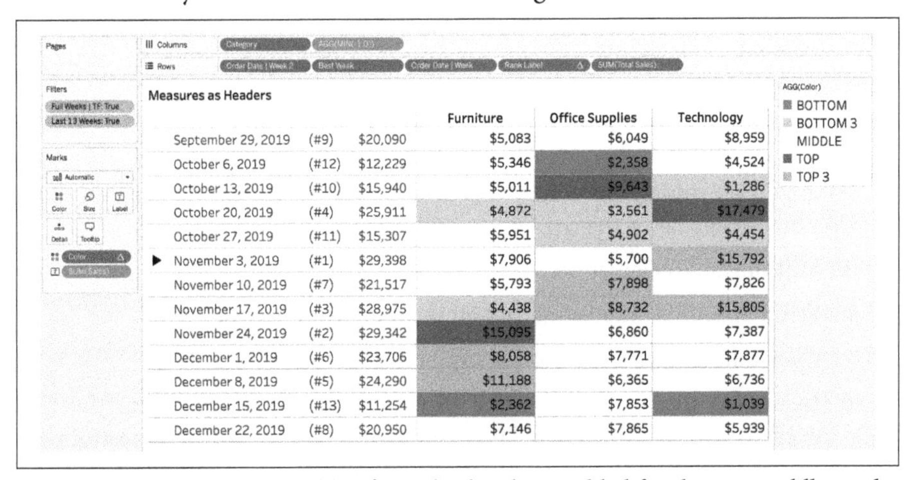

Figure 6-34. A view of the table after color has been added for the top, middle, and bottom values

13. Create your rank by using the `INDEX()` function and an ad hoc calculation, as follows:

    a. Double-click the whitespace underneath SUM(Sales) on the Marks card. This should create an ad hoc calculation.

    b. Type **`INDEX()`** and then press Enter. You should have a second calculation on the Marks card now: INDEX().

c. Click the little icon to the left of INDEX()—it's a mini detail icon—and change the type to Text.

d. Depending on the size of the spacing on your view, you may see the symbols #####, indicating you don't have enough space for the text. To fix this, click and edit the text on the view.

e. Format your text so the INDEX() calculation is a little smaller than the SUM(Sales); see Figure 6-35.

f. Also set the color to be a slightly lighter tint. We used the second-to-bottom color in the black column on the color editor.

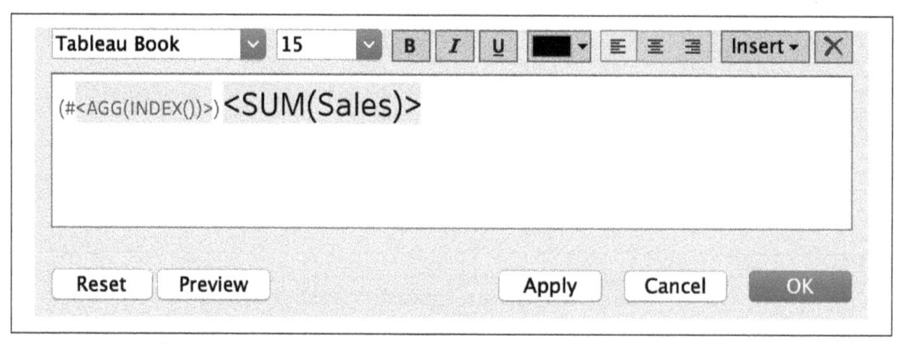

Figure 6-35. The text editor for the tooltips

14. After you set your formatting, you'll notice your "rankings" via the `INDEX()` function do not appear in the correct order. Fix them as follows:

    a. Edit the table calculation.

    b. Select Specific Dimensions and all the dimensions listed.

    c. Click and drag Category to the top of the list of dimensions.

    d. In the "At the level" list, make sure Deepest is selected. From the "Restart every" list, select Category. Then complete a custom sort in descending order, on the sum of sales.

    Your result should look like Figure 6-26.

Problem-Solving with Subsets and Hiding the Dreaded Abc: Office Essentials Case Study

OE has *lots* of products. In fact, within one sub-category, there might be hundreds of products. Knowing the top-selling products is important, of course, but Ali and the data team need more details: they want to build a table to display the top products *within each sub-category*. They also want to compare those products' sales numbers with all other products in that category combined. They tried using a set, but it didn't work. What should they do?

Oh, and they have one more problem—some of their tables keep producing a weird column that just reads "Abc" on every row. The team needs to get rid of that "Abc" column before they can present their results, but it won't budge. What's the deal?

Oh yes, your authors know both of these problems well. Don't worry, we'll get these tables sorted out.

Strategy: Show the Top Products Within Sub-Categories

Let's start with those sub-category totals. The OE data team tried to use sets, with no luck. That's because sets apply at only the highest level of the data source, and they're looking to break out that data by individual products within a category. You can't use a set at the product level. Follow along here as we show you a workaround:

1. To build the top sub-categories set, create a new integer parameter called **`[Top N Sub-Category]`**. Set the value to 4.

   Now create a Top N set with a sub-category called **`[Sub-Category | Profit]`**. Use the [Top N Sub-Category] parameter to set the number of sub-categories to be within the set and choose the sum of [Profit] as the measure.

2. Build a top sub-categories calculation:

   ```
   // Sub-Category Header
   IF [Sub-Category | Profit]
   THEN [Sub-Category]
   ELSE "All Others"
   END
   ```

   Place [Sub-Category Header] on Rows.

3. Write a sub-category sorting calculation:

   ```
   // Sub-Category Sort
   IF MIN([Sub-Category | Profit])
   THEN COUNTD([Order ID])
   ELSE 0
   END
   ```

   Right-click [Sub-Category Header] on Rows, and sort descending on Sub-Category Sort (Figure 6-36).

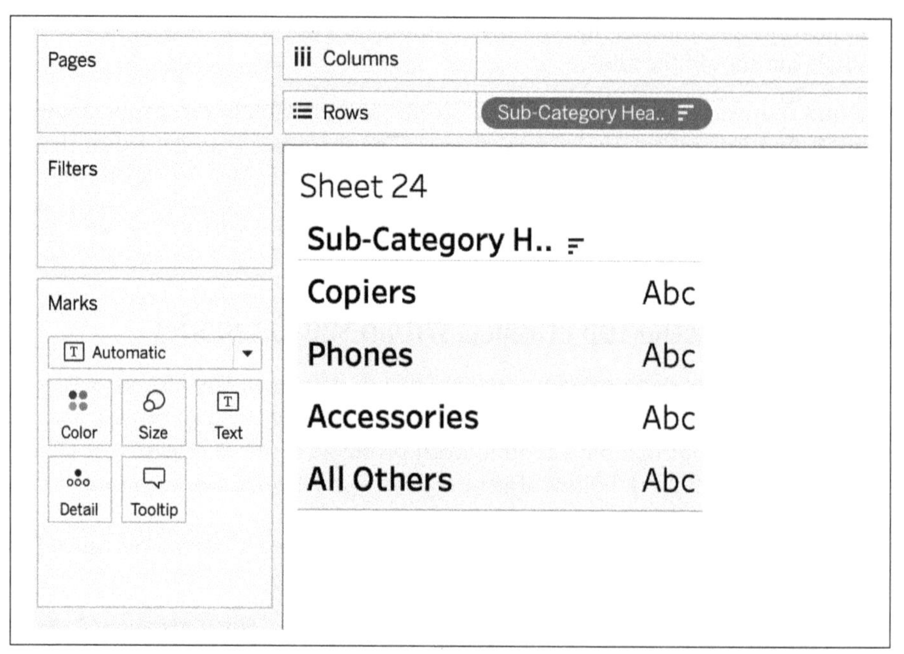

Figure 6-36. The [Sub-Category Header] calculation on Rows

4. Create an integer parameter for the number of products to show. Call it **`[Top N Products]`**. Set the integer to 5. This parameter can be used to dynamically update the number of products that will show details.

5. Create a function called **`[Product Index]`**:

```
// Product Index
INDEX()
```

   This calculation will work as a ranking of each product within its sub-category (we'll discuss that more in a bit). In the meantime, you can use this calculation to help you build your remaining calculations.

6. Build a calculation to help you sort the products into our Top N (in our example, Top 5) per sub-category. If a product falls outside the Top N, it still gets a ranking, just outside the top:

```
// Product Sort
IF [Product Index] > [Top N Products]
THEN [Top N Products] + 1
ELSE [Product Index]
END
```

   Place [Product Sort] to the left of [Sub-Category Header] on the Rows shelf (make sure it is Discrete). Add [Product Name] to the right of [Sub-Category

Header] on the Rows shelf. Edit the sort of [Product Name] to be in descending order by your key metric—in this case, sum of [Profit].

Now right-click to edit the table calculation for [Product Sort]. Select Specific Dimensions, and select only Product Name. You should see something like Figure 6-37.

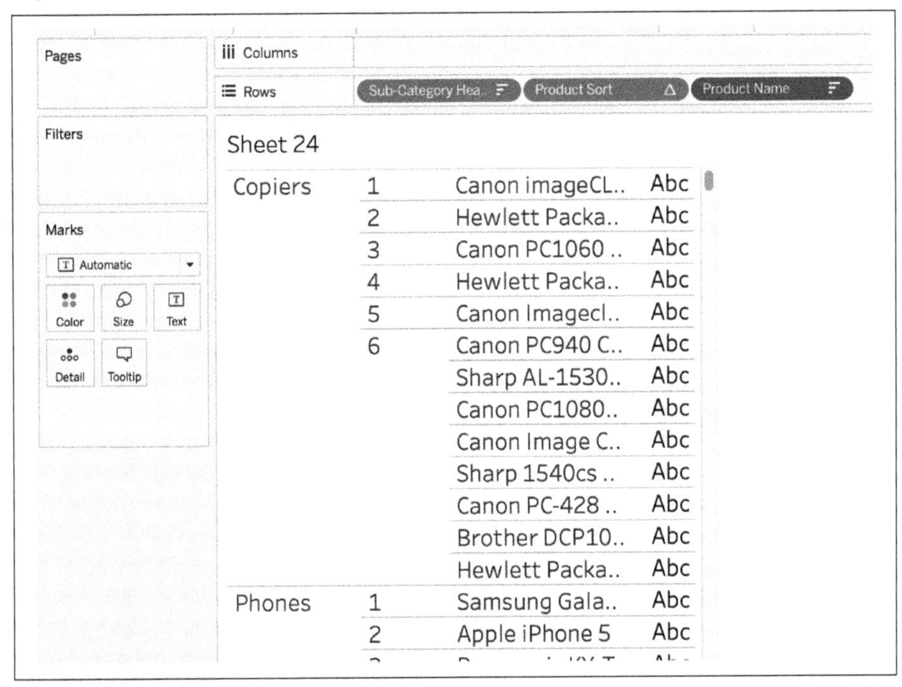

Figure 6-37. In this table, the top 5 products are separated from the remainder of the products

7. Create a calculation called **`[Product Row Labels]`**. This calculation will be the label you see on rows:

```
// Product Row Labels
IF [Product Index] <= [Top N Products]
THEN ATTR([Product Name])
ELSE "All Others"
END
```

Did you notice that we're using the Attribute function, `ATTR()`, here? Since the index calculation is an aggregate, dimensions within the calculation also need to be aggregated. `ATTR()` evaluates the minimum and maximum values of the dimension, and if they're the same, returns the value; if they're different, it will return an asterisk (*).

Add this dimension to the Rows shelf to the right of [Product Name]. Edit the table calculation for the value on the Rows shelf, select Specific Dimensions, and select only Product Name.

8. Create the row header for top products. Start by creating a calculation called **`[Product Header]`**. You will use this calculation as a filter to show the top products and the first product in the [Other Products] sub-category:

```
// Product Header
[Product Index] <= [Top N Products] + 1
```

After you have saved the calculation, click and drag this dimension onto the Filters shelf. Click OK (don't worry about selecting values yet). Edit the table calculation; within select Specific Dimensions, choose only Product Name. Then edit the filter and select only values that are True. Figure 6-38 shows what this should look like.

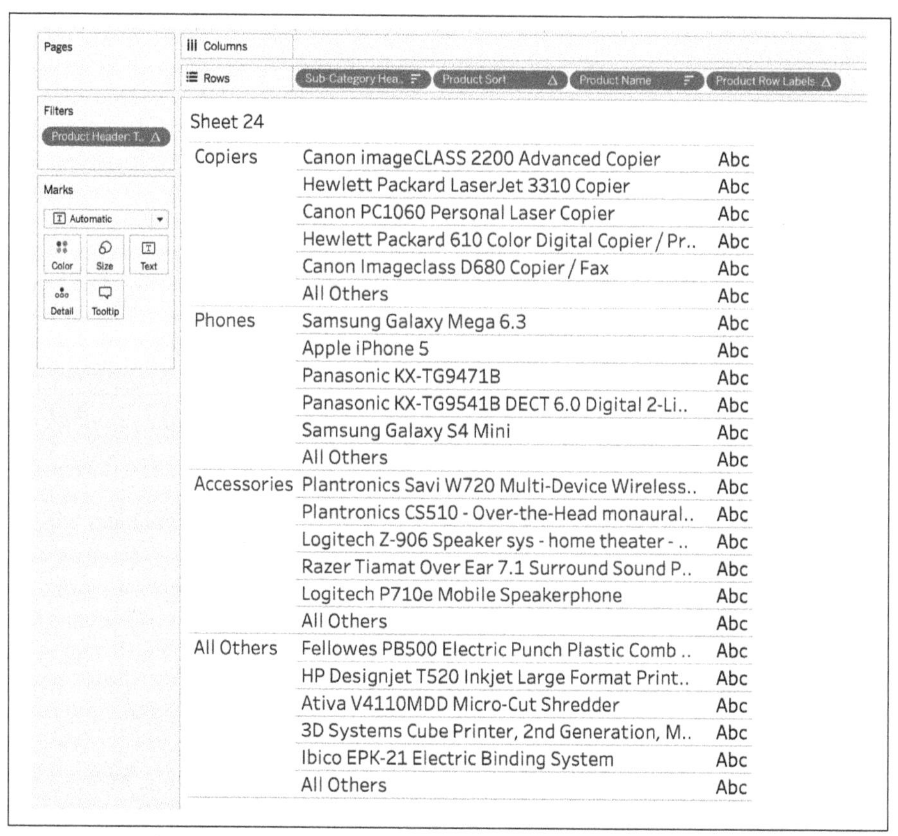

Figure 6-38. The results after adding a filter and creating product row labels

9. Once you've added [Product Row Labels] to the Rows shelf, you can then hide [Product Sort] and [Product Name] by right-clicking each dimension on the Rows shelf and unchecking Show Header.

10. Now you must create all the measures you will use on your view. For every measure, you'll use the same format. For values that are outside the Top N, you need to calculate the `TOTAL()` of some aggregation. If the value is inside the Top N, you can just use the standard aggregation.

    Here, place `SUM(Profit)` inside the `TOTAL()` function, but only when it's outside the top 5. When a [Product Name] is inside the top 5, take the standard `SUM(Profit)`:

    ```
    // Profit | Top N Products
    IF [Product Index] > [Top N Products]
    THEN WINDOW_SUM(
      IF [Product Index] > [Top N Products]
      THEN SUM([Profit])
      END
    )
    ELSE SUM([Profit])
    END
    ```

    To include other aggregated measures, you could repeat the same steps. Here are calculations for [Sales]:

    ```
    // Sales | Top N Products
    IF [Product Index] > [Top N Products]
    THEN WINDOW_SUM(
      IF [Product Index] > [Top N Products]
      THEN SUM([Sales])
      END
    )
    ELSE SUM([Sales])
    END
    ```

    And here are calculations for [Total Orders]:

    ```
    // Total Orders | Top N Products
    IF [Product Index] > [Top N Products]
    THEN WINDOW_SUM(
      IF [Product Index] > [Top N Products]
      THEN COUNTD([Order ID])
      END
    )
    ELSE COUNTD([Order ID])
    END
    ```

    When you place all three measures on the visualization, you'll need to edit the nested table calculations within each calculation. Be sure to edit both. Select Specific Dimensions and select Product Name for both.

Because the names of the calculations that display on the header might be less than intuitive for your audience, we recommend editing their aliases. For the measures, right-click [Measure Names] on your Columns shelf and select Edit Aliases; then rename the dimension aliases. For the dimensions on Rows, you can right-click the header names and select Hide Field Labels for Rows. Be sure to add appropriate table formatting too.

The final visualization is shown in Figure 6-39.

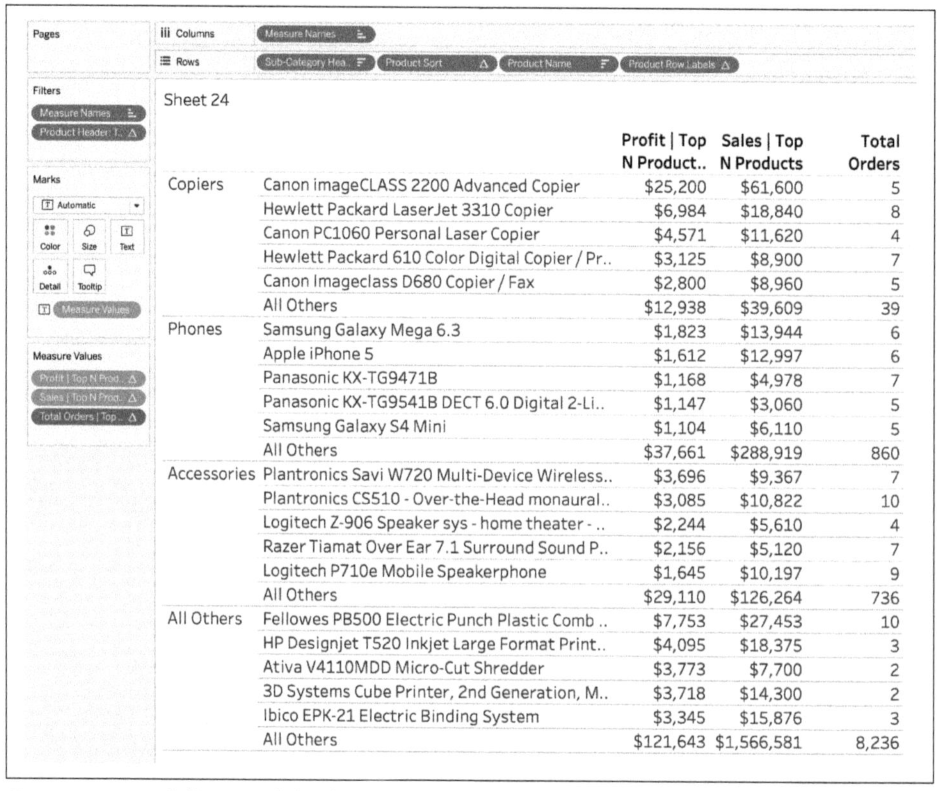

		Profit | Top N Product..	Sales | Top N Products	Total Orders
Copiers	Canon imageCLASS 2200 Advanced Copier	$25,200	$61,600	5
	Hewlett Packard LaserJet 3310 Copier	$6,984	$18,840	8
	Canon PC1060 Personal Laser Copier	$4,571	$11,620	4
	Hewlett Packard 610 Color Digital Copier / Pr..	$3,125	$8,900	7
	Canon Imageclass D680 Copier / Fax	$2,800	$8,960	5
	All Others	$12,938	$39,609	39
Phones	Samsung Galaxy Mega 6.3	$1,823	$13,944	6
	Apple iPhone 5	$1,612	$12,997	6
	Panasonic KX-TG9471B	$1,168	$4,978	7
	Panasonic KX-TG9541B DECT 6.0 Digital 2-Li..	$1,147	$3,060	5
	Samsung Galaxy S4 Mini	$1,104	$6,110	5
	All Others	$37,661	$288,919	860
Accessories	Plantronics Savi W720 Multi-Device Wireless..	$3,696	$9,367	7
	Plantronics CS510 - Over-the-Head monaural..	$3,085	$10,822	10
	Logitech Z-906 Speaker sys - home theater - ..	$2,244	$5,610	4
	Razer Tiamat Over Ear 7.1 Surround Sound P..	$2,156	$5,120	7
	Logitech P710e Mobile Speakerphone	$1,645	$10,197	9
	All Others	$29,110	$126,264	736
All Others	Fellowes PB500 Electric Punch Plastic Comb ..	$7,753	$27,453	10
	HP Designjet T520 Inkjet Large Format Print..	$4,095	$18,375	3
	Ativa V4110MDD Micro-Cut Shredder	$3,773	$7,700	2
	3D Systems Cube Printer, 2nd Generation, M..	$3,718	$14,300	2
	Ibico EPK-21 Electric Binding System	$3,345	$15,876	3
	All Others	$121,643	$1,566,581	8,236

Figure 6-39. A full view of the final visualization

Note that the bottom product always shows as All Others and is the total of all other products in the category.

Strategy: Hide the Abc

Sometimes you want to make a table just to explore its dimensions, or to output data to a *.csv* file or other format. If you use only dimensions on rows without any measures on columns or dimensions on text, though, you will end up with the "Abc" in the dimension value on your view: literally, just the letters "Abc" displayed as a value. You can see it in Figure 6-40, as well as in some of the figures in the previous strategy.

Tables with no Measures

Order ID	Customer Name	Country/Region	State	City	Ship Mode	
CA-2016-100006	Dennis Kane	United States	New York	New York City	Standard Class	Abc
CA-2016-100090	Ed Braxton	United States	California	San Francisco	Standard Class	Abc
CA-2016-100293	Neil Französisch	United States	Florida	Jacksonville	Standard Class	Abc
CA-2016-100328	Jasper Cacioppo	United States	New York	New York City	Standard Class	Abc
CA-2016-100363	Jim Mitchum	United States	Arizona	Glendale	Standard Class	Abc
CA-2016-100391	Barry Weirich	United States	New York	New York City	Standard Class	Abc
CA-2016-100678	Kunst Miller	United States	Texas	Houston	Standard Class	Abc
CA-2016-100706	Laurel Elliston	United States	Virginia	Springfield	Second Class	Abc
CA-2016-100762	Nat Gilpin	United States	Michigan	Jackson	Standard Class	Abc
CA-2016-100860	Cindy Stewart	United States	California	Pomona	Second Class	Abc
CA-2016-100867	Eugene Hildebrand	United States	California	Lakewood	Standard Class	Abc
CA-2016-100881	Daniel Raglin	United States	New Mexico	Albuquerque	Standard Class	Abc
CA-2016-100895	Stewart Visinsky	United States	Georgia	Roswell	Standard Class	Abc

Figure 6-40. The dreaded Abc column results from having no values on text in a table

Aesthetically, this Abc column is a nightmare. It's also a distraction to your end users. You should get rid of it. Here's how:

1. Use the Sample - Superstore dataset and place [Order ID], [Customer Name], [Country/Region], [State], [City], and [Ship Mode] on Rows.
2. Double-click the whitespace below the Marks card button and type `""`. These are *empty quotes*—two quotation marks with nothing between them.
3. Click the icon to the left of the empty quotes and change the type to Text. This will then place the empty quotes on Text and remove the Abc.
4. On your keyboard, press Ctrl-right-arrow three times to make the label wider.
5. Manually size the column to be as small as possible by clicking and dragging the right-border to the Ship Mode column.
6. On your keyboard press Ctrl-left-arrow five times.

Step 6 adjusts the row width of your table. This also works for row height: instead of using the left and right arrows, you can use the up and down arrows.

This will leave you with a table that is export-ready and no longer shows "Abc" (Figure 6-41).

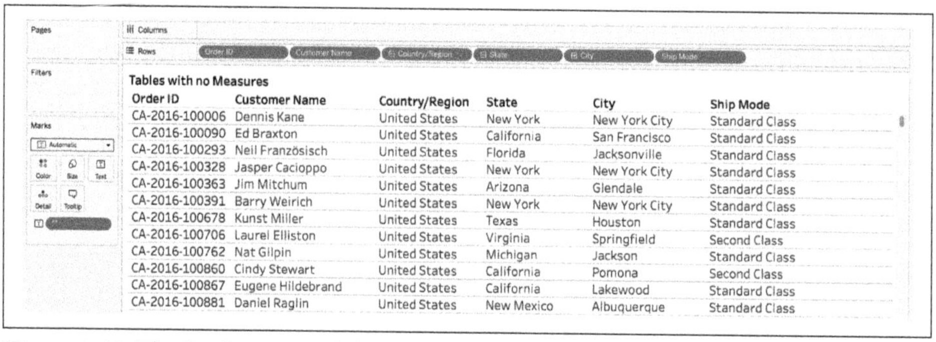

Tables with no Measures

Order ID	Customer Name	Country/Region	State	City	Ship Mode
CA-2016-100006	Dennis Kane	United States	New York	New York City	Standard Class
CA-2016-100090	Ed Braxton	United States	California	San Francisco	Standard Class
CA-2016-100293	Neil Französisch	United States	Florida	Jacksonville	Standard Class
CA-2016-100328	Jasper Cacioppo	United States	New York	New York City	Standard Class
CA-2016-100363	Jim Mitchum	United States	Arizona	Glendale	Standard Class
CA-2016-100391	Barry Weirich	United States	New York	New York City	Standard Class
CA-2016-100678	Kunst Miller	United States	Texas	Houston	Standard Class
CA-2016-100706	Laurel Elliston	United States	Virginia	Springfield	Second Class
CA-2016-100762	Nat Gilpin	United States	Michigan	Jackson	Standard Class
CA-2016-100860	Cindy Stewart	United States	California	Pomona	Second Class
CA-2016-100867	Eugene Hildebrand	United States	California	Lakewood	Standard Class
CA-2016-100881	Daniel Raglin	United States	New Mexico	Albuquerque	Standard Class

Figure 6-41. The final version of the table with Abc removed

Conclusion

For dashboard developers, creating tables can be a tough pill to swallow. We're capable of so much more! But tables aren't going away—and for some audiences, they really are the best way to take in and evaluate data. It's our job to use our analytical tools to provide them with insights that will make their jobs easier and their decision making faster. Sometimes that means designing tables.

Remember that creating a table always starts with a purpose. Whether your audience is a supply-chain team trying to understand individual routes and loads, or an executive team trying to understand spending, if you design around their purpose, you will be able to help them.

Don't underestimate the difference it makes when you take the time to improve the overall look and feel of a table. By following the simple steps we outlined at the beginning of this chapter, you can easily clean up a table's design.

Consider adding a pop of color too. By using color intermittently and thoughtfully, you can emphasize the critical information in a table—just as you would in a dashboard. We showed you how to limit the amount of color in your tables. We also showed you how to change the encoding from a highlight to dots, which simplifies the way your audience interacts with the data.

Another way to emphasize the insights in your data is by using different encodings in each column. Instead of only dots or colored cells, for example, you can use arrows or additional text. We've shown you how to create custom tables with different mark types for different columns, even ones that don't use measure names or measure values as headers, as well as how to re-create the headers when you build your dashboard. We also showed you how to use measures in your headers. You also learned multiple ways to add context to values in a table, an important topic that we'll return to in Chapter 12.

Finally, we did a little problem-solving for two common issues with Tableau tables. Since there is no out-of-the-box method for highlighting the top products within various subsets of the data, we showed you a workaround. You also learned how to get rid of the pesky "Abc" column that shows up in some very large tables.

In the next chapter, we'll shift gears and look at another powerful way to provide context to your data: with maps.

CHAPTER 7

Working with Geospatial Data

When was the last time you looked—and we mean *really looked*—at a map? Have you ever stopped to admire and acknowledge the amount of data and information they house?

Maps are one of the most powerful visualization types and perhaps the one we encounter earliest in life. Try to remember when you first understood how to read a map—it's hard to recall, right? Many people understand maps at an almost instinctual level, which makes them useful tools for communicating data.

It's important to remember that geographic proximity and distribution are central to any map analysis. That is really the key distinguishing factor when you're considering making a map. Just because you *can* map something doesn't mean you *should*.

Throughout this chapter, we will show you techniques for creating maps in Tableau. We'll also help you see when it's appropriate to create a map instead of something simpler, like the trusty bar chart.

In This Chapter

You'll need a basic understanding of US geography, including the hierarchy of geographic elements like zip codes (postal codes) as they relate to cities and states. You'll also need to download the Office Supply Store, US State Populations, and US State Attainment datasets, and the Tile Map Template.

In this chapter, you'll learn how to do the following:

- Map geographic data in Tableau, starting with a basic choropleth map
- Use custom color palettes to add clarity to your visualizations
- Normalize data
- Create symbol maps
- Create cartograms/tile maps and know when they are useful
- Create custom polygon territories and use data elements
- Use built-in geographic functions to perform additional analysis

Choropleth (Filled) Maps

One of Tableau's key differentiators is its capability to map data. Mapping is one of the most robust data professions out there—in fact, there is even a specialized domain called geographic information systems (GIS). While this level of specialization has led to some amazing tools, it has also meant that, until recently, data mapping and business intelligence data visualization have mostly been done separately, and with different tools for creating maps than for dashboards and bar charts.

Tableau distinguishes itself by allowing users to take common geographic data elements housed in datasets, like postal code, city, state/province, region, and country, and automatically attribute the correct latitude and longitude coordinates to them. It can also assign them the appropriate shape or geographic polygon (spatial object). This means that you don't have to know the coordinates for Seattle, Washington; Tableau will provide them by using its built-in geographic database. Similarly, you don't need to create the polygon object or a list of coordinates to construct the state of Washington; giving Tableau the state's name is enough for it to recognize and plot the polygon. A *choropleth map* takes a numeric value and plots it as a color scale on a map.

Displaying Customer Penetration with Choropleth Maps: Office Essentials Case Study

OE is assessing demographic data to better understand its core customers. The executives want to answer the seemingly simple question, "Where are our customers from?" How would you build a visualization to demonstrate this?

To start, you will build a choropleth map. (In Tableau this is called a *filled map*; the terms are synonymous. You'll shade each state in the United States according to the number of purchasing customers who live there (Figure 7-1).

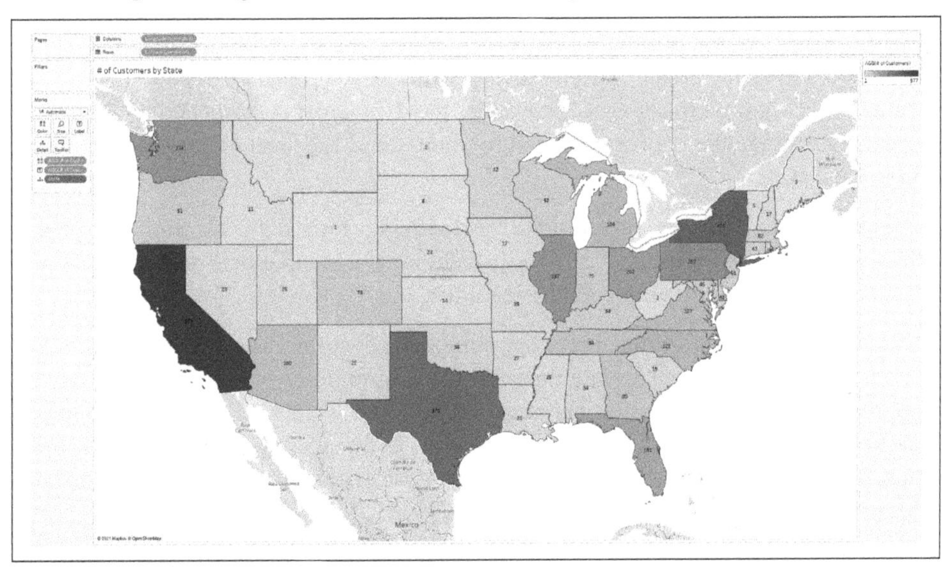

Figure 7-1. Basic choropleth map showing number of customers by US state

Strategy: Create a Simple Choropleth Map

Let's get started building our map:

1. Double-click the [State] field within your Data pane. Tableau will see that this is a geographic value and automatically create a map. This is one of the rare occasions when we recommend double-clicking a field and letting Tableau decide how it should be presented in the visualization.
2. Create a calculated field called **`[# of Customers]`**. This will be the Distinct Count of Customer Names:

```
//# of Customers
COUNTD([Customer Name])
```

3. Drag this newly created measure onto Color on the Marks card. Tableau automatically changes the mark type.
4. Click into Label on the Marks card and turn on labels.

Stop and inspect the chart. What insights can you immediately glean? California, Texas, and New York have the highest number of customers compared to the other states. Anything else? If we're being honest, it's hard to distinguish the states in the middle of the pack. Beyond the top few states, everything looks about the same.

Do you have a nagging feeling that this map isn't telling a complete or accurate data story? If so, we're right there with you. If all the states had the same population, we could just compare the number of customers. But while it is true that the three states mentioned have the most customers, they also have the largest populations. So, although it is accurate to say that these states have the most customers, this statement neglects to provide the context we need to understand the level of customer penetration in each state.

To improve this map, we recommend normalizing your data. *Normalizing* means adjusting the scale of a number so that it is no longer independent for each reading, but instead is consistent among all the readings.

For this map, you need to normalize the number of customers against the population of each state. A more appropriate measure to communicate where our office supply store may have deeper customer penetration would be to use a *per capita* (per person) measurement. This relays our customer numbers as a percentage of each state's population. It could also be phrased as "How many customers do we have per 1,000 state citizens?"

Strategy: Normalize a Choropleth Map

We're now going to introduce a new dataset to normalize the data:

1. Enhance the data by including the population data of each state:
   a. Go to Data Source and add a new connection.
   b. Navigate to the file that contains the census population and estimates by US state.
2. Using Tableau's logical data modeling, create a relationship between the census data and the customer order data. This relationship will be based on selecting State = Geographic Area, as shown in Figure 7-2. We will delve into more detail on data modeling in Chapter 10. For now, it is enough to know that Tableau is relating the two datasets together by the name of the state.

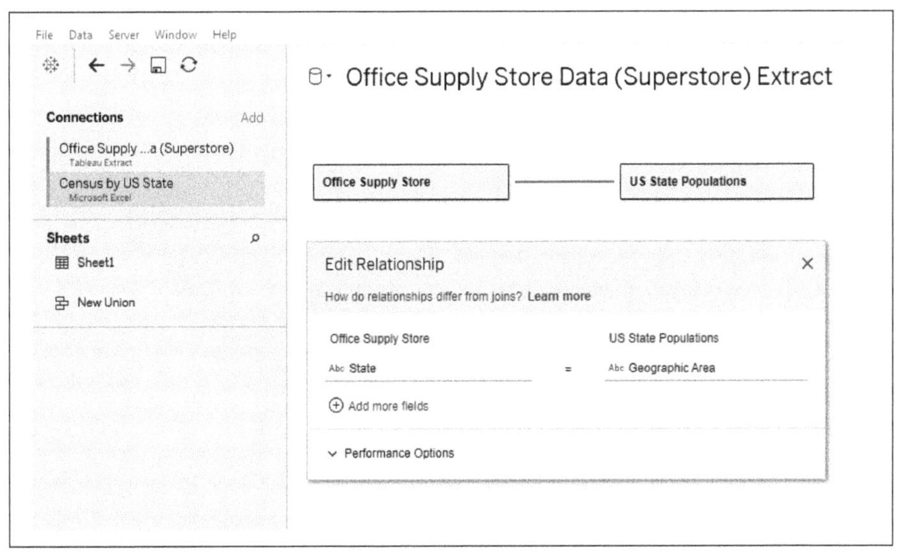

Figure 7-2. Creating a relationship using Tableau's logical data modeling

3. Build a calculated field that is **`[% of Customers per State]`**. The numerator will be **`[# of Customers]`** and the denominator will be **`[SUM(Census)]`**. Set the default format of the field to Percentage, with four decimal places showing:

```
//% of Customers per State
[# of Customers]
/
SUM([Census])
```

What do you think of the updated map in Figure 7-3? Does it change your understanding of the customer data?

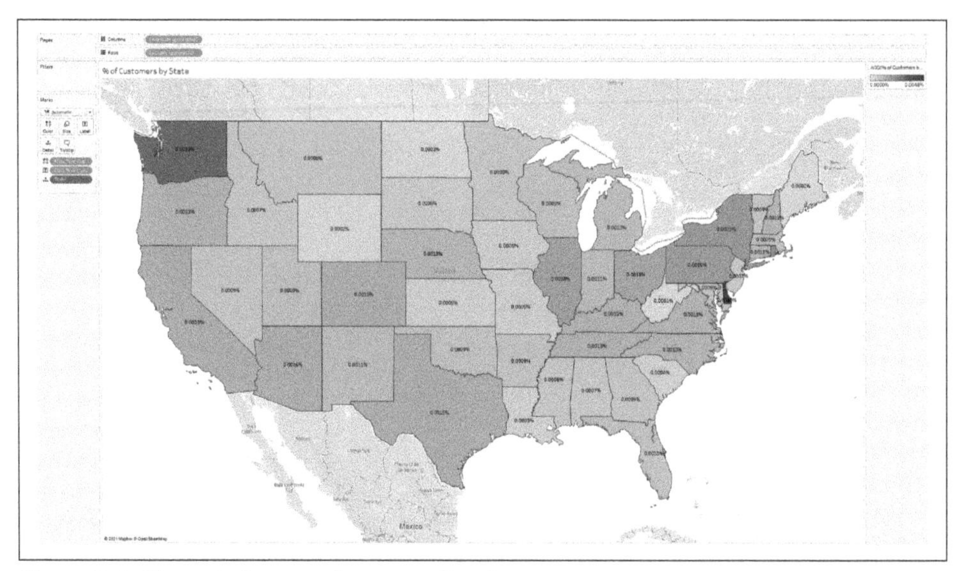

Figure 7-3. Choropleth map after normalization

We think so. The first map made it look like California, Texas, and New York were the high performers. Now that we've normalized the data, though, it's clear that our store has a strong presence in Washington and Delaware. Summative measures don't tell the full story.

But we still think this map needs more clarity. The next item to tackle is the color gradient. As you saw in Chapter 1, color encoding isn't the most precise method of distinguishing data points. Particularly with mapping, it is important that you take time to consider how to best display the data.

Fortunately, research on color perception has led to the development of color palettes that take into account accessibility issues like uniformity of perception, vibrancy, and colorblindness. We particularly recommend four gradient color palettes: Viridis, Magma, Plasma, and Inferno. All four started out in R and are easy to use in Tableau.

Strategy: Add a Custom Color Palette

You can code a few types of color palettes into your Tableau preferences. For this exercise, we will use an ordered-sequential palette, which means that each color is listed in order and will be available when using continuous fields on Color:

1. Choose File → Repository Location to determine the default save location of Tableau files.
2. Find the *preferences.tps* file and open it in a text editor.

3. Go to this color palettes dashboard (*https://oreil.ly/Cfj8k*) on Tableau Public built by Jacob Olsufka and copy and paste in the color palette hex codes for Viridis.
4. Save and close your Preferences file. Restart Tableau Desktop to activate the new palettes.
   a. Click Color on the Marks card.
   b. Change the palette to Viridis.

   This minor enhancement to the visualization, shown in Figure 7-4, improves viewers' understanding dramatically. All the central states, once slightly different shades of blue, are easier to distinguish. You can now clearly see that Delaware has the highest customer penetration.

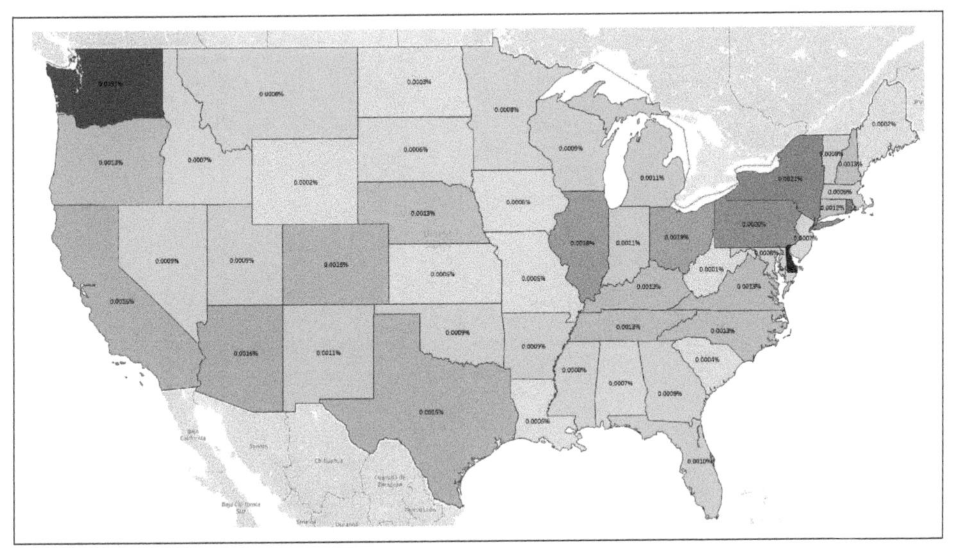

Figure 7-4. Normalized choropleth map with Viridis color palette

You can further simplify this visualization by using another derivative measure on color. Let's say your audience wants to know which states have the highest customer penetration. They want names of states for an answer; the numeric percentage is secondary. Using a percentile can make the picture even clearer.

Strategy: Make a Decile Choropleth Map

For this strategy, you will create a quantile map. The colors of the states in this map will indicate the percentile of value, between 0% and 100%. A *quantile* is any evenly distributed number of proportional pieces. The two most common quantiles are the *quartile* (numbers are separated into four even distributions: 0 to 25%, 25 to 50%, 50% to 75%, and 75% to 100%) and *deciles* (data is evenly distributed into 10 portions). Here, you'll make a decile map:

1. Right-click the [% of Customers per State] measure on Color and choose Quick Table Calculation → Percentile (Figure 7-5). Now data will be represented in percentiles, with a minimum at 0% and a maximum at 100%.

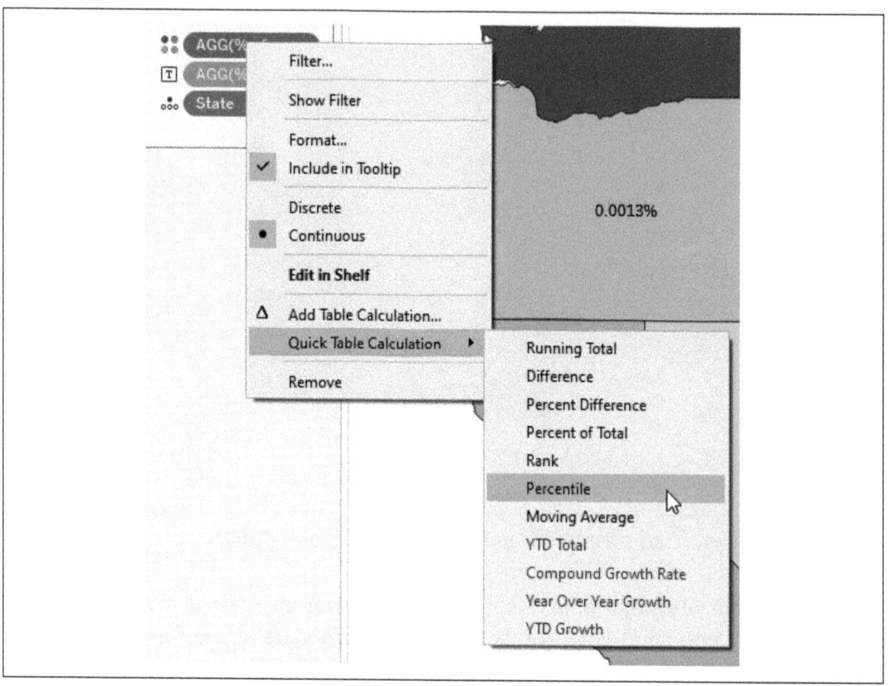

Figure 7-5. Representing data in percentiles

2. Click Color → Edit Colors and select Stepped Color. Change the number of steps to 10.

3. Fix the Start and End of the palette range at 0 and 1, respectively, by clicking Advanced, checking each box, and entering the value (Figure 7-6).

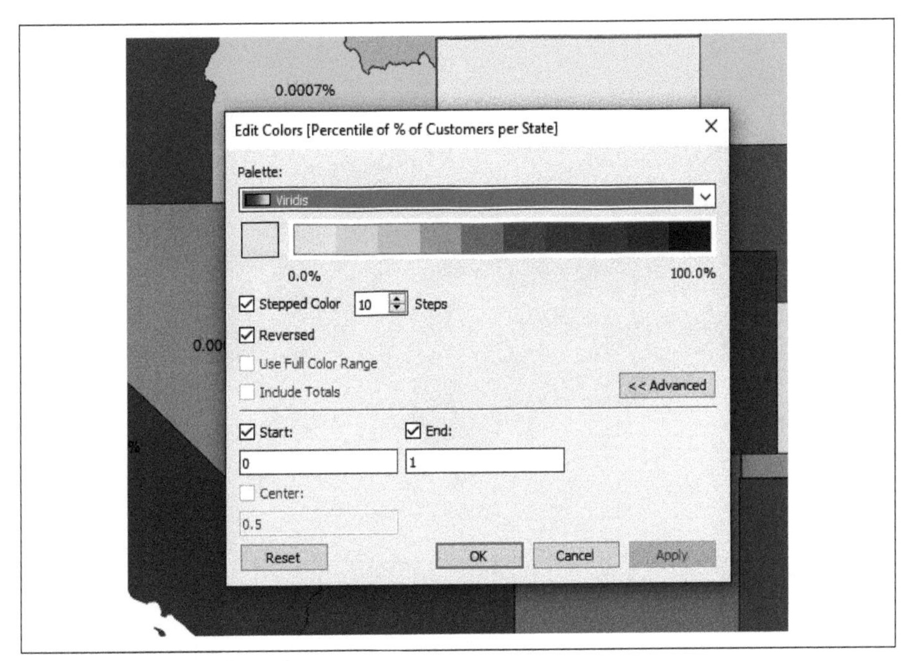

Figure 7-6. Setting the palette range

4. Change the label back to [% of Customers per State] by dragging the measure from the Data pane onto Label.

   The benefit of the color palette is clear: you can distinguish between the different colors effortlessly and mentally map the states to the different deciles (Figure 7-7). This lets you tell your audience that the top 10% of states for customer penetration are Washington, Pennsylvania, New York, Delaware, and Rhode Island.

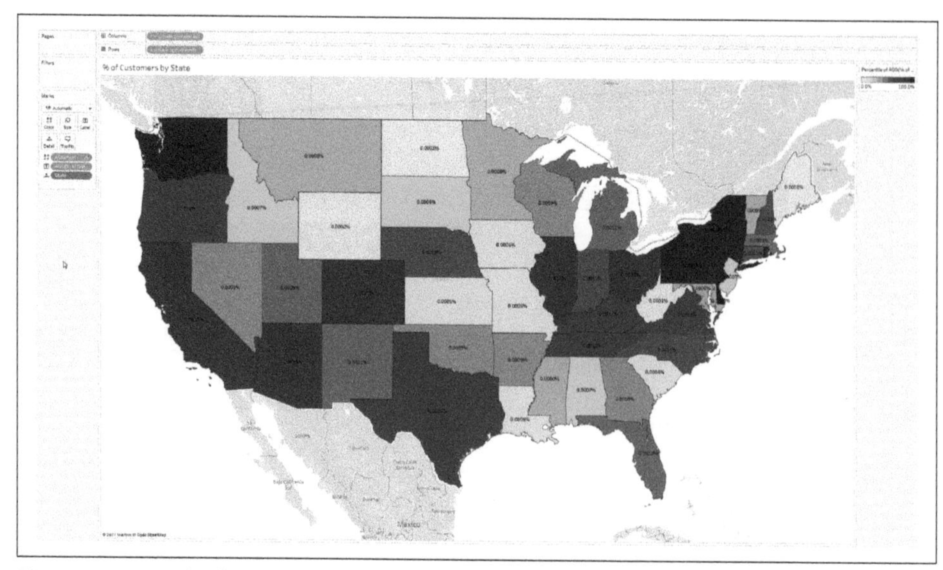

Figure 7-7. Decile choropleth map with Viridis color palette

We started with a basic metric (# of Customers) and advanced to percentiles. The drastic difference in readability underscores how important it is to think critically about the way you display data and communicate it to your audience. This simple evolution has gone a long way toward creating a compelling and insightful visualization.

Symbol Maps

The second most common map type you're likely to create in Tableau is a *symbol map*. It takes a discrete or continuous field and encodes it on a shape. The shape can have additional encoding on both size and color to distinguish it further.

Using Symbol Maps to Show Profitability and Channel Distribution: OE Case Study

Now OE wants to know which city is generating the most profit for the company. Using the same dataset, how would you construct a map that shows two measures at the same time: sales and profit ratio? Remember, this question is multifaceted: while profit ratio is the primary metric your audience is trying to understand, it needs context, like total sales. Which do you think a sales manager cares about more: a million-dollar city with a 10% profit ratio, or a $500 city with a 90% profit ratio?

Strategy: Create a Symbol Map

Starting with a new worksheet, you'll first create a Tableau default Symbol Map and then enhance the formatting for your audience:

1. Double-click [City] within the Data pane to automatically create a symbol map showing a circle for each city present in the dataset.

Depending on which version of the dataset you're working with, you may want to first create a hierarchy with Country, Region, State, City, and Postal Code. When you create a hierarchy with geographic information, Tableau will automatically add the higher levels of the hierarchy to the view.

2. Drag [Sales] onto Size. Adjust the size of the marks to be the largest recommended by Tableau, as shown in Figure 7-8.

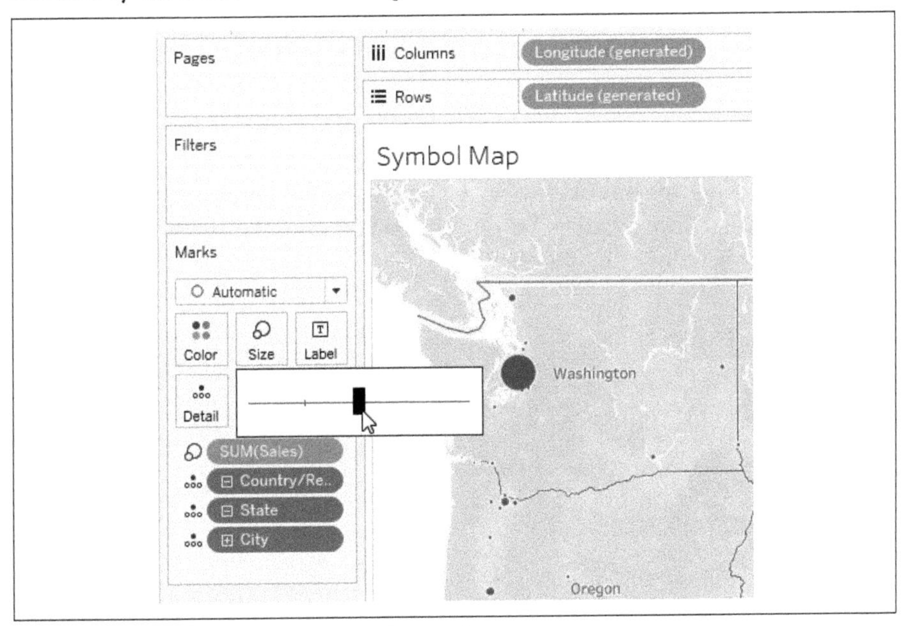

Figure 7-8. Adjusting mark size to show sales

3. Now create a calculated field called **`[Profit Ratio]`**:

```
//Profit Ratio
SUM([Profit])
/
SUM([Sales])
```

4. Drag [Profit Ratio] to Color. Tableau automatically creates a diverging color palette centered at 0, recognizing that this measure has both positive and negative values.
5. To visually distinguish overlapping marks, adjust the color opacity down slightly and add a border (Figure 7-9).

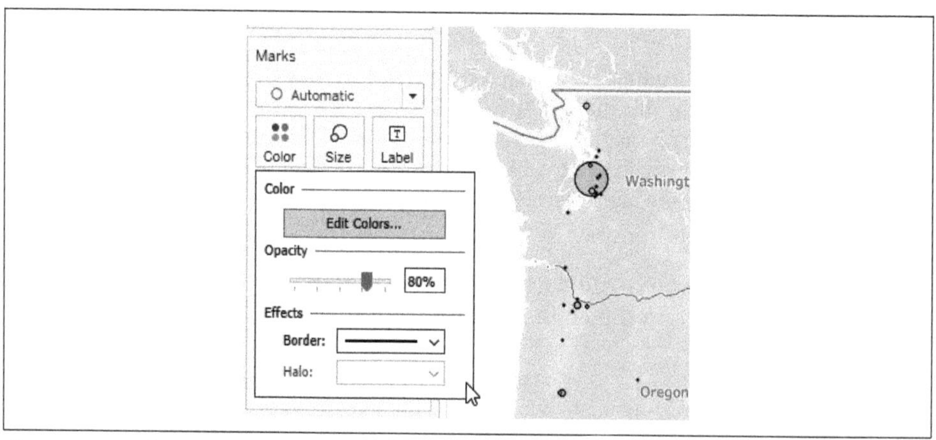

Figure 7-9. Adjusting opacity and adding a border

The classic symbol map, as shown in Figure 7-10, will always have a place in data analysis. But what if we could go beyond the default thinking and use symbol maps to create something unique and different?

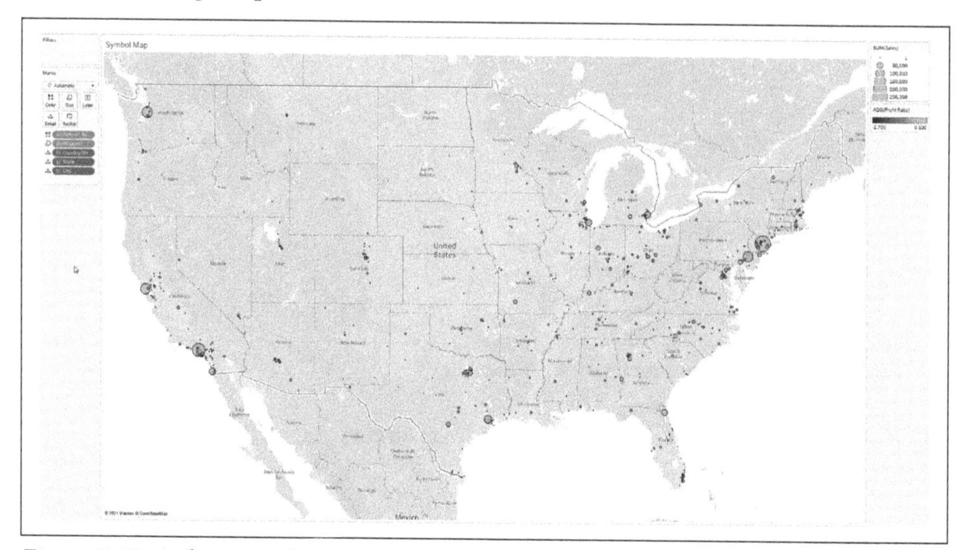

Figure 7-10. A classic symbol map

Let's say our office-supply executives ask you to provide an overview of sales distribution by product channel for each state. (The product channels are in our data segments.) To answer this question, you'll need to show a measure separated out by dimensions. You could do this with a stacked bar chart, but instead let's take a mapping approach.

Strategy: Create a Map with Donut Charts

In this strategy, you'll use a donut chart on top of a map to answer the team's queries:

1. Double-click [State] to construct a map.
2. Change your mark type to Pie and drag [Sales] onto Angle.
3. Drag [Segment] onto Color and adjust the size of your marks to the largest recommended size.
4. Create a dual axis to change the pie chart to a donut chart:
   a. Drag a copy of [Latitude] to the Rows shelf.
   b. Right-click the field and select Dual Axis.
5. Click the copy of [Latitude] to access the Marks card; remove [Segment] from Color and change your mark color to white. Change the size of the mark to be smaller, so the segment-separated pie chart is still visible.
6. Drag [Sales] onto Label and format it to be single-decimal currency expressed in Thousands (K) units.

   Notice that the map in Figure 7-11 provides two levels of information for more insight. First, it tackles the main analytical question: "What are our sales by segment per state?" Then it provides overview information: the total sales for each state. While your audience can see the visual proportion of the sales, adding the denominator (total sales per state) allows them to do mental math to process whether the visual distribution they are seeing is important information to retain.

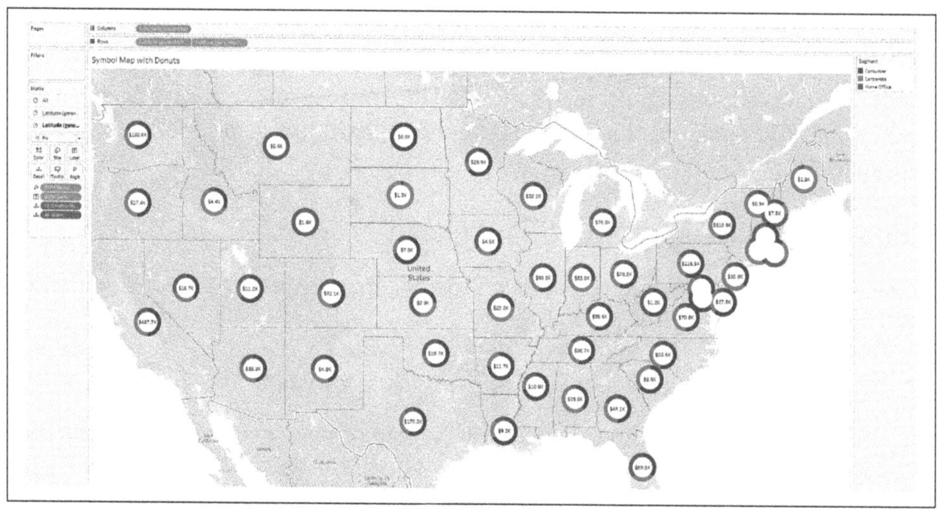

Figure 7-11. Sales by segment per state, using a map with donut charts

Tile Maps (Cartograms)

Maps are useful when there is an inherent need to understand geography along with data. But it's also important to recognize that the size of geographic areas can bias an audience. Larger geographies will naturally dominate the space of any visualization. When color encoding is added, the audience's eyes are naturally drawn to the large, colorful areas, and it's easy for them to miss some of the nuances in the map.

To compensate, we can use a *cartogram*, or *tile map*. A tile map takes geographic areas, strips out their complicated real shapes, and replaces them with more uniform shapes. Most often this shape is a square, a hexagon, or another shape that will tessellate. In *tessellation*, a pattern uses shapes that fit together without any gaps or spaces between them. Cartograms can use other measures to construct shapes too; for example, a map of the world with the area of each shape represented by the total population.

Showing Education Level with a Tile Map: Aloft Educational Services Case Study

In this next use case, Aloft Educational Services, an organization whose mission is to increase the educational attainment of adults in the US, is looking at education demographics in this country. They want to know what percentage of each state's adult population has attained a bachelor's degree and show the locations of the highest and lowest rates. How would you create a visualization to communicate this?

Strategy: Build a Square Tile Map

You'll connect to the US State Attainment dataset and then follow these steps to enrich the data and build a tile map:

1. Create a calculated field called **`[% College Attainment]`** from the dataset. Set the default number format to a percentage with one decimal point:

    ```
    //% College Attainment
    SUM([Total Population with Bachelor's Degree 25 Years and Over (Estimate)])
    /
    SUM([Total Population 25 Years and Over (Estimate)])
    ```

2. Enrich the dataset by adding a file that maps each state to a column or row (shown in Figure 7-12). More-sophisticated shapes might map each state to an *x,y* coordinate, but we'll keep it simple here.

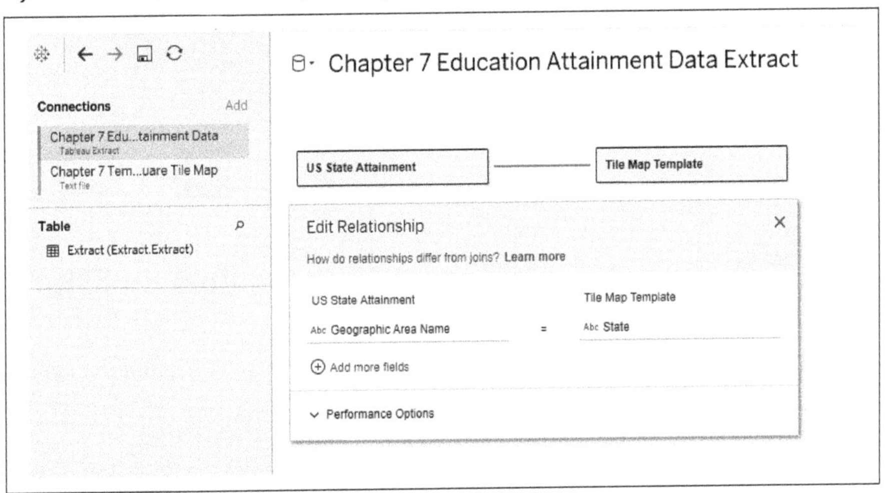

Figure 7-12. Mapping states to columns and rows

3. Drag [Row] to Rows and [Column] to Columns; change the mark type to Square. (Make sure they are both discrete dimensions.) Set the view sizing to Entire View.

If [Row] and [Column] are continuous measures (green), you can drag them above the horizontal line in the Data pane to change each to a discrete dimension.

4. Drag [% College Attainment] to Color and to Label. Drag [Abbreviation] to Label. You may need to change your mark type back to Square. Change the color palette to a custom palette, as you did in the first section of this chapter. (Figure 7-13 uses Color Brewer Greens.) Add a white border as well.

5. Right-click [Row] and [Column] and uncheck Show Header.

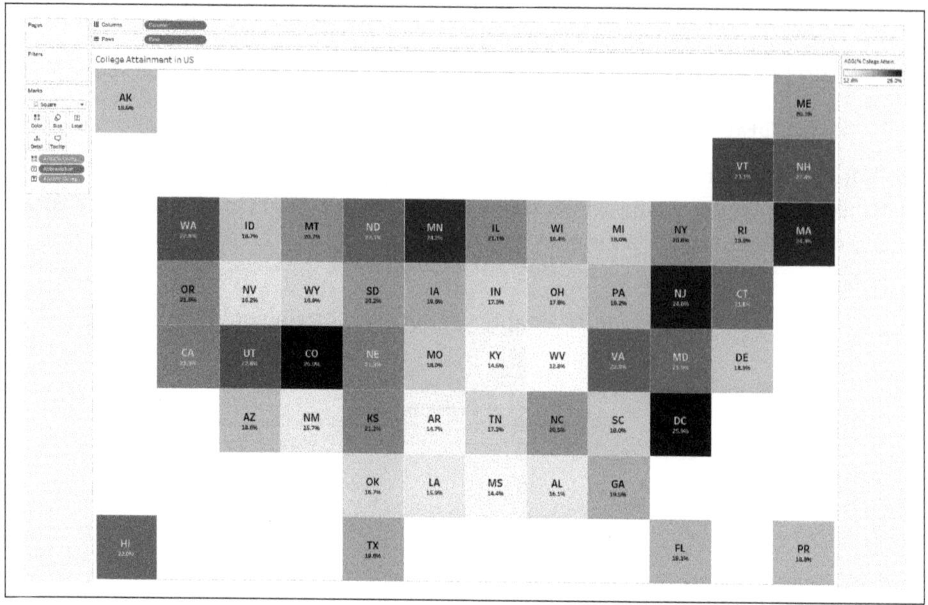

Figure 7-13. Educational attainment rates shown in a tile map (top), and the same data shown in a choropleth map (bottom)

As you compare the tile map to the original, what changes about where the eye is drawn? While Colorado remains noticeable, in the tile map it is much easier to see the pockets of lower attainment in the Southeast, and higher attainment on the East Coast. Alaska no longer dominates the visualization because of its size, and smaller areas like Hawaii, the District of Columbia, and Puerto Rico all get equal space.

Having seen the tile map you just made, your nonprofit stakeholders might well ask how they can tell if a particular state's rate is above or below the national average. How can we make this display more insightful? By using something other than the direct measure on color.

Strategy: Compare Parts Against the Whole with a Tile Map

Starting with our chart from the last strategy, we'll add formatting, text, and color enhancements:

1. Create a new calculated field called **`[Total US College Attainment]`**.

   This will be an LOD expression that computes the percentage of college attainment for the entire dataset (encompassing all states):

   ```
   //Total US College Attainment
   {FIXED : [% College Attainment]}
   ```

2. Using this new calculated field, create another calculated field called **`[Delta from US Overall]`**. This will be the percentage point difference in attainment between the state's rate and that of the US overall:

   ```
   //Delta from US Overall
   [% College Attainment] - MAX([Total US College Attainment])
   ```

   Here, you are re-aggregating the LOD expression to the maximum. LODs are naturally row-level. Since the other metric is aggregated, [Total US College Attainment] must also be aggregated. It is conventional to use Maximum, Minimum, or Average when aggregating LODs for calculated fields; mathematically, they will all return the same result.

3. Drag the new measure onto Color and Label.
4. For advanced formatting, change the Default Number format for [Delta from US Overall] by using the symbols ▼▲▬, as shown in Figure 7-14.

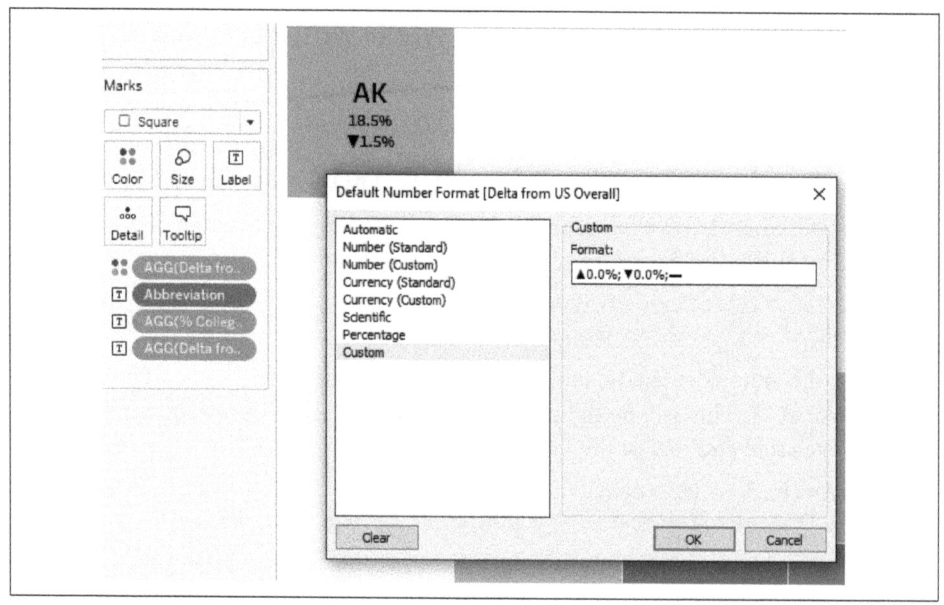

Figure 7-14. Changing the default number format

In the final iteration of the visualization (Figure 7-15), the emphasis is now on comparing against the overall rate, and it tells a compelling story. Every southern state is below the average, as is a group of states in the Midwest.

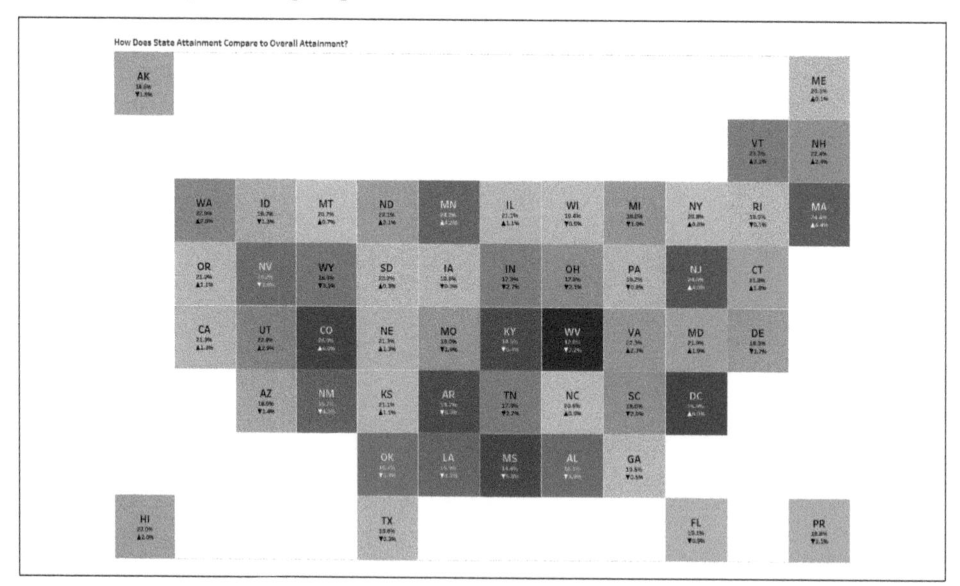

Figure 7-15. Tile map showing how each state's college attainment rate compares to the national average

Built-in Features and Functions

You've now seen quite a few examples of how to make default and nondefault maps in Tableau. We've guided you through basic examples and provided valuable options for maximizing the data you're presenting geographically. However, that's only one part of geospatial analysis. In this section, we'll show you some of Tableau's built-in features and analytical computations.

Using Built-in Features and Functions with Marketing Data: SAGE Digital Marketing Case Study

In this example, Savvy Actionable Gains Executed (SAGE) is a digital marketing agency that provides its clients with leads through paid ads in search results. The agency's client is a local landscaping company. To the client, the less they pay for a lead, the more profit they get from each converted customer—so the goal is to give them as many consistent leads as possible, at the lowest possible cost.

The client has three physical locations and divides its customers into three service areas based on home address. How would you understand how well the ads in each service area are performing? The available data shows how much it costs to display a paid search advertisement and how many people in any given zip code clicked the ad.

Tableau lets you build custom geographic areas, which it calls *custom polygons*. For the first strategy, you'll set up your client's three service areas by using the zip code data they've provided. As with most things in Tableau, there are multiple ways to do this. We will walk you through two methods.

Strategy: Build Custom Polygons Using Groups

While your data provides zip codes, you don't know exactly which zip code falls in the three service areas (shown in Figure 7-16). You may have only a rough approximation of where those areas are.

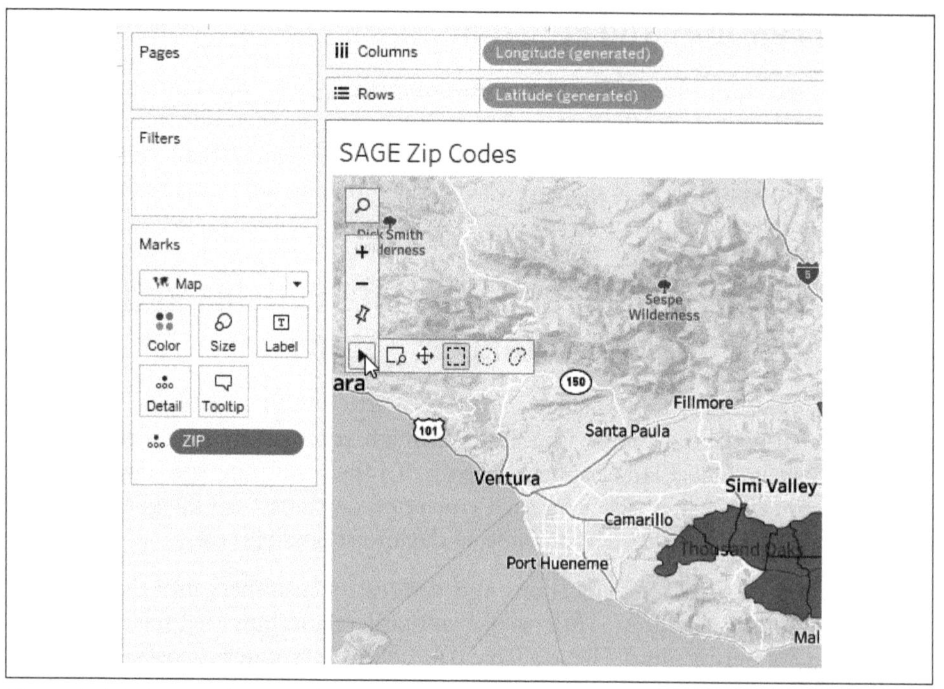

Figure 7-16. A map showing the approximate locations of your client's service areas

You can easily create groups in Tableau to generate your different areas:

1. Double-click [ZIP] to create a map showing available data by zip code in your dataset. Change your mark type to Map.
2. Modify the Map background by going to Map on the toolbar and selecting Map Layers at the top. Change the Style to Streets.
3. Hover over the upper-left area of the map image to view the toolbar; hover over the bottom icon (a right arrow) to see various ways data can be selected. Click the lasso option on the far right (Figure 7-17).

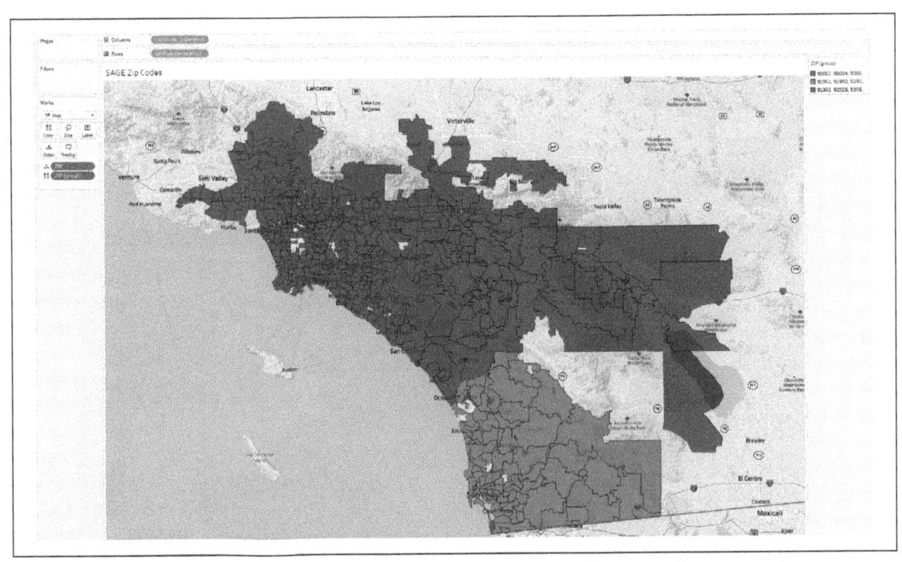

Figure 7-17. The map toolbar automatically displays when you hover in the upper-left corner of a worksheet

4. With the lasso option chosen, select zip codes that are part of the North service area shown in Figure 7-16. If you miss one, you can hold down your Ctrl key afterward to manually select additional zip codes. The selection does not have to be precise.

5. After you have selected the appropriate zip codes, click the group icon (paperclip) at the top toolbar; then click the word ZIP. Tableau will automatically construct a group consisting of the zip codes you selected. The new group is automatically placed on Color.

6. Repeat this process two more times, until you have three areas roughly resembling the service areas (Figure 7-18). You can rename them by editing the Group you just created.

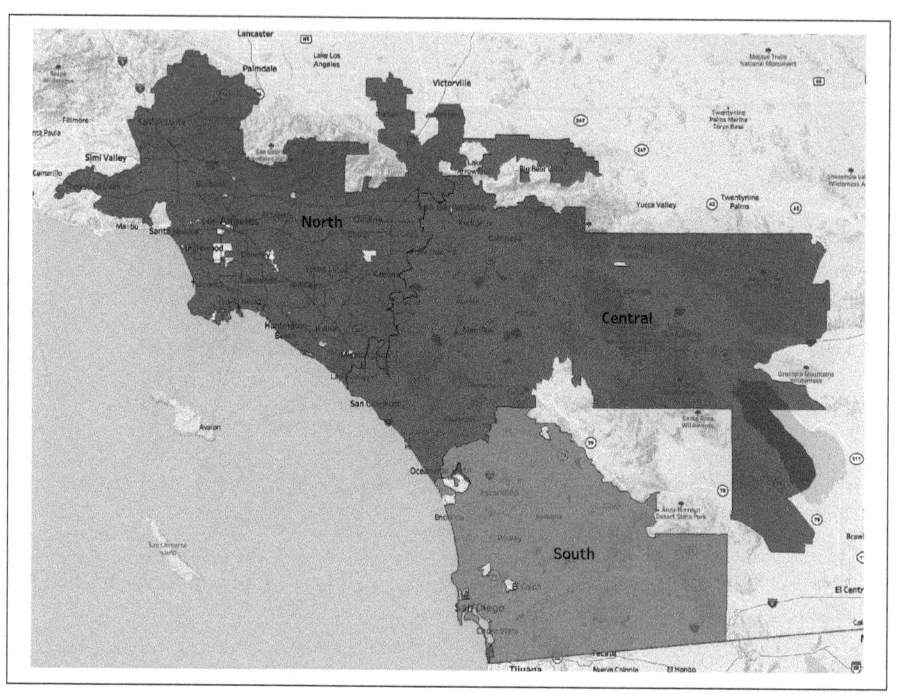

Figure 7-18. Each zip code is now colored by service area

7. Remove [ZIP] from the Detail property. This is the most important step. You have now created a custom polygon. Using this information, you can see aggregate values for metrics by service area.

8. Create a calculated field called **`[Cost per Click]`**. Add it to the label, along with the service area name:

```
//Cost per Click
SUM([Cost])
/
SUM([Clicks])
```

Figure 7-19 shows the result.

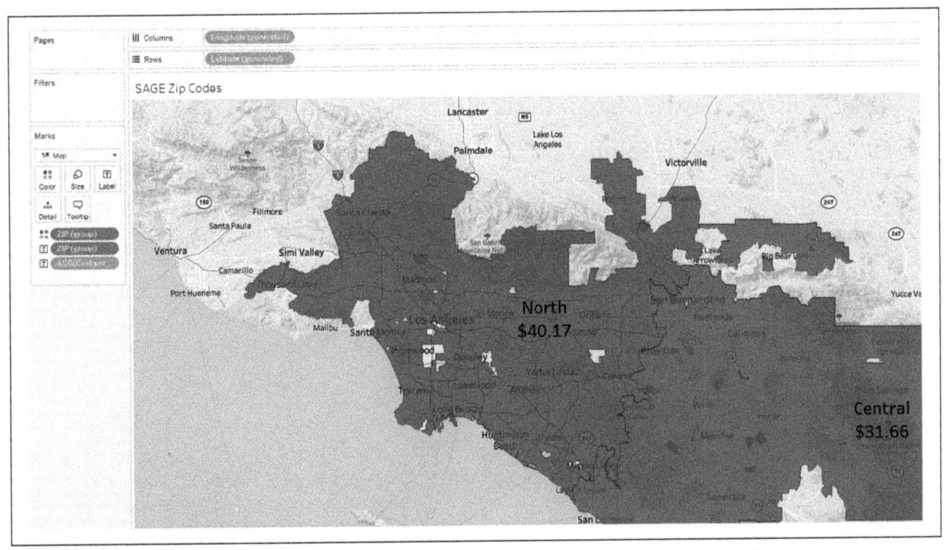

Figure 7-19. Cost per click to the client by service area, created using ad hoc groups

The capability to aggregate geographic data this way is extremely powerful. The original data did not provide what you needed to know—cost per click for each area—but aggregating it provides clarity on that key metric.

In this scenario, we used an ad hoc technique to construct the areas. If you have a territory, area, or region encoded or mapped to your data, you can use the underlying geography to compute the same result, instead of building ad hoc groups.

Strategy: Build Custom Polygons Using Underlying Geographic Data

In this strategy, you'll create a map similar to the previous exercise, but this time using data fields that exist within the dataset:

1. Right-click the field named [Service Area] and choose Geographic Role → Create from → ZIP.

   This feature aligns geographic data to the service areas (territories) in your dataset. As long as there is a one-way relationship between a zip code and a territory, your data will be combined into the new groupings.

   Notice that a new hierarchy appears, showing the relationship between service area and zip code (Figure 7-20).

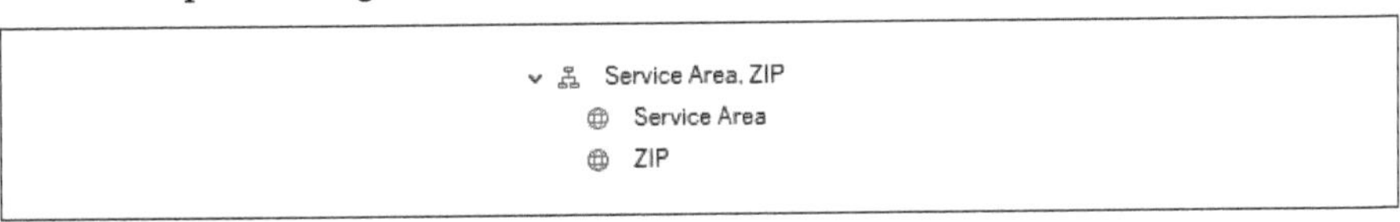

Figure 7-20. The Service Area/ZIP code hierarchy

2. Double-click [Service Area]. Change your mark type to Filled Map.
3. Add [Cost per Click] to Label along with [Service Area]. Add [Service Area] to Color. You may have to change the colors to match the previous strategy.

   Figure 7-21 shows the result.

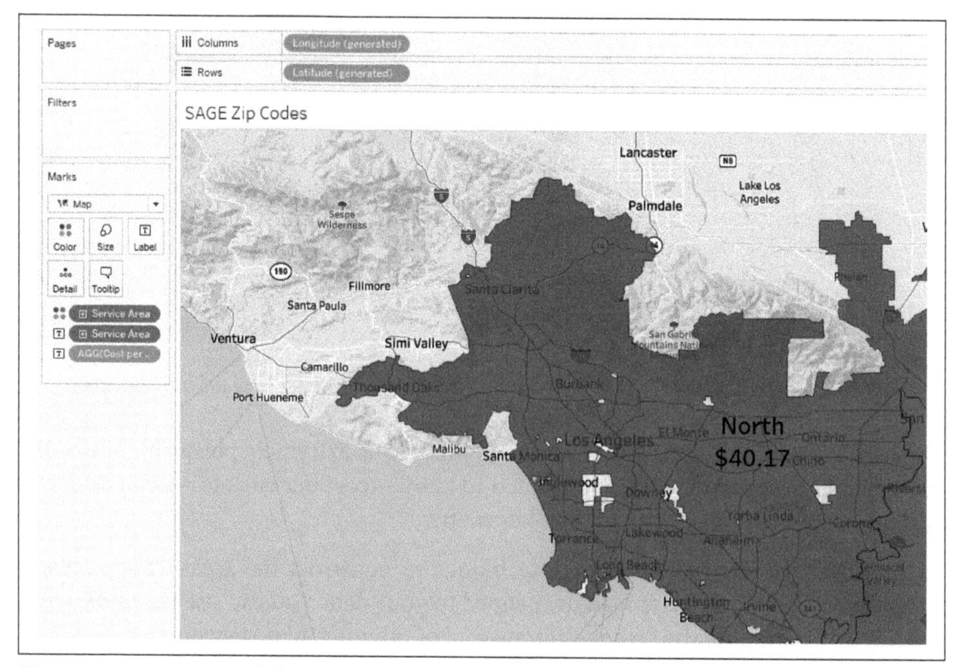

Figure 7-21. Cost per click to the client by service area, created using underlying geographic data

Strategy: Use the Distance Function to Show Zip Codes' Distance from a Central Location

Now your clients at the landscaping company want to know how far their business locations are from the people who view their ads. They want to test their theory that the closer an ad viewer is to a store (which we'll name *Service Area Center*), the more likely they are to purchase services. You can use built-in spatial functions to answer this question.

To do this, we'll explore Tableau's four built-in functions related to geography:

`DISTANCE()`
: Uses geospatial fields (marked with a globe icon) as inputs to compute the distance between two objects.

`MAKELINE()`
: Creates a line between two objects.

`MAKEPOINT()`
: When you provide coordinates (latitude and longitude or *x,y* coordinates), Tableau identifies the coordinate or GIS system they represent and issues a *spatial reference identifier (SRID)*.

`BUFFER()`
: Takes a geospatial point object and computes a radius around it with a unit distance that you can specify.

Follow these steps to find the distance from zip codes to the Service Area Center:

1. Use the `MAKEPOINT()` function with latitude and longitude to create a point called **`[Zip Code Point]`**:

    ```
    //Zip Code Point
    MAKEPOINT([Latitude],[Longitude])
    ```

2. Repeat this process, using the [Service Area Latitude] and [Service Area Longitude] to construct a field called **`[Service Area Point]`**:

    ```
    //Service Area Point
    MAKEPOINT([Service Area Latitude], [Service Area Longitude])
    ```

3. Using the two calculated fields, construct a final calculated field called **`[Distance from Service Area Center]`**. Within it, set the unit (the third parameter) to **`mi`**:

    ```
    //Distance from Service Area Center
    DISTANCE([Service Area Point], [Zip Code Point], "mi")
    ```

4. Right-click [Distance from Service Area Center] and set the default aggregation to Average. This lets you display the distance between a zip code and the Service Area Center, even if there are multiple rows of data for an individual zip code. You can also use this measurement to compute the distance at a higher level of aggregation, like [Service Area].

5. Construct a dot plot:

    a. Drag [Service Area] onto the Rows shelf.

    b. Drag [Distance from Service Area Center] onto the Columns shelf. Change the mark type to a Circle.

    c. Drag [ZIP] onto Detail on the Marks card.

6. Drag [Conversions] onto Size. Adjust the color opacity to 60%.

   What conclusions can you draw from the visualization in Figure 7-22?

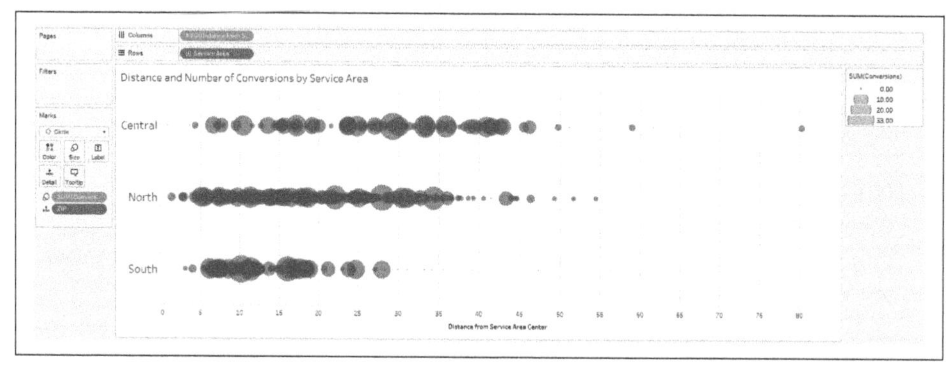

Figure 7-22. The distance of each zip code within a service area from its central point, along with the number of conversions

First, there are certainly some distinctions among each service area's results: in the North, it seems that more conversions come from zip codes within 30 to 35 miles. For South, that distance is a little bit tighter, with most coming in closer than 25 miles. And with Central, it's apparent that there aren't many zip codes near the center point.

Thanks to your insights, your landscaping clients now know where most conversions in the North service area are coming from. They decide to adjust their strategy and target only customers who live within 25 miles of each center point. They ask you for an impact analysis: which zip codes will they retain and which will they drop? How many will they lose? You can do this by working with the `BUFFER()` function. This process creates a second map layer within your visualization, which is useful if you want to display different layers of geographic detail on one map.

Strategy: Use the Buffer Function to Map a Trade Area Using Radial Distance

Using the same dataset from the previous strategy, you'll construct a new map layer with a circle of a specified radius:

1. Create a calculated field called **`[Service Area Trade Area]`**:

```
//Service Area Trade Area
BUFFER([Service Area Point], 25, "mi")
```

2. Double-click this new calculated field to plot the areas on a map. Add [Service Area] to Color and Label. Change the opacity of the mark to 20% and add a black border.

3. Drag [ZIP] to the map in the upper-left corner; this should generate a drop zone for adding another Maps Layer.
4. On the secondary Marks card just created, add [Service Area] to Detail. Reduce the opacity of the points to 60% and set both the Border and Halo to None.

   Now your clients can clearly see which zip codes they will be targeting in their revised paid search efforts, and which ones will be left out (Figure 7-23).

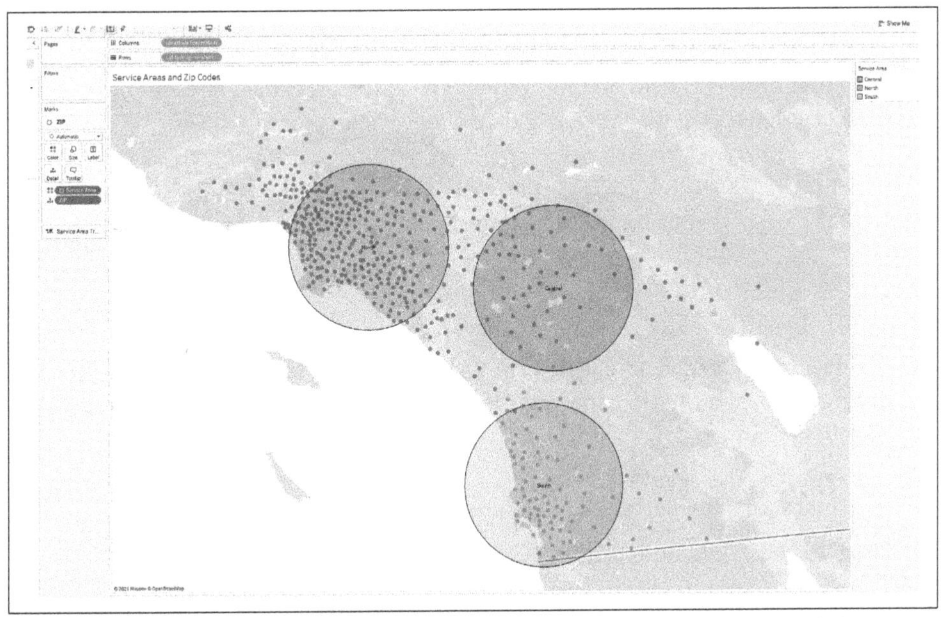

Figure 7-23. Map showing trade areas constructed in a 25-mile radius around the center of each service area, plotted with the zip codes currently being targeted with marketing efforts

In our final visualization, you'll create a map that explains the whole story to your client: which zip codes are part of each service area, how far each one is from the center, which ones fall into the trade areas targeted by the revised ad strategy, and how many conversions are associated with each zip code. You will do this by constructing lines from the Service Area Center to each zip code.

Strategy: Create Paths Between a Starting Point and Destinations with MakePoint and MakeLine

In this last strategy of the chapter, you'll pull everything together and use the remaining two spatial functions introduced earlier:

1. Create a calculated field called **`[Line from Zip to SA Center]`**:

```
//Line from Zip to SA Center
MAKELINE([Service Area Point],[Zip Code Point])
```

2. Drag and drop this field on top of the [Zip Code Point] field already on your Zip Code layer Marks card. *The dots should change to lines.*

When you drag and drop a field on top of another field within a visualization, the original field is immediately replaced with the field you drop.

3. Create a calculated field called **`[In Service Area]`** that identifies whether the distance between the zip code and the Service Area Center is less than or equal to 25 miles:

```
//In Service Area
[Distance from Service Area Center]<=25
```

4. Drag this new field onto the [Zip Code] Marks card.

   a. Click the icon to the left of the field, changing it from Detail to Color. You should now have two different colors for each zip code within a Service Area, one darker and one lighter. If necessary, adjust the colors using the Tableau 20 palette.

The color legend can represent multiple dimensions, but only if you manually assign each one to the Color property on the Marks card. If you drag and drop a field onto Color, it will automatically replace the one that was previously there and instead set it as Detail.

5. Finally, on the [Line from Zip to SA Center] Marks card, drag [Conversions] onto Size. This will size each line by the number of conversions associated with the zip code.

   This visualization in Figure 7-24 helps your client see the whole picture.

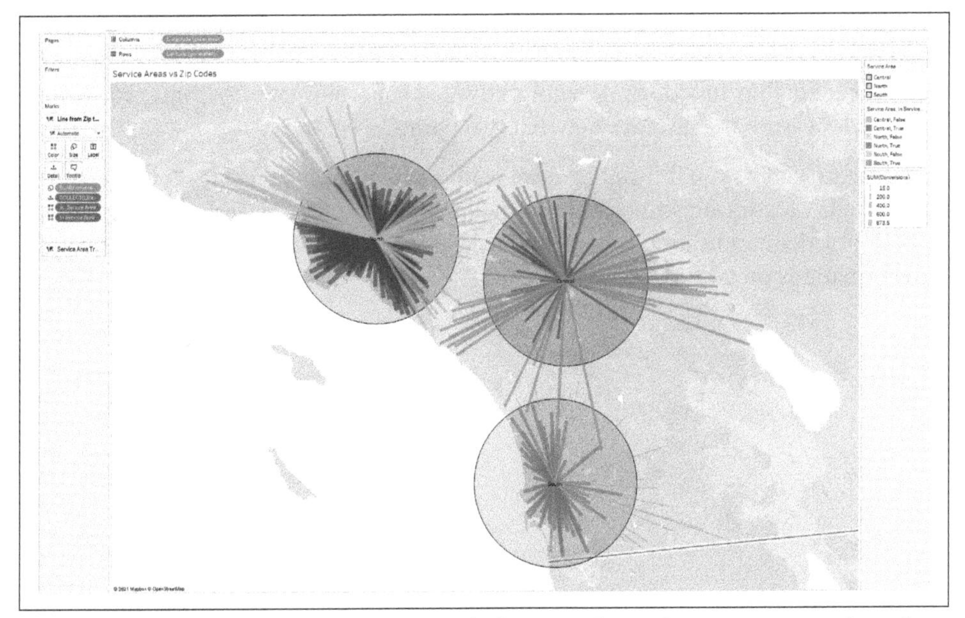

Figure 7-24. The new trade areas along with the zip codes and conversions inside and outside each trade area

It also suggests that, while the revised targeting strategy works well for the North and South territories, it isn't as successful in the Central service area. You might suggest that your clients move Central's center or separate the service area into two smaller sections.

Conclusion

You've now seen several ways to work with maps and geographic data in Tableau, starting with a basic choropleth map that showed the penetration of OE by each state. You also learned how to normalize your data to provide a more accurate numerical result to your audience.

Additionally, you've seen the importance of color encoding when working with maps. A color palette like Viridis has intentionally significant differences in intensity and hue, which helps your audience distinguish between numerical values represented as color.

You also created a cartogram by way of a tile map, in which the geographic area of each state was stripped from the map and replaced by a uniform square. This allowed for easier comprehension of the data values and added equity for smaller states.

You've also worked with spatial functions to perform additional analysis on your geographic data—which doesn't always have to be a map. First you used built-in features to construct completely new spatial polygons, and then you constructed polygons based on other geographic data points available in your data. You also explored the built-in geographic functions, learning how to make new points, construct service areas, and measure the distance between two points.

In the next chapter, you'll focus on advanced mathematical concepts. You'll learn the ins and outs of working with advanced mathematical functions and concepts, as well as best practices indicating when and where to do more-advanced mathematical transformations on your data.

CHAPTER 8

Advanced Mathematical Concepts

The purpose of this chapter is to explore relationships among your data. It's very common to want to explore the relationship between two or more numerical metrics, use your time-bound data to determine what may come in the future, or to use recurring relationships among dimensions to further categorize and segment. The aim of this chapter is to explore all of these concepts.

We think this chapter is of utmost importance and designed for the individual most likely to be exploring data—that is, looking for deeper insights or uncovering patterns, which is different from some of the other concepts we've explored that are aimed at presenting data with evident insights. In this chapter, some of the charts you will build may not make it to a large audience but will lead to deeper understanding of that data you have to work with.

In the first case study, you will revisit our call center and use Tableau to forecast call volumes to help inform staffing levels.

In the second case study, you will work with digital marketing data to help marketing teams make optimization decisions on which ads should stay and which should go. You'll also explore the relationship between two marketing metrics to determine if you can estimate an outcome or provide a required minimum to reach a goal.

In the third case study, you'll explore alternative axes and visualize COVID-19 data. You'll see how working with different axes types can lead to better data stories.

Finally, in the last section of this chapter, we'll discuss when to and when not to use other analytics add-ins to extend your types of analyses even further. We'll provide two generalized case studies, one on A/B testing and one on text analysis that can easily be ported over to your own use cases.

In This Chapter

Each strategy in this chapter uses Tableau Desktop as a starting point. To explore more-advanced mathematical modeling, you'll need to have a local TabPy (Python) server or a local Rserve (R). Both tools are free to install. In this chapter, you'll learn how to do the following:

- Use built-in forecasting features to predict the future of time-bound data
- Work with two metrics to explore relationships
- Determine whether there is statistical significance or correlation between two metrics
- Organize and categorize the relationship between two metrics into a chart that executives can understand
- Work with multiple data attributes (dimensions) to segment your data
- Use alternative ways to display data, including dynamically built trellis charts and nonlinear numerical axes
- Perform advanced mathematical modeling, using external statistical languages

Forecasting

Forecasting is the process of fitting models by using historical data to predict future observations. It is also one of the most common use cases for time-bound data. Many industries, like retail, rely on historical data that displays seasonality to predict or provide direction as to what's coming in the following months. And while that may seem unsophisticated, many processes do follow seasonal trends that can be well predicted. In retail, you know that each year an uptick in sales will occur during the last quarter, driven by holiday shopping. Similarly, in the US healthcare industry, a large increase in utilization often occurs as Americans' insurance benefits reset each calendar year—allowing them to avoid paying out-of-pocket costs for capping out on their healthcare benefits.

That said, forecasting is not foolproof and cannot react to unforeseen events like natural disasters, rapid economic changes, social unrest, or a global pandemic. The COVID-19 pandemic is a perfect example: no forecasting model in existence could have accurately predicted the dramatic changes in consumer habits that resulted. You can bet that any forecasting models for retail, major sports, food service, travel, and recreational industries were rendered completely useless during this global pandemic.

When you use forecasting, think of it as guidance of what is to come, and specifically think of it in terms of "what's to come if it's business as usual." If you present it in that lens to your audience and in your visualization practice, it can be a useful tool in planning for future outcomes.

Using Forecasting to Predict Staffing Needs: CaB Call Center Case Study

One industry for which it is critical to predict future events is *customer service*. Many organizations have minimum service-level requirements to keep their customers satisfied, and often the business-to-business (B2B) world has service-level agreements (SLAs). So, in these situations it is critical to have the correct number of employees staffed.

The call center of automotive parts manufacturer CaB can average 3,000 calls or more each month. Calls can spike around standing order deliveries and new product releases. The company wants to be able to more accurately forecast its call loads to manage staff hours, reduce caller wait time, and improve customer satisfaction. How can you take advantage of the vast amount of time-bound data available to predict future call volumes at many levels of time (monthly, weekly, daily)?

Strategy: Create a Monthly Forecast

The manager of the CaB call center wants to know what monthly call volumes are going to be in the future. She wants to use this information to estimate the number of additional employees she may need to hire, given current employee attrition. You'll create a monthly forecast as follows:

1. Create a line chart that trends Calls by Month. Do this by dragging `SUM([Calls])` onto the Rows shelf and Month(Start Date Time) onto Columns.

We chose to limit data to between 1/1/18 and 3/31/20. This was done to keep whole months of data and to ensure that the forecasting model wasn't incorrectly influenced by partial-data months. When you are forecasting, always choose the most relevant time periods for your analysis.

2. On the Analysis pane, drag and drop [Forecast] onto your chart (Figure 8-1). The data in your chart should extend and show forecast values for the future (Figure 8-2).

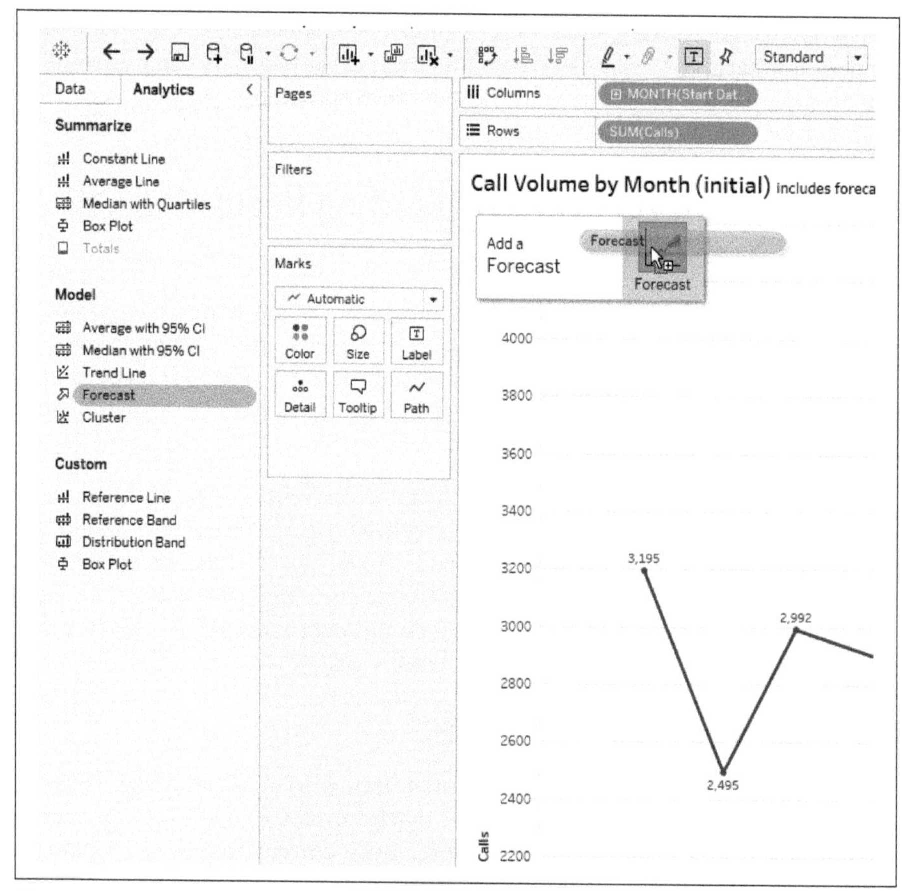

Figure 8-1. Dragging and dropping a forecast line onto a line chart

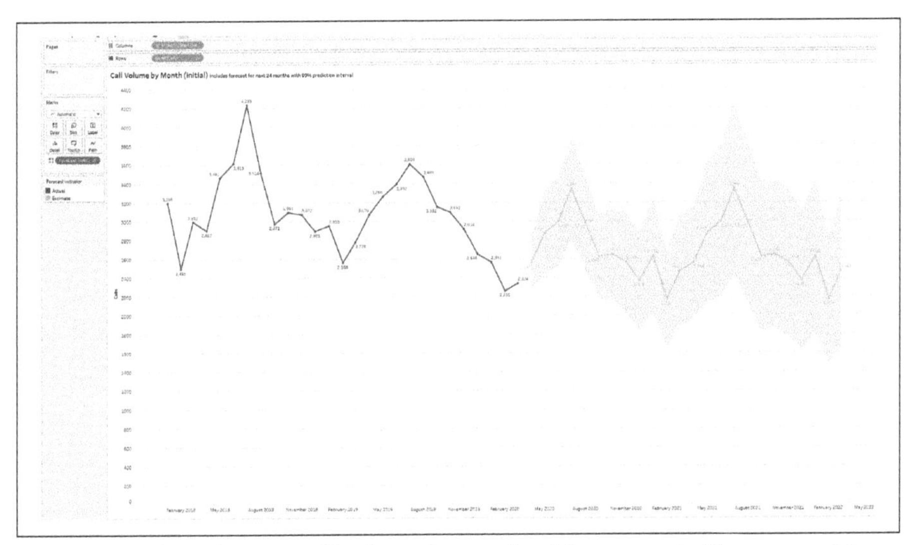

Figure 8-2. Chart showing call volume by month with automatically generated forecast

3. Remove the break in the line between actual data and estimated data by removing the automatically generated Forecast indicator from Color on the Marks card.
4. Customize the forecast by choosing Analysis → Forecast → Forecasting Options. Change the length to 24 months. Adjust the prediction intervals to 99%. Your chart should now match Figure 8-3.

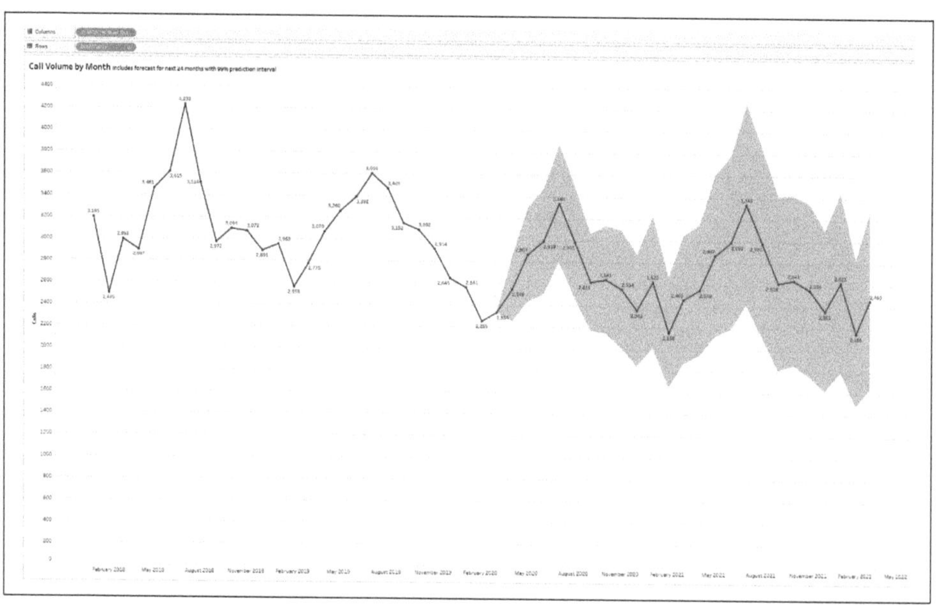

Figure 8-3. Customized forecast of monthly call volume that extends for 24 months and includes a 99% prediction interval

5. View the summary of the Forecast output by choosing Analysis → Forecast → Describe Forecast to open the Describe Forecast dialog box (Figure 8-4).

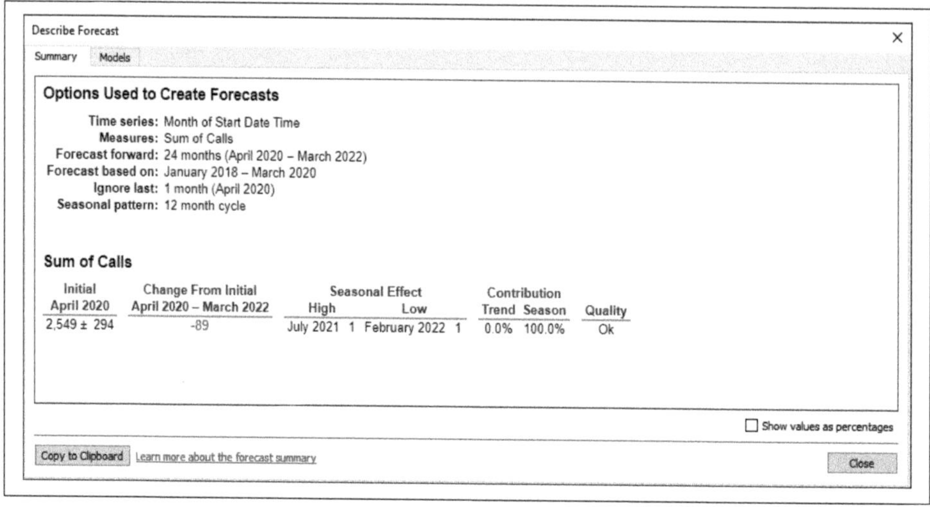

Figure 8-4. Forecast summary description generated by Tableau

After creating forecasts in Tableau, it is critical to review the summary. Here you will find key information, including the quality of the model (valued at Good, Ok, or Poor). Mastering what's in the Summary and Models sections will help you to gain confidence in describing and using the model generated.

In the previous strategy, we just scratched the surface on forecasting in Tableau. It is worth mentioning you can access detailed documentation on how forecasts are generated and how to interpret all of the statistical components that are output by clicking the "Learn more about the forecast summary" link at the bottom of the Describe Forecast dialog box. It also offers guidance on customizing your forecast model. When explaining the key properties of forecasting in Tableau to others, we narrow our focus of key features as follows:

- Uses exponential smoothing, a method that decreases the weight of older observations
- Uses both trend and seasonality to generate a forecast, and both can be customized to be additive or multiplicative, based on your existing understanding of the data

You can visit Tableau Desktop's help article to learn more about describing a forecast output (*https://oreil.ly/cONgi*).

Relationships Between Two Numerical Values

Another powerful type of analysis is exploring the relationship between two metrics. Most data points are likely to have more than one numerical metric, so it is natural to want to explore those properties and see what, if any, relationship or influence the metrics have on one another.

A practical example is looking at the life expectancy of a country's population against the percentage of population living in poverty (Figure 8-5). You or your audience may have a natural inclination that a relationship exists, but by using the concepts explored in this chapter, you can present the findings along with statistical significance.

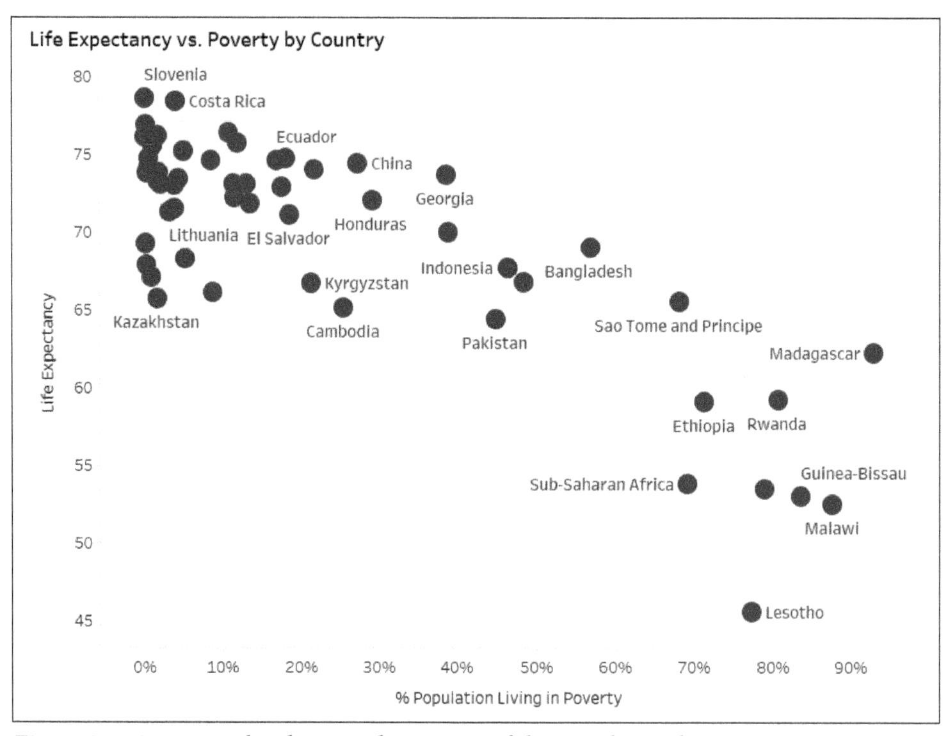

Figure 8-5. A scatter plot showing the percent of the population living in poverty compared to the average life expectancy

After inspecting Figure 8-5, it would certainly be appropriate to say that, in general, countries with less poverty have longer life expectancy—but the next level of this analysis, and the scope of this chapter, would be to explore it mathematically. This eliminates overestimation and intuition-based inferences to describe the data presented to the audience.

It is also worth noting that you don't always have to apply additional statistical processing when exploring two metrics. It is enough to use these methods to gain a deeper understanding of your data and how the distributions of two metrics relate to one another. You can also use this type of analysis as a simple way to organize data points for your audience in an actionable way.

Using Scatter Plots to See Relationships Between Spend and Conversions: SAGE Digital Marketing Case Study

In marketing, there is always a drive to optimize—for example, optimizing each ad to achieve the greatest results, frequently measured by the number of consumers purchasing versus the amount of money invested. Achieving this optimization can take the form of changing the medium of the ad or the location (think social media or search results).

To address this ever-pressing need, a practical tool is the scatter plot. A *scatter plot* positions one metric on the x-axis and the other on the y-axis. The dots plotted depend on the level of data you want to display and can range from individual records (think a specific ad) or more aggregated results (like a whole marketing campaign).

In this case study, SAGE Digital Marketing is seeking to quickly identify which campaigns in a given marketing channel were most effective at driving consumers to purchase. How would you design a scatterplot to visualize this data?

Strategy: Create a Scatter Plot

Starting with a new worksheet, you'll construct a scatter plot to compare two measures:

1. Drag SUM(Spend) to the Columns shelf and drag SUM(Conversions) to the Rows shelf.
2. Drag [Channel] to Color and Label. Drag [Platform] and [Campaign] to Detail.
3. Adjust the marks to be circles, make the size larger, and adjust the opacity of the dots to 75%.

   Stop here and take a moment to review the plot shown in Figure 8-6.

In general, what conclusions can you draw?

- OOH (Out of Home aka Billboards) is extremely expensive, yet yields relatively few conversions.
- There's one Paid Search that performed tremendously well, netting almost three times the number of conversions as the other initiatives.
- The rest of the data points are dispersed relatively evenly in the bottom-left sector of the chart—having similar spend and conversion outcomes.

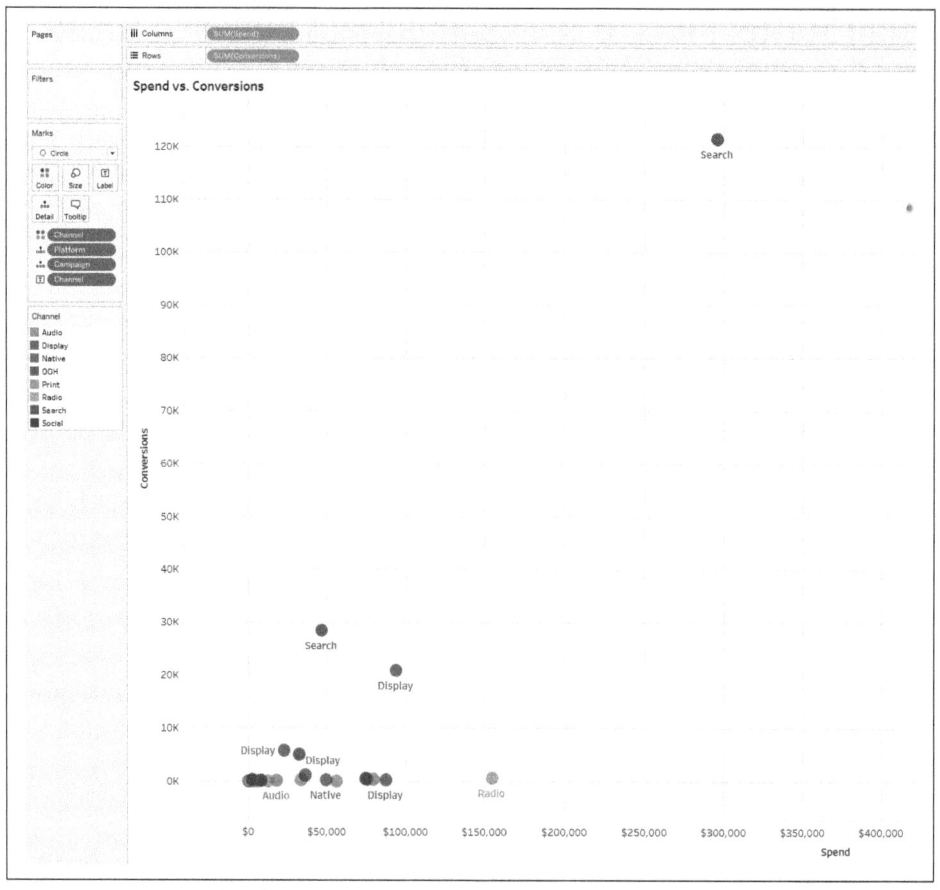

Figure 8-6. Scatter plot showing Spend versus Conversions for marketing data

This scatter plot by itself has a wealth of insights that the audience can now describe. But what if we want to take this view and make it even more actionable? In the next strategy, we'll explore how to turn your scatter plot into a chart that categorizes the outcome of each data point.

Strategy: Create a Quadrant Chart

As you just saw, the scatter plot constructed in the previous strategy tells a compelling story around two critical metrics. To make this dialog of insight more visible and understandable to your audience, you can turn the scatter plot into a quadrant chart. A *quadrant chart* is a transformed scatter plot, in which average lines are used on both the x-axis and y-axis metrics to segment your data into four sections, or quadrants. Once the data is sectioned into quadrants, you can label them into groups that generalize their performance.

For our marketing optimization scenario, we want to end up with four categories:

- Low impact, high cost
- High impact, low cost
- Low impact, low cost
- High impact, high cost

Stakeholders who are tasked with making optimizations under these categories face a simpler process of deciding which initiatives to keep or invest in further, and which ones may need to be removed. The steps for this strategy are as follows:

1. From the Analysis pane, click and drag an Average Line for both `SUM(Spend)` and `SUM([Conversions])`, as shown in Figure 8-7. Both should be for the Table.

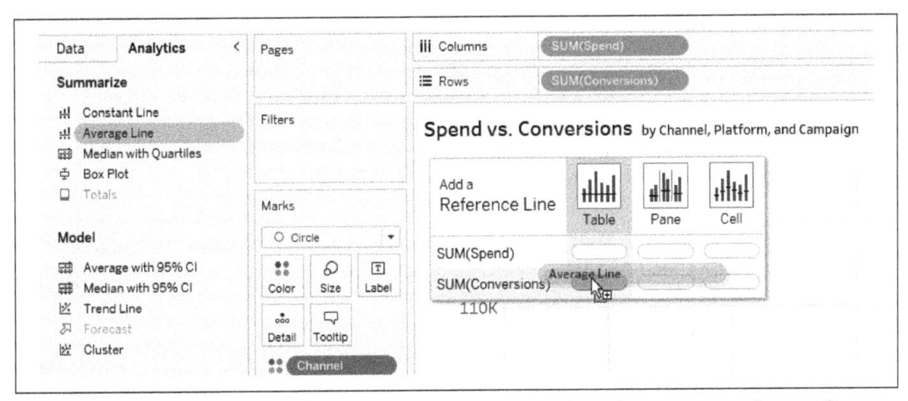

Figure 8-7. The drag-and-drop menu that appears when dragging a reference line to a view

When adding a reference line, there are three ways to address your data based on the dimensions on Rows and Columns. *Table* reads across all data (think left to right); *Pane* will construct a line that is for the subset of data delineated by a header; and *Cell* will construct a line for the innermost header or individual data point.

2. Right-click each Average line, select Edit, and modify settings so you have a dashed line with the value displayed along with the average. Uncheck "Show recalculated line for highlighted or selected data points." Figure 8-8 shows the result.

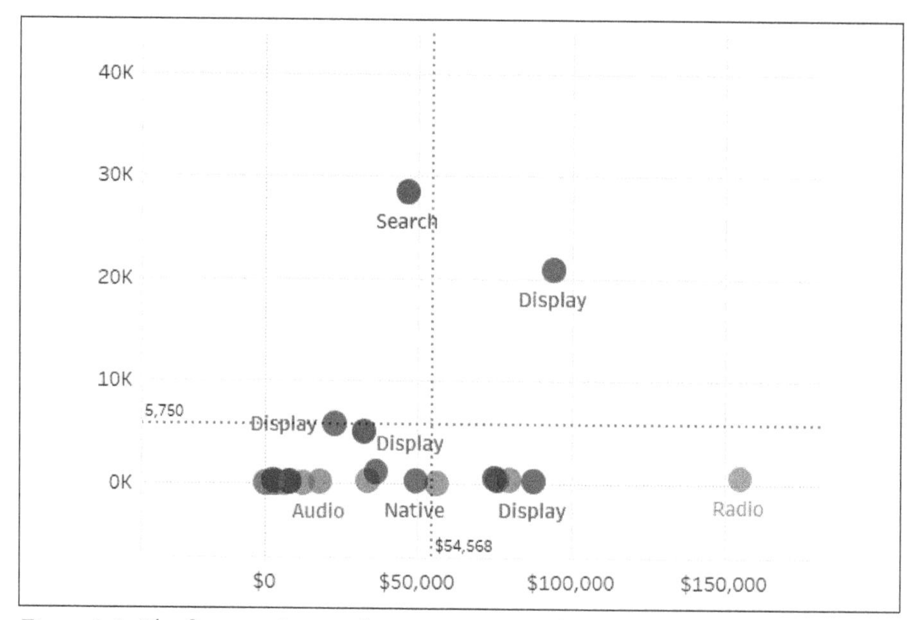

Figure 8-8. The first step in transforming a scatter plot to a quadrant chart; average lines have been added for each axis.

Next, you'll complete two items: one is a calculated field that will categorize the data points into the four labels we've mentioned, and the other is to annotate the chart to emphasize those categories.

3. Create a calculated field called **`[Average Spend]`** and a field called **`[Average Con versions]`**. These two fields will match the value of the average lines on the chart:

```
//Average Spend
 WINDOW_AVG(SUM([Spend]))
//Average Conversions
WINDOW_AVG(SUM([Conversions]))
```

We're using table calculations to construct the same values as the reference lines. Working with reference lines and table calculations together is an easy way to validate that your table calculation is accurate.

4. Create a calculated field called **`[Optimization Category]`**:

```
//Optimization Category
IF SUM([Spend])>=[Average Spend] AND
SUM([Conversions]) >=[Average Conversions]
THEN 'High Cost, High Impact'
ELSEIF SUM([Spend])>=[Average Spend] AND
SUM([Conversions])<[Average Conversions]
THEN 'High Cost, Low Impact'
ELSEIF SUM([Spend])<[Average Spend] AND
SUM([Conversions])<[Average Conversions]
THEN 'Low Cost, Low Impact'
ELSEIF SUM([Spend])<[Average Spend] AND
SUM([Conversions])>=[Average Conversions]
THEN 'Low Cost, High Impact'
END
```

5. Drag [Optimization Category] onto Color, replacing [Channel]. Right-click [Optimization Category] and ensure that it is computed across all dimensions at the deepest level.
6. Right-click the whitespace of a quadrant and select Annotate → Area. Enter the Optimization Category that represents the quadrant. Adjust formatting as desired.
7. Filter your data to August 2020, September 2020, and October 2020—fiscal quarters for SAGE. Figure 8-9 shows the plot at this stage.

Now you have a completed quadrant chart, one that you can present on a dashboard or to stakeholders. Executives will now be able to derive insights from this two-dimensional chart, which makes it easy to focus on a subset of items for optimization.

The final concept we'll cover with scatter plots is trend lines. A *trend line* is added to a chart to show the general direction of data or the relationships among numerical data. Constructing and including trend lines is useful when you explicitly want to demonstrate a correlation between variables or to estimate/predict a variable.

When you create trend lines with data, you're generating a mathematical equation, with an x-axis metric, or the *independent variable*, and an output that is the y-axis variable, or *dependent variable*. The type of equation you generate depends on the relationships among your data (line of best fit), and similarly in Tableau you can generate various line types. The most commonly known type of trend line is a *linear trend* that follows the pattern $y = mx + b$, where m is the slope or angle of the line, and b is the y-intercept, or the point where the line crosses the y-axis.

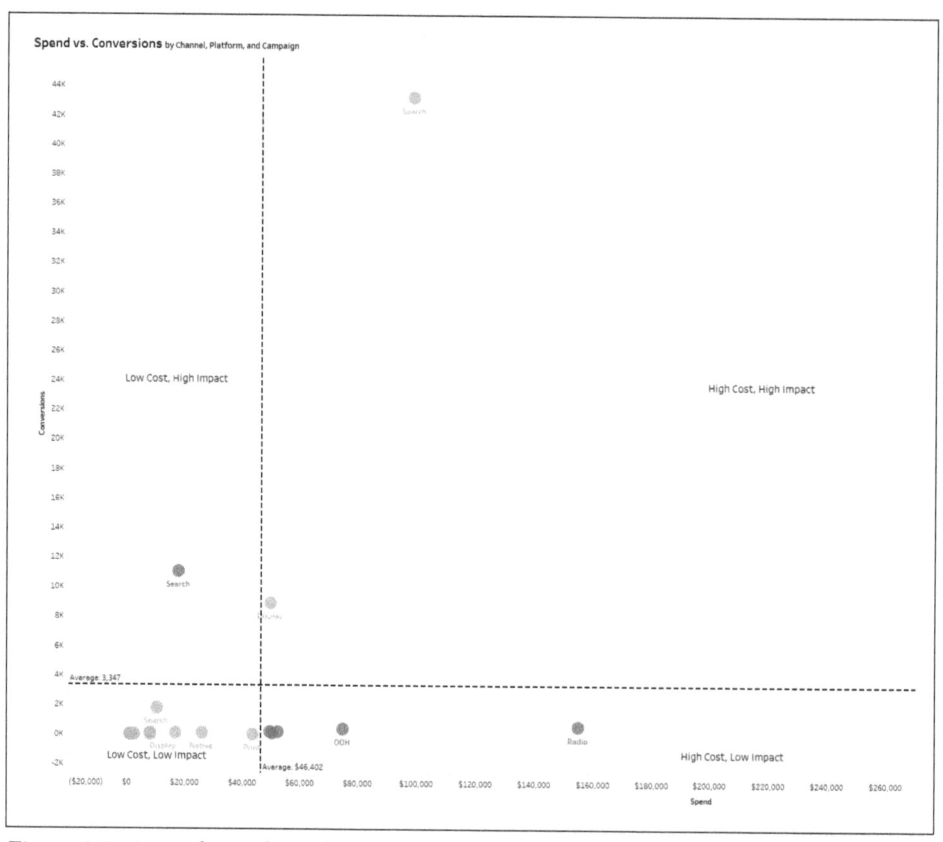

Figure 8-9. A quadrant chart showing the four categories of optimization the data falls into

It is worth noting that you can always place a trend line on a chart, selecting the line of best fit, but drawing/constructing a line does not guarantee a statistical or mathematical relationship among your metrics. You'll need to use the outputs provided from the trend line to determine whether the line is trustworthy for predictions or explanations.

Strategy: Add a Trend Line to a Scatter Plot

In this strategy, you are going to continue working with marketing data, but change the two metrics being examined. This time, you are going to use impressions and conversions, and instead of breaking apart the data by marketing channel and campaign, we'll break the data apart by day. This will facilitate answering the question, "How many conversions can we anticipate receiving based on the number of impressions for a piece of marketing content?" The relevance of this exercise is being able to predict and have an estimated outcome if you know the number of impressions something will receive:

1. Drag SUM(Impressions) onto the Columns shelf and SUM(Conversions) onto the Rows shelf.
2. Right-click and drag the date onto Detail, dropping it with the selection on the continuous date value.
3. Filter the data to include only the Search Channel (Figure 8-10).

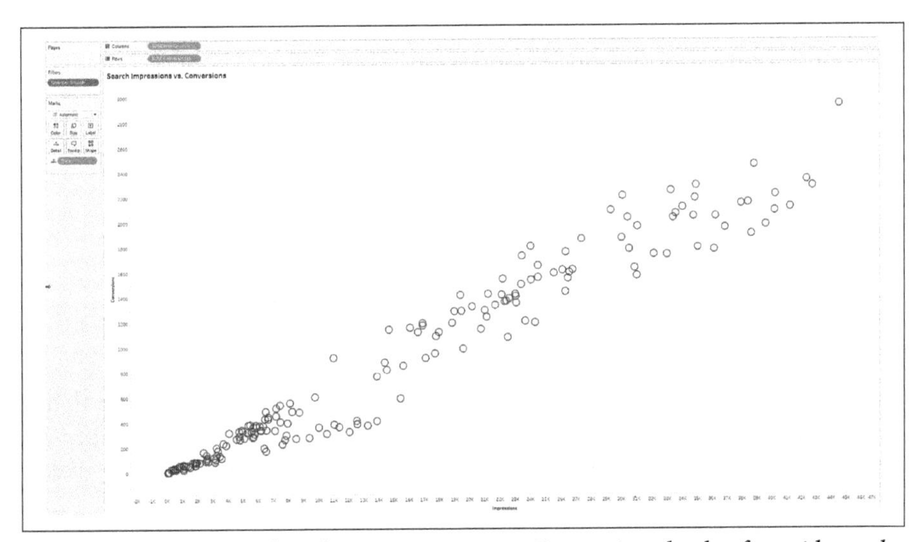

Figure 8-10. A scatter plot of Impressions versus Conversions by day for paid search marketing

From this visual you've just created, it might already be obvious that a strong relationship exists between these two metrics (variables). But next, you will add a trend line to express the relationship mathematically and get an output that demonstrates the confidence you can have with the model.

4. From the Analytics pane, drag the Trend Line and drop it on Linear.

Tableau provides five equations for generating a trend line (Figure 8-11):

- *Linear* follows the format $y = mx + b$.
- *Logarithmic* generates a curved line that plateaus at one end.
- *Exponential* also creates a curved line, often used when there is rapid increase or decrease.
- *Polynomial* generates curved lines that bend one or more times.
- *Power* creates a curved line with a slope that is changing at a known rate.

Since there are many options, we encourage you to explore them and find the one that fits best.

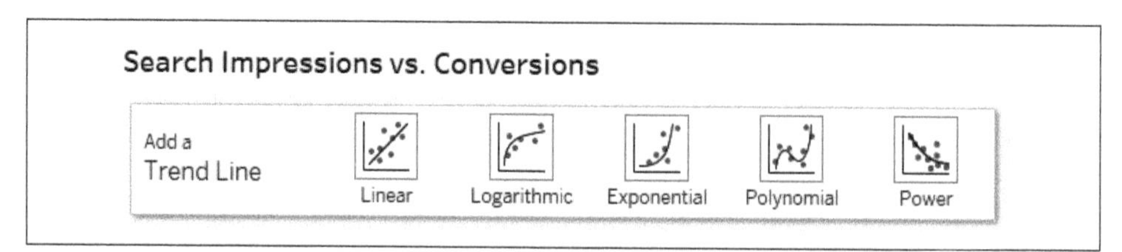

Figure 8-11. The types of trend line models that are built into Tableau

Now that you've built the trend line (Figure 8-12), here comes the important part—evaluating the model.

Anytime Tableau generates a trend line, it includes the equation of the line along with a menu you can access that describes the trend model. You can access this menu by right-clicking the line and selecting Describe Trend Model. This opens the dialog box shown in Figure 8-13.

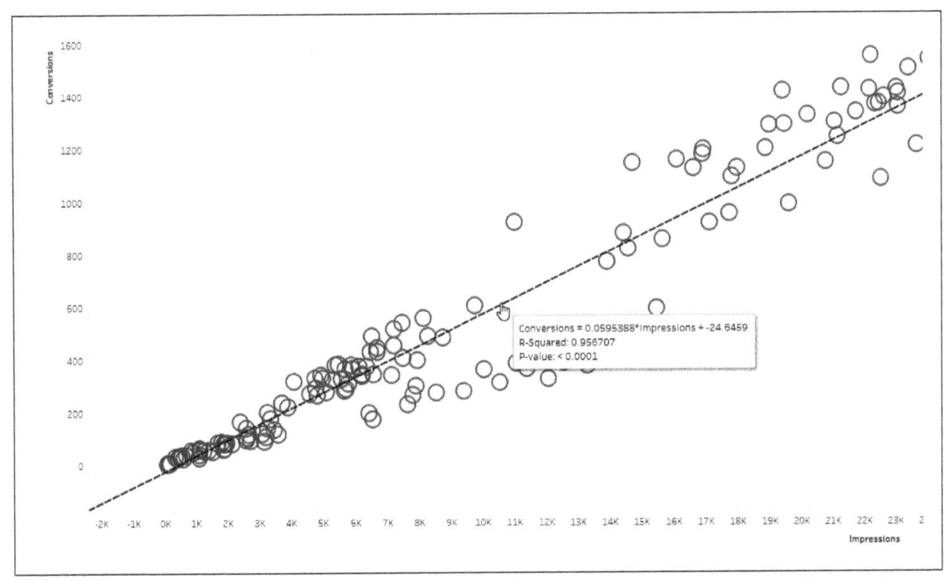

Figure 8-12. A linear trend line fit to the chart showing Impressions versus Conversions

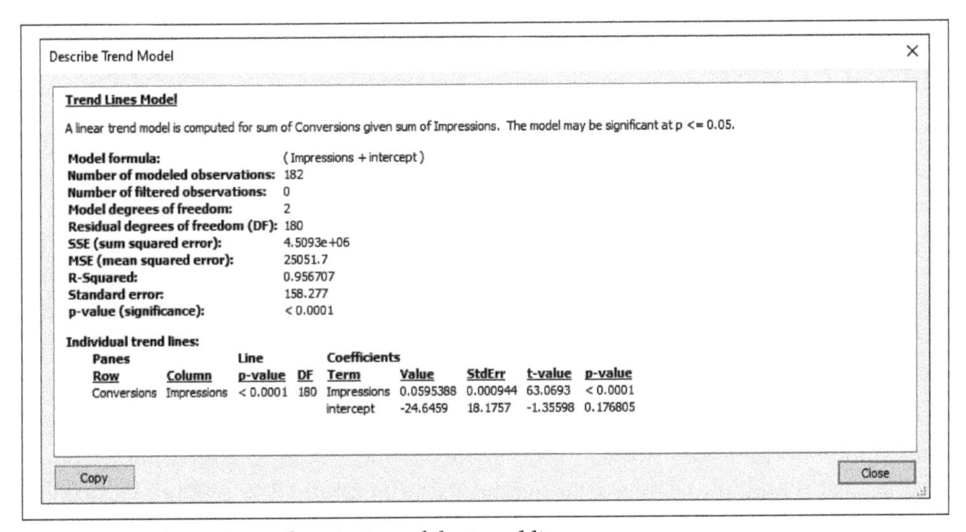

Describe Trend Model

Trend Lines Model

A linear trend model is computed for sum of Conversions given sum of Impressions. The model may be significant at p <= 0.05.

Model formula:	(Impressions + intercept)
Number of modeled observations:	182
Number of filtered observations:	0
Model degrees of freedom:	2
Residual degrees of freedom (DF):	180
SSE (sum squared error):	4.5093e+06
MSE (mean squared error):	25051.7
R-Squared:	0.956707
Standard error:	158.277
p-value (significance):	< 0.0001

Individual trend lines:

Panes		Line		Coefficients				
Row	Column	p-value	DF	Term	Value	StdErr	t-value	p-value
Conversions	Impressions	< 0.0001	180	Impressions	0.0595388	0.000944	63.0693	< 0.0001
				intercept	-24.6459	18.1757	-1.35598	0.176805

Copy Close

Figure 8-13. Accessing a description of the trend line

When we review these models, we hone in on two key components: the p-value and R-Squared value. The *p-value* is the probability that the output is caused by random chance. The smaller this value, the more likely it is that the output is statistically significant. Said in laymen's terms, the closer the p-value is to 0, the more likely it is that a statistical relationship exists between the two variables. The *R-Squared* value is a number between 0 and 1 used to describe the amount of variability in the trend line. The closer R-Squared is to 1, the better the line fits the data.

If you are getting very small p-values, but not high R-Squared values, test a different model type to see if you can get a better line of fit.

The model generated in Figure 8-12 is fantastic—the p-value is very small, and R-Squared is very close to 1. If you knew the number of impressions that were going to be served, you could confidently predict or estimate the number of conversions. For the strategy, the specific model equation is Conversions = 0.0595388 × Impressions – 24.6459. Stated more clearly, on any given day, you're likely to receive 19 conversions for every 100 impressions.

Bringing it back around to our marketing case study, this means that we can also conversely figure out the number of ads to serve to reach a specific conversion value, equating to revenue generation. Essentially, if we want to sell 100,000 units of a product, we will need to generate about 1.68 million impressions.

Cluster Analysis

With the explosion of technology surrounding data storage and data processing, the domain of data analytics has extended to include more statistical and mathematical methods of processing data. Data science is increasingly playing a common role in the world of business. Data science takes advanced mathematical and statistical concepts, along with machine learning, and applies it to data models. The algorithms used in data science rely heavily on some form of programming, since the way they are generated is usually complex, involves many iterations, or has an element of recursion.

In the last part of this chapter, we will explore integrating two data science applications into your Tableau visualizations, but for now, we want to start with a type of analysis that is built into Tableau: cluster analysis. *Cluster analysis* is the task of taking similar data points and grouping them together. Different from manual grouping or categorization, cluster analysis uses an algorithm to determine similarities and differences among data points. Tableau uses the k-means clustering algorithm to generate clusters.

Cluster analysis has many practical applications in business analytics. It can be used for population segmentation, development of customer personas, fraud and anomaly detection, and optimization. It can be used when you have numerical data values, categorical data values, or both. It also works well because it works with many variables or fields, when it may not be feasible to plot four or five different data points on a single chart and assess for similarities.

Just like forecasting, cluster analysis isn't a panacea for categorizing your data. Instead, consider it a suggestion of a way to group, categorize, or segment your data. Because it relies on mathematical outputs, business acumen and domain knowledge should be applied in conjunction to the output to determine whether the results will be useful to your organization and use case.

Creating Segmentation Among Employees to Assess Differences in Attrition: Banco de Tableau Case Study

Banco de Tableau is considering offering remote employment opportunities to some of its customer support and IT staff. Company executives are aware that advances in technology are having a broad impact on the landscape of office work and want to understand whether a link exists between an employee's proximity to the office and staff attrition rates. In particular, they are seeking to understand whether they are losing top talent because of concerns such as a long commute.

The HR team has narrowed its focus to three key contributors to work-life balance and happiness at work: tenure, monthly income, and distance from the office. How would you use cluster analysis to take these three variables and segment them into groups?

Strategy: Add Cluster Analysis to a Scatter Plot and Use Analysis in Another Chart

Before you get started, we want to note that most often when you see cluster analysis done, it starts with a scatter plot. That's not by coincidence; since you'll generally be working with three or more variables, it is natural to start with a scatter plot to begin to explore relationships among the variables. That said, it is not required for cluster analysis, and specifically won't work if you have a series of dimensions you want to group together instead of numerical and categorical data. Let's get started:

1. Create a scatter plot that has [Years at Company] versus [Monthly Income] by [Employee Number].
2. Limit the data to the Sales Department.
3. Include the third metric, [Distance from Home], by adding it to Size; change the mark type to Circle and set the opacity to 70%. Figure 8-14 shows the resulting scatter plot.

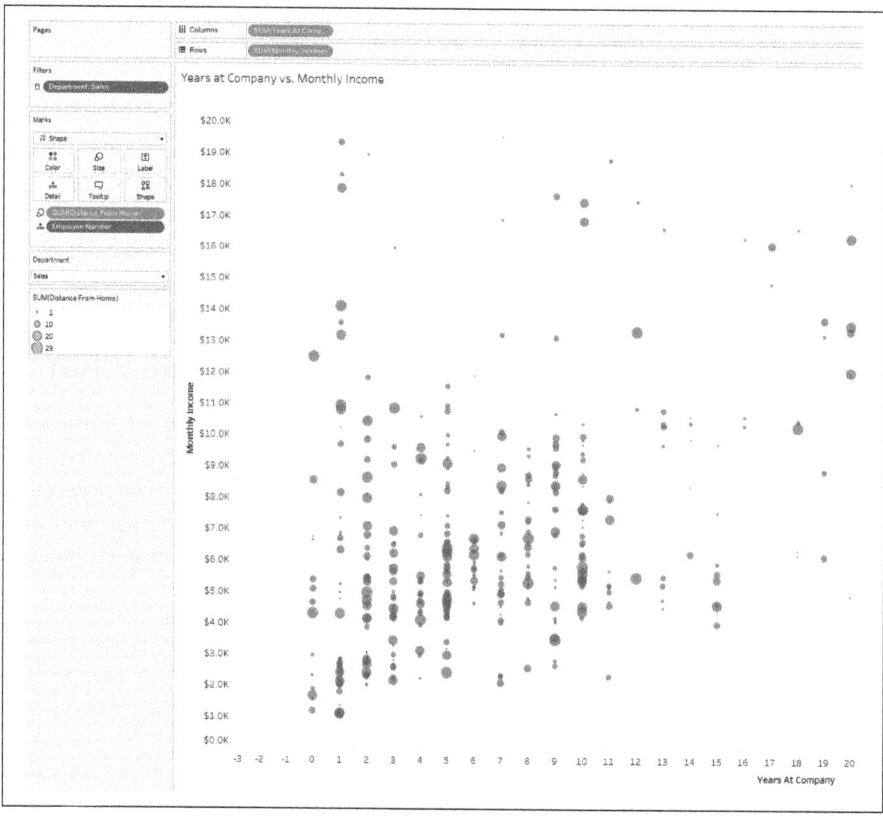

Figure 8-14. A scatter plot that has an additional variable, Distance from Home, on Size

Now stop and review this visualization. If you had to group similar dots, where would you begin? Unfortunately, for most humans, once we introduce a third dimension, it becomes increasingly difficult to complete the task. But that's exactly where cluster analysis comes into play.

From the Analytics pane, drag [Clusters] onto the worksheet. Figure 8-15 shows the result.

When you create clusters in Tableau, you have control over what is used in the analysis and the number of clusters. There's no right or wrong answer, so work with variables that you think may have an impact and refine the number of clusters for your use case. Just know that Tableau defaults to an automatic, or optimized, number of clusters (meaning the greatest number of clusters that have the most dissimilarity to the other clusters and similarity to other members within a cluster).

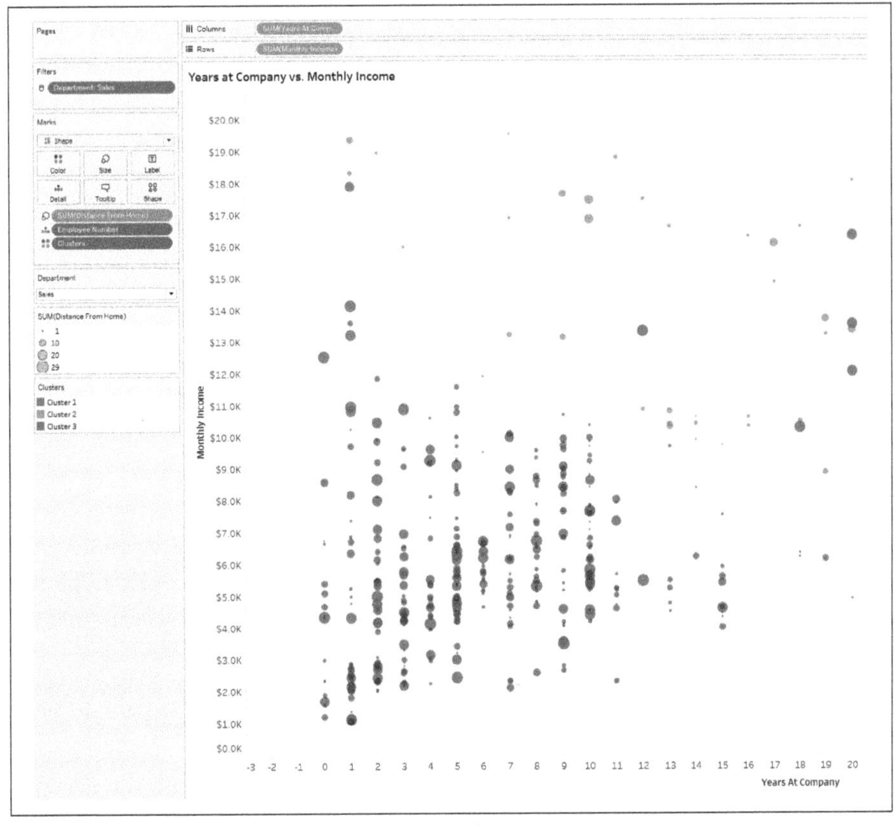

Figure 8-15. The result of adding cluster analysis to Figure 8-13

4. Right-click the [Clusters] field on the Marks card and select Describe Clusters. You'll see the summary of details in Figure 8-16.

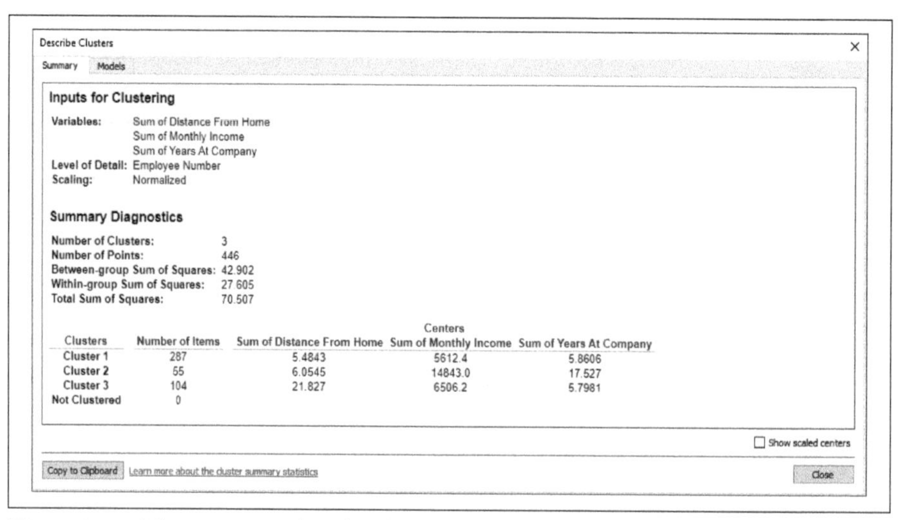

Figure 8-16. The summary details related to the cluster analysis of HR data

This step is important. This is where you will provide a semantic description and interpretation to the clusters generated. The values shown for each metric in the description represent the mathematical center of each cluster—not quite the average. Instead, imagine the cluster as a circle or oval and the data point at the center. In this example, we have three clusters, which we would define as follows:

Cluster 1
: Short Commute, Low Income, Low Tenure

Cluster 2
: Short Commute, High Income, High Tenure

Cluster 3
: Long Commute, Low Income, Low Tenure

5. Now hold down Ctrl and drag the [Clusters] field to your Data pane to save it as a field. Name it **`[Employee Groupings]`**.
6. Right-click and edit the group to rename the clusters to match the descriptions we've chosen. As we said before, this is just the beginning of using cluster analysis. The fun comes when we apply these new segmentations to other variables.
7. In a new sheet, construct a bar chart showing the number of employees by [Employment Status].
8. Drag [Employee Groupings] to Color and change the metric to percentage of total for each status. Figure 8-17 shows the completed chart.

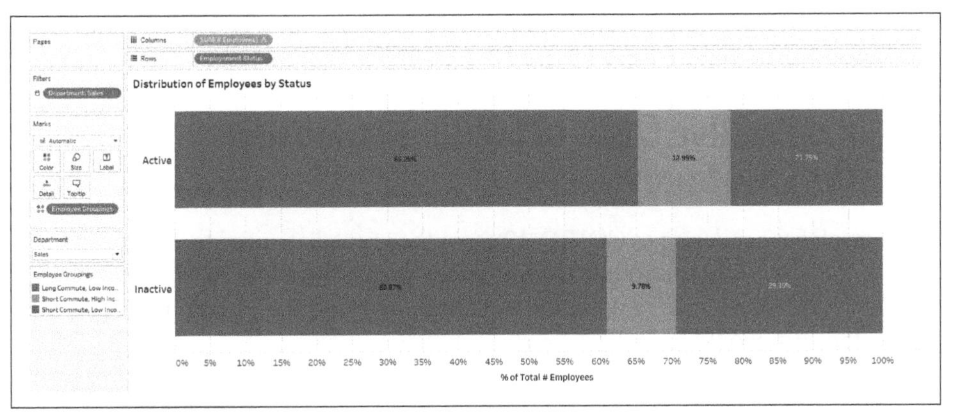

Figure 8-17. Stacked bars showing the distribution comparison among active and inactive employees

Now you have some fascinating results. Clearly there is a difference of the proportions between the active and inactive employees. Most strikingly, there's a larger proportion of employees who are no longer with the company and who had a long commute, low income, and low tenure. And what's even more useful is that you can use the centers produced by the analysis as guidelines to identify which employees, both current and future, may be at higher risk of leaving the organization.

Alternative Axis and Many Multiples

If you've ever encountered a dataset that is tough to visualize, this section is made just for you. Some datasets simply don't conform to conventional visualizations. Maybe the data is changing exponentially, making it difficult to plot over time, or there are too many attributes to make a compelling comparison.

When we encounter these situations, it's even more important to step back and try to find the right visualization to communicate the data. Because the properties of the dataset are complex, it's probably even more important to lean on visualizations to synthesize and communicate it. And as we write this, there's no better example than COVID-19 data. Unlike traditional data you're likely to encounter, it has unique properties that must be mastered. Depending on the given month, the number of positive people can be doubling or tripling daily. And the headline and major insight is exactly that point—clearly demonstrating how rapidly the virus is spreading.

To deal with rapidly changing numerical values, and especially those that have a massive range (think 1 to 1 million), we recommend a logarithmic scale. A *logarithmic scale*, or *log scale*, is a nonlinear scale with tick marks that aren't spaced out by the same amount. Instead, they are spaced out by multiples, or factors, of a base scale. The beauty of a log scale is that it can display a massive data range quite compactly and can aid in providing a chart narrative; the language of visual math becomes "*x* times more" or "*x* times greater" instead of an additive amount.

Tracking Positive Cases of COVID-19 Globally: Logistics Case Study

With the global pandemic of COVID-19 came the immediate need to understand and analyze the virus. Organizations, world leaders, media, health experts, and individuals around the globe were looking for ways to understand the situation and potential impact the virus would have.

The first instinct most global leaders had when they learned of the severity of COVID-19 was to start assessing the total number of positive cases in a given country compared to its peers. This was useful for figuring out where the greatest need was and which preventative measures were contributing to positive outcomes. How would you design a logarithmic scale to visualize daily infection rates by country?

Strategy: Use a Logarithmic Axis to Understand Rate of Change

In this strategy, you will use a log axis to plot eight countries across the globe—Argentina, China, France, Italy, South Korea, Spain, United States, and Zambia:

1. Create a trend line showing SUM(People Positive Case Count) by day. *Note: this measure is a cumulative total.*
2. Filter the data to the eight countries mentioned and place [Country Short Name] on Color.
3. Right-click the y-axis and change the Scale to Logarithmic; leave it set to Positive. You can also hide the indicator at the bottom right noting that there are negative values. Figure 8-18 shows the resulting visualization.

Logarithmic scales typically work only with positive numbers, but Tableau also allows users to select Symmetric, which includes negative numbers also displayed logarithmically.

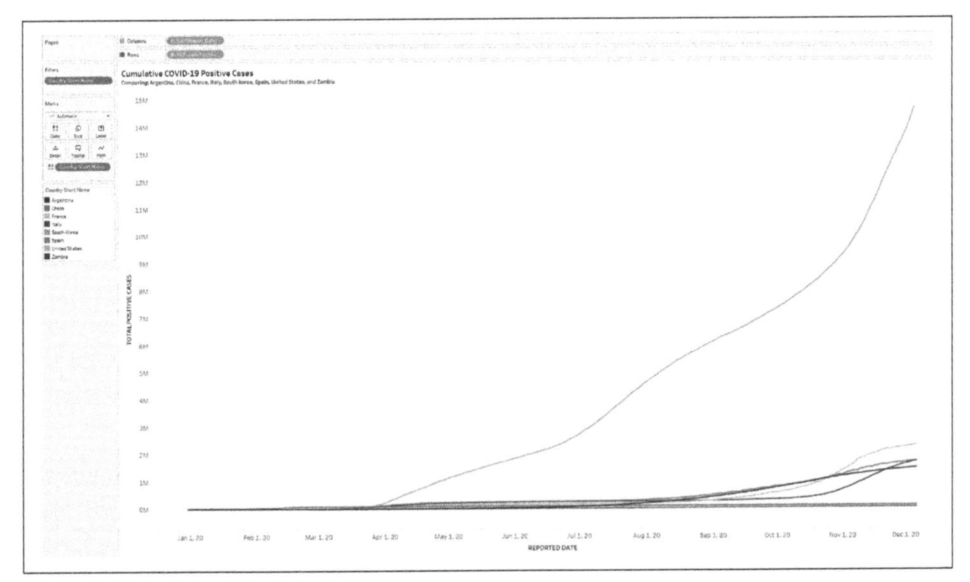

Figure 8-18. Eight countries, plotted on a linear axis

Now compare the outcome to a chart with a linear axis, as in Figure 8-19. How does your visual interpretation and the story change?

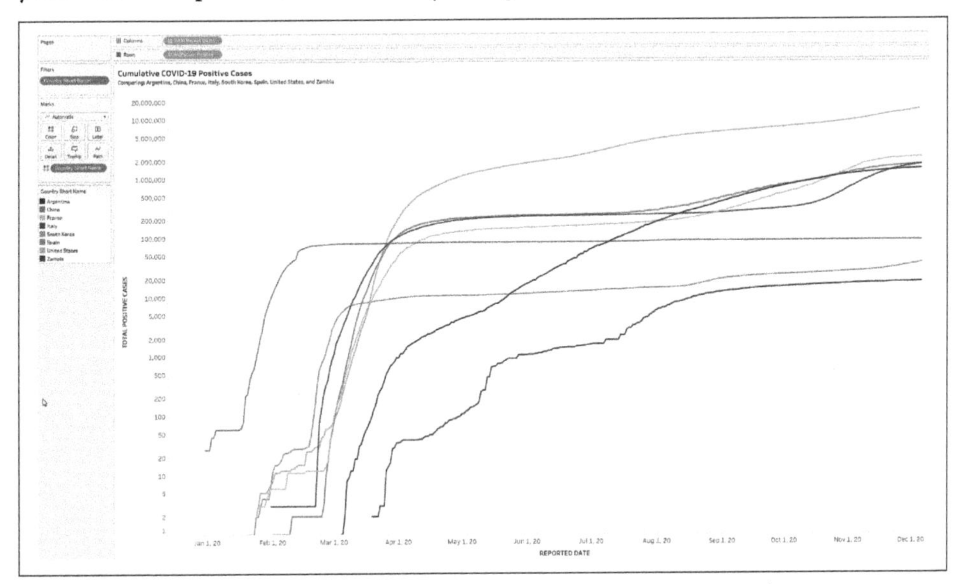

Figure 8-19. Total COVID-19 cases by day for eight countries, displayed on a logarithmic scale

With the log axis, it is immediately clear to the audience how quickly cases surged in the eight target countries. It also does a great job of displaying nuances among the countries. China, for example, hits a peak reported cumulative total around 75,000 and then tapers out to a completely flat line. Zambia, on the other hand, has seen two or three surges since it has begun reporting. And the US in the linear plot dwarfs the other countries, but in the log axis view can more clearly be compared to peers. At the time of this writing, the US has almost seven times more total positive reported cases than France.

While here we're focused on COVID-19, logarithmic axes are useful anywhere you have a metric with exponential growth. Other common subject areas that are typically plotted logarithmically include growth and adoption (of technology or a product), money interest and savings, and (uncontrolled) population growth.

As we move on, there's one additional tactic we employed to really help demonstrate COVID-19 data—and that is normalizing the x-axis, or time. Remember we said our stakeholders wanted to know how effective various mitigation strategies were in the countries? In the current view, it would be difficult to see whether a full shutdown order in China in February had the same impact as a partial stay-at-home order in the US. But by now you're an expert in Tableau, so this should be an easy solve!

Strategy: Normalize a Date Axis

In this strategy, we will normalize the reporting date so that each day is in reference to when the virus was first reported in a given country:

1. Create an LOD expression called **[First Date per Country]**; this calculation will be the first reported date for each country:

```
//First Date per Country
{FIXED [Country Alpha 3 Code]: MIN(IF [People Positive Cases Count]>0
  THEN [Report Date] END)}
```

This data contains records for every country going back to 12/31/19, regardless of whether any cases were reported. Adding the IF statement requiring positive cases to be greater than 0 ensures that the minimum date indicates when the first case was reported.

2. Create another calculated field called **[Days Since First Case]**:

```
//Days Since First Case
DATEDIFF('day',[First Date per Country],[Report Date])
```

3. Before dragging [Days Since First Case] onto Columns, turn it into a continuous dimension by right-clicking and selecting Convert to Dimension and then Convert to Continuous.

Remember, you can change the property of a field in Tableau at any time. When you set this field as a dimension, you tell Tableau you don't want it to be aggregated when it is used in a worksheet. By making it continuous, you're also telling Tableau that you want an infinite number line or axis when it is plotted.

4. Drag [Days Since First Case] on top of DAY(Report Date) to replace it.

As we review this final transformation (Figure 8-20), we ask again, how has the data story changed?

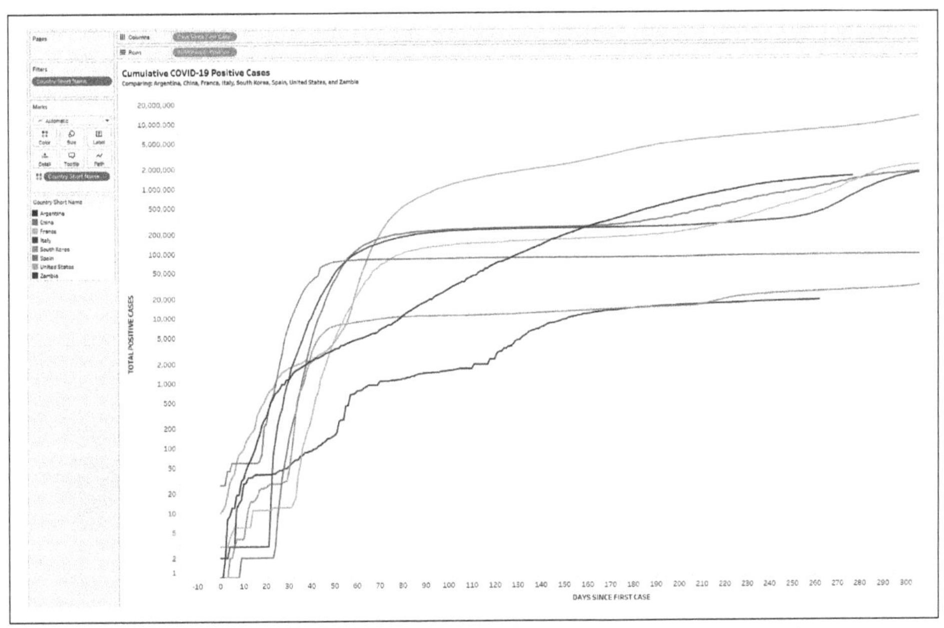

Figure 8-20. COVID-19 data on a log axis with reporting dates normalized

With each country set to the same timescale, it becomes easier to compare and contrast how actions taken have impacted outcomes. Within 50 days of the first reported case, South Korea was able to completely flatten out the total number of cases, whereas the US experienced two initial large climbs, one up through day 40, and another up through day 90. This chart also gives guidance to countries like Argentina and Zambia, which are behind others, on where to look globally for success stories as well as on pitfalls that should be avoided.

Before we close the case study on COVID-19, we also want to address dealing with the many members (like countries, states, and regions) that come with this dataset. With eight members in the previous two strategies, our line charts are already at capacity for the amount of information that can be shared without becoming spaghetti. So, we want to address one final item: how to dynamically display many dimensions in one chart.

In Chapter 1, we created small multiples when we had five members of the dimension, but that tactic won't always work, especially if you have ten or more members. Instead, we want to introduce a method to dynamically create many small multiples, a tactic that is configurable based on the screen space you may have available.

Strategy: Create a Trellis Chart

As you know from previous chapters, a *trellis chart* is a fancy term to describe small multiples. Imagine a trellis that supports vines climbing up the side of a wall. That same structure describes the axes lines and grid pattern created by having many small multiples in one chart. In this strategy, you'll create a trellis chart for our COVID-19 data:

1. Use the same principles from the former strategy to construct a chart that shows total COVID-19 positive cases for each US State and Region by Report Date using a logarithmic axis (Figure 8-21).

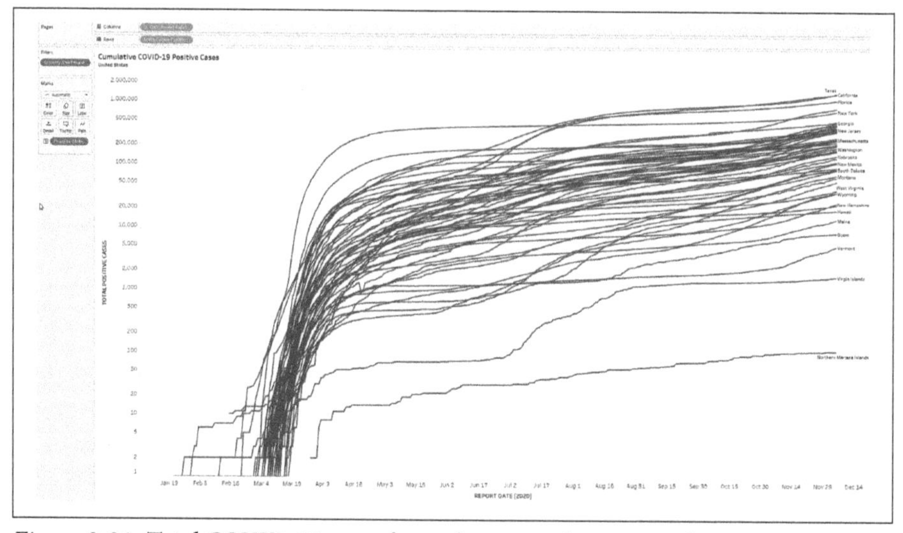

Figure 8-21. Total COVID-19 cases for each state and region in the US

2. Create a calculated field called **`[Index]`**:

```
//Index
INDEX()
```

Index is our favorite table calculation because of its utility. It allows you to number data in your view, something that can be done at the level of data you assign it to.

3. Create an integer parameter called **`[Total Columns]`**; set it to 8.
4. Create two additional calculated fields, one called **`[Column]`** and one called **`[Row]`**:

```
//Column
([Index]-1)%[Total Columns]
//Row
(([Index]-1)-[Column])/[Total Columns]
```

Bookmark these calculations! The parameter, Column, and Row calculations are all you need to create small multiples whose size you can dynamically adjust by simply changing a parameter value.

5. Convert both calculated fields to be Discrete and then drag [Row] and [Column] to their respective shelves.
6. Right-click [Column] and choose Edit Table Calculation. Set the Compute Using option to Specific Dimensions and select Province State Name. Repeat this for [Row].

Here is where you're setting how Tableau will number the data in the view. By specifying Province State Name and not including the Day of Report Date, Tableau will number the Province State Names from 0 up through *N*, ignoring the date.

7. Right-click and hide the Columns and Rows headers. Add formatting to have the latest date show the total number of cases and the name of the State/Region (Figure 8-22).

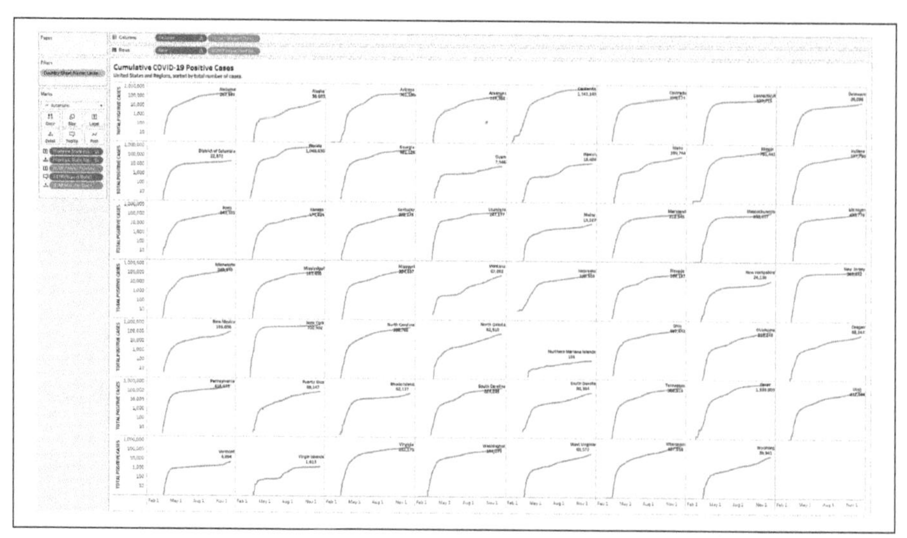

Figure 8-22. COVID-19 total cases by day for every state and region in the US

As a last step, we want to add just a tiny bit more value to this chart by sorting the states/regions based on their total case count.

8. Create a calculated field called **`[Max per State]`** that returns the maximum total number of cases per state:

```
//Max Per State
{FIXED [Province State Name], [Country Alpha 3 Code]: SUM({FIXED [County Fips Number
  [Province State Name], [Country Alpha 3 Code]:
    MAX([People Positive Cases Count])})}
```

We're using a nested LOD here because data is reported at the County level. So the LOD first finds the maximum People Positive Case Count per County and then sums up that value at the State level.

9. Right-click [Province State Name] on the Marks card and Sort it in Descending order by the Maximum of Max per State. Figure 8-23 shows the resulting chart.

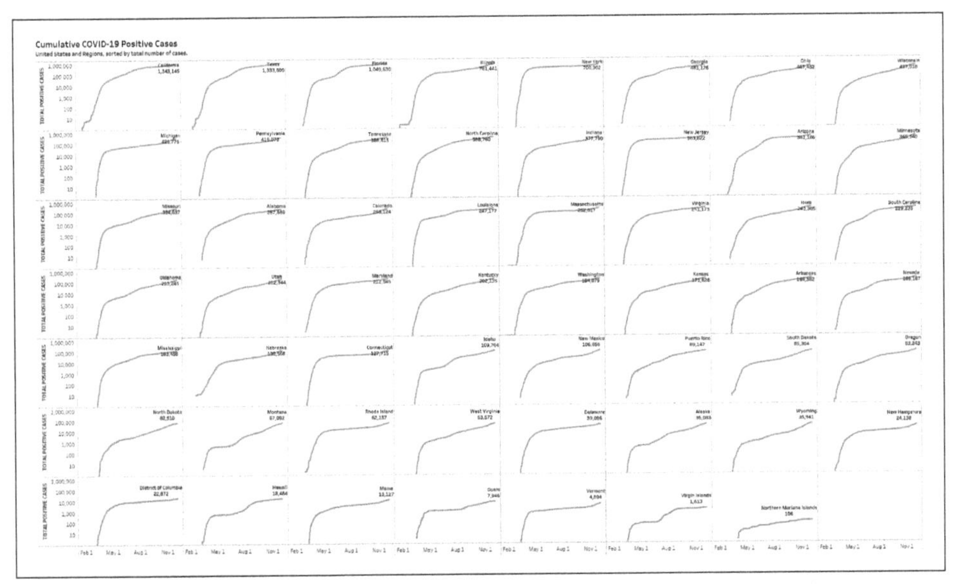

Figure 8-23. COVID-19 total cases by date for every US state, sorted by total case count

While the topic is grim, the chart produced is effective. Your audience can now see the shape of the COVID-19 virus in all regions of the US and assess the impact the virus has had on each area's community.

Advanced Modeling Using Statistical Add-Ons

For the last section of this chapter, we want to discuss even more ways that you can extend the types of analysis of your data. As we touched on, data science is becoming a more common practice in the analyst's everyday life. Fortunately, Tableau makes it easy to integrate two of the most common languages with advanced statistical and mathematical modeling, Python and R. Tableau calls these *Analytics Extensions*.

Before you begin using these tools, it's important to know when to use them. Tableau considers the outputs of the integration of these tools as table calculations. This is a critical notion, because it means that you'll be passing the data available in your view (worksheet) through to be processed and then further outputted. Also, by virtue of Tableau's Order of Operations, this step will happen after everything else, immediately before visualization rendering.

We recommend using these integrations under the following scenarios:

- You or your audience need to change your algorithm output on the fly. A great example is calculating a forecast output after filtering other fields. If you have many combinations of filtering, or you aren't quite sure which combination will yield the best result, computing a forecast output on the fly is your best option.
- You want to do advanced statistical significance tests, like the test of two proportions, and need to display the p-value. Although you've seen that it is possible to view the p-value of a trend line, it is not possible to visualize it in a chart; it simply remains as a tooltip. So, if you are testing many combinations of proportions and want to generate a table of p-values, your best bet is to let Python or R render them as a calculated field.
- Your data will be changing or growing. If your model will constantly be taking in new inputs or using data as it becomes available, it is best to do the processing in Tableau Desktop. Since Tableau will read all available data and pass it through, you can use the same algorithm to continuously generate new outputs.
- You are doing data transformations that are beyond the scope of normal Tableau functions. Because both Python and R have a myriad of libraries and functions, it may be easy to connect and use one of those packages to transform your data. A practical use case we've encountered is translating IPv4 IP addresses to IPv6. There's an algorithmic way to do it that has many steps but is just one function in Python.

Conversely, here are some scenarios of when *not* to use Analytics Extensions:

- You want to work with your data to find the best algorithm. If you are trying to find the best algorithm to fit your data, consider working completely in Python or R. After finding the best fit, use Tableau to visualize your output.
- You are doing a one-time processing step, like sentiment analysis, on a dataset. If you are relying on a library or package to do text analysis or scoring, there's no need to process that data on the fly with Tableau. Instead, process all of your data beforehand and then visualize in Tableau.
- You want to create a trend line or forecasting with minimal feature changes to the model. If you can achieve results by using the built-in Tableau functionality, save yourself some time and avoid these add-ons. As you saw in "Strategy: Add a Trend Line to a Scatter Plot" on page 305, we created a linear trend line with an exceptional R-squared value. We recommend you test out these statistical tools in Tableau first before investing additional time.

- You are using machine learning to create a model and have a training dataset. Machine learning models rely on first inputting a training dataset and then applying the algorithm on your actual dataset. Since you'll be limited to using data in a view, it may become overly complicated to try to pass through a training dataset and the actual dataset in a single script-calculated field.
- You are using any type of recursion or looping. If you are iterating over your data at a variable length, this preprocessing should be done outside Tableau. Imagine counting the number of words for every sentence in a book.
- Your visualizations need to be super-fast for end users. This may sound like a silly reason, but if you have an impatient audience, consider preprocessing your data. Passing through data subsets to another application, having that application process the data, pushing back the results, and finally rendering the visualization can be costly—especially if it is a complicated task or done over many rows of data.

Now that we've walked through a few best practices, let's take a closer look at statistical add-ons.

Using Python Analytics Extension for Web Page A/B Testing: Squeaks Pet Supply Case Study

A/B testing is a common practice for anyone with a website. It is a great way to test multiple iterations of content, get results back, and determine which version generates more engagement. For ecommerce organizations, it can be an invaluable tool for finding the perfect fit of content that will cause the greatest conversion (and ultimately purchase) from a customer. That said, to evaluate A/B tests, you must use hypothesis testing for two sample proportions, otherwise known as a *t-test*.

Squeaks Pet Supply is a pet product company looking to launch several A/B tests across many products. It's testing two layouts of its website to determine whether one version increases likelihood of a customer purchasing a product. Company executives are interested in seeing the results for all products in a single visualization. How would you represent this? Note, while the content format was the same across the respective versions, the content team would like to know if there are impact variations among the individual products versus the overall test results.

To accomplish this task, we turned to using the Python Analytics Extension (TabPy). The high-level workflow to using TabPy in this scenario is as follows:

1. Install TabPy locally on your machine or on a standalone server. You can find documentation on how to install TabPy in its GitHub repo (*https://github.com/tableau/TabPy*). If you install using the default settings, you should see that it is active by going to **`localhost:9004`** in a browser, as shown in Figure 8-24.

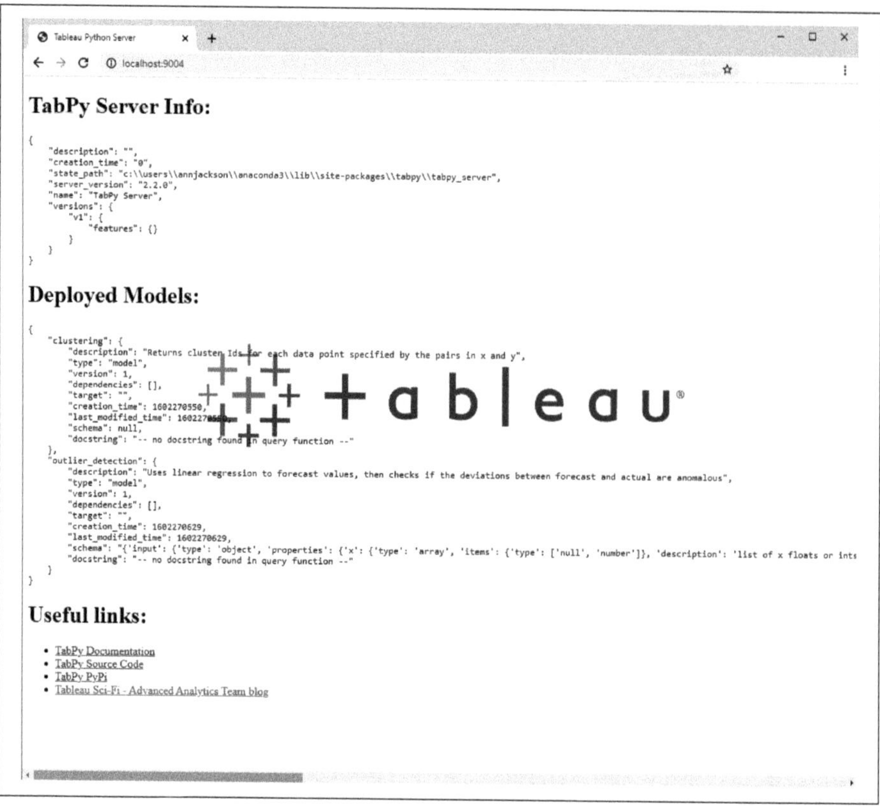

Figure 8-24. After TabPy is installed locally on a Windows machine, it can be accessed via port 9004 on the localhost

2. After installing TabPy, we enable connection to the TabPy server in Tableau Desktop; see Figures 8-25 and 8-26.

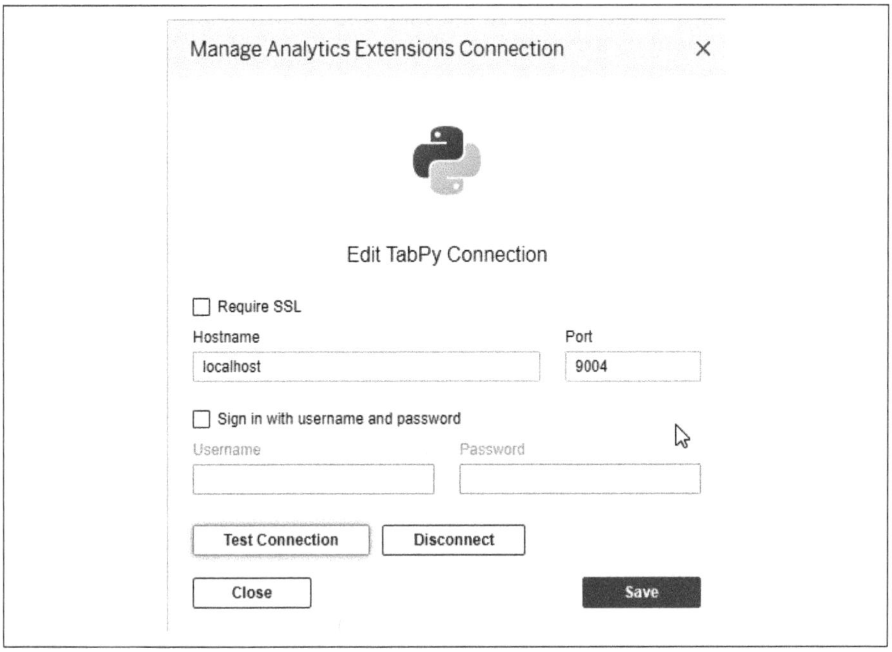

Figure 8-25. Accessing the Analytics Extension connection

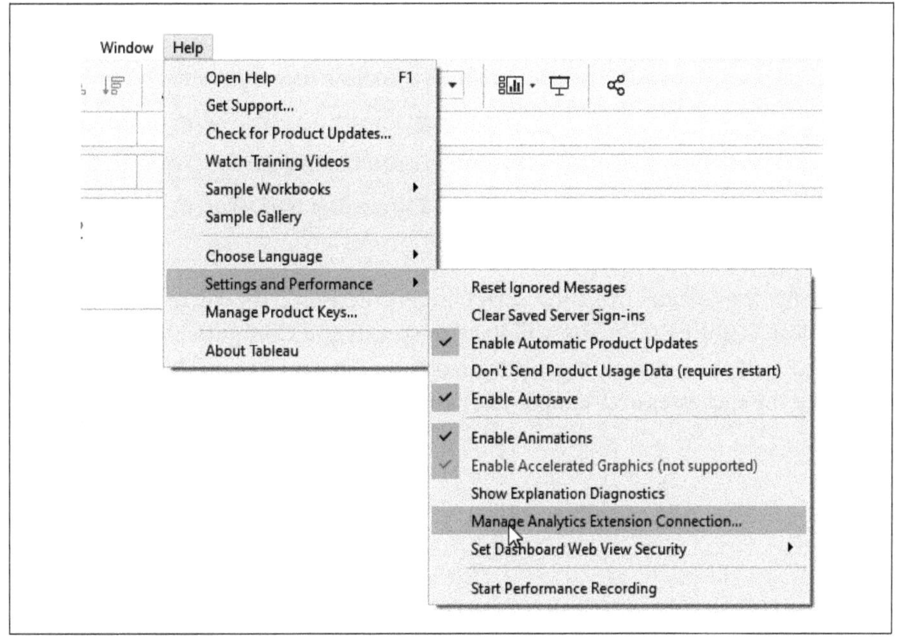

Figure 8-26. Adding connection information for a local instance of TabPy

3. Script code is written in Tableau Desktop and passed to the TabPy server. Both the Python code and the fields, which are called *arguments*, are used in the function:

```
//p-value
SCRIPT_REAL("import numpy as np
from statsmodels.stats.proportion import proportions_ztest
test_success  = _arg1[0]
test_sample = _arg2[0]
control_success = _arg3[0]
control_sample = _arg4[0]
success = np.array([test_success,control_success])
sample = np.array([test_sample,control_sample])
value = 0
alternative = 'two-sided'
stat, pval = proportions_ztest(success, sample, value, alternative)
return(pval)", SUM([Test Success]), SUM([Test Sample]), SUM([Control Success]),
SUM([Control Sample]))
```

You can test this code yourself by modifying the four arguments that are referencing fields with constant integers.

When working with Analytics Extensions, you will pass in fields as arguments. You will also use Script functions to return the values from the Analytics Extension. There are four Script functions:

- `SCRIPT_BOOL()` returns a Boolean data type value.
- `SCRIPT_INT()` returns an integer data type value.
- `SCRIPT_REAL()` returns a real number (or float) data type.
- `SCRIPT_STR()` returns a string data type value.

4. Finally, the output from the script is returned and shown in a visualization in Tableau (Figure 8-27). You will have to specify the Compute Using option for the `SCRIPT_REAL()` function table calculation. For this example, we're using Cell since the successes and sample sizes are aggregated per row in the visualization.

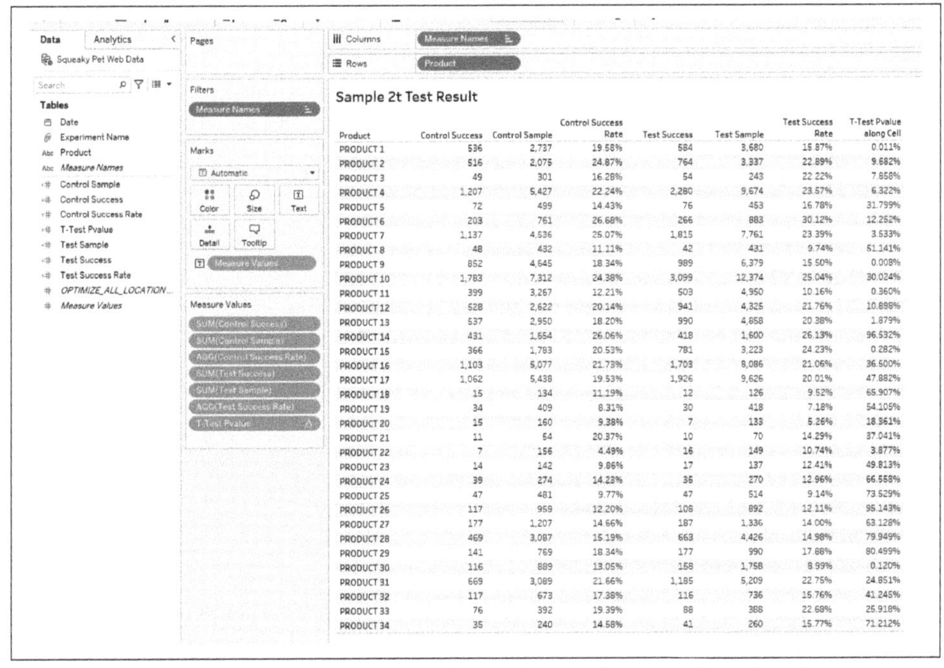

Product	Control Success	Control Sample	Control Success Rate	Test Success	Test Sample	Test Success Rate	T-Test Pvalue along Cell
PRODUCT 1	536	2,737	19.58%	584	3,680	15.87%	0.011%
PRODUCT 2	516	2,075	24.87%	764	3,337	22.89%	9.682%
PRODUCT 3	49	301	16.28%	54	243	22.22%	7.858%
PRODUCT 4	1,207	5,427	22.24%	2,280	9,674	23.57%	6.322%
PRODUCT 5	72	499	14.43%	76	453	16.78%	31.799%
PRODUCT 6	203	761	26.68%	266	883	30.12%	12.252%
PRODUCT 7	1,137	4,536	25.07%	1,815	7,761	23.39%	3.533%
PRODUCT 8	48	432	11.11%	42	431	9.74%	51.141%
PRODUCT 9	852	4,645	18.34%	989	6,379	15.50%	0.008%
PRODUCT 10	1,783	7,312	24.38%	3,099	12,374	25.04%	30.024%
PRODUCT 11	399	3,267	12.21%	503	4,950	10.16%	0.360%
PRODUCT 12	528	2,622	20.14%	941	4,325	21.76%	10.898%
PRODUCT 13	537	2,950	18.20%	990	4,858	20.38%	1.879%
PRODUCT 14	431	1,654	26.06%	418	1,600	26.13%	96.532%
PRODUCT 15	366	1,783	20.53%	781	3,223	24.23%	0.282%
PRODUCT 16	1,103	5,077	21.73%	1,703	8,086	21.06%	36.500%
PRODUCT 17	1,062	5,438	19.53%	1,926	9,626	20.01%	47.882%
PRODUCT 18	15	134	11.19%	12	126	9.52%	65.907%
PRODUCT 19	34	409	8.31%	30	418	7.18%	54.105%
PRODUCT 20	15	160	9.38%	7	133	5.26%	18.361%
PRODUCT 21	11	54	20.37%	10	70	14.29%	37.041%
PRODUCT 22	7	156	4.49%	16	149	10.74%	3.877%
PRODUCT 23	14	142	9.86%	17	137	12.41%	49.813%
PRODUCT 24	39	274	14.23%	35	270	12.96%	66.558%
PRODUCT 25	47	481	9.77%	47	514	9.14%	73.529%
PRODUCT 26	117	959	12.20%	108	892	12.11%	95.143%
PRODUCT 27	177	1,207	14.66%	187	1,336	14.00%	63.128%
PRODUCT 28	469	3,087	15.19%	663	4,426	14.98%	79.949%
PRODUCT 29	141	769	18.34%	177	990	17.88%	80.499%
PRODUCT 30	116	889	13.05%	158	1,758	8.99%	0.120%
PRODUCT 31	669	3,089	21.66%	1,185	5,209	22.75%	24.851%
PRODUCT 32	117	673	17.38%	116	736	15.76%	41.245%
PRODUCT 33	76	392	19.39%	88	388	22.68%	25.918%
PRODUCT 34	35	240	14.58%	41	260	15.77%	71.212%

Figure 8-27. The far-right column shows the p-value of the 2t test based on the inputs

Because there are relatively few inputs for a test of two proportions, and the resulting p-value was output from a single function in Python, it saved our team members valuable time and allowed them to accrue data over time to see whether the test outcomes changed as time went on. They were also able to slice and dice data by customer demographic information to determine whether the behavior was consistent across all customers or resonated with a subset.

Sentiment Analysis on Customer Reviews: Amazing Products Case Study

Have you ever read customer reviews for a product? For companies that sell products, customer endorsements or complaints can be the difference between product success and failure. But with the proliferation of the internet, reading, interpreting, and classifying the vast amount of reviews can be very difficult.

Amazing Products is considering refreshing its digital streaming device, the Amazing Fire TV. First, however, company executives want to know more about the pain points and how customers feel about the existing version. They've asked for an analysis of online reviews and want to see this data quantified. What role could sentiment analysis play as you approach this task?

Sentiment analysis scores words, phrases, or sentences based on a preset knowledge dictionary that provides positive or negative (emotion) scoring. When working with sentiment analysis, you can choose default dictionaries or customize your own. When we did this practice with Amazing Products (see Figure 8-28), we started with a base library. Because the output was a number between –1 and 1, we passed the script through TabPy and plotted the results in a jitterplot:

```
//Review Sentiment
SCRIPT_REAL("
import nltk
from nltk.sentiment import SentimentIntensityAnalyzer
reviews =_arg1
scores = []
sid = SentimentIntensityAnalyzer()

for text in reviews:
    ss = sid.polarity_scores(text)
    scores.append(ss['compound'])
return scores
",MAX([Review Text]))
```

You can test the code yourself by replacing `Review Text` with a sentence in quotes.

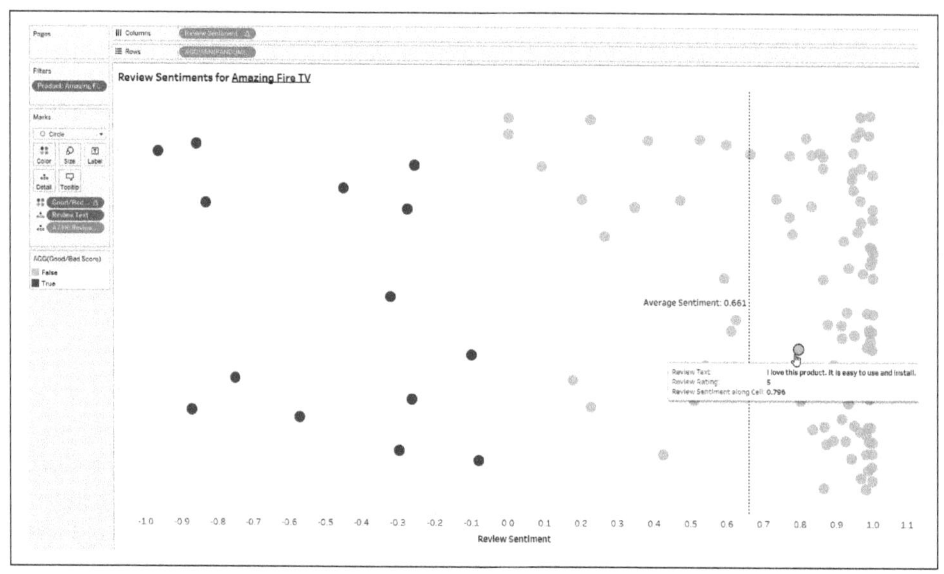

Figure 8-28. Jitterplot showing individual product reviews and their sentiments

Conclusion

While the topics explored in this chapter are a little more mathematically intense than others, we hope you've seen the virtue of the examples presented. You've learned how to create forecasts and follow forecasting best practices when working with time series, a technique to use when you have reasonably predictable data, like call center volume. You've also learned how to create both scatter plots and quadrant charts, tools critical for evaluating the relationship between two numerical variables. And specifically, with the quadrant chart, you've learned to further segment your data to communicate it clearly to the audience.

You have also learned more-advanced mathematical analysis techniques built into Tableau: constructing trend lines and creating clusters via k-means clustering. Quickly dragging and dropping trend lines is a good way to statistically evaluate whether a legitimate statistical relationship exists between two variables. If it does, you also know how to use that output to further inform business decisions. And with k-means clustering, you've learned to work with relationships among more than two variables (often three or more) to segment your data. You used cluster analysis to segment employees as an organization to compare whether certain attributes led to more likelihood of attrition.

Additionally, we tackled COVID-19, a complicated data topic that doesn't conform to normal chart standards. You used alternative axes to transform both the total number of cases and dates to tell clear data stories. With the transformed axes, you could construct a tough, but clear comparison among various countries of the world. You were also able to enumerate COVID-19 cases for every state and region in the US in a compact and dynamic trellis chart. With the technique covered, you can quickly change the shape of the trellis scaffolding to best fit the space the visualization will reside in.

And finally, we discussed additional statistical concepts and the best times to use analytics extensions and when not to. You've learned the two most popular ways to integrate statistical languages with Tableau for visualizations, including a use case that allows for on-the-fly processing.

At this point, you've also reached the conclusion of chapters designed to teach you how to build charts. You now have a broad foundation of data visualization best practices and analytical techniques designed in Tableau Desktop, foundations that you will be able to work with in concert to create dynamic analyses, which is the focus of the next chapter.

CHAPTER 9

Constructing Dynamic Analyses

Up to this point, we've been focused on identifying the correct visualizations to answer key questions. You've learned when it is appropriate to use pie charts, how to make effective and varied bar charts, and many ways to work with different data types. Each chapter has been foundational to your understanding of good data visualization and has given you the tools necessary to build the right charts. But now, you're going to take that knowledge one step further with the introduction of dynamic elements.

We named this chapter "Constructing Dynamic Analyses" because you will be using a variety of features within Tableau to manipulate visualizations, changing the way data is displayed, providing user input, and creating charts that are customized to the audience's interests. Unlike the other strategies in this book, everything discussed in this chapter will result in several (sometimes infinite) permutations of a visual. We'll be providing you baseline ideas and the building blocks for incorporating dynamic elements into your own data projects. We hope your takeaway from this chapter is how to effectively add interactivity and customization for your audience and use case.

In the first case study, you will explore how to build several chart variations to satisfy your audience with parameters. In the second case study, you will work with parameter actions to create dynamic charts that change based on user interaction. Finally, you will use set actions to drill down to various date levels and create expanding and collapsing tables.

In This Chapter

This chapter relies on your understanding of the prior chapters, or on your knowledge of building a variety of charts within Tableau. Each strategy will focus on the task at hand, and not the build of the initial chart to which the dynamic analysis will be applied. In this chapter, you'll learn how to do the following:

- Use parameters inside Tableau
- Create sets
- Use set actions
- Use parameter actions
- Use animations

Parameters

We've introduced parameters a few times throughout the book (namely, in Chapter 1), when we used a simple parameter to limit a bar chart to a Top N (in our case, 10) number of merchant categories. Before we dive deeper, let's walk through a brief refresher.

Parameters in Tableau are dynamic entry fields that you can use in your visualizations. You can set them as the following field types: float, integer, string, Boolean, date, and datetime. They allow you or your audience to pick a variable that exists outside a dataset. These variables can be used to change the way visualizations are displayed, what field values are set to, or as complex filters, just to name a few.

Additionally, you can define default behaviors associated with them. You can set them as a static list—either from manual entry or derived from a field, or as a range of values (for dates and numbers), or allow them to be free for any entry of the same data type. Over the following strategies, we'll explore examples that take advantage of parameters in straightforward as well as less obvious ways.

Using Parameters to Change Measures and Dimensions: Office Essentials Case Study

One of the most popular ways to work with parameters in Tableau is to use them as a way for the audience to define which measures they want to view as well as to display charts by a dimension of their choosing.

As an office supply store, OE is managing tens of thousands of SKUs across many demographics—from individuals making purchases in local stores to corporate accounts with standing orders. As such, company executives are interested in tracking many key metrics. If you were to build a chart for each one, you would end up with many dashboards and reports. How can you avoid report proliferation?

Let's take a closer look at leveraging parameters so you can have the same type of chart represent many types of analysis.

Strategy: Use a Parameter to Change a Measure in a Bar Chart

Delving further into our scenario, imagine the Sales team at OE wants to be able to compare many metrics—namely Sales, Profit Ratio, Number of Orders, and Quantity for different Sub-Categories of product types. This request may seem benign, but imagine that this request is supposed to live on a Sales dashboard. If four bar charts for each metric were put on one dashboard, very little space would remain for anything more. Sure, you could switch the charts to a table, but as we've discussed, tables don't allow for the same visual comparisons that bar charts afford. So instead, a parameter will serve as the control for the audience to select which measure they want displayed.

Basic parameters can be used in visualizations without being a part of calculated fields, but most parameters, especially in more sophisticated techniques, will require use within one or many calculated fields.

1. Create a parameter called **`[Select A Metric]`** and make it a data type of Integer. Set the "Allowable values" option to List. Enter the following values in the list:

Value	Display As
1	`Sales`
2	`Profit Ratio`
3	`# Orders`
4	`Quantity`

We recommend creating these types of parameters as integers. When you do this, the integer is aliased as the Display As option in the parameter controller, but calculations will evaluate based on the Value. Integers are faster to process than strings within calculations and will also make your calculated fields shorter and faster to create.

2. Create a calculated field called **`[Selected Metric]`**. Using a case statement, enter the following:

```
//Selected Metric
CASE [Select A Metric]
WHEN 1 THEN SUM([Sales])
WHEN 2 THEN SUM([Profit])/SUM([Sales])
WHEN 3 THEN COUNTD([Order ID])
WHEN 4 THEN SUM([Quantity])
END
```

Remember that calculated fields need to either be row level or aggregated. Since [Profit Ratio] and [# Orders] will need to be pre-aggregated, the other two measures, [Sales] and [Quantity], are included as aggregates.

3. Drag [Selected Metric] onto the Columns shelf and drag [Sub-Category] onto the Rows shelf. Right-click [Sub-Category] and sort by [Selected Metric] in Descending order.

4. Edit the title of the sheet by double-clicking the title, and then entering the following by using the Insert feature in the Edit Title dialog box:

```
<Parameters.Select A Metric> by Subcategory
```

5. Right-click the [Select A Metric] parameter in the lower left and select Show Parameter. Use the parameter control and notice that the bar chart changes depending on the metric and that the sub-categories stay sorted in descending order (Figure 9-1).

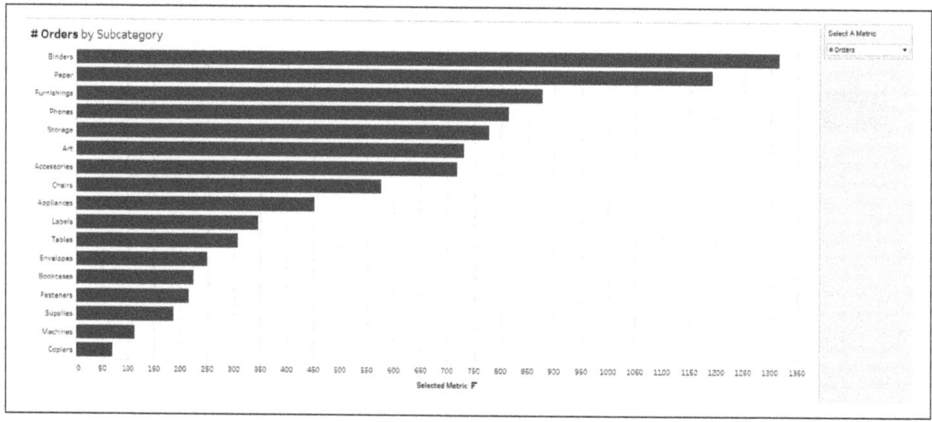

Figure 9-1. A bar chart whose measure changes with a parameter

You've now created four bar charts, with the use of a parameter and a single calculated field. This possibility gives your audience the ability to choose which chart is most impactful for them as they navigate through the data shared in dashboards.

When you use a parameter to toggle between many metrics, you'll notice that the number format changes, but not precisely for each metric. Within Tableau, each field can have only one default number type, so a common workaround is to create another calculated field that matches the desired number format.

Strategy: Use Regex Functions to Dynamically Format a Parameterized Metric

To make your visualizations user-friendly and intuitive, we always recommend making numbers match the data types and formatting that they represent. Think back to what we've said earlier; as the audience reads the chart, there will be internal dialogue. Why not help them out, by including a dollar sign or a percentage when appropriate?

One approach to creating a formatting calculation that matches the metric type is by using regular expressions (regex). These are used to find and sometimes extract or replace patterns within a string of text. Regex functions are popular in many programming languages, and several exist within Tableau as well. For this example, we will use `REGEX_REPLACE()` along with `ROUND()` (the function that rounds numbers), and `STR()` (the function that turns numbers into strings) to construct well-formatted numeric labels:

1. Create a calculated field called **`[Formatted Label]`**:

```
//Formatted Label
CASE [Select A Metric]
WHEN 1 THEN "$" +
(REGEXP_REPLACE(STR(ROUND([Selected Metric],0)),"(\d)(?=(\d{3})+$)","$0,"))
WHEN 2 THEN STR(ROUND(ROUND([Selected Metric],4)*100,2))+"%"
WHEN 3 THEN REGEXP_REPLACE(STR(ROUND([Selected Metric],0)),
  "(\d)(?=(\d{3})+$)","$0,")
WHEN 4 THEN REGEXP_REPLACE(STR(ROUND([Selected Metric],0)),
  "(\d)(?=(\d{3})+$)","$0,")
END
```

While we don't have time to go into regular expressions in great detail, it is enough to describe the repeated pattern shown in the preceding code. `(\d)(?=(\d{3})+$)` looks for any patterns of three digits in a row. `$0` inserts a comma after that pattern is found. This is done to take a number that has been converted to a string and add comma separators back in.

2. Drag the calculated field to Label and change the measure by using the parameter control. Your chart should now match Figure 9-2.

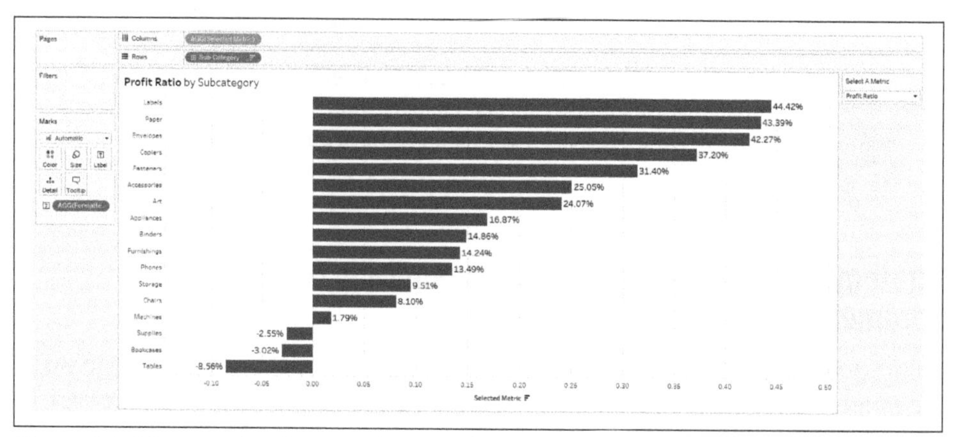

Figure 9-2. The regex has been implemented on the bar chart labels to format numbers based on the chosen metric

Strategy: Use a Parameter to Change the Dimension in a Bar Chart

So what if your audience also wants a way to change the bar chart from comparing Sub-Categories to States, Customers, or Regions? You can apply a similar technique, using parameters to create a dynamic dimension:

1. Create an integer parameter called **[Slice By]** along with a calculated field called **[Selected Dimension]**:

Value	Display As
1	Sub-Category
2	State
3	Customer
4	Region

```
//Selected Dimension
CASE [Slice By]
WHEN 1 THEN [Sub-Category]
WHEN 2 THEN [State]
WHEN 3 THEN [Customer Name]
WHEN 4 THEN [Region]
END
```

2. Double-click [Sub-Category] on the Rows shelf to edit the field displayed. Then erase Sub-Category and replace it with [Selected Dimension], and press Enter.

When you change which field is used by editing it on the visualization, it will retain any formatting and sorting that has been placed on it. Here, we are using this technique to retain the field sort by Selected Metric.

3. Double-click the title and replace [Sub-Category] with [Slice By]. Show the parameter and toggle between the different dimensions. You can also update the title with this new parameter. Your final visualization should match Figure 9-3.

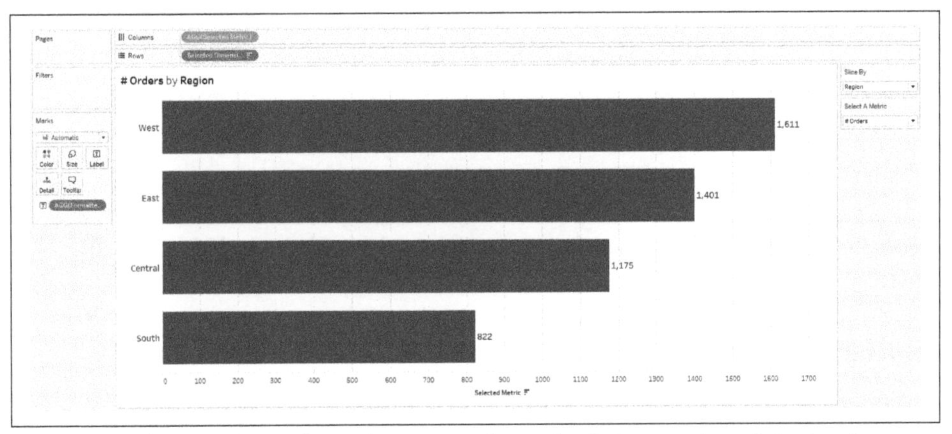

Figure 9-3. This bar chart allows the user to change both the metric and the header

Congratulations—you've now created what could be 16 different bar charts by using two parameters and a few calculated fields. And you've also offered a multitude of options to your audience for analyzing their data.

Using Parameters to Set Time Periods and Date Aggregations in Line Charts: Office Essentials Case Study

Now that the OE executives have highly customizable bar charts to better understand their sales, they've surfaced additional questions. They want to see their data trended over time, but aren't quite sure how to ask for what they're looking for, because it depends on what they're using the data for. Sometimes they want to see weekly data for their biweekly reports, sometimes they want to see it daily when a specific promotion is going on, and sometimes they want to see it monthly.

All are reasonable scenarios and ways to analyze the data, but each one sounds like a chart of its own. Are there any simplified solutions? With parameters, you can again simplify all this data down to one chart, as shown in Figure 9-4, allowing the audience to select which time part they want to use to analyze their data.

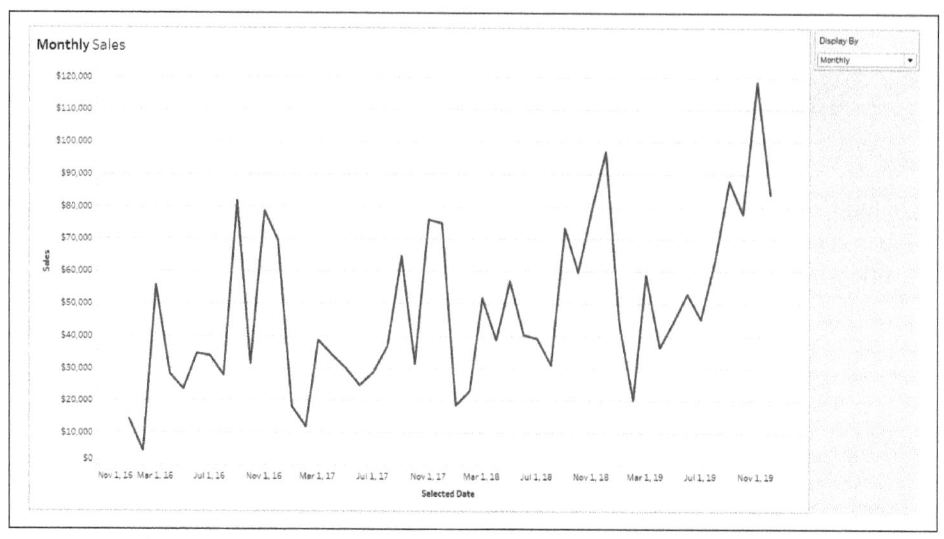

Figure 9-4. This line chart allows the user to change the level at which dates are displayed

Strategy: Use a Parameter to Change the Date in a Line Chart

In this strategy, you'll create the line chart in Figure 9-4:

1. Create an integer parameter called **`[Display By]`**:

Value	Display As
1	`Daily`
2	`Weekly`
3	`Monthly`

2. Create three custom date fields, from the field Order Date: [Days Date Value], [Week Numbers Date Value], and [Months Date Value]. Remember, you can create these by right-clicking the [Order Date] field, using the drop-down list for the date precision, and selecting Date Value.

3. Create a calculated field called **`[Selected Date]`**:

```
//Selected Date
CASE [Display By]
WHEN 1 THEN [Order Date (Days)]
WHEN 2 THEN [Order Date (Week numbers)]
WHEN 3 THEN [Order Date (Months)]
END
```

4. Right-click and drag [Selected Date] onto the Columns shelf, choosing the first option, Selected Date (Continuous). Drag `SUM([Sales])` to Rows. Show the [Display By] parameter and toggle between different time values. Make a matching title by using the parameter, shown in Figure 9-5.

At this point, you may be questioning why we aren't using the built-in date hierarchy features in Tableau to allow the audience to drill down and up, through various date levels. We have found that the experience often is less than intuitive, leading to over-drilling, too many options, or periods of time during which data is truncated and shows a less-than-accurate-picture.

Indeed, in this view, when we switch to weekly, the week of 12/29/2019 appears to be down a significant amount, but that is only because the data for this week includes just three days. Typically, when we build these charts, we will also build in hidden date filters to ensure that no matter what time part is selected, a full amount of time is displayed.

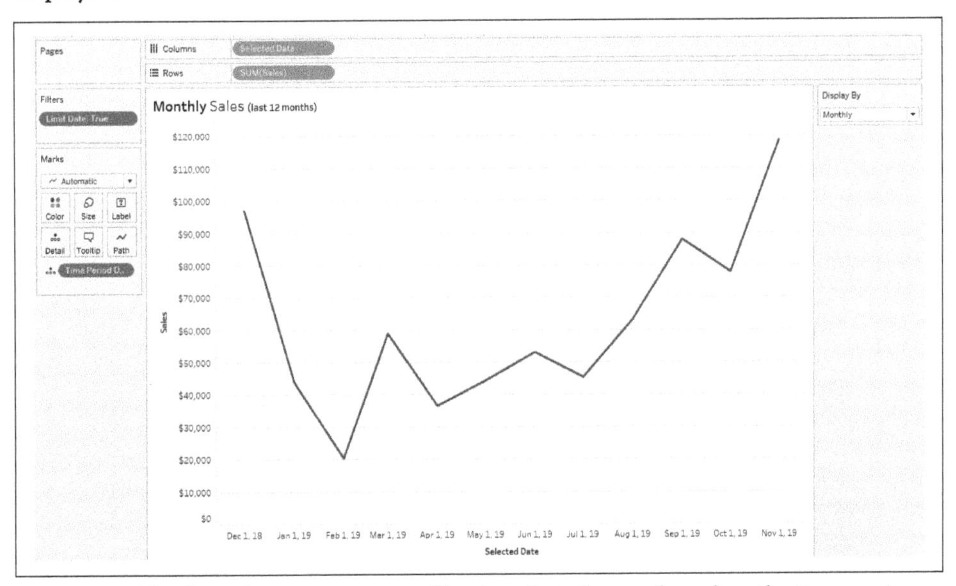

Figure 9-5. This line chart incorporates filtering that changes based on the time part selected

Strategy: Add Hidden Date Filters to Prevent Truncated Time Parts

To prevent truncated date parts, you can add more calculated fields that serve as filters. These will limit data appropriately, ensuring that the audience is looking at full time periods. And after asking around, you've found the most popular time window for each cut of time (daily, weekly, monthly). For daily, most end users want to see at most the last 30 days. For weekly, they're looking back 13 weeks, and for monthly, they're looking at the last 12 months.

Continuing with the chart from the previous strategy, you'll now implement this dynamic filtering:

1. Create a calculated field that identifies the latest date in the dataset:

```
//Maximum Date
{MAX([Order Date])}
```

When writing fixed LOD expressions, you don't have to include the modifier `fixed` when applied to the entire dataset. This is a handy way to save time and easily identify what the LOD is computing.

2. Make a calculated field called **`[Limit Date]`**:

```
//Limit Date
CASE [Display By]
WHEN 1 THEN [Order Date (Days)]>= DATEADD('day',-14,[Maximum Date])
  AND [Order Date]<=[Maximum Date]
WHEN 2 THEN [Order Date (Week numbers)]
  >=DATEADD('week',-13,DATETRUNC('week',[Maximum Date]))
AND [Order Date (Week Numbers)]
  <=DATEADD('week',-1,DATETRUNC('week',[Maximum Date]))
WHEN 3 THEN [Order Date (Months)]
  >=DATEADD('month',-12,DATETRUNC('month',[Maximum Date]))
  AND [Order Date (Months)]
  <=DATEADD('month',-1,DATETRUNC('month',[Maximum Date]))
END
```

Whenever you're building calculated fields that limit dates, we always recommend identifying the most recent date programmatically. An LOD is an easy way to do this, but you could also switch this up and make it a user input (aka another parameter).

3. Drag the [Limit Date] field onto your Filters shelf and select True.
4. As a final touch, build a calculated field that explains how much data is being shown for each time part selected. Drag that onto Detail and then add it to your title:

```
//Time Period Description
CASE [Display By]
WHEN 1 THEN 'last 30 days'
WHEN 2 THEN 'last 13 weeks'
WHEN 3 THEN 'last 12 months'
END
```

Now you've provided the audience with charts (as shown in Figure 9-5) that are customizable to different time parts and popular time periods. You've also built in guardrails for the audience. They won't find themselves comparing truncated weeks of data to each other, and when they look back at the data at a later date, the charts will have updated in time.

Using Parameter Actions to Change Trended Metrics: SAGE Digital Marketing Case Study

SAGE Digital Marketing wants to take the concept of allowing end users the flexibility to choose their metrics a step further. Now, instead of just allowing them to select from a drop-down list, they want to integrate that functionality directly into the visualization itself. How would you build a visualization enabling the user to click a metric to automatically display more information?

To get this functionality to work, you'll be relying on *parameter actions*. First introduced in Tableau 2019.2, these allow end users to inject (or set) a parameter value by using a dashboard or worksheet action. Dashboard and worksheet actions are described in greater depth in Chapter 12, so for now it is enough to know that an *action* is a way to specify a condition (like clicking a bar within a chart) and having something happen as a result. The most basic action is a *filter action*: a user clicks East Region, and the surrounding charts in the dashboard filter to East. With the inclusion of parameter actions, the user can now click East Region and have a parameter set to East Region.

Strategy: Change a Metric Dynamically Using Parameter Actions and Measure Names

In this strategy, we'll build a visualization with a metric the user can click to display information:

1. Create a single sheet with KPIs by using [Measure Names] and [Measure Values] that includes Clicks, Conversions, Conversion Rate, Cost, Cost per Click, Cost per Conversion, Revenue, Revenue per Conversion, and Return on Ad Spend (ROAS); see Figure 9-6.

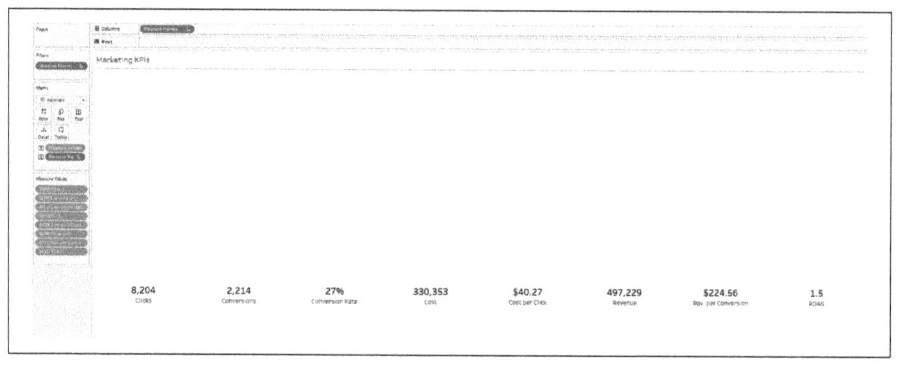

Figure 9-6. A basic set of KPIs

2. Now create a String parameter called **`[Marketing Metric]`**. Leave the "Current value" set to 1 and "Allowable values" set to All.

While we just recommended using integers for metric selection parameters, when using Measure Names along with parameter actions, you must use a String data type. This is because the name of the measure (Clicks) is set to the current value.

3. Create a calculated field that displays when a marketing metric is chosen:

```
//Selected Marketing Metric
CASE [Marketing Metric]
WHEN 'Clicks' THEN SUM([Clicks])
WHEN 'Conversions' THEN SUM([Conversions])
WHEN 'Conversion Rate' THEN [Conversion Rate]
WHEN 'Cost' THEN SUM([Cost])
WHEN 'Revenue' THEN SUM([Revenue])
WHEN 'Cost per Click' THEN [Cost per Click]
WHEN 'ROAS' THEN [ROAS]
WHEN 'Rev. per Conversion' THEN [Rev. per Conversion]
END
```

4. Create a horizontal bar chart showing the Selected Marketing Metric by Service Area. Add both the KPI worksheet and bar chart to a new dashboard.

5. Create the dashboard action to change the parameter by choosing Dashboard → Actions. In the Actions dialog box, click Add Action → Change Parameter.

6. In the Edit Parameter dialog box, shown in Figure 9-7, set the functionality such that Marketing KPIs is the source—that is, the worksheet that when interacted with will cause a change. Set Marketing Metric as the target parameter, whose value will change. The field value that will be put into the parameter is Measure Names.

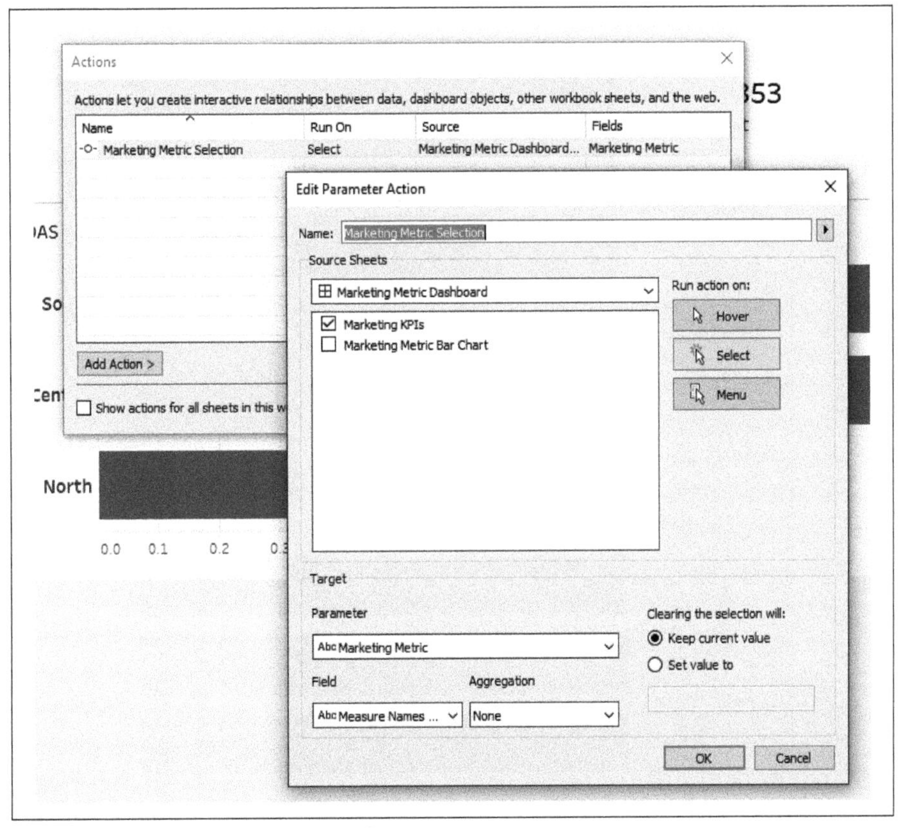

Figure 9-7. Configuration screen for parameter action

Now when a user wants to see a bar chart based on any of these metrics, instead of a drop-down list, they can click the metric directly, as shown in Figure 9-8.

Figure 9-8. The output when the user selects Clicks (top), and when the user selects Conversion Rate (bottom)

Using Parameter Actions to View Daily and 7-Day Average Value: SAGE Digital Marketing Case Study

One of the interesting things about the world of digital marketing is that every day a significant amount of fluctuation occurs in the data that's read out. Depending on the activity of your competitors, potential customers, and other external factors, your business metrics can display a wide range of values.

To help compensate for this, SAGE Digital Marketing has requested that a moving average be added to its report, to smooth out the data. A *moving average* takes a data point and a certain number of preceding data points to calculate a value. This adds a smoothing effect to data, since the highs and lows of the past *x* data points are averaged. But the SAGE executives also want to see the actual value.

How would you represent the actual value and the moving average at the same time?

Strategy: Compute a 7-Day Moving Average Reference Line That Dynamically Changes with a Date Selection

Using the same dataset from the previous strategy, you'll introduce parameter actions to dynamically change the moving average:

1. Create a line chart showing Revenue by Day.
2. Create a parameter called **`[Selected Date]`** and make it a date field with all allowable values:
3. Create a calculated field called **`[Time Window]`**:

```
//Time Window
[Date]>=DATEADD('day',-7,[Selected Date])
AND [Date]<=[Selected Date]
```

4. Create another calculated field called **`[Time Window Revenue]`**:

```
//Time Window Revenue
IF [Time Window]
THEN [Revenue]
END
```

5. Drag [Time Window Revenue] to Detail as a sum. Create a reference line by using this field, showing the Average value. Change the label to Custom. Unselect "Show recalculated lien for highlighted or selected data points."
6. Create another calculated field called **`[Time Window Date]`**:

```
//Time Window Date
IF [Time Window]
THEN [Date]
END
```

7. Drag [Time Window Date] onto Detail as the Minimum.

Here we are dragging the date as aggregated to avoid the lines on the line chart from turning into dots. This method can also be helpful when trying to add dimensions to a chart that has table calculations that you don't want to interfere with.

8. Create a reference band that goes from the Minimum Time Window Date to the Maximum Time Window Date and set the shading to light green.
9. Create a worksheet action called **`Change Date`** that sets Selected Date to DAY([Date]) when the user hovers on the chart.

   Now you've built a chart that satisfies both conditions from the audience: a daily readout and a smoothed average amount for a chosen time period (Figure 9-9).

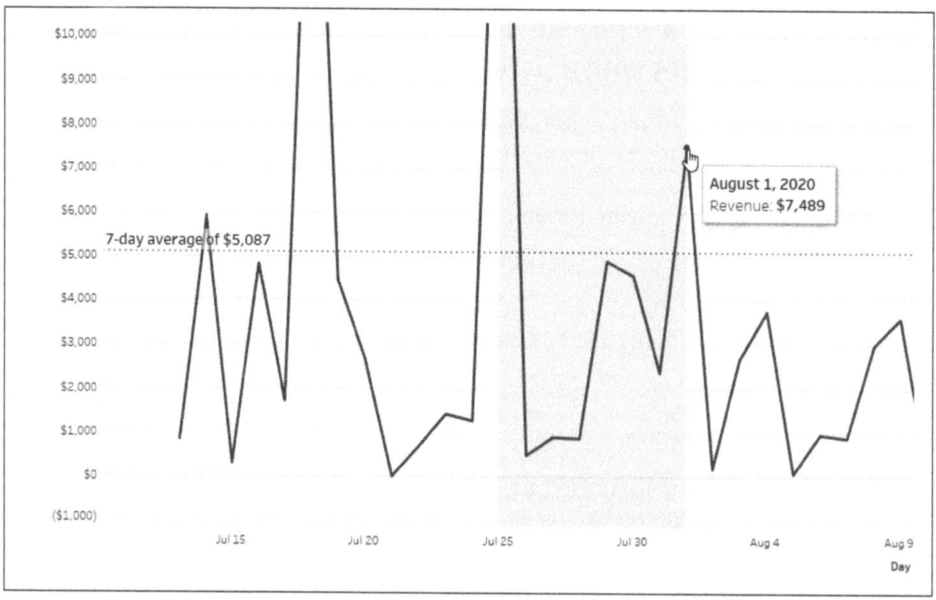

Figure 9-9. A line chart with a reference line for the seven-day moving average related to the point selected

Sets and Set Actions

You've had some basic firsthand experience with set actions early on in the book, but we haven't yet taken the time to dive deep. In this section, we'll start by explaining what a *set* is. In Tableau, it's a computed field based off a dimension (or dimensions) where you can define the members based on conditions. Those conditions can be a manual selection like a list, they can be based on a computed condition or value, or they can be a top or bottom grouping based on a field or formula. In fact, the options for creating a set match those that are available when you use a dimension (blue field) as a filter. Sets can also be combined, meaning you can use two conditions to create a subgroup of a dimension.

A practical example of combining two sets and their outcomes is to construct a set of the Top 10 Customers by Sales and a set of Top 10 Customers by Quantity Sold. Those two sets can be combined in three ways:

All members in both sets
: If a Customer is either in Top 10 by Sales or in Top 10 by Quantity, that Customer will be included.

Shared members in both sets
: The Customer is in both Top 10 sets.

Except shared members
: This subtracts one set from the other. In this scenario, if the Top 10 Customers by Quantity Sold set is subtracted from Top 10 Customers by Sales, you would be left with those customers with high sales, but not high quantity.

As a row-level calculation, when you use a set, you can think of it as creating a Boolean value behind the scenes to categorize your data. In Tableau, sets have unique functionality in the way they can be displayed. First, you can use the set as dimensions that will construct an In and Out header and have the members in the respective group represented. You can also use sets as filters, filtering the data either to the entire In or Out group, or to the members within the In group. This means you can use sets to dynamically display either the dimension values that meet the set condition or the entire set itself. Finally, you can expose the set control and have your end users use it like a pick list.

Now, on to *set actions*, first introduced in Tableau 2018.3. With a set action, you're taking the first type of set, from which members are manually selected from a pick list, and instead of that being a one-time task, members are added and subtracted from the pick list based on user interactivity. This has powerful implications because you are giving your end users the ability to create ad hoc groupings of data. The end user can also dynamically define members that are either In or Out of the set, which is useful for filtering. As you saw in Chapter 1, a common use case for set actions is to dynamically change the granularity of data based on interactivity.

In the next two strategies, we will expand on drilling into data with set actions and combine them with filtering and custom formatting to generate a compelling and intuitive user experience.

Using Set Actions to Drill Down and Filter Data: Digital Marketing Case Study

SAGE is looking back at its digital marketing data after receiving an interesting request from end users. The executives typically like to look at weekly performance, but sometimes, especially if there is a particularly great week in terms of user engagement, they like to drill into daily volumes and isolate any content that could have potentially contributed. Unlike the prior concepts we've discussed, in this scenario they want to have flexibility to drill down to what is interesting to them instead of being confined to showing all the daily data or a pre-filtered subset.

This presents an interesting problem. How would you approach it? Let's take a closer look at how this can be solved by implementing set actions, which allow the user to select a data point and see more detailed data within the point.

Strategy: Drill from Week to Date in a Line Chart

Starting with a new worksheet, follow these steps to introduce drill-down interactivity:

1. Create two custom date fields, one that is a Date at the Day level, and one that is a Date at the Week level. Drag the week-level calculation to the Marks card.
2. Create a set based on the week-level date. Click the All button to select all the values and name the set **`Included Dates Set`**.
3. Create a set action called **`Drill Date`** that runs on select, configure it to "Add values to set" when run, and when clearing "Remove all values from set."
4. Create a calculated dimension based on the set, and place it as the [Date] field on the Columns shelf:

```
//Date to Show
IF [Included Dates Set]
THEN [Date (Days)]
ELSE [Date (Week numbers)]
END
```

   At this point, the interactivity you have built will cause the week date marks to change to daily. However, both will still be visible in the view. The last part of this strategy will be to add a filter for the dates, to ensure that you're drilling *and* filtering to the daily view.

5. Create a calculated field called **`[Filter Date]`**:

```
//Filter Dates
IF {COUNTD(IF [Included Dates Set] THEN [Date (Days)] END)}>=1
THEN [Included Dates Set]
ELSE TRUE
END
```

This calculation determines the number of dates that are included in the set. If there are one or more dates in the set, the calculation will filter to include only those in the set. Otherwise, if there are 0 or NULL members in the set, the entire line chart will show.

6. Add some final formatting to the sheet by building a calculated field to dynamically change when the granularity of the data changes; this will use the same logic as the date filter. Bring it on to Detail as Attribute.

```
// Title Date Part
IF {COUNTD(IF [Included Dates Set] THEN [Date (Days)] END)}>=1
THEN 'Day'
ELSE 'Week'
END
```

Bringing the formatting field on as an attribute will ensure that the mark selections still stay highlighted when you change from one date level to another. Otherwise, after you select a mark, the value will no longer be a part of it and will cause the mark to deselect.

Using this logic, you've now deployed a line chart that can dynamically change from weeks to days, as shown in Figure 9-10. This technique and interactivity follows the flow of analysis of the end user. Someone can come in, inspect the weekly data, and then quickly drill into a particular week to see more-granular daily data.

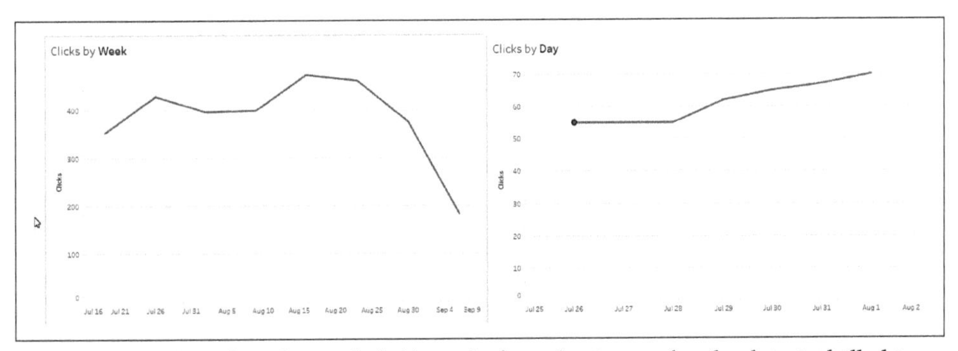

Figure 9-10. A line chart by week (left), and after selecting weeks, the data is drilled to the day (right)

In addition to using set actions to dynamically drill down on time parts, they can also be used to drill into tables. Has anyone ever said to you, "I want to see more detail for this area?" When they do, it is often a signal that you have hierarchical data and they're looking to see it disaggregated. What they'd love to have is a table where they can "expand" a category to see the child members as they see fit. Unlike a built-in hierarchy within Tableau, set actions enable you to provide your users the opportunity to choose a category that presents more detail, giving them the ability to see both aggregated information for categories of less concern and disaggregated data of a particular category, all in the same view.

With our marketing data, this problem presents itself in the scenario of having users who want to drill into a service area of interest and then visually review the metrics associated with each zip code. They don't want to be inundated with extra zip code detail for territories that are of less interest.

Strategy: Expand a Section of a Data Table

In this strategy, you'll build interactivity that allows a data table to expand and collapse using set actions:

1. Create a data table that shows Cost, Clicks, Cost per Click, Conversions, Conversion Rate, and Rev. per Conversion by Service Area.
2. Create a set called **`[Selected Service Area]`** based on [Service Area], and include all values in the set.
3. Create a calculated field called **`[Show Zip Code]`** and place it to the right of [Service Area] on the Rows shelf:

    ```
    //Show Zip Code
    IF [Selected Service Area]
    THEN [ZIP]
    ELSE [Service Area]
    END
    ```

4. Create a set action based on the new set, and configure the action to run on Select. Configure it such that running the action will "Assign values to set" and when clearing the selection it will "Remove all values from set."

    Now when a user clicks a service area, the table will dynamically expand to show the zip codes that are a part of that service area. To finish up, we recommend adding the following final touches.

5. Create a calculated field called **`[Drill Symbol]`**:

    ```
    //Drill Symbol
    IF [Selected Service Area]
    THEN '▼'
    ELSE '►'
    END
    ```

6. Create a calculated field called **`[Zip Code Header]`**:

    ```
    //Zip Code Header
    IF [Selected Service Area]
    THEN [ZIP]
    ELSE 'Total'
    END
    ```

7. Drag the new calculated fields onto the Rows shelf, placing [Drill Symbol] after [Service Area], and [Zip Code Header] after [Show Zip Code]. Right-click [Show Zip Code] and hide the header. Right-click the headers and select Hide Field Label Rows. Adjust row banding and formatting as desired. Your visualization should now resemble Figure 9-11.

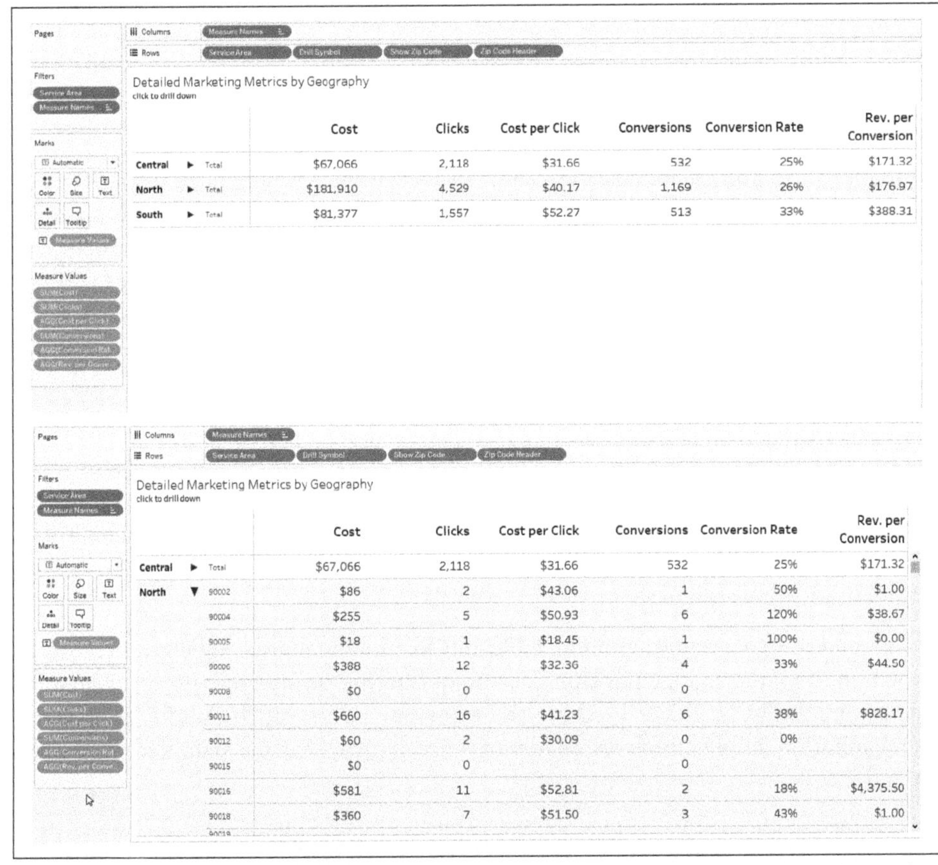

Figure 9-11. The marketing metrics aggregated by Service Area (top), and the chart after clicking the expand button for North to show the zip codes (bottom)

While it may take a little bit of time to add extra formatting details, trust us when we say it's worth it. These small design decisions tally up to a good user experience (something we will explore in more detail in Chapter 12). And you've added functionality that is now more familiar to your audience.

The type of functionality and interactivity that set actions bring to visualizations is something both familiar and popular with end users. Implementing these types of features and functionality will ultimately result in your audience responding more positively to the content you develop. They're also very likely to be more engaged because of the added interactivity.

Animation

Animation is a newer concept in the scheme of data visualization. This technique was arguably first popularized by Hans Rosling's Gapminder, which showed how the relationship between number of births and life expectancy has changed among all countries in the world over the past 50 years. Animation has become increasingly more prevalent as software tools for creating visualizations have evolved.

Animations are unique in that they allow visualization developers to display more-complex themes, like change over time, transitions, change in rank/order, or an emotional/narrative context. By literally "playing" through the data, the changes are offloaded from the audience and placed directly into the visual. A word of caution here: when using animations, we strongly recommend having someone there to describe the story as it unfolds. Because animation typically requires a narrator, we think it is best suited in the context of a presentation as opposed to a self-service data product.

Tableau provides two ways that animation can be used. First is with the Pages shelf. This feature allows users to "page through," or construct different views, based on the dimension on the shelf. Imagine you have a book, and each page represents a day. With the Pages shelf, you can add a controller to move through each page, or even automatically flip through them.

The second way to use animation in Tableau is to animate marks. When this feature is turned on, if marks change position because of filtering or interactivity, an animation will automatically be made to show the transition. This can be useful at drawing users' attention as selections are made, or to recognize that the data itself has been filtered.

Competitor Analysis Study: Using Animations to Compare Ranking Performance

For most brands or products, it is natural to have competition. On any given day, your company is vying for the most attention from consumers. One popular way to measure customer awareness is to compare an engagement metric like web page views. A traditional way to analyze this may be to plot the number of page views by day, seeing how your company performs, and then also your competitors. But what if you want to glean the shuffling for market share that is happening in the day-to-day? If that's the case, animation can help demonstrate the motion and movement among peers and tell the story of how your organization is performing.

In the next strategy, we're going to work with another dataset from SAGE. This dataset has a few competitors and a target company (You) with metrics defining their daily rank based on page views.

Strategy: Build an Animated Bump Chart

In Chapter 3, you learned how to build a bump chart, which takes dimensions ranked by a metric and plots the change in rank over time. In this example, you'll take the bump chart a step further by introducing custom highlighting and animation:

1. Build a bump chart, showing the Rank by Day for each competitor.

Remember, a bump chart is typically a dual-axis chart; the first axis marks are circles, and the secondary axis marks are lines. If your measure is continuous, reverse the axis, so that 1 is at the top of the chart.

2. Create a parameter called **`[Comparison]`** based on the [Competitor] field, and remove You from the list.
3. Create a calculated field called **`[Highlight Color]`** that will color You, the Comparison Competitor, and all Others in different colors. Put this on Color on the Marks card:

```
// Highlight Color
IF [Competitor]='You'
THEN 'You'
ELSEIF [Comparison]=[Competitor]
THEN 'Comparison'
ELSE 'Other'
END
```

4. Right-click and drag [Date] as a discrete field onto the Pages shelf. Configure the controller to show history for all marks, both trails and marks without fade.

   With animations, you can now tell the story of your brand, starting at position 6, making some headway and peaking at position 4 about halfway through the time period, and ending at position 5. Likewise, you can visually see Competitor 5 making a strong push to the top spot toward the end of the time period (as seen in Figure 9-12).

An alternative to the bump chart is commonly called a *racing bar chart*. This takes the rank of a metric and animates it over time to see the position change day by day. It's called a racing bar chart because each member is vying for position, like racers on a track.

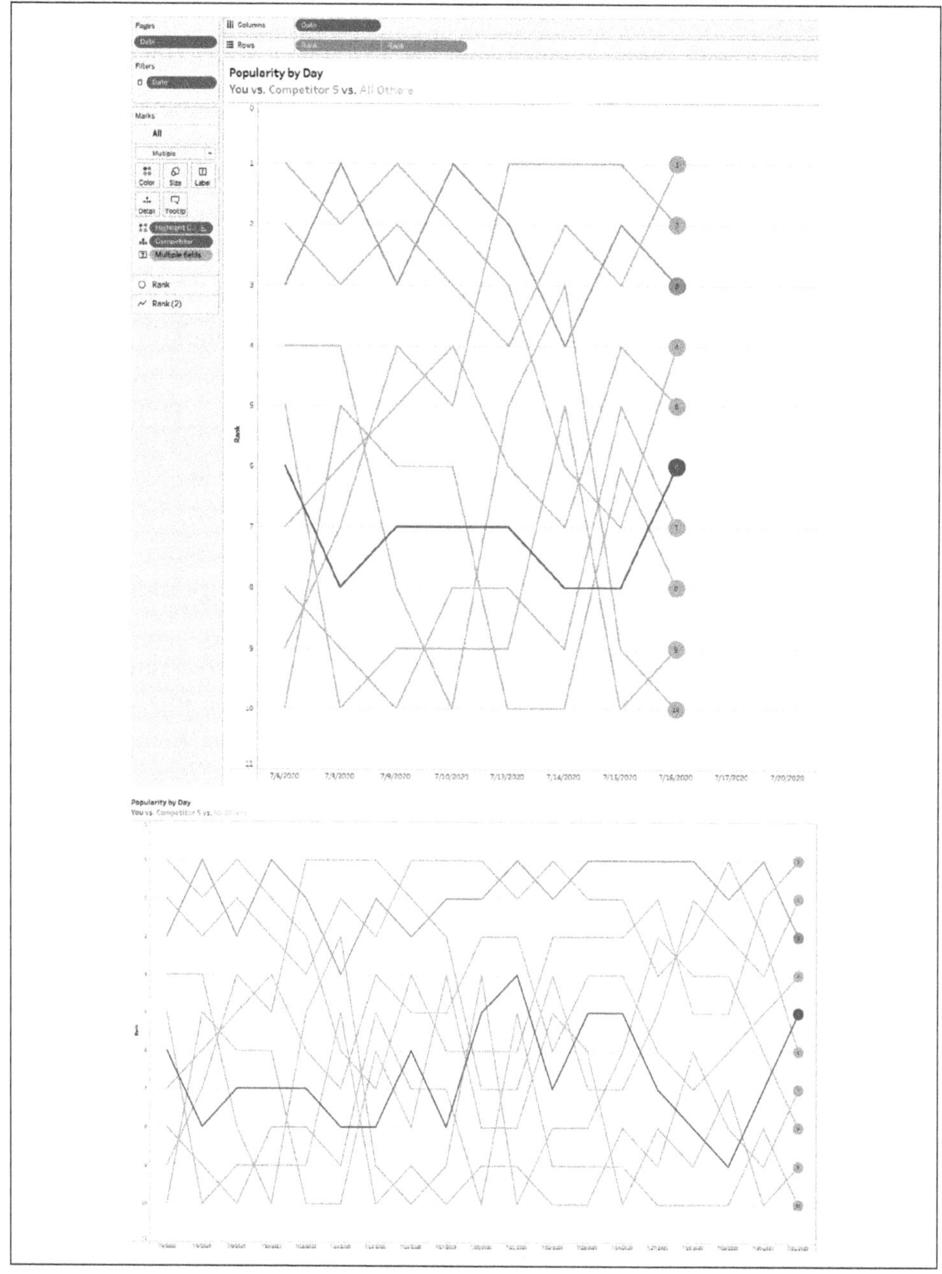

Figure 9-12. The initial ranking as compared to competitors (top) and the rest of the data (bottom)

Strategy: Create a Racing Bar Chart

Using the same dataset, create a new worksheet and do the following:

1. Create a horizontal bar chart with Rank and Page Views.
2. Use the Pages shelf to page through each day.
3. Put the [Highlight Color] calculated field from the previous strategy onto Color.
4. Do not enable history for the Pages shelf.

   This type of animated chart takes advantage of changing positions over time, while still retaining the magnitude of the measurement. Figure 9-13 shows that while your brand was ranked sixth in terms of page views, all competitors were pretty close. However, Competitor 5 later dominated everyone, leaving you at the back of the pack.

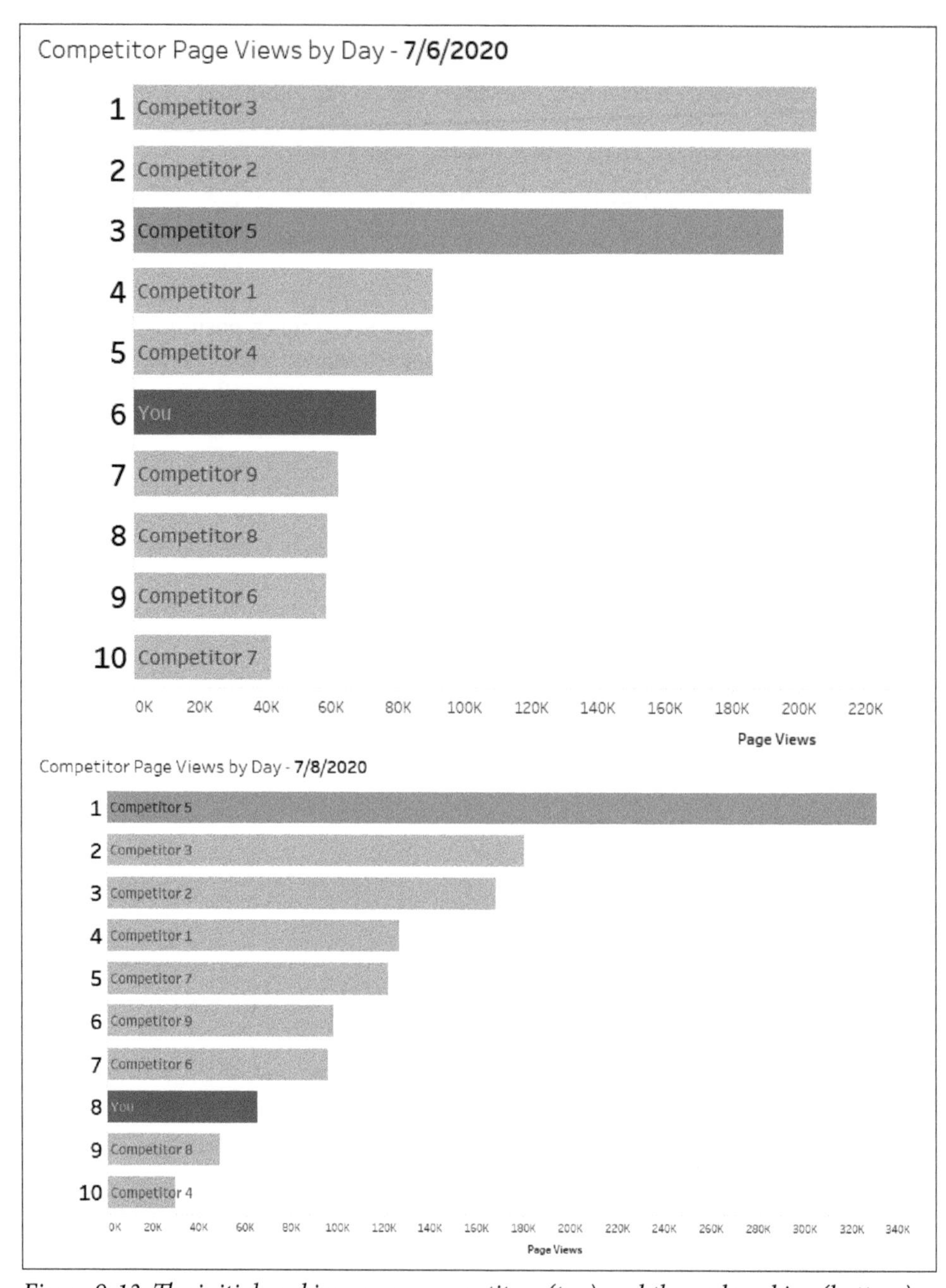

Figure 9-13. The initial ranking versus competitors (top) and the end ranking (bottom)

Conclusion

In this chapter, you've seen how powerful parameters can be. They allow for direct user input, and when combined with calculated fields, allow for complex filtering and interactivity. They're extremely useful for reducing chart proliferation, as well as for providing flexibility in chart design (remember how you made a line chart change from monthly to weekly data with a drop-down). *Direct user input* means that the actions of your audience can define the output of the visualization—making for a highly customized ending visual.

You've also worked with sets more directly and seen the enhanced interactivity they bring. If your users are asking for drill capability, use *drill* as a keyword to consider how a set action could enhance your charts. And finally, you've seen how animation can take static charts and make them more compelling. By using movement and motion, data stories become easier to tell and more exciting to explore. The suite of techniques in this chapter will serve you well in leveling up your data products, both analytically and from a user-experience perspective. Refined data products built in Tableau are likely to have one or more of these tools employed for maximum effect.

In the next chapter, you'll learn how to enrich your existing datasets to remove gaps for dates and values and to construct sophisticated visualizations.

CHAPTER 10
Advanced Data Modeling

Sometimes you want to make a chart type that's just out of grasp for out-of-the-box Tableau. Don't fret: you can make any chart in Tableau with a little patience and some out-of-the-box thinking.

In this chapter, you will tackle different approaches to data modeling in Tableau Desktop. When we use the term *data modeling*, we are referring to the inclusion of additional data that might not be necessary for standard visualizations. We will discuss several techniques to data modeling, including using table calculations, blends, unions, and joins. Each technique has a different time and place. If you aren't yet familiar with these techniques, no worries; we will articulate these differences in this chapter's scenarios.

In every case, you are trading dashboard performance for dashboard design. Some of these trade-offs will not compromise speed. Others can slow things significantly, particularly when you're adding additional marks to your visualizations or rows and columns to your datasets.

Before you dive in deep, we'll lay the foundations for understanding with some practical examples.

All but two of the strategies use Tableau's Sample – Superstore dataset. The other examples use sales pipeline data and a modified version of the superstore dataset. This chapter won't have a single-threaded story for the use cases. Each is independent. As a result, we'll set up the scenario in each of the strategies.

In This Chapter

For this chapter, you need Tableau Desktop and three data sources: the standard Sample - Superstore, the sales funnel dataset, and the Sample - Superstore - Modified dataset. Additionally, this chapter will require you to use a text editor to create data sources; don't worry, these data sources are fairly straightforward.

In this chapter, you'll learn how to do the following:

- Identify domain padding and domain completion
- Fill a calendar with missing values
- Build a rounded bar chart by using a data union
- Build an accordion table by using a join
- Build an accurate funnel chart by using a join
- Create a basket analysis by using a join
- Create a multiple dimension waterfall by using a blend
- Create a multilayered visualization using map features

Data Modeling

Data modeling might be the most advanced concept in Tableau. *Data modeling*—sometimes called *data densification* (which is a subset)—is a technique for increasing the amount of data showing in your visualization.

Data densification is most commonly used for creating curved charts, but there are many other practical uses for data modeling. We will discuss only a handful of possibilities. The most common type of data modeling is filling in potentially missing values (*domain padding*). It's so common that you've already used this method in Chapter 6! In addition to filling in missing values, you can also add more values to the beginning or end of a series (*domain completion*). Both domain completion and domain padding do not have to be difficult; in fact, in our first strategy we'll show you how to do both.

As you advance your skills, you will find yourself trying to develop unique solutions that reach beyond the base capabilities of Tableau Desktop. Remember, at its core, any visualization is just a compilation of dots, lines, and polygons. If you can't make a base chart in Tableau, you can use data modeling to redefine your visualization.

We can execute data modeling in at least three ways: unioning data sources, joining data sources, or bin/range densification. For most visualizations, you can use any of these methods. Each comes with trade-offs, but all add more data to your existing data sources to create your visualization.

The most simplistic approach to data modeling is *unioning*, or appending, a data source to itself. This duplicates the data you are working with—all of the data—but the process of unioning data to itself is the easiest of all the approaches. For instance, if you want to double your data and are working with the Sample - Superstore dataset, you will just union orders together. The downfall of this method is that you have to load all the data at least a second time for the union. If you are working with lots of data (a million records or more), this is probably not the best method to use.

The second approach is to simply take the data source of interest and join the data with a placeholder dataset. This placeholder dataset is often straightforward—usually a single column with rows counting up from 1 to whatever integer is necessary for analysis. This join will also effectively double your data. However, by using an extract, Tableau will just take a single snapshot of the data rather than duplicate as it would with a union. This method is the most efficient, in terms of performance, of the three methods.

In this case, if you were working with the Sample - Superstore dataset and wanted to replicate it five times over, you would first create a new placeholder dataset that looks like this:

placeholder
1
2
3
4
5

Then you would join the dataset with Sample - Superstore by using a custom join, where 1 = 1 (we'll show you how to do this later in this chapter). If you needed to duplicate the data 100 times, you would just have 101 rows with the header and then each row counting by 1s from 1 to 100. When we are data modeling, this is our approach.

The last method requires you to create a join to a supplementary data source containing two rows: the first row being the first value, and the last value being the maximum number of times you want the data source duplicated on a view. You then use bins and table calculations to pad the missing values. This method is often the least efficient in terms of performance because the data modeling occurs through table calculations. These types of calculations are completed on the rendering of your visual and are typically the slowest to return.

In this case, as in the preceding join example, if you were working with the Sample – Superstore dataset and wanted to replicate it five times over, you would first create new placeholder data. This time, the placeholder data is much simpler:

placeholder
1
5

You'll then join to the data as you would with the join example. This is where the methods diverge. From there, you need to fill in values 2, 3, and 4 by using bins and a few more steps that we won't discuss. If you are interested in this technique, you can read about it in a great blog post (*https://oreil.ly/5miVm*) by Ken Flerlage.

As noted earlier, we'll tackle the union and join methods in this chapter. It's worth noting at this juncture that the bin/table calculation method is extremely prevalent in the Tableau community and remains the way many individuals perform data modeling. The reason we mostly move away from the technique is because Tableau has significantly improved the way it creates extracts, and the original method allowed for a workaround without significantly increasing data sources. We've found that using a simple join over the bin/table calculation method increases performance as much as 20 times when replicating data over 100 times.

Strategy: Create a Calendar with Data Densification

Your manager wants a heatmap that shows the profitability by day of the year. To do this, you decide you are going to make the visualization in the shape of a calendar. The only challenge: you don't have sales for every day of the year, causing many blanks in your calendar.

With this strategy, we'll show you how to create a calendar showing profit by day for four years on a calendar. In this example, you'll learn how to do basic data densification using the Sample – Superstore dataset that fills in the missing values. Let's start by looking at what Tableau would produce by default; take a look at Figure 10-1.

Calendar with Standard Dates

Order Date

	2017							2018							2019							2020						
	S	M	T	W	T	F	S	S	M	T	W	T	F	S	S	M	T	W	T	F	S	S	M	T	W	T	F	S
January			3	4	5	6	7			2	3	4	5	6				2	3	4	5				1	2	3	
		9	10	11		13	14			9	10		12	13		7	8	9	10	11			6	7	8	9		
	15	16		18	19	20	21				17		19			14	15	16	17			12	13	14	15	16		
		23			26	27	28			23	24		26	27		21	22	23	24	25		19	20	21	22	23	24	
		30	31					28		30	31					28		30	31			26	27	28	29	30		
February				1	2	3	4							3						1	2							
		6	7	8			11			6	7	8	9	10	3	4	5	6	7	8	9	2	3	4	5	6		
	12		14	15	16	17	18				14	15	16			11	12	13	14	15	16	9	10	11		13		
		20	21	22	23	24		18		20	21	22	23				19	20	21	22	23	16	17	18	19	20	21	
		27						25		27	28					25		27	28			23	24	25	26		28	
March				1	2	3	4					1	2							1								
	5		7			10	11		5	6	7	8	9	10	3	4	5	6	7	8	9		2	3	4	5	6	7
			14	15	16	17	18		12	13	14	15	16	17	10	11	12	13	14	15		8	9	10	11	12	13	14
	19		21	22	23	24	25		19	20	21	22	23	24	17	18	19	20	21	22			16	17	18	19	20	21
	26		28	29	30	31			26	27	28	29	30	31	24	25	26	27	28	29	30		23	24	25	26	27	28
															31							29	30	31				
April							1																					
	2	3	4	5	6	7	8		2		4	5	6	7		1	2	3	4	5	6				1	2	3	4
			11	12	13		15		9	10	11	12	13	14	7	8	9	10		12	13		6	7	8	9	10	11
	16		18	19	20	21	22		16	17	18	19	20	21	14	15	16	17	18	19		12	13	14	15	16	17	
	23		25	26		28	29	22		24	25	26	27	28	21	22	23	24	25	26			20	21	22	23	24	25
	30							29	30						28		30					26	27	28	29	30		

Figure 10-1. A calendar without data modeling

You can see that Tableau returns cells only where data exists. If we look at January 1 or in 2017, for example, they don't show up. This is because there is no data in our dataset for these dates. Our goal with this strategy is to fill in those missing values so that our visualization renders as we would expect:

1. Build the calendar as follows:
   a. Add Year of [Order Date] and Weekday of [Order Date] to Columns.
   b. Add Month of [Order Date] and Week of [Order Date] to Rows. Right-click WEEK(Order Date) on Rows.
   c. Add Day of [Order Date] to Text and SUM(Profit) to Color, and then change the mark type to Square. The result is Figure 10-2.

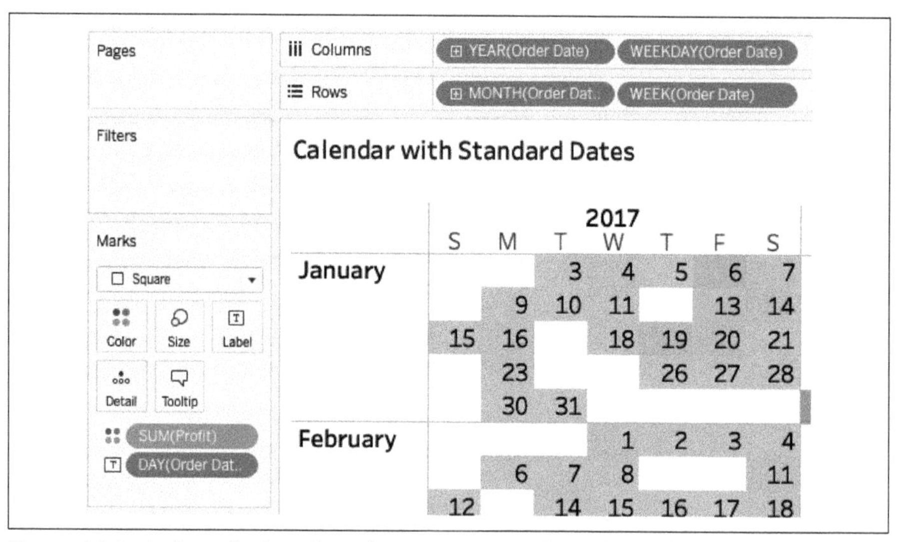

Figure 10-2. A closer look at the columns, rows, and Marks card

2. This step is very simple. Right-click any dimension on the Rows or Columns shelf and select Show Missing Values (Figure 10-3). This fills in all the values, from the minimum to the maximum.

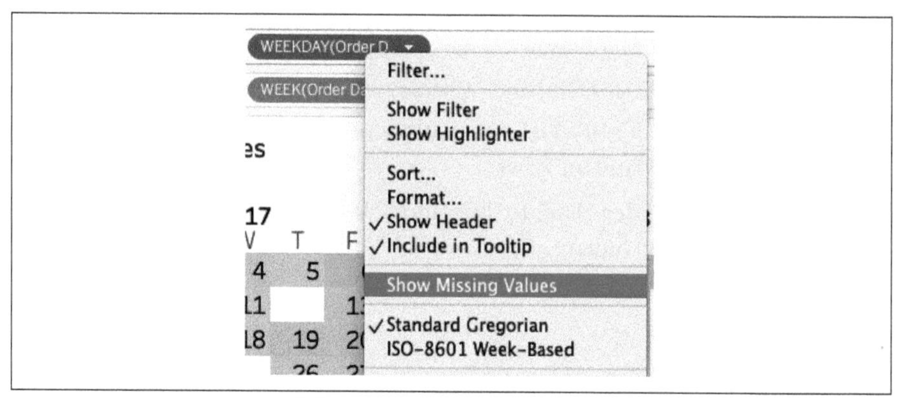

Figure 10-3. The Show Missing Values option

As you can see in Figure 10-4, all values from January 3, 2017 through December 30, 2020 are filled. But if you are looking to show January 1, 2017, January 2, 2017, and December 31, 2020, you need to extend our methodology. We can do this through some Tableau trickery that layers sheets on top of each other.

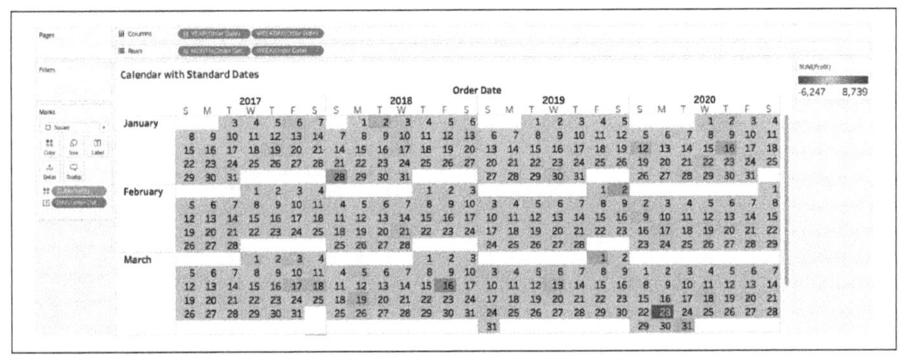

Figure 10-4. A filled calendar from the minimum value through the maximum value

3. To format the existing visualization, right-click it and set the background fill of the worksheet to None. Also change your text color to a dark gray.
4. Build the background sheet as follows:
   a. Create a calculated field called **`[Order Date Domain]`**. If the date in the dataset is equal to the earliest date, it will return January 1 of the earliest date; and if the date is equal to the latest date, it will return December 31 of the year of the latest date:

```
// Order Date Domain
IF [Order Date] = {MIN([Order Date])}
THEN DATETRUNC("year", [Order Date])
ELSEIF [Order Date] = {MAX([Order Date])}
THEN DATEADD("year", 1, DATETRUNC("year", [Order Date]))-1
ELSE [Order Date]
END
```

   b. Create a new sheet that will be used for the background. Follow steps 1 and 2 for the [Order Date Domain] date field. After showing the missing values, you will notice that values now go from January 1, 2017 through December 31, 2020.
   c. Edit the color by choosing a custom diverging palette and setting the color to #E1E1E1 for both options. Then set the stepped color to 2 units. This will create an all-gray sheet (Figure 10-5).

Order Date Domain																													
		2017							2018							2019							2020						
Month ..	Week ..	S	M	T	W	T	F	S	S	M	T	W	T	F	S	S	M	T	W	T	F	S	S	M	T	W	T	F	S
January	Week 1	1	2	3	4	5	6	7		1	2	3	4	5	6			1	2	3	4	5				1	2	3	4
	Week 2	8	9	10	11	12	13	14	7	8	9	10	11	12	13	6	7	8	9	10	11	12	5	6	7	8	9	10	11
	Week 3	15	16	17	18	19	20	21	14	15	16	17	18	19	20	13	14	15	16	17	18	19	12	13	14	15	16	17	18
	Week 4	22	23	24	25	26	27	28	21	22	23	24	25	26	27	20	21	22	23	24	25	26	19	20	21	22	23	24	25
	Week 5	29	30	31					28	29	30	31				27	28	29	30	31			26	27	28	29	30	31	
Februa..	Week 5				1	2	3	4					1	2	3						1	2							1
	Week 6	5	6	7	8	9	10	11	4	5	6	7	8	9	10	3	4	5	6	7	8	9	2	3	4	5	6	7	8
	Week 7	12	13	14	15	16	17	18	11	12	13	14	15	16	17	10	11	12	13	14	15	16	9	10	11	12	13	14	15
	Week 8	19	20	21	22	23	24	25	18	19	20	21	22	23	24	17	18	19	20	21	22	23	16	17	18	19	20	21	22
	Week 9	26	27	28					25	26	27	28				24	25	26	27	28			23	24	25	26	27	28	29
March	Week 9				1	2	3	4					1	2	3						1	2							
	Week ..	5	6	7	8	9	10	11	4	5	6	7	8	9	10	3	4	5	6	7	8	9	1	2	3	4	5	6	7
	Week ..	12	13	14	15	16	17	18	11	12	13	14	15	16	17	10	11	12	13	14	15	16	8	9	10	11	12	13	14
	Week ..	19	20	21	22	23	24	25	18	19	20	21	22	23	24	17	18	19	20	21	22	23	15	16	17	18	19	20	21
	Week ..	26	27	28	29	30	31		25	26	27	28	29	30	31	24	25	26	27	28	29	30	22	23	24	25	26	27	28
	Week ..															31							29	30	31				

Figure 10-5. The background sheet

With the calculation, you completed domain padding by "adding" dates to your dataset. You then used Show Missing Values to execute domain completion.

5. For this calendar, you will layer the two visualizations on top of each other:

   a. If you do not have a dashboard yet, create a new dashboard and add a vertical container. Add the background sheet to the container and fit the sheet to the entire container.

   b. Add a floating container on top of the background view. Add the top layer to the top container. Hide all the headers and then size the top container to match the background layer. You can match the size by matching the left, top, width, and height between the two sheets.

The final result is two visualizations that look like one full calendar (Figure 10-6).

Order Date Domain																													
		2017							2018							2019							2020						
Month ..	Week ..	S	M	T	W	T	F	S	S	M	T	W	T	F	S	S	M	T	W	T	F	S	S	M	T	W	T	F	S
January	Week 1	1	2	3	4	5	6	7		1	2	3	4	5	6			1	2	3	4	5				1	2	3	4
	Week 2	8	9	10	11	12	13	14	7	8	9	10	11	12	13	6	7	8	9	10	11	12	5	6	7	8	9	10	11
	Week 3	15	16	17	18	19	20	21	14	15	16	17	18	19	20	13	14	15	16	17	18	19	12	13	14	15	16	17	18
	Week 4	22	23	24	25	26	27	28	21	22	23	24	25	26	27	20	21	22	23	24	25	26	19	20	21	22	23	24	25
	Week 5	29	30	31					28	29	30	31				27	28	29	30	31			26	27	28	29	30	31	
Februa..	Week 5				1	2	3	4					1	2	3						1	2							1
	Week 6	5	6	7	8	9	10	11	4	5	6	7	8	9	10	3	4	5	6	7	8	9	2	3	4	5	6	7	8
	Week 7	12	13	14	15	16	17	18	11	12	13	14	15	16	17	10	11	12	13	14	15	16	9	10	11	12	13	14	15
	Week 8	19	20	21	22	23	24	25	18	19	20	21	22	23	24	17	18	19	20	21	22	23	16	17	18	19	20	21	22
	Week 9	26	27	28					25	26	27	28				24	25	26	27	28			23	24	25	26	27	28	29
March	Week 9				1	2	3	4					1	2	3						1	2							
	Week ..	5	6	7	8	9	10	11	4	5	6	7	8	9	10	3	4	5	6	7	8	9	1	2	3	4	5	6	7
	Week ..	12	13	14	15	16	17	18	11	12	13	14	15	16	17	10	11	12	13	14	15	16	8	9	10	11	12	13	14
	Week ..	19	20	21	22	23	24	25	18	19	20	21	22	23	24	17	18	19	20	21	22	23	15	16	17	18	19	20	21
	Week ..	26	27	28	29	30	31		25	26	27	28	29	30	31	24	25	26	27	28	29	30	22	23	24	25	26	27	28
	Week ..															31							29	30	31				
April	Week ..							1																					
	Week ..	2	3	4	5	6	7	8	1	2	3	4	5	6	7		1	2	3	4	5	6				1	2	3	4
	Week ..	9	10	11	12	13	14	15	8	9	10	11	12	13	14	7	8	9	10	11	12	13	5	6	7	8	9	10	11
	Week ..	16	17	18	19	20	21	22	15	16	17	18	19	20	21	14	15	16	17	18	19	20	12	13	14	15	16	17	18
	Week ..	23	24	25	26	27	28	29	22	23	24	25	26	27	28	21	22	23	24	25	26	27	19	20	21	22	23	24	25
	Week ..	30							29	30						28	29	30					26	27	28	29	30		

Figure 10-6. The final result: a two-layered sheet

With this strategy, you applied domain padding and domain completion by using a single calculation and Show Missing Values. This is the most simplistic version of data densification. In the rest of the chapter, you will learn more-complicated versions of data modeling.

Strategy: Create Rounded Bar Charts

For this strategy, you'll create a summary of sales by sub-category. At first, you may think, "This seems easy," but what if your stakeholder wants to see a very specific type of chart: a rounded bar chart? You know that this isn't ideal for interpretation but agree it's a break from the mundane bar charts everyone typically gets.

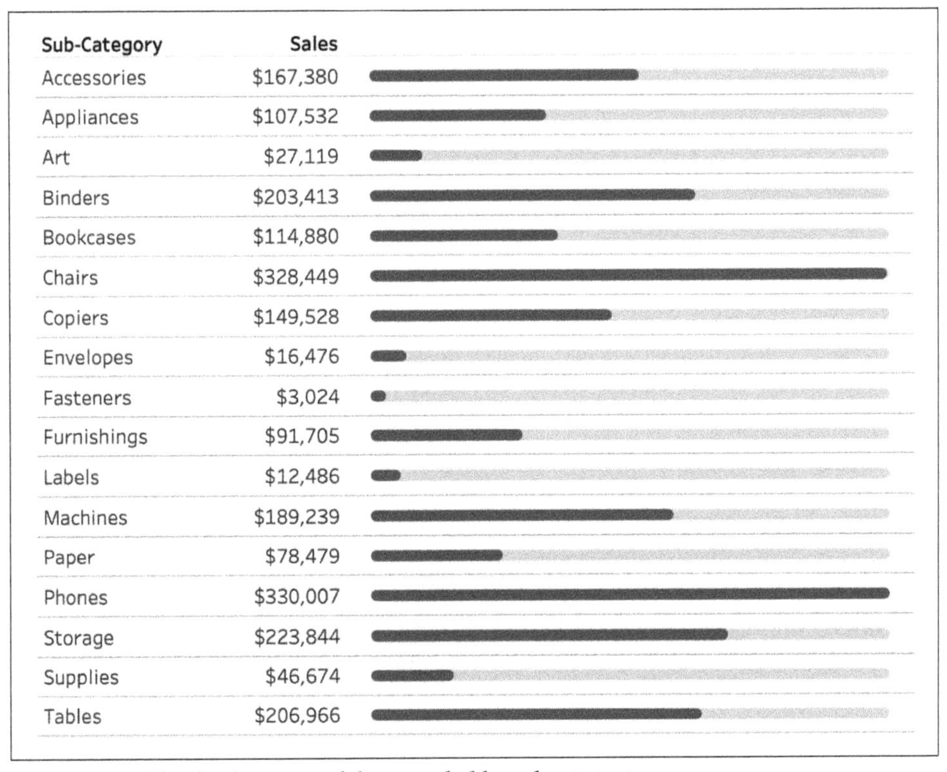

Sub-Category
Sales
Accessories
$167,380
Appliances
$107,532
Art
$27,119
Binders
$203,413
Bookcases
$114,880
Chairs
$328,449
Copiers
$149,528
Envelopes
$16,476
Fasteners
$3,024
Furnishings
$91,705
Labels
$12,486
Machines
$189,239
Paper
$78,479
Phones
$330,007
Storage
$223,844
Supplies
$46,674
Tables
$206,966

Figure 10-7. The final output of the rounded bar chart strategy

In this strategy, you'll learn how to make rounded bar charts, as shown in Figure 10-7, but more importantly, you will begin to learn the essentials of data modeling. With data modeling, you add more data to your dataset to build out-of-the-box solutions. As mentioned earlier, this can be done one of three ways: unions, joins, and table calculations.

For this rounded bar chart strategy, and the next strategy that refers to lessons learned in Chapter 6, we will use a union. Again, we could accomplish this using the other techniques, but we will start with the option we find the easiest:

1. Let's begin by connecting to the Sample – Superstore dataset. Instead of using any quick connections you have, select Microsoft Excel and find Sample – Superstore.xls. Once you have connected, click and drag the Orders table to the data source. The result is shown in Figure 10-8.

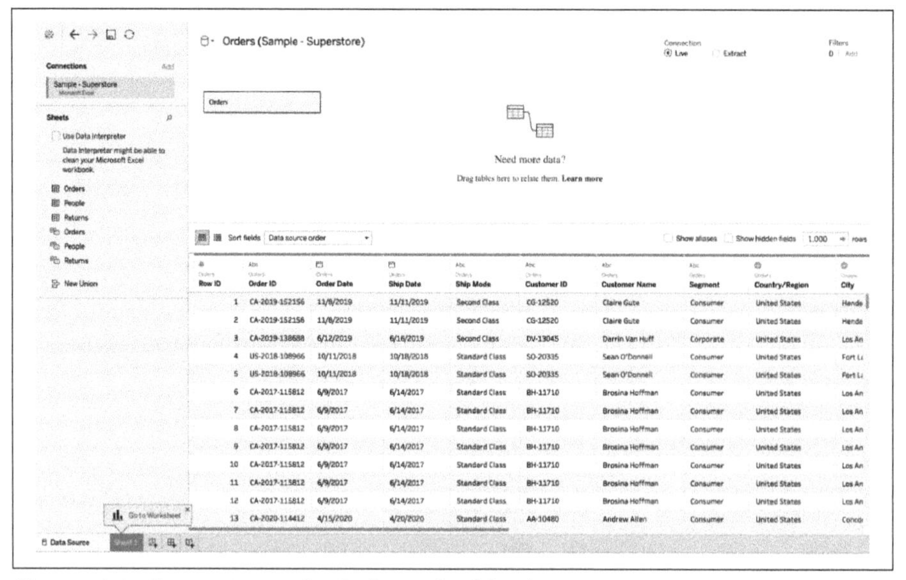

Figure 10-8. Connecting to the Orders tab of the dataset

   What you have done is very typical. But we're not done with the Data pane yet.

2. Let's take the same Orders table you selected from the Connections pane and create a union with the Orders data you already added to the data source window, shown in Figure 10-9.

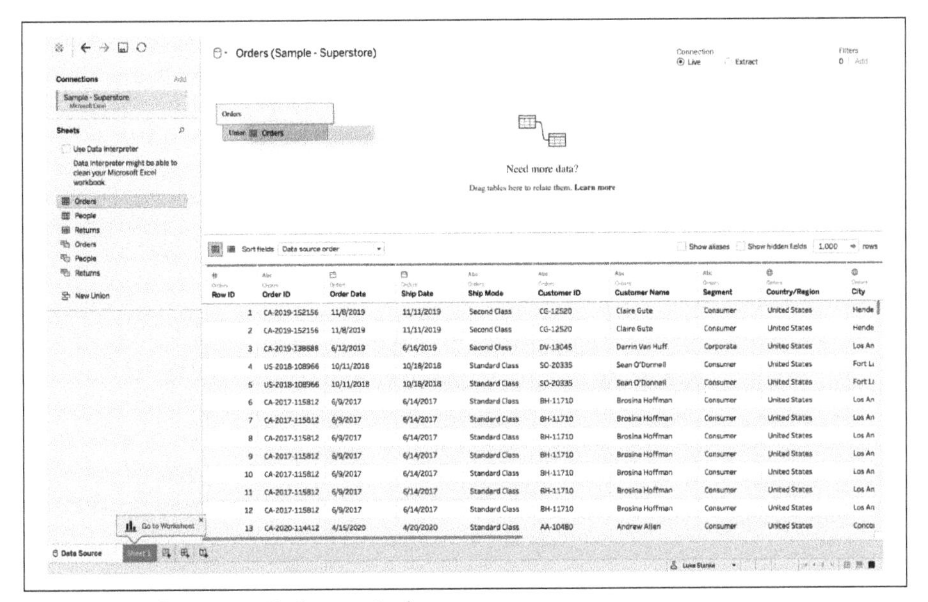

Figure 10-9. Unioning Orders to Orders

By unioning the Orders data to itself, we are essentially creating a second layer of data to work with. This duplicate layer of data provides increased flexibility for working in Tableau, but it also requires us to track each layer all the time. You constantly have to ask yourself, What is happening with the first layer of data? What's happening with the second layer of data? If you can track what you are doing with both, you are well on your way to becoming a next-level professional of Tableau Desktop.

So how do you track both layers of data? When you union data sources together, Tableau creates two new fields in your dataset to help track the layers: [Sheet] and [Table Name]. As you are building out our data, you can see the [Sheet] and [Table Name] by scrolling all the way to the right (Figure 10-10).

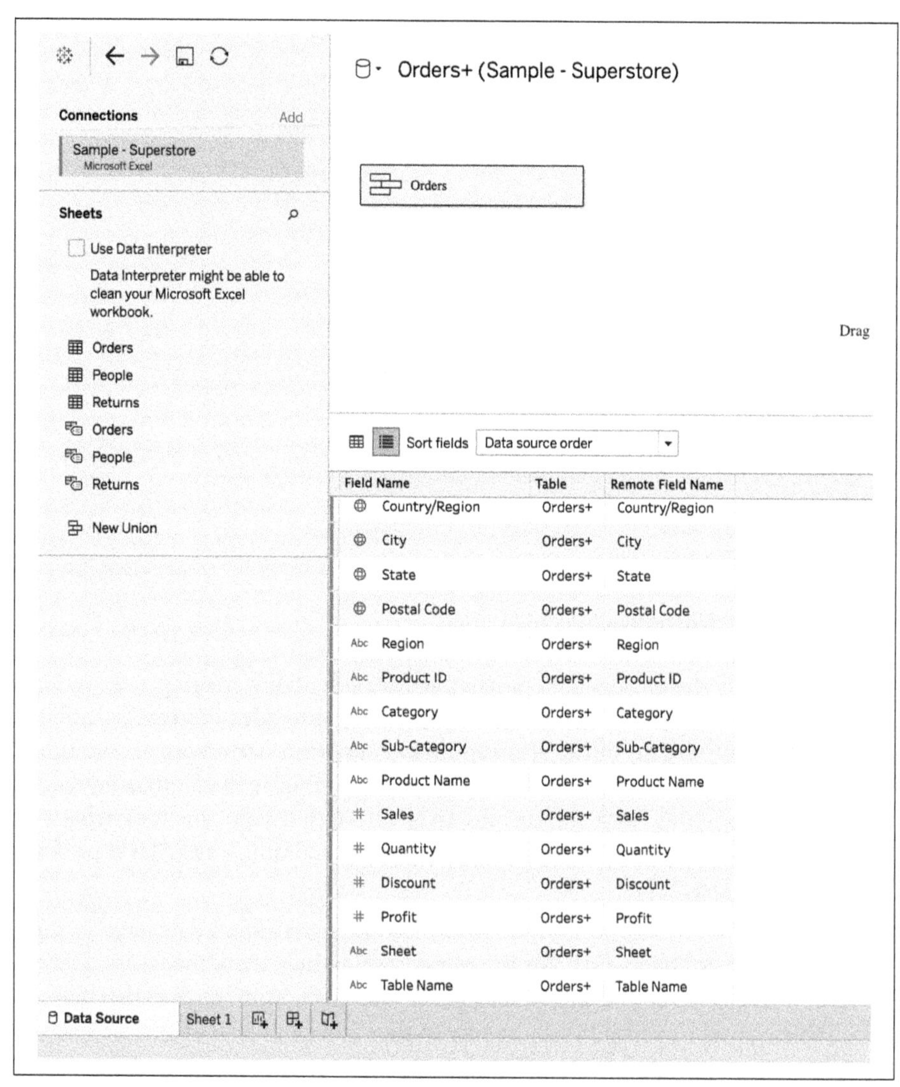

Figure 10-10. After a union, the [Sheet] and [Table Name] dimensions are added

For our strategies, you will use the [Table Name] dimension to track what is happening with each layer.

3. Click on Sheet 1. Before we start building, add [Sales] to Columns and [Table Name] to Rows. You will see that we have two members of [Table Name]: Orders (our first layer), and Orders1 (our second layer, whose name Tableau updated by appending a 1 to the end). You'll also notice that the sales are equal. That's because our layers are identical.

   Half the battle of data modeling is getting the data in order. The other battle: tracking where the data is on your visualization.

4. Now that we have two layers of data, we have to think about what our rounded bar chart really is: it's just two overlapping lines. We need to control the start and end of each lines. And we can do that with our two data layers. Let's consider the final product in Figure 10-6. The start of our main bars is at zero. We'll need one layer to be equal to zero. The other layer of our data needs to go to the total sales.

   Let's build out a calculation called **[Bars]** to do exactly this:

   ```
   // Bars
   IF [Table Name] = "Orders"
   THEN 0
   ELSE [Sales]
   END
   ```

   You also have our background bar, whose overall length is equal to the maximum sales across each [Sub-Category]. Create a calculation called **[Background]**:

   ```
   // Background
   IF MIN([Table Name]) = "Orders"
   THEN MIN(0)
   ELSE WINDOW_MAX(SUM([Sales]))
   END
   ```

5. To build the visualization, create a new sheet and change the view from Standard to Entire View.

   a. Add [Sub-Category] to Rows.

   b. Add the sum of [Sales] to Rows. Convert the data to discrete. Then format the number to currency and right-align the text. Figure 10-11 shows the view after this step.

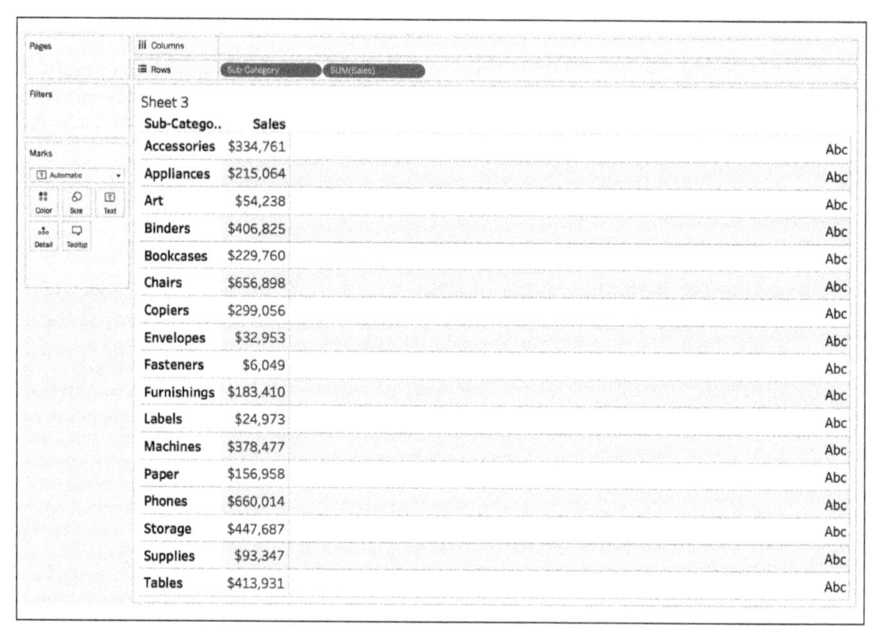

Figure 10-11. The view after formatting the discrete sum of sales

c. Add [Table Name] to Detail on the Marks card.

d. Add the [Bars] calculation to Columns, change the mark type to Line, and adjust the size to be about 75% of the maximum value. Figure 10-12 shows the view after this step.

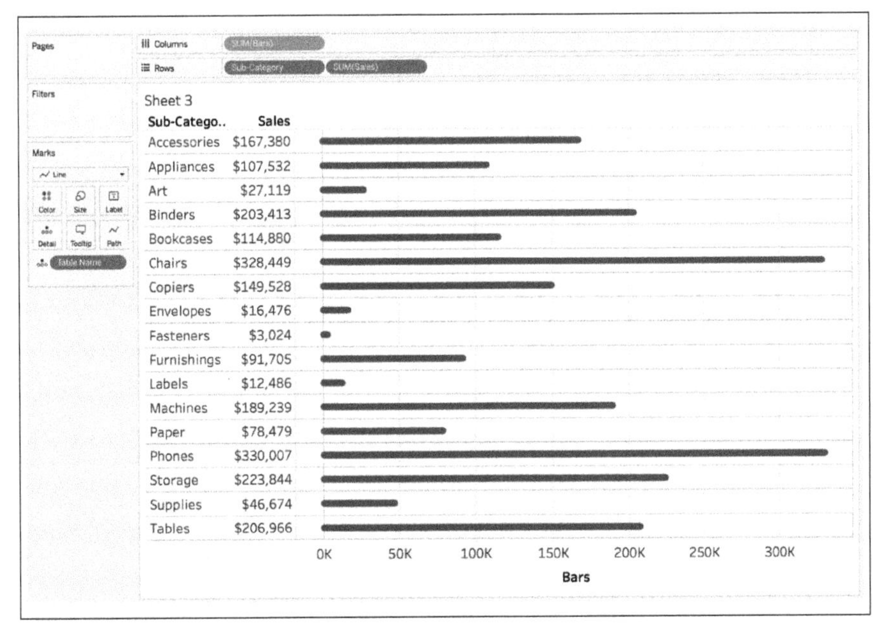

Figure 10-12. The view after adding the Bars calculation to rows

e. Add the [Background] calculation to Columns. Make sure the mark type is set to Line. Create a dual axis between [Bars] and [Background]. Right-click the [Background] axis and select "Move marks to back." (Depending on your version of Tableau, you may need to change the mark type of [Table Name] to Line instead of Detail.)

f. Remove [Measure Names] from both Marks cards. Edit the opacity of the color on the [Background] Marks card to 25%. Figure 10-13 shows the visualization at this point.

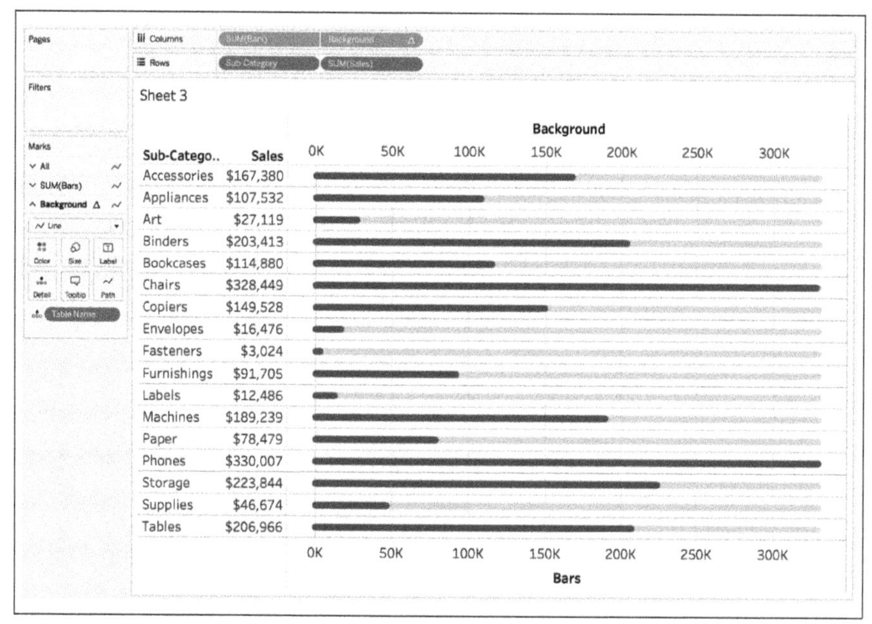

Figure 10-13. The view after formatting the color of the background marks card.

g. Format the view by hiding the axes and removing grid lines, zero lines, axis rulers, and column dividers. Format your tooltip.

Once you've done these steps, you've created your final product (Figure 10-14).

It's important to know that not all data sources can create unions in Tableau. For a full list, see Tableau's help doc on unions (*https://oreil.ly/t5N03*).

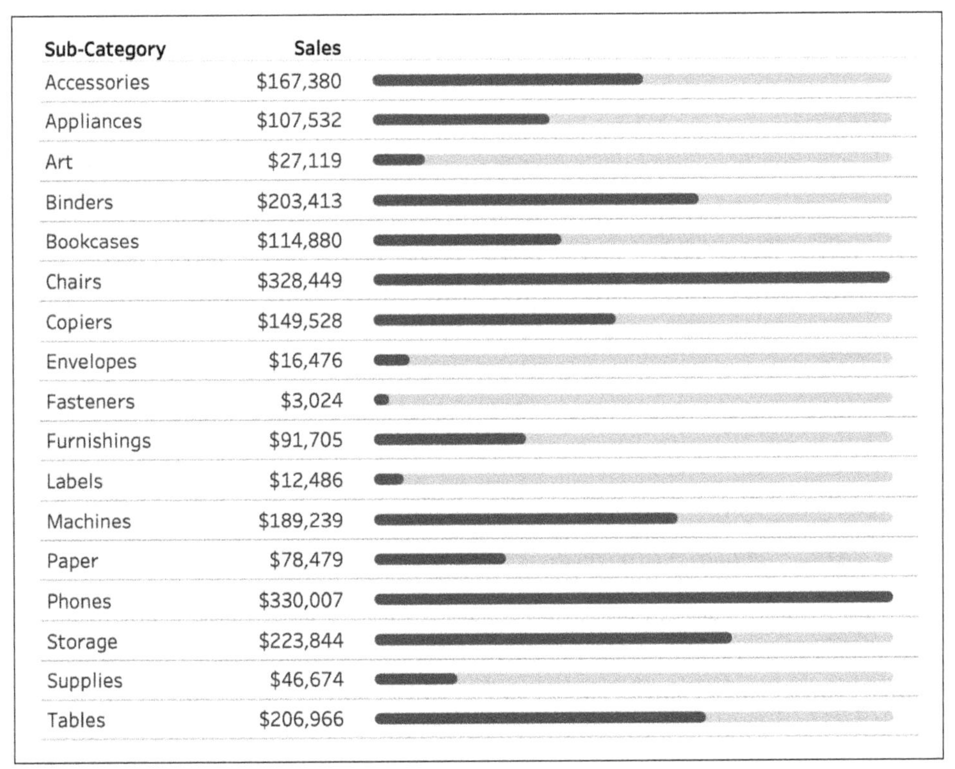

Figure 10-14. A formatted rounded bar chart

Strategy: Create an Accordion Table

You build lots of tables because your audience demands that they can just get to the numbers quickly. You use hierarchies to allow your users to drill up and down through the data. But one of the common complaints you hear is that when a user expands the table, they see too much information and can't compare one level to the next.

For example, say a user is looking for the profit ratio in Figure 10-15 for Office Supplies. The user would have to drill back up the hierarchy to get the number.

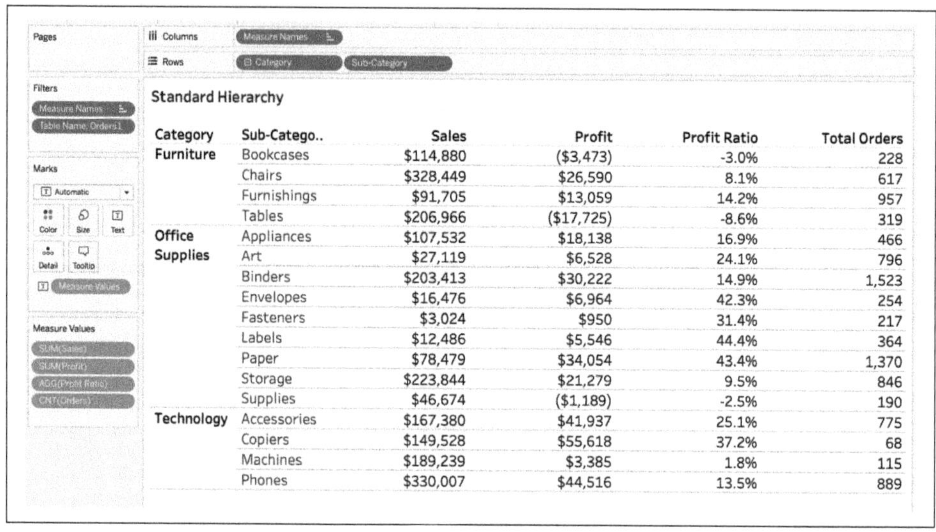

Standard Hierarchy

Category	Sub-Catego..	Sales	Profit	Profit Ratio	Total Orders
Furniture	Bookcases	$114,880	($3,473)	-3.0%	228
	Chairs	$328,449	$26,590	8.1%	617
	Furnishings	$91,705	$13,059	14.2%	957
	Tables	$206,966	($17,725)	-8.6%	319
Office Supplies	Appliances	$107,532	$18,138	16.9%	466
	Art	$27,119	$6,528	24.1%	796
	Binders	$203,413	$30,222	14.9%	1,523
	Envelopes	$16,476	$6,964	42.3%	254
	Fasteners	$3,024	$950	31.4%	217
	Labels	$12,486	$5,546	44.4%	364
	Paper	$78,479	$34,054	43.4%	1,370
	Storage	$223,844	$21,279	9.5%	846
	Supplies	$46,674	($1,189)	-2.5%	190
Technology	Accessories	$167,380	$41,937	25.1%	775
	Copiers	$149,528	$55,618	37.2%	68
	Machines	$189,239	$3,385	1.8%	115
	Phones	$330,007	$44,516	13.5%	889

Figure 10-15. A standard hierarchy in Tableau

One alternative to help reduce the cognitive complexity is an *accordion table*. At a high level, you see only top-level information. In Figure 10-16, this means seeing values only at the category level.

	Sales	Profit	Profit Ratio	Total Orders
▶ Furniture	$742,000	$18,451	2.5%	2,121
▶ Office Supplies	$719,047	$122,491	17.0%	6,026
▶ Technology	$836,154	$145,455	17.4%	1,847

Figure 10-16. An overview at the category level

However, if your users want to see details about office supplies, they could open the accordion and understand details one level lower without revealing too much information, as shown in Figure 10-17.

	Sales	Profit	Profit Ratio	Total Orders
▶ Furniture	$742,000	$18,451	2.5%	2,121
▼ Office Supplies	$719,047	$122,491	17.0%	6,026
Appliances	$107,532	$18,138	16.9%	466
Art	$27,119	$6,528	24.1%	796
Binders	$203,413	$30,222	14.9%	1,523
Envelopes	$16,476	$6,964	42.3%	254
Fasteners	$3,024	$950	31.4%	217
Labels	$12,486	$5,546	44.4%	364
Paper	$78,479	$34,054	43.4%	1,370
Storage	$223,844	$21,279	9.5%	846
Supplies	$46,674	($1,189)	-2.5%	190
▶ Technology	$836,154	$145,455	17.4%	1,847

Figure 10-17. Using the accordion table to drill into the Office Supplies category and see the performance by sub-category

When we are building drillable tables, we often use accordion tables. So how is this done?

1. For this strategy, we will be using the same Samples – Superstore data source as in our preceding strategy. If you are not connected to this data source, follow steps 1 and 2 of "Strategy: Create Rounded Bar Charts" on page 367 to connect.
2. Rather than selecting Standard or Entire View, set your view size to Fit Width. Build the visualization as follows:
   a. Add [Category] to Rows.
   b. Add [Table Name] to the right of [Category] on Rows. Make sure that the Orders member occurs before Orders1.
   c. Right-click [Category] in the Data pane and create a new set. Call the set **`[Category Set]`**. Do not select any values yet.
   d. Create a calculation for the arrows called **`[Arrow]`**:

```
// Arrow
IF [Table Name] = "Orders"
AND NOT [Category Set]
THEN "►"
ELSEIF [Table Name] = "Orders"
AND [Category Set]
THEN "▼"
ELSE ""
END
```

This will show a down arrow when an accordion is open and a right arrow when an accordion is closed. Place [Arrows] to the right of [Table Name] on Rows.

e. Create a calculation called **[Level]**. This calculation will show either [Category] or [Sub-Category], depending on which layer of data we are working with:

```
// Level
IF [Table Name] = "Orders"
THEN [Category]
ELSE [Sub-Category]
END
```

Place [Level] to the right of [Arrow] on Rows.

f. Add [Measure Names] to Columns and add [Measure Values] to Text on the Marks card. Include only the metrics you are interested in. For this example, we are using SUM(Sales), SUM(Profit), [Profit Ratio], and COUNTD(Orders).

If [Profit Ratio] and COUNTD(Order ID) don't exist, you can create the following calculated fields:

```
// Profit Ratio
SUM([Profit])/SUM([Sales])

// Total Orders
COUNTD([Order ID])
```

You now have a fully built-out table.

g. Hide the [Category] and [Table Name] dimensions on Rows by right-clicking each and deselecting Show Header.

h. Format your table by removing row banding, zero lines, axis rulers, and grid lines. Add row dividers so they match the view in Figure 10-18.

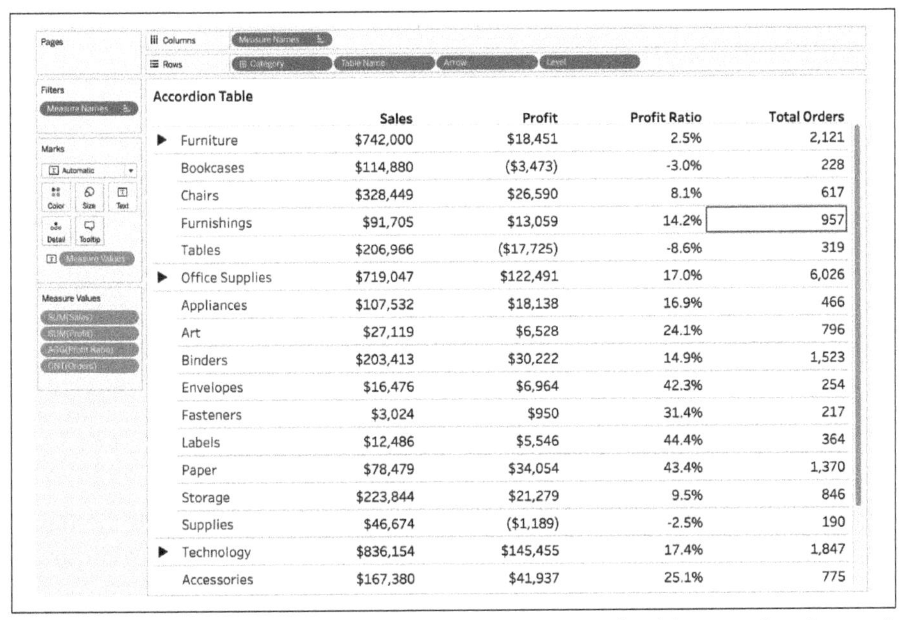

	Sales	Profit	Profit Ratio	Total Orders
▶ Furniture	$742,000	$18,451	2.5%	2,121
Bookcases	$114,880	($3,473)	-3.0%	228
Chairs	$328,449	$26,590	8.1%	617
Furnishings	$91,705	$13,059	14.2%	957
Tables	$206,966	($17,725)	-8.6%	319
▶ Office Supplies	$719,047	$122,491	17.0%	6,026
Appliances	$107,532	$18,138	16.9%	466
Art	$27,119	$6,528	24.1%	796
Binders	$203,413	$30,222	14.9%	1,523
Envelopes	$16,476	$6,964	42.3%	254
Fasteners	$3,024	$950	31.4%	217
Labels	$12,486	$5,546	44.4%	364
Paper	$78,479	$34,054	43.4%	1,370
Storage	$223,844	$21,279	9.5%	846
Supplies	$46,674	($1,189)	-2.5%	190
▶ Technology	$836,154	$145,455	17.4%	1,847
Accessories	$167,380	$41,937	25.1%	775

Figure 10-18. The accordion after hiding the Category and Table Name headers and format

3. You want to show data only when it is selected. To do so, you need to create a filter. Create a calculation called **`[Filter]`**:

```
// Filter
[Table Name] = "Orders"
OR [Category Set]
```

Add Filter to the Filters shelf and select True. Figure 10-19 shows the visualization at this stage.

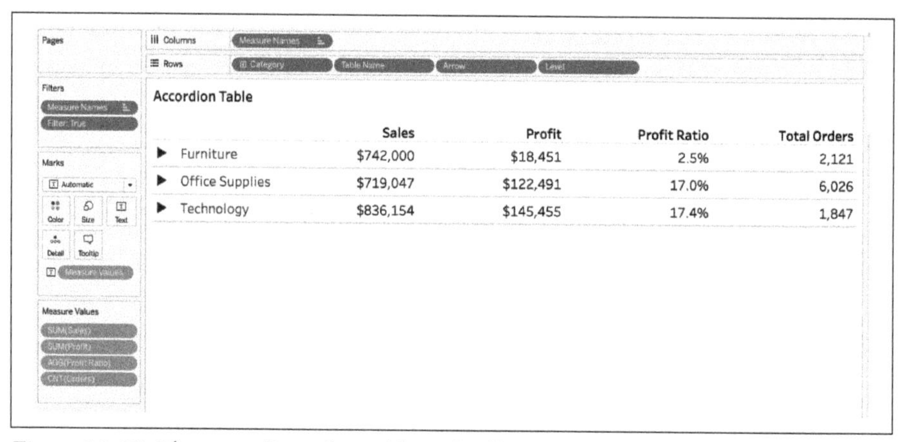

Figure 10-19. The accordion after adding the filter calculation

4. After you have added your visualization to the dashboard, from the top menu choose Dashboard → Actions and add an action. Alternatively, you can add an action on the sheet by choosing Worksheet → Actions from the top menu.

   Select to Change Set Values to add a set action. On select, assign a value to Category Set, and when the selection is cleared, remove all values. Click OK. Then click OK again. These options are shown in Figure 10-20.

Go ahead and test your table by selecting Furniture. You will notice that your table expands to show the four sub-categories of information, as shown in Figure 10-21. You will now be able to drill in and out of your data without revealing an overwhelming amount of information to your audience.

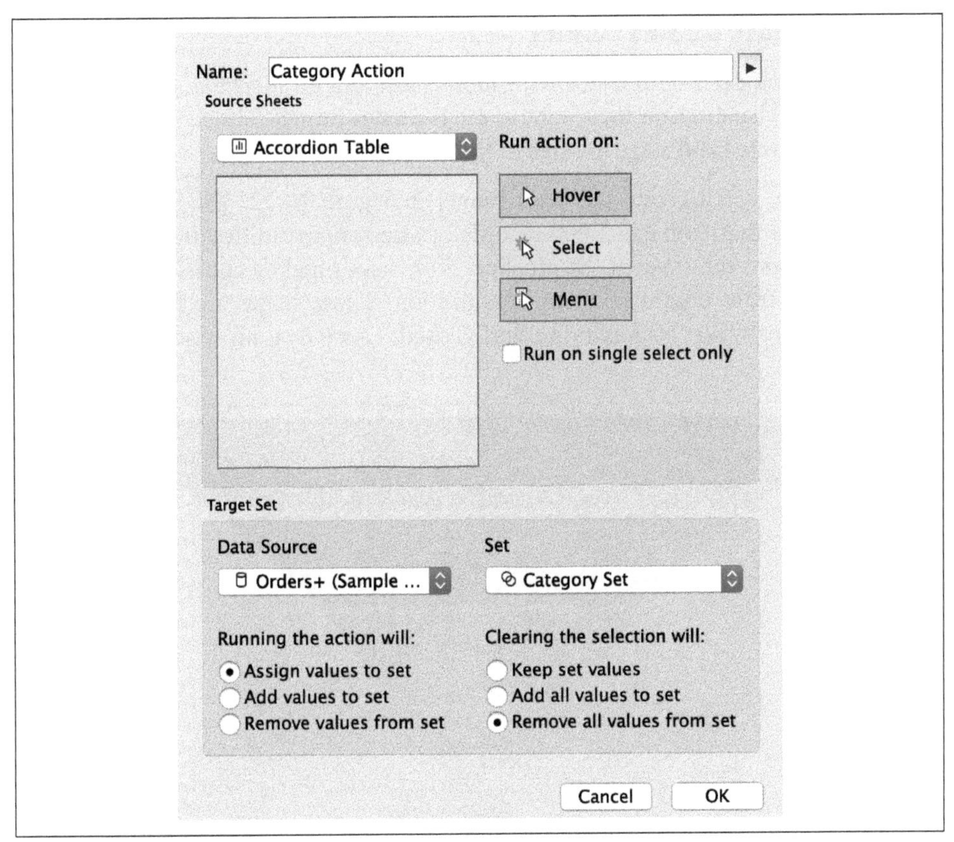

Figure 10-20. Adding the set action to the Category Set calculation

Accordion Table

	Sales	Profit	Profit Ratio	Total Orders
▼ Furniture	$742,000	$18,451	2.5%	2,121
Bookcases	$114,880	($3,473)	-3.0%	228
Chairs	$328,449	$26,590	8.1%	617
Furnishings	$91,705	$13,059	14.2%	957
Tables	$206,966	($17,725)	-8.6%	319
▶ Office Supplies	$719,047	$122,491	17.0%	6,026
▶ Technology	$836,154	$145,455	17.4%	1,847

Figure 10-21. The accordion table after drilling into the Furniture category

Strategy: Create a Sales Funnel

Data provides modern sales teams with motivation and an incentive to sell, and the most common visualization they want to see is a sales funnel. The *sales funnel* shows the conversion rate from stage to stage in the sales cycle.

In this strategy, you'll create the sales funnel in Figure 10-22. You will work with mock sales data that contains 1,000 records of sales opportunities, the values of the opportunities, and the stage of the pipeline. Your sales pipeline consists of six stages, but this varies from organization to organization. These stages are Prospect, Lead, Qualified, Opportunity, Negotiations, and Closed. Every deal must go through each of these stages.

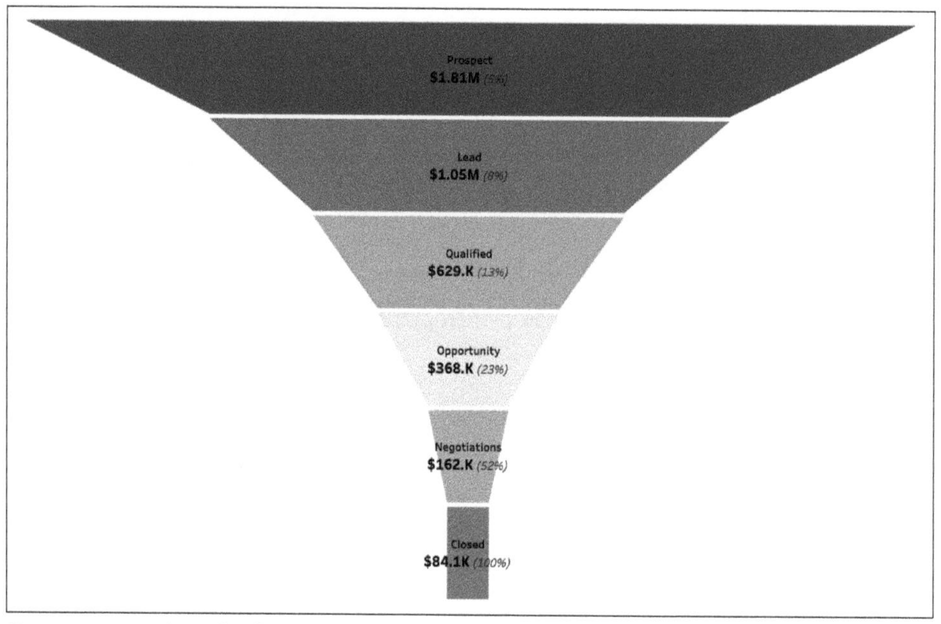

Figure 10-22. The sales funnel

The sales funnel is a nonstandard data visualization. While there are some ways to create a funnel without data modeling, the biggest issue with those techniques is that areas in each stage of the funnel end up misrepresented. Additionally, a common retort against the sales funnel is that a funnel is a metaphor that could be represented by a series of bar charts instead. This is absolutely true. However, when we think about our audiences and how they engage with data, sometimes a sales funnel metaphor is the exact visualization they need to understand their pipeline conversion. And a sales pipeline with areas that are proportionally correct can't be built in Tableau without a little data modeling.

This will be the first data model to use a join. We will use a join to duplicate the data four times over: one for each corner of the sales polygons. Here are the steps:

1. In Notepad or any text editor, create your modeling data source by typing the following:

   ```
   Model
   1
   2
   3
   4
   ```

   Save the file as *funnel_model.csv*.

2. When modeling the data, you need to connect to two data sources: the sales data source and the modeling data. You also need to join the two sources together with a custom join:

   a. Connect to the *funnel_model.csv* data source. Click and drag *funnel_model.csv* to the Drag Tables Here section.

   b. Add another data source connection to the sales stage data. Click and drag *sales_stage_data* to the *funnel_model.csv* data source.

   c. To bring the two data sources together, you need to edit the join. Create custom relationship calculations for both data sources and set the value for each to 1 (Figure 10-23).

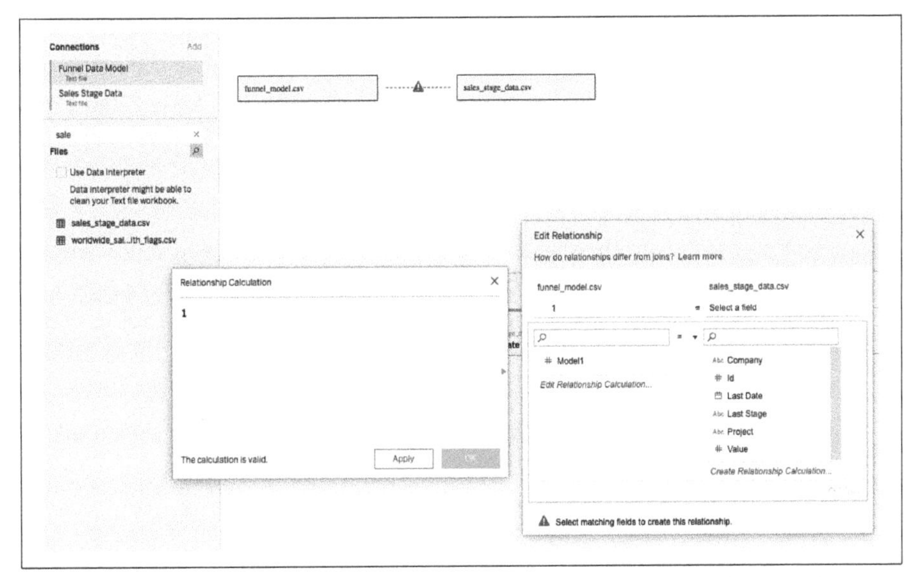

Figure 10-23. Creating the custom relationship for the sales_stage_data.csv data source

Your data source connection will change from a red warning sign to an orange line. This line indicates the relationship between the two data sources. Data is ready for analysis. Create a new sheet to begin building our sales funnel.

3. To determine and fill in the values of the funnel, we need to code the order of the funnel sections. Use a case statement to put stage names into a numeric order by creating a calculation called **[Sort]**:

```
// Sort
CASE [Stage]
WHEN "Prospect" THEN 1
WHEN "Lead" THEN 2
WHEN "Qualified" THEN 3
WHEN "Opportunity" THEN 4
WHEN "Negotiations" THEN 5
WHEN "Closed" THEN 6
END
```

   In this example, our stages are ordered as Prospect, Lead, Qualified, Opportunity, Negotiations, and then Closed.

4. Calculate the total value of the sales opportunities as follows:

   a. Create a calculation called **[Total Value]**:

   ```
   // Total Value
   RUNNING_SUM(SUM([Value]))
   ```

   We will use this calculation to determine the total value of opportunities throughout the pipeline.

   b. As part of creating the funnel chart, you need to calculate the value of each stage, but you also need to calculate the value of the next stage. Create a calculation called **[Next Value]**:

   ```
   //Next Value
   IFNULL(LOOKUP([Total Value] ,1), [Total Value])
   ```

   c. You may not want the stages of your funnel touching one another. You can fix this by adding padding. Rather than hardcoding the padding between stages, create a float parameter called **[padding]**. We've assigned a value of 1.05 to our parameter. We suggest a value from 1 (no padding) to 1.3 (a lot of padding).

   d. Create a calculation called **[size]** that assigns values to your data model. This will help create the sides of your funnel:

   ```
   //size
   IF [Model] = 1
   OR [Model] = 2
   THEN 2
   ELSE 1
   END
   ```

5. Create two calculations, one for the *x* coordinates and one for the *y* coordinates, as follows:

   a. For the *x* coordinates, create a calculation called **[x]**:

   ```
   //x
   CASE MIN([Model])

   WHEN 1 THEN [Total Value]
   WHEN 2 THEN -[Total Value]
   WHEN 3 THEN -[Next Value]
   WHEN 4 THEN [Next Value]
   END
   ```

   b. For the *y* coordinates, create a calculation called **[y]**:

   ```
   //y

   FLOAT([size]) - ([Sort]*[padding])
   ```

   In this case, `[size]` is just the overall height of the funnel, and the `[sort] * [padding]` will specify the start of each section of the sales funnel.

6. Now that you have the calculations to create a funnel, you can begin to build the funnel chart. You will eventually create a dual-axis chart, so this is only part of the visualization, but if you can build the funnel, then the final steps will be a piece of cake:

   a. Change the mark type to Polygon. Then add [Stage] to Color and [Model] as a dimension to Path.

   b. Add [y] to Rows. Change the aggregation to Average. Don't forget the average aggregation; this is critical later!

   c. Add [x] as a dimension to Columns. Right-click to edit the table calculations. Because you are working with nested calculations, you have two table calculations to edit. For the [Next Value] calculation, set the Compute Using option to Specific Dimensions and then select only the Stage checkbox. From the "Sort order" list, select Custom Sort, and do a custom ascending sort on the [Sort] calculation you created, using Minimum as the aggregation type. Use sort ascending for Next Value and use sort descending for Total Value. Use the exact same settings for the [Total Value] table calculation, which is shown in Figure 10-24.

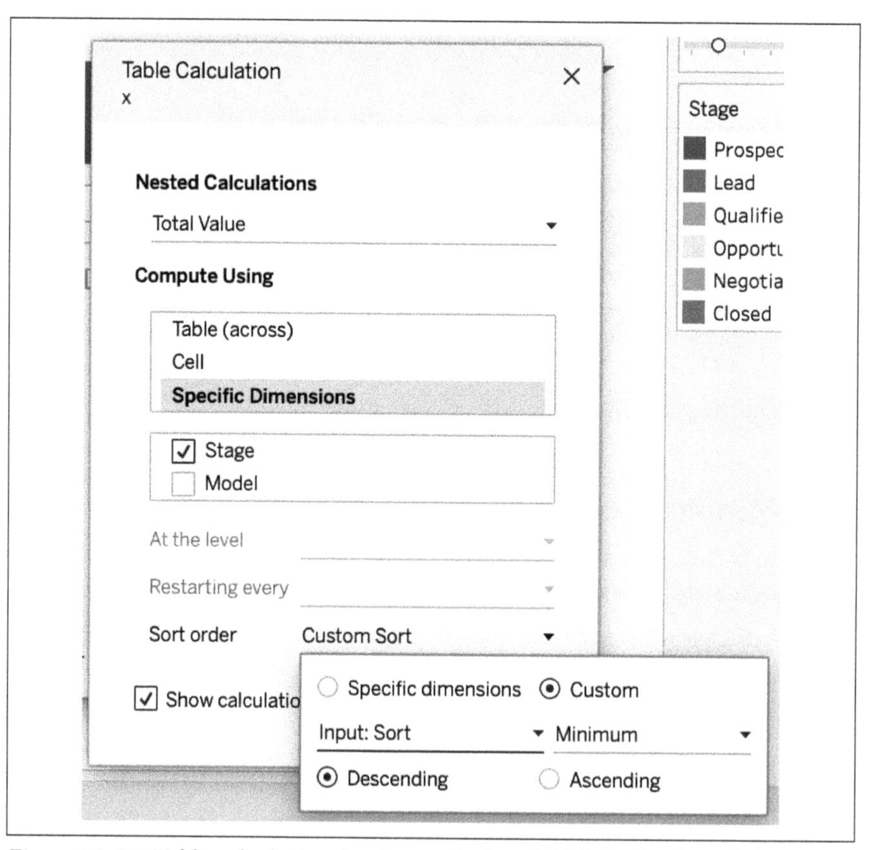

Figure 10-24. Table calculation for the x calculated field for the funnel chart

d. Remove all lines and dividers, and hide the headers. Then click the color and remove the border around the polygons. Edit the colors of the stages.

   You have three options for coloring the stages: keep the color the same because you are already using the shape of the funnel to encode change; use discrete colors to emphasize that each stage is separate; use a custom sequential color palette (which, as mentioned, acts as a double encoding of the information). The result is Figure 10-25.

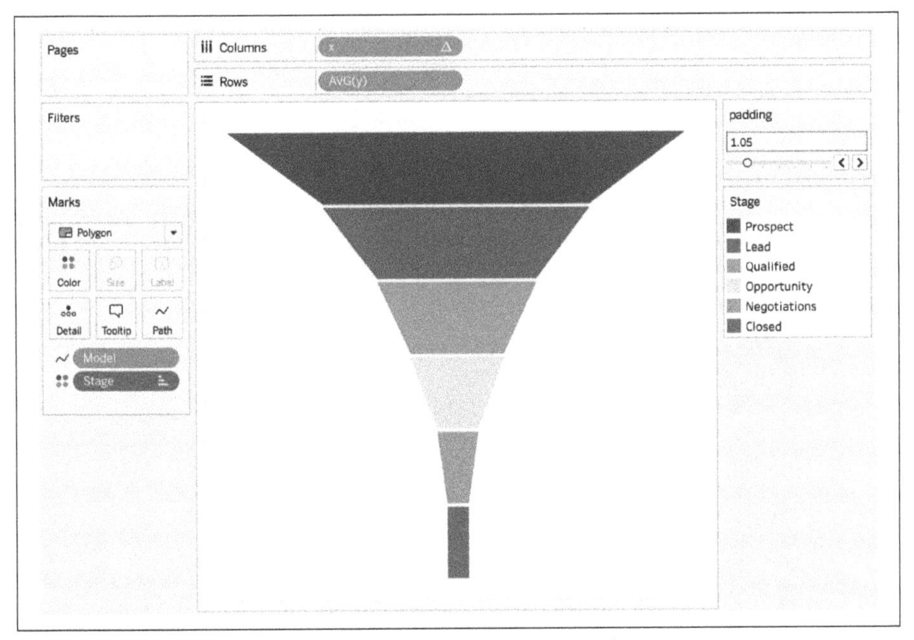

Figure 10-25. The first half of the funnel chart, formatted

7. You will add context to the funnel by adding text and labels. To do so, you need to build a calculation that centers labels on the chart:

   a. Create a calculation called **`[Center Label]`**:

   ```
   //Center Label
   IF [Model] = 1 THEN 0 END
   ```

   Because you modeled your data four times over, you need to place labels to calculate the totals for only a single point in the data model. In this case, you are going to calculate them for when the model dimension is equal to 1.

   Add [Center Label] to Columns. The default aggregation is sum, and you won't need to update this. Create a synchronized dual axis with [x] and sum of [Center Label]; don't forget to remove [Measure Names] from both Marks cards. When you are done with the axes, change the mark type to Text.

   Add [Stage] to Label.

   b. You will want to show the value of each stage in the pipeline. Remember that regardless of which stage any opportunity is at, it will have passed through each of the previous stages. This means that we need to roll back the values of each stage into the prior stage—even though the data does not technically exist in our dataset. We can do this with table calculations.

Create a calculation called **[Stage Value Label]**:

```
//Stage Value Label
RUNNING_SUM(SUM(IF [Model] = 1 THEN [Value] END))
```

This will add labels to model values that are equal to 1.

Since your values will be in both thousands of dollars and millions of dollars, there is no need to show precision to the nearest dollar. You might prefer to show a *K* for thousands and an *M* for millions. You can do that with a custom calculation:

```
//Stage Value Suffix
IF [Stage Value Label] > 1E6
THEN "$" + LEFT(STR(ROUND([Stage Value Label]/1E6,2)),4) + "M"
ELSEIF [Stage Value Label] > 1E3
THEN "$" + LEFT(STR(ROUND([Stage Value Label]/1E3,0)),4) + "K"
ELSE "$" + STR(ROUND([Stage Value Label]))
END
```

This calculation will round to values below one billion.

Add [Stage Value Suffix] to Text on the Marks card. Edit the table calculation and select Specific Dimensions and then select the Stage checkbox. You will also want to add a custom sort. Use a descending sort on the Sort calculation, using Minimum as the aggregation. This table calculation is shown in Figure 10-26.

The result of this table calculation is actually a great hidden trick of Tableau. We are doing a table calculation up the table! Tableau regularly suggests "Table (across)" and "Table (down)." But there is never a Table (up) option. This proves you can do it; you just have to use a custom sort.

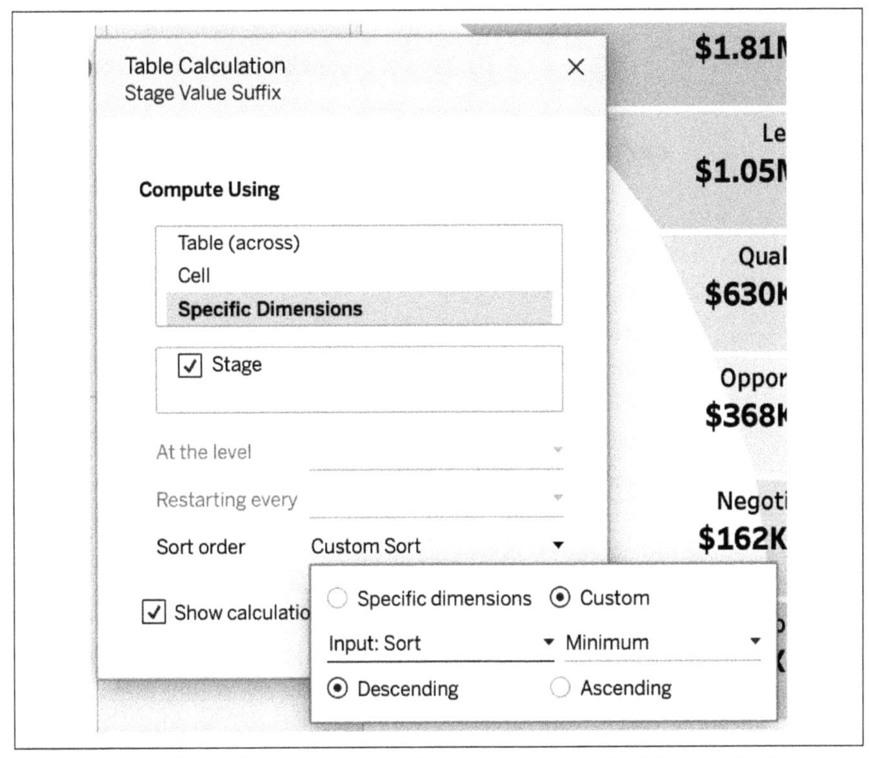

Figure 10-26. The table calculation and custom sort for the labels in the funnel calculation

c. Create a calculation called **`[% Closed]`** that will show the percentage of opportunities that closed at each stage:

```
//% Closed
WINDOW_MIN([Stage Value Label])/[Stage Value Label]
```

Add [% Closed] to Text and then edit the table calculations. Set both of the [% Closed] table calculations to match the table calculation in step 7b, and sort minimum descending.

d. Format the text to match Figure 10-27, including changing [% Closed] to display as a percentage. [Stage] should be a smaller font than [Stage Value Suffix]. Format [% Closed] to be the smallest, lightest-colored font.

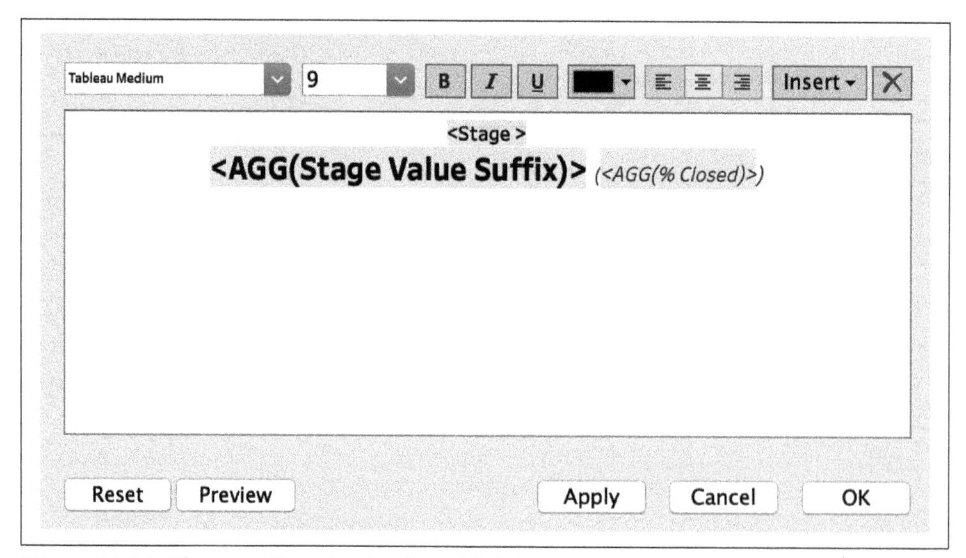

Figure 10-27. The text editor for the funnel labels

Figure 10-28 shows the resulting funnel. It shows the change from stage to stage while maintaining the look and feel of an actual funnel. You put a lot of effort into this chart. You probably don't need to put this sort of effort into every chart you create—but in some cases, your audiences will want more-polished, infographic-like charts. In these cases, you'll probably need to use data modeling.

In addition to learning how to apply principles of data modeling, you learned how to execute a table calculation that will go up (rather than down) a table. Remember that when you are modeling data, you need to think about what's happening at every level. In this example, you have to think about what is going to happen for all four members of the Model dimension. You've modeled data for four corners of each stage. You've also controlled where the labels display by using `IF` statements.

This method of using placeholder data to model your visualization could have been used in the previous strategy. To do so, you would have had to model only two rows. But sometimes our data modeling doesn't involve creating a placeholder dataset for the join; sometimes we just need to join back to our existing data source.

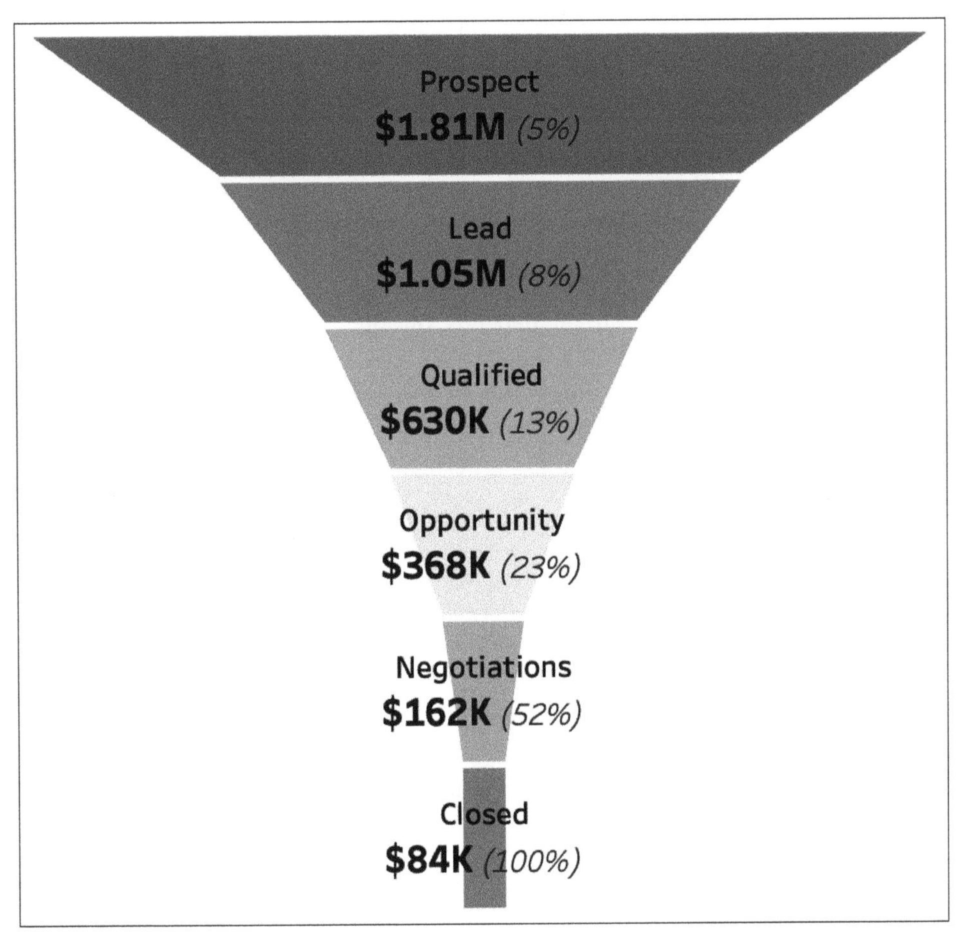

Figure 10-28. The final funnel chart

Market Basket Analysis

Market basket analysis is a technique used by retailers, restaurants, and manufacturers to uncover associations among products. Market basket analysis looks for combinations of items that occur together frequently in transactions. By understanding transactions, you can understand the behaviors of customers.

Basket analysis is not an out-of-the-box analysis in Tableau. To complete this analysis, we need to use data modeling. But for this data model, we will relate one instance of the data source back to a duplicate version of the data. When you model data to itself, you will be increasing your data exponentially. Good news: Tableau's logical layer makes the relationships simple, so your data doesn't get too out of hand. Logical layers are one of two ways to work with data in Tableau. Starting with Tableau Desktop

2020.2, Tableau has allowed its users to build interactions at both the logical level and the physical level.

With logical levels, which Tableau calls *noodle relationships*, you can drag any data source onto the canvas and allow the data sources to have flexible relationships depending on the level of analysis on a worksheet. You do not need to specify join types for relationships, and instead Tableau will automatically select appropriate join types based on the fields and type of analysis on each worksheet. In some cases, Tableau may use the relationships to join the data types together; other times it might not use those relationships. It all depends on the analysis in your workbook. Figure 10-29 shows an example of a logical table.

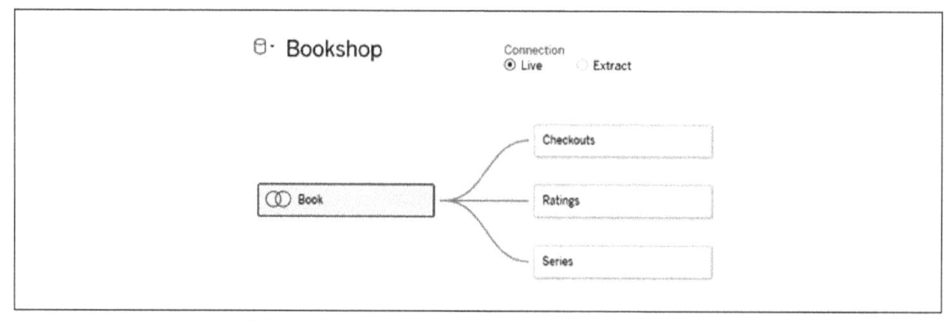

Figure 10-29. From Tableau (https://oreil.ly/hTOcx), this is an example of a logical table

For more detail on how the relationships work, we highly recommend this Tableau Help article (*https://oreil.ly/n2sVV*).

Think of a relationship as a contract between two tables. When you are building a visualization with fields from these tables, Tableau brings in data from the tables by using that contract to build a query with the appropriate joins.

Unlike logical tables, *physical tables* use joins to merge data from two tables into a single table before your analysis begins. Unlike the logical model, merging tables can cause data to be duplicated, filtered from one or both tables, or cause NULL rows to be added to your data. Joins provide you greater control over your data, but also require you to understand exactly what you are doing to your data and where consequences could occur. For many of our examples, we are going to use physical tables. Figure 10-30 shows an example.

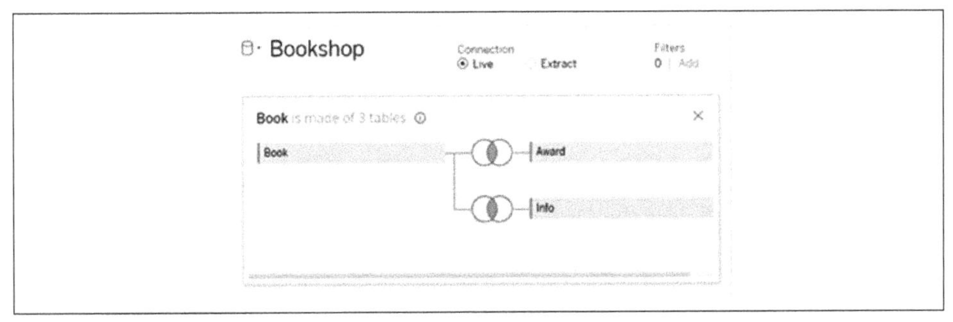

Figure 10-30. From Tableau (https://oreil.ly/DCVR7), this is an example of a physical table

Strategy: Basket Analysis

When doing basket analysis, you often need to calculate three metrics: support, confidence, and lift. Each is somewhat dependent on the next. Of these metrics, lift is the most important. *Lift* is a ratio that represents the odds of two items occurring within a single transaction over the odds of the two items showing up independently in transactions.

Let's consider an example. Imagine we have a dataset with 100 transactions. In 25 of the transactions, customers bought binders; 10 of the transactions were for fasteners; and 5 transactions included both binders and fasteners. Let's say our hypothesis is that Binders lifts (increases) Fasteners sales.

Support is calculated by finding the percentage of occurrences in which both items were purchased—in this case, 5/100, which is 5%.

Confidence is simply the number of purchases including both items divided by the number of transactions for fasteners—in this case, 5/10, or 0.5. Confidence can also be calculated as the support percentage divided by the percent of transactions for fasteners—in this case, again, 5%/10%, which is 50%.

Finally, we can calculate *lift*, which is confidence divided by the percent of transactions for binders—in this case, 0.5/0.25, which is 2.0. So, for the lift of binders on fasteners—or vice versa—our final number is 2.0. This means if you know a customer is purchasing binders, the probability of purchasing fasteners is 2.0, or double.

A lift value greater than 1 means the items depend on each other. Values equal to 1 mean there is no relationship. Values less than 1 mean there is a negative effect between the two and could mean that the items are substitutes for each other.

For this strategy, we will use logical tables—though the process would be nearly identical for physical tables. To complete a basket analysis, typically you need a lot of information to compare products. The Sample - Superstore dataset has 9,994 rows of data, 1,849 products, and 5,009 transactions. This is not enough to do a basket analysis at the product level. For 1,849 we would probably need closer to 100,000 transactions to make the analysis worthwhile. Instead, we can roll our basket analysis to a higher level: the sub-category level. In our basket analysis, we will determine which sub-categories provide a lift for other sub-categories:

1. Open Tableau and add a data source. Connect to the Sample – Superstore dataset and then click and drag the Orders table onto the data view (Figure 10-31).

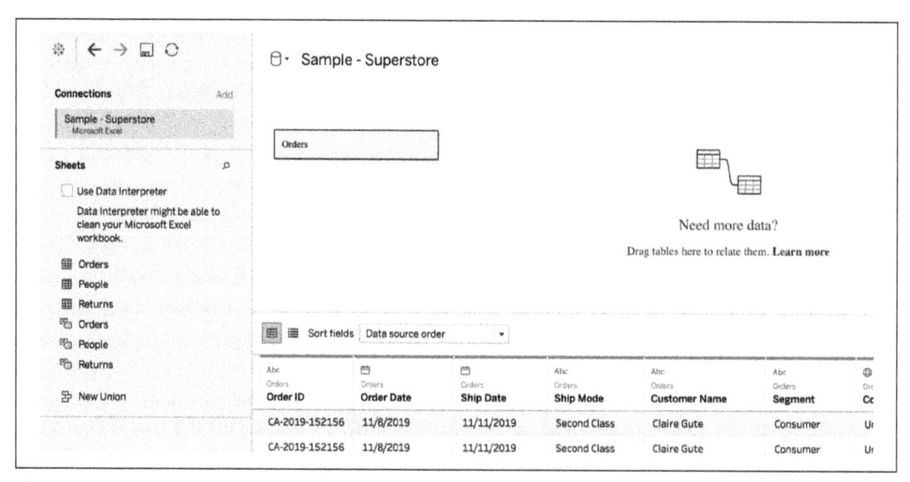

Figure 10-31. Connecting to the Orders table in Sample – Superstore

Next, take the same Orders table from Sheets on the Connections pane and click and drag the same Orders table onto the Data pane. These sources will attempt to connect with a noodle, shown in Figure 10-32. Because you are using the same dataset twice, Tableau will require you to select fields to join on. Let's join [Order ID] on Orders to [Order ID] on Orders1. Let's also join [Sub-Category] to [Sub-Category]—but change the sign from equal-to to less-than-or-equal-to. Without giving too much away, this sign change will create a cleaner final visualization showing just a cross-section of analysis in the matrix, rather than the full matrix.

Remember, we are going to do our basket analysis at the Sub-Category level. If we were doing analysis at the product level, you would join on Product. You do not need to make any other joins—even though the data is identical.

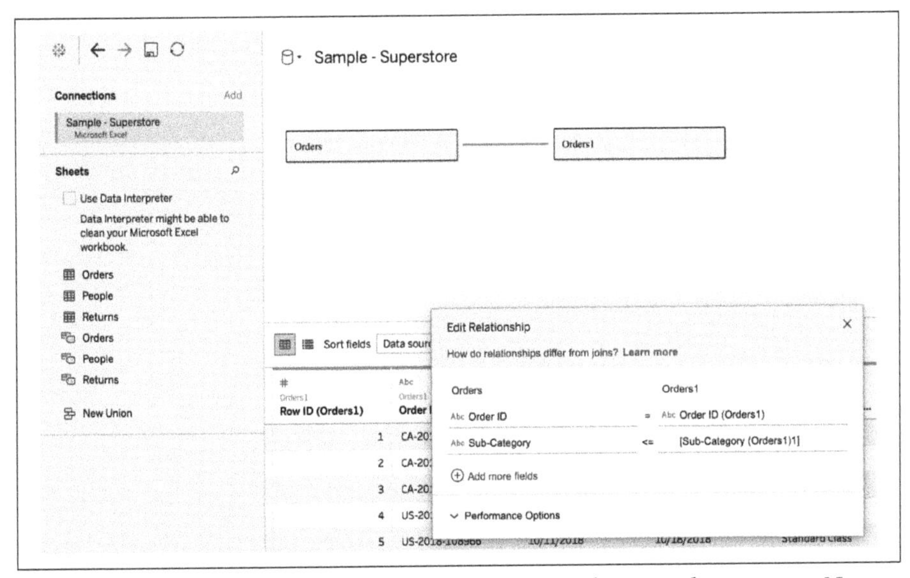

Figure 10-32. Editing the logical relationship between the same data sources. Note the less-than-or-equal-to sign on Sub-Category

When you are done creating these joins, you can go to a new sheet.

2. You'll notice with this data connection you have each field duplicated on your Data pane—one for the initial Orders dataset and one for the Orders dataset you joined in. Tableau renamed the Orders dataset you joined in to Orders1, to help clarify the data sources. This is also why you see half of your measures and dimensions with (Orders1) as a suffix in the field name.

   a. Add [Sub-Category] to Columns and [Sub-Category (Order1)] to Rows. You will notice that Abc shows up on only half of the fields separated on the diagonal (Figure 10-33).

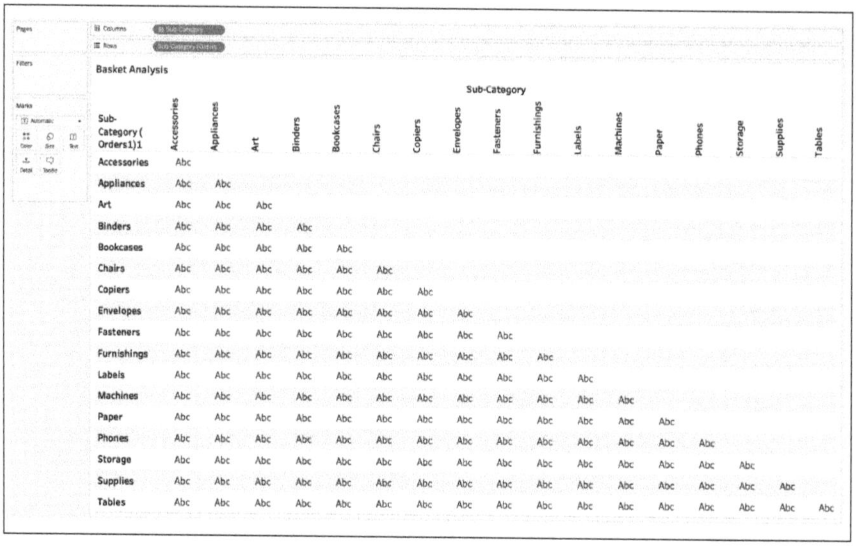

Basket Analysis

Sub-Category (Orders1)1	Accessories	Appliances	Art	Binders	Bookcases	Chairs	Copiers	Envelopes	Fasteners	Furnishings	Labels	Machines	Paper	Phones	Storage	Supplies	Tables
Accessories	Abc																
Appliances	Abc	Abc															
Art	Abc	Abc	Abc														
Binders	Abc	Abc	Abc	Abc													
Bookcases	Abc	Abc	Abc	Abc	Abc												
Chairs	Abc	Abc	Abc	Abc	Abc	Abc											
Copiers	Abc	Abc	Abc	Abc	Abc	Abc	Abc										
Envelopes	Abc	Abc	Abc	Abc	Abc	Abc	Abc	Abc									
Fasteners	Abc	Abc	Abc	Abc	Abc	Abc	Abc	Abc	Abc								
Furnishings	Abc	Abc	Abc	Abc	Abc	Abc	Abc	Abc	Abc	Abc							
Labels	Abc	Abc	Abc	Abc	Abc	Abc	Abc	Abc	Abc	Abc	Abc						
Machines	Abc	Abc	Abc	Abc	Abc	Abc	Abc	Abc	Abc	Abc	Abc	Abc					
Paper	Abc	Abc	Abc	Abc	Abc	Abc	Abc	Abc	Abc	Abc	Abc	Abc	Abc				
Phones	Abc	Abc	Abc	Abc	Abc	Abc	Abc	Abc	Abc	Abc	Abc	Abc	Abc	Abc			
Storage	Abc	Abc	Abc	Abc	Abc	Abc	Abc	Abc	Abc	Abc	Abc	Abc	Abc	Abc	Abc		
Supplies	Abc	Abc	Abc	Abc	Abc	Abc	Abc	Abc	Abc	Abc	Abc	Abc	Abc	Abc	Abc	Abc	
Tables	Abc	Abc	Abc	Abc	Abc	Abc	Abc	Abc	Abc	Abc	Abc	Abc	Abc	Abc	Abc	Abc	Abc

Figure 10-33. The market basket analysis after adding dimensions to Rows and Columns

b. In market basket analysis, *support* is the total transactions containing *x* and *y* divided by the total transactions. Create a calculation called **`[Support]`**:

```
// Support
COUNTD([Order ID])/MIN({COUNTD([Order ID])})
```

If you are asking yourself which Order ID you should be using: it doesn't matter. Because we joined our datasets on [Order ID], they will be identical. We personally prefer the shorter version because it makes our functions easier to read.

Note that we are using an LOD calculation to return the total orders in the entire dataset. In one quirk of Tableau, it is not possible to calculate the total orders by using a table calculation because of the use of the `COUNTD()` function.

c. *Confidence* is calculated by counting the total transactions containing *x* and *y* and dividing it by the transactions containing only *x*. There are several ways to write this calculation, but for this example, we'll use the [Support] calculation to create the **`[Confidence]`** calculation:

```
// Confidence
[Support]/
MIN(
    {FIXED [Sub-Category] : COUNTD([Order ID])}
    /
    {COUNTD([Order ID])}
)
```

d. *Lift* is calculated by taking the confidence calculation and dividing it by the proportion of orders containing *y*. Create a calculation called **`[Lift]`**:

```
// Lift
[Confidence]
/
(COUNTD([Order ID])/MIN({FIXED [Sub-Category (Orders1)]
  : COUNTD([Order ID])}))
```

After you have created the [Lift] calculation. Add [Lift] to Text on the Marks card. Then change the mark type to Square and add [Lift] to Color as well. The result is Figure 10-34.

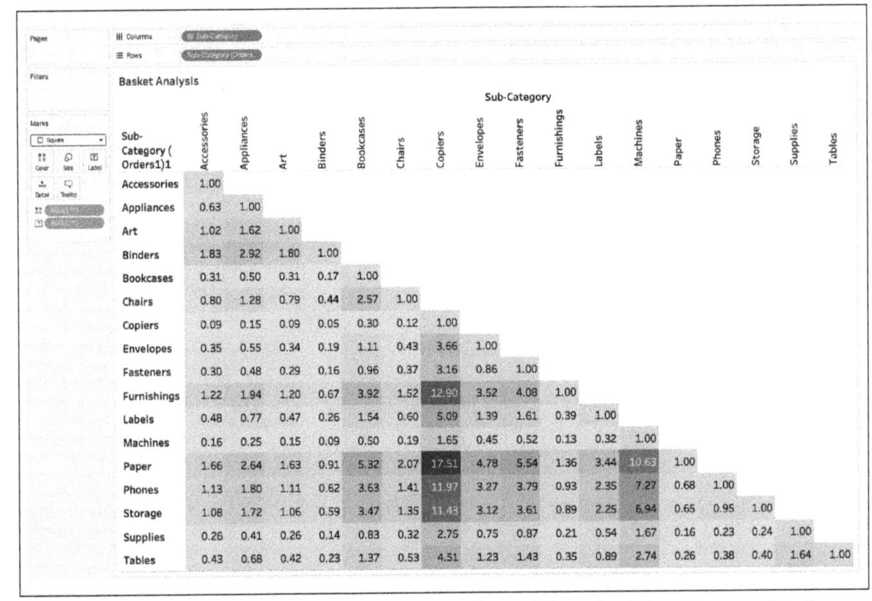

Basket Analysis

Sub-Category (Orders1)1	Accessories	Appliances	Art	Binders	Bookcases	Chairs	Copiers	Envelopes	Fasteners	Furnishings	Labels	Machines	Paper	Phones	Storage	Supplies	Tables
Accessories	1.00																
Appliances	0.63	1.00															
Art	1.02	1.62	1.00														
Binders	1.83	2.92	1.80	1.00													
Bookcases	0.31	0.50	0.31	0.17	1.00												
Chairs	0.80	1.28	0.79	0.44	2.57	1.00											
Copiers	0.09	0.15	0.09	0.05	0.30	0.12	1.00										
Envelopes	0.35	0.55	0.34	0.19	1.11	0.43	3.66	1.00									
Fasteners	0.30	0.48	0.29	0.16	0.96	0.37	3.16	0.86	1.00								
Furnishings	1.22	1.94	1.20	0.67	3.92	1.52	12.90	3.52	4.08	1.00							
Labels	0.48	0.77	0.47	0.26	1.54	0.60	5.09	1.39	1.61	0.39	1.00						
Machines	0.16	0.25	0.15	0.09	0.50	0.19	1.65	0.45	0.52	0.13	0.32	1.00					
Paper	1.66	2.64	1.63	0.91	5.32	2.07	17.51	4.78	5.54	1.36	3.44	10.63	1.00				
Phones	1.13	1.80	1.11	0.62	3.63	1.41	11.97	3.27	3.79	0.93	2.35	7.27	0.68	1.00			
Storage	1.08	1.72	1.06	0.59	3.47	1.35	11.43	3.12	3.61	0.89	2.25	6.94	0.65	0.95	1.00		
Supplies	0.26	0.41	0.26	0.14	0.83	0.32	2.75	0.75	0.87	0.21	0.54	1.67	0.16	0.23	0.24	1.00	
Tables	0.43	0.68	0.42	0.23	1.37	0.53	4.51	1.23	1.43	0.35	0.89	2.74	0.26	0.38	0.40	1.64	1.00

Figure 10-34. The market basket analysis after adding [Lift] to Text and Color

If you take a look at the analysis, you will notice a very high association between Paper and Copiers. This probably makes sense! What about Tables and Copiers? Does every person need to buy a table to go with a copier? Maybe to put the copier on a table. Or maybe we don't have enough data to make that kind of inference. What might make sense is that we have a minimum number of orders in each cell before we show data.

3. Let's say we need at least 10 orders before we want to show a cell. We could add a filter, but that would just make that cell disappear. It would be better if the cell was a different color and the text was omitted. We can do that with a parameter and a few different colors:

   a. Create an integer parameter called **`[Min Orders]`** and set the value to 10.

b. Create a calculation called **`[Lift | Color]`**:

```
// Lift | Color
IF COUNTD([Order ID]) >= [Min Orders]
THEN [Lift]
ELSE -1
END
```

Click and drag [Lift | Color] to replace Lift. Edit the color to change the palette to Green-Blue-White Diverging (or whatever palette you prefer). Click Apply. Then edit the blue to be #D3D3D3. Select the Use Full Color Range and Reversed checkboxes. Under your advanced settings, change the range to be hardcoded from –1 to 5 (which is a lot in real life). Fix the center to 0. Figure 10-35 shows the color settings.

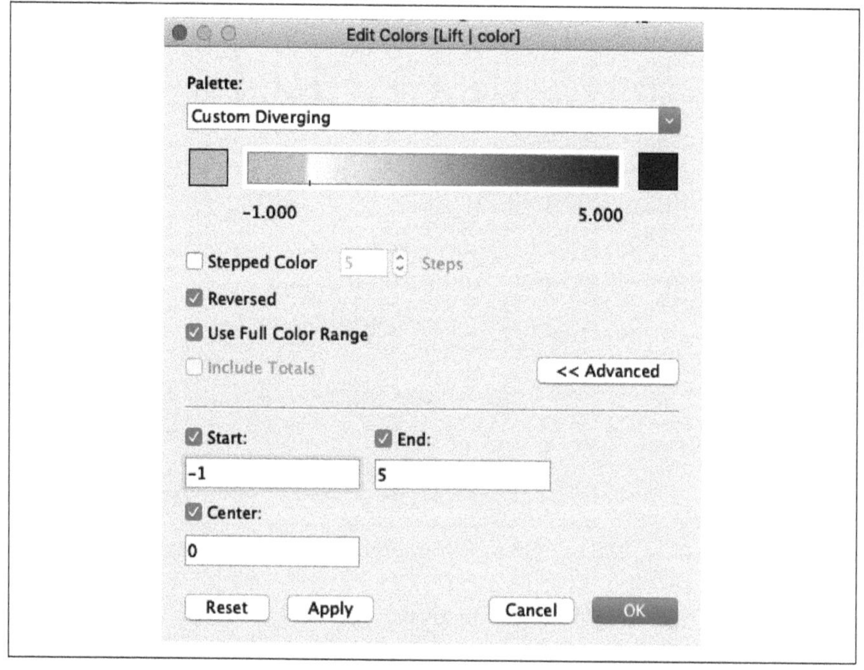

Figure 10-35. Color settings for [Lift | Color] on the market basket analysis

c. Create a new label calculation for lift called **`[Lift | Label]`**:

```
// Lift | Label
IF COUNTD([Order ID]) >= [Min Orders]
THEN [Lift]
END
```

Click and drag [Lift | Label] to replace [Lift]. The resulting visual is shown in Figure 10-36.

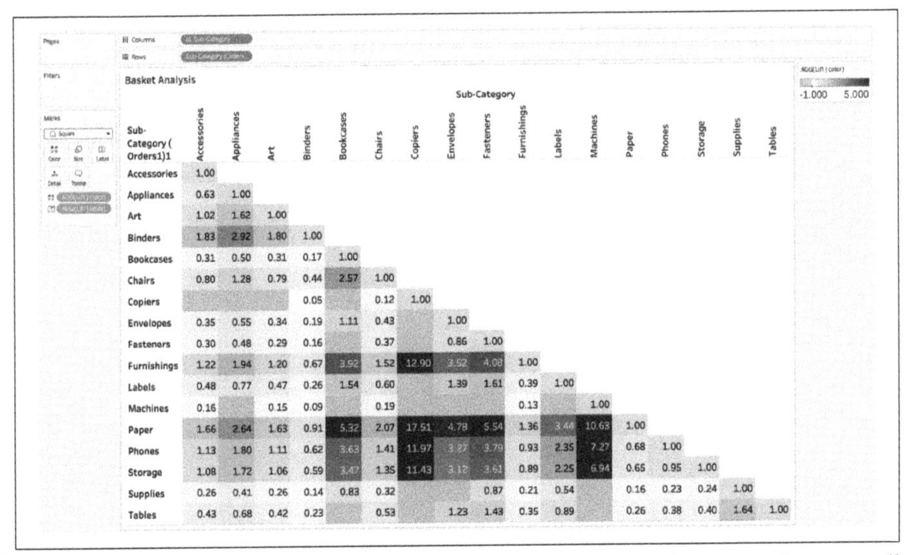

Figure 10-36. The market basket analysis, adjusted to show at least 10 orders per cell

You have completed a market basket analysis and are showing only cells with enough orders. You now have enough orders to say the lift between Paper and Copiers is extremely high! (PS. While we used a threshold of only 10 with our example, you might want this to be closer to 100 or even 300 in reality—there's no reason to return a false positive on any relationship.)

If you wanted to call your analysis complete here, you could. And you would probably want to add row and column dividers for every cell. But we're not going to stop here.

4. One frustration with this chart is that the labels are at the top! We want the labels to be on the left (and right-aligned) and on the bottom (and top-aligned). How can we do this? Well, we have to trick Tableau just a little:

   a. Double-click Rows and type **`MIN(0.0)`** to create an ad hoc calculation. This is going to significantly change our chart type. Don't panic! First, notice that our chart now has squares on each cell; this matches our mark type selection of Square! Next, notice we've added an axis next to the members of [Sub-Category]; see Figure 10-37.

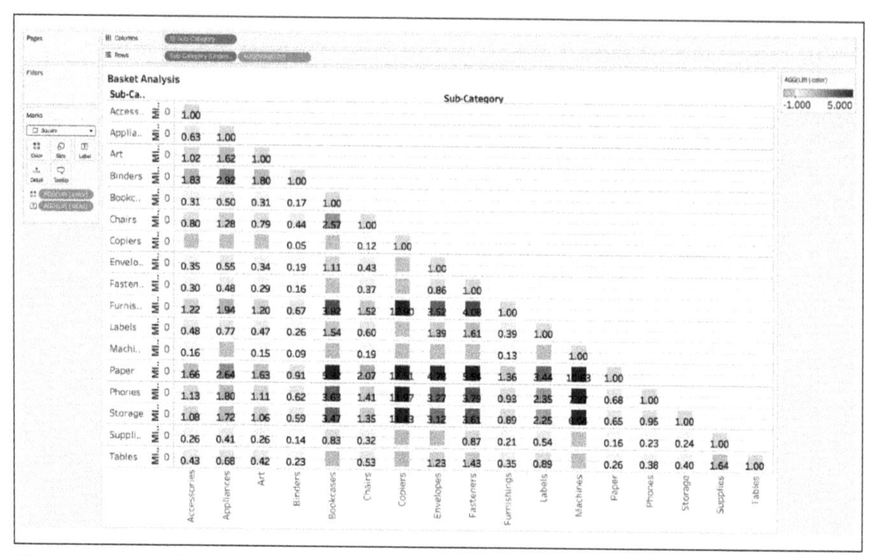

Figure 10-37. If you see this, don't panic! This is what your chart should look like after step 4a.

b. Change your mark type from Square to Gantt. Create an ad hoc calculation on the Marks card by typing **`MIN(1.0)`**. Then add this calculation to Size. This will change the depth of the bars. Edit the width by changing the slider to the largest size possible.

c. Edit the MIN(0.0) axes and set the range of the axes from 0 to 1; then hide the axes.

d. Click the Text button and then align the text to the center and middle.

e. Remove all the grid lines, axis lines, and axis rulers.

f. Add row and column dividers that match your background. Adjust the size on the Marks card down ever so slightly to match the row dividers.

g. Right-align the members of [Sub-Category]. Top-align members of [Sub-Category (Orders1)]. Finally, hide label fields for rows and columns to get the visual shown in Figure 10-38.

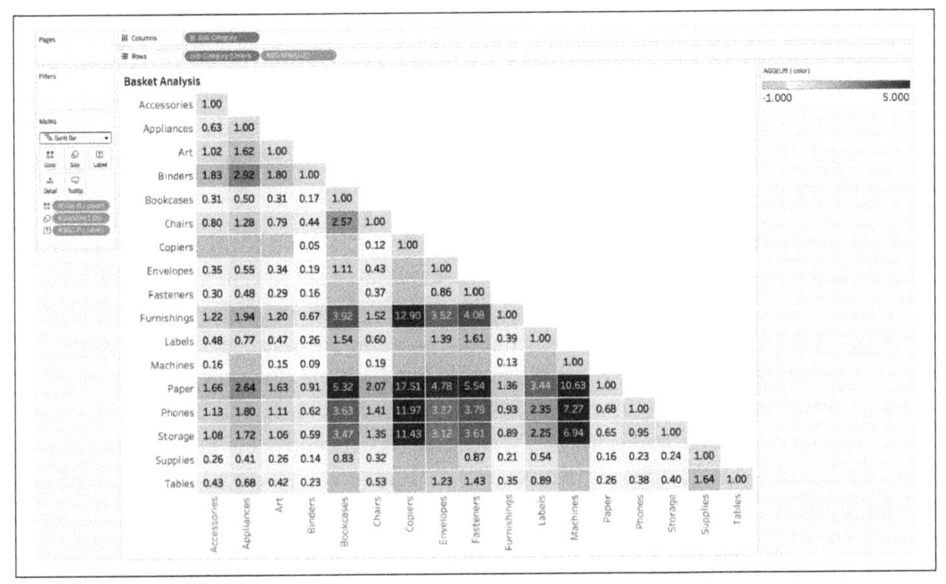

Figure 10-38. Our market basket analysis for Sub-Category using the Sample – Superstore dataset

In this strategy, you learned how to model data to itself. This allowed you to do a market basket analysis. In addition to building out the matrix, you were able to calculate lift for each sub-category. To do this, you couldn't use a table calculation and had to use an LOD calculation. You also learned how to place labels at the bottom of a visualization that looks like a grid by using a Gantt chart.

In the end, our final product is a very clean-looking matrix displaying lift for all categories. With the addition of the spacing between each cell, the eyes of our audience will be able to quickly track which two sub-categories they are comparing—and identify the total lift.

Strategy: Build a Multidimension Waterfall

In this chapter, we've tackled data modeling through table calculations, unions, and joins. You will be able to get to 90% of your solutions through these techniques. One of the last unique, but consistent, problems we've encountered are with financial statements that use multiple measures to calculate a final metric.

For instance, if you are creating a profit-and-loss statement, revenue minus cost of goods is gross profit. Gross profit minus operating expenses is earnings before interest and taxes (EBIT). If you consider the flowing nature of the metrics, you can think of them almost as a waterfall. And many people like to visualize these as a waterfall too. Waterfall charts are used only in accounting; they are used to analyze everything from supply chains to human capital.

For this strategy, we will work with a modified version of Sample – Superstore. This modified data source has several additional columns including MSRP (Manufacturer Suggested Retail Price), Promotional Cost, Manufacturing Cost, Transportation Cost, and Administrative Cost. Unlike most tutorials, where you create a waterfall chart with a single measure and a single dimension, we want you to visualize this waterfall with multiple dimensions.

The use of multiple dimensions puts your data in a less-than-ideal state for Tableau. And ideally, you would pivot all of your measures into a set of dimensions and values by using a tool like SQL, Tableau Prep, Alteryx, Python, or R. But sometimes you don't get that luxury because of your data source. So we want to show you how to create a waterfall chart when you are stuck using multiple dimensions. There's good news with this method, though: even though it's harder to use, you can actually build more flexible visualizations. With this technique. your waterfall is fully customizable: you can create totals at any point in your waterfall, place labels on specific bars, or update labels with very little effort.

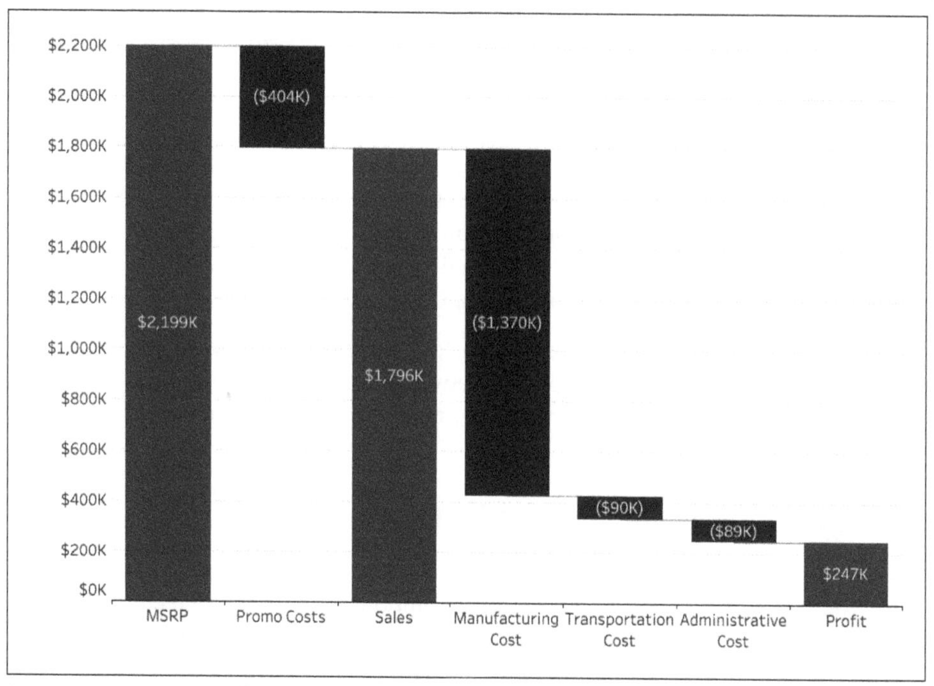

Figure 10-39. A waterfall chart using multiple dimensions

For instance, if you take a look at Figure 10-39, you will notice that we have full bars not just for the starting and ending values of MSRP and Profit, but we've also included a total for Sales. In this strategy, we will re-create this example and add a filter for the region of sales. Remember, we're going to do some data modeling, so you'll need your text editor handy to build a custom dataset.

The reason you need to create a placeholder dataset is that your only option for a sustainable single-sheet solution would be to use [Measure Names] and [Measure Values.] And under our current constraints, you're going to have to apply a table calculation to [Measure Values].

Before we start building our solution, let's take a look at Figure 10-40 to see how far you would have gotten by using [Measure Names] and [Measure Values] with no additional calculations.

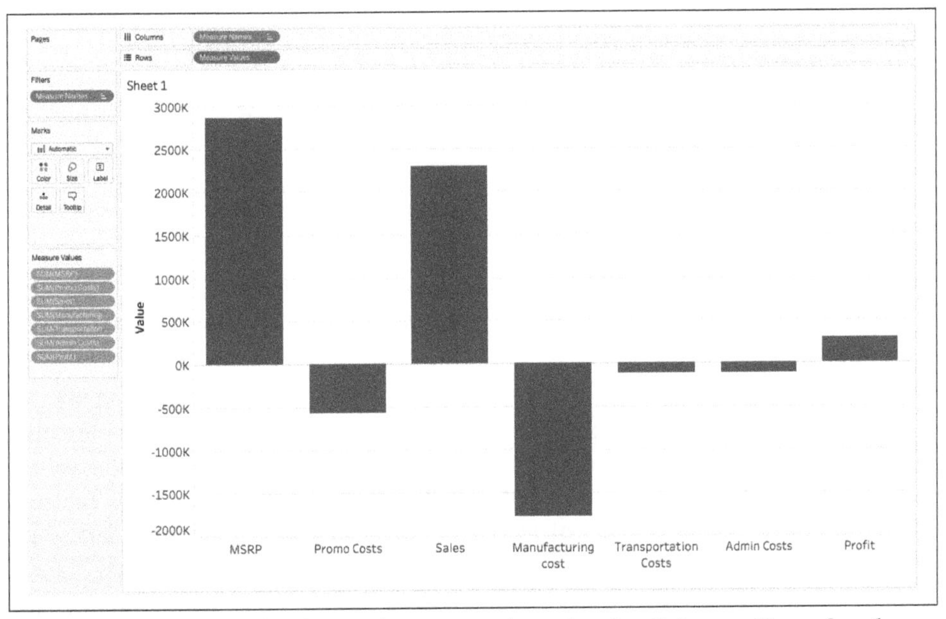

Figure 10-40. A waterfall solution that just won't work using [Measure Names] and [Measure Values]

To make a waterfall with multiple measures, you need to create a very simple secondary data source:

1. Find and connect to the Sample – Superstore – Modified data source. Nothing special, yet.
2. To build our solution, we are going to do a data blend. But we don't need a particularly interesting dataset for the blend.

A *blend* allows you to place measures or dimensions from multiple data sources on the same sheet. If you want to read more, we suggest you start with Tableau's help section on blends (*https://oreil.ly/jz9Bk*).

For our example, we are working with seven separate measures. We need to build a new data source to help support these measures. In your text editor, type the following:

```
Values
1
2
3
4
5
6
7
```

Then save the file as *values.csv*. These seven values will eventually be assigned to each of our measures of interest.

3. Create a new data source connection and connect to *values.csv*. You do not need to join your data sources together. In, fact we want them as two completely different data sources.

4. Before we start building our calculations, we just want to tell you that we are going to do something slightly unorthodox; we are going to include measures from our modified superstore data and add them to calculations in the Values data source. This was far more commonplace in the earlier versions of Tableau, before the existence of LOD calculations or being able to join multiple data sources together.

   With this visualization, we are going to create a Gantt mark type. With a Gantt chart, we need to identify the start point for each Gantt bar and to specify the length of the bar. Additionally, we're going to use a line to connect all the bars together. To do this, we need three separate calculations. In addition to calculations specific to the Gantt bars, we'll use calculations that control the labels of our chart and the color of the bar—technically, this color could apply to any chart type.

   a. Click the Values data source. Then click and drag [Values] to Detail on the Marks card. Make sure [Values] is set to be a dimension and not an aggregation. Change the mark type to Gantt.

   b. Using the Values data source, not the Sample - Superstore — Modified data source, create a new calculation called **`[Labels]`**. Type and save the following:

```
// Labels
CASE MIN([Values])
WHEN 1 THEN "MSRP"
WHEN 2 THEN "Promo Costs"
WHEN 3 THEN "Sales"
WHEN 4 THEN "Manufacturing Cost"
WHEN 5 THEN "Transportation Cost"
WHEN 6 THEN "Administrative Cost"
WHEN 7 THEN "Profit"
END
```

Since we will be completing a blend on our data sources, the key is that we use an aggregation, hence the `MIN()` function wrapped around [Values]. We will build a series of calculations around each of the members of the [Values] dimension.

Click and add [Labels] to Columns. Right-click [Labels] on the Columns shelf and edit the sort. Select Manual and adjust the values to match the order shown in [Labels].

c. Using the Values data source, not the Sample – Superstore — Modified data source, create a new calculation called **`[Start]`**. Type and save the following:

```
// Start
CASE MIN([Values])
WHEN 1 THEN 0
WHEN 2 THEN SUM([Orders (Sample - Superstore - Modified)].[MSRP])
WHEN 3 THEN 0
WHEN 4 THEN SUM([Orders (Sample - Superstore - Modified)].[Sales])
WHEN 5 THEN SUM([Orders (Sample - Superstore - Modified)].[Sales])
  + SUM([Orders (Sample - Superstore -
Modified)].[Manufacturing cost])
WHEN 6 THEN SUM([Orders (Sample - Superstore - Modified)].[Sales])
  + SUM([Orders (Sample - Superstore -
Modified)].[Manufacturing cost]) + SUM([Orders (Sample - Superstore
  - Modified)].[Transportation Costs])
WHEN 7 THEN 0
END
```

The first thing you should notice is that we have measures that start with Orders (Sample – Superstore – Modified).

For [MSRP], [Sales], and [Profit], these bars are just the total values. Set those values to zero. All other values are part of the waterfall.

Add this calculation to Rows. You might get an error prompt asking how you should blend your data (you won't need to worry about your blend) since it's at a global level. At this point, your visualization will look similar to Figure 10-41.

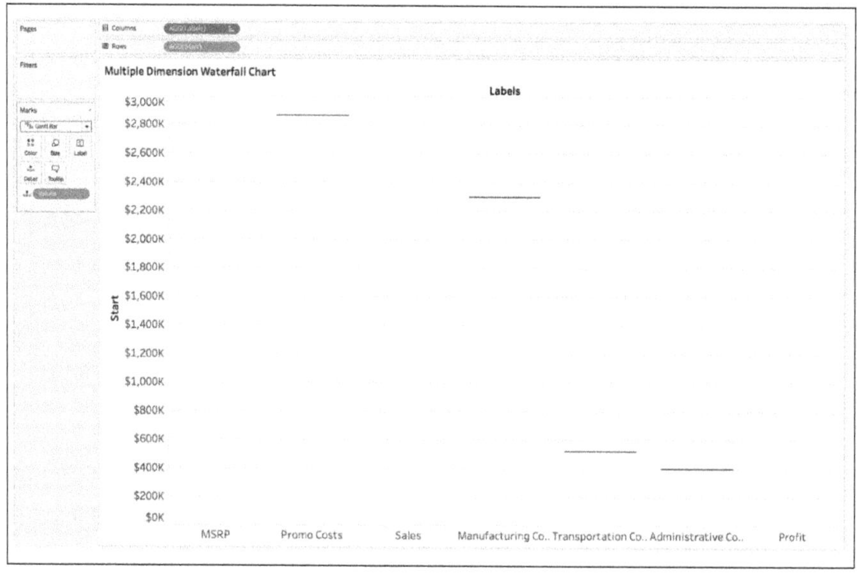

Figure 10-41. Our waterfall chart after adding Start to the Rows shelf

d. Using the Values data source, create a new calculation called **`[Bars]`** to determine the length of the Gantt bars. Type and save the following:

```
// Bars
CASE MIN([Values])
WHEN 1 THEN SUM([Orders (Sample - Superstore
  - Modified)].[MSRP])
WHEN 2 THEN SUM([Orders (Sample - Superstore
  - Modified)].[Promo Costs])
WHEN 3 THEN SUM([Orders (Sample - Superstore
  - Modified)].[Sales])
WHEN 4 THEN SUM([Orders (Sample - Superstore
  - Modified)].[Manufacturing cost])
WHEN 5 THEN SUM([Orders (Sample - Superstore
  - Modified)].[Transportation Costs])
WHEN 6 THEN SUM([Orders (Sample - Superstore
  - Modified)].[Admin Costs])
WHEN 7 THEN SUM([Orders (Sample - Superstore
  - Modified)].[Profit])
END
```

Add this calculation to Size. The chart will start looking correct.

Add [Bars] to Label on the Marks card. Align the text to the center and middle.

e. Once again, using the Values data source, create a new calculation called **`[Color]`**. Type and save the following:

```
// Color
CASE MIN([Values])
WHEN 1 THEN SIGN(SUM([Orders (Sample - Superstore
  - Modified)].[MSRP]))
WHEN 2 THEN SIGN(SUM([Orders (Sample - Superstore
  - Modified)].[Promo Costs]))
WHEN 3 THEN SIGN(SUM([Orders (Sample - Superstore
  - Modified)].[Sales]))
WHEN 4 THEN SIGN(SUM([Orders (Sample - Superstore
  - Modified)].[Manufacturing cost]))
WHEN 5 THEN SIGN(SUM([Orders (Sample - Superstore
  - Modified)].[Transportation Costs]))
WHEN 6 THEN SIGN(SUM([Orders (Sample - Superstore
  - Modified)].[Admin Costs]))
WHEN 7 THEN SIGN(SUM([Orders (Sample - Superstore
  - Modified)].[Profit]))
END
```

With Color, we will use the sign calculation. If a value is positive, it returns 1; if it is negative, it returns –1; and if it is 0, it returns 0. With our blend calculation, we need to do this for each value, but the results will be consistent.

Add [Color] to Color on the Marks card. Your visual will now look like Figure 10-42.

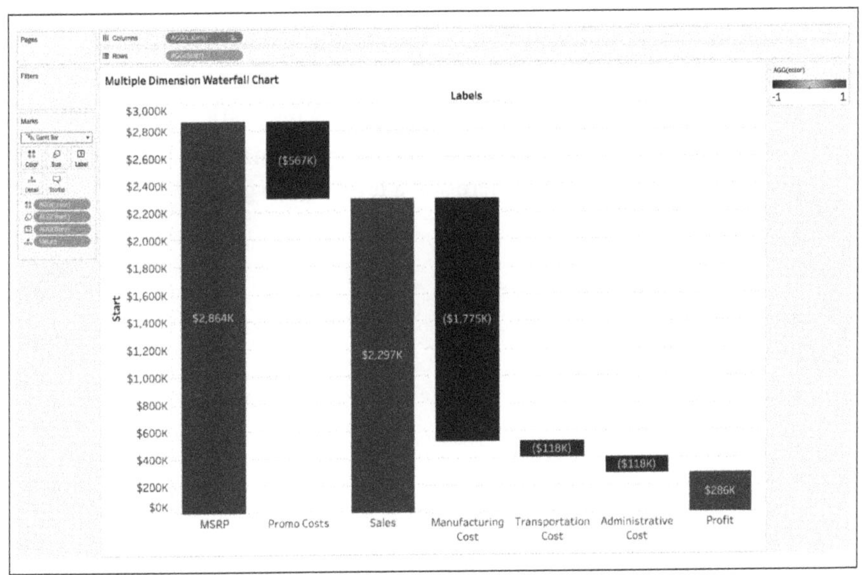

Figure 10-42. Our waterfall chart after adding bar length and labels

We're almost there: we just need to add a line to help the eye coordinate all the values.

f. One last time, using the Values data source, create a new calculation called **[Line]**. Type and save the following:

```
// Line
CASE MIN([Values])
WHEN 1 THEN SUM([Orders (Sample - Superstore - Modified)].[MSRP])
WHEN 2 THEN SUM([Orders (Sample - Superstore - Modified)].[MSRP])
  + SUM([Orders (Sample - Superstore -
Modified)].[Promo Costs])
WHEN 3 THEN SUM([Orders (Sample - Superstore - Modified)].[Sales])
WHEN 4 THEN SUM([Orders (Sample - Superstore - Modified)].[Sales])
  + SUM([Orders (Sample - Superstore -
Modified)].[Manufacturing cost])
WHEN 5 THEN SUM([Orders (Sample - Superstore - Modified)].[Sales])
  + SUM([Orders (Sample - Superstore -
Modified)].[Manufacturing cost]) + SUM([Orders (Sample - Superstore
  - Modified)].[Transportation Costs])
WHEN 6 THEN SUM([Orders (Sample - Superstore - Modified)].[Sales])
  + SUM([Orders (Sample - Superstore -
Modified)].[Manufacturing cost]) + SUM([Orders (Sample - Superstore
  - Modified)].[Transportation Costs])
+ SUM([Orders (Sample - Superstore - Modified)].[Admin Costs])
WHEN 7 THEN SUM([Orders (Sample - Superstore - Modified)].[Profit])
END
```

You will use this calculation to help your audience track the values across the waterfall. Add this measure to the left of the start bars. Remove [Color] and [Bars] from this [Lines] Marks card. Change size to be about one-quarter of the maximum value. Click Path and change the mark type to Jump.

Right-click the Start axis and create a synchronized dual-axis chart. This gets you to Figure 10-43, but there still a few steps to go.

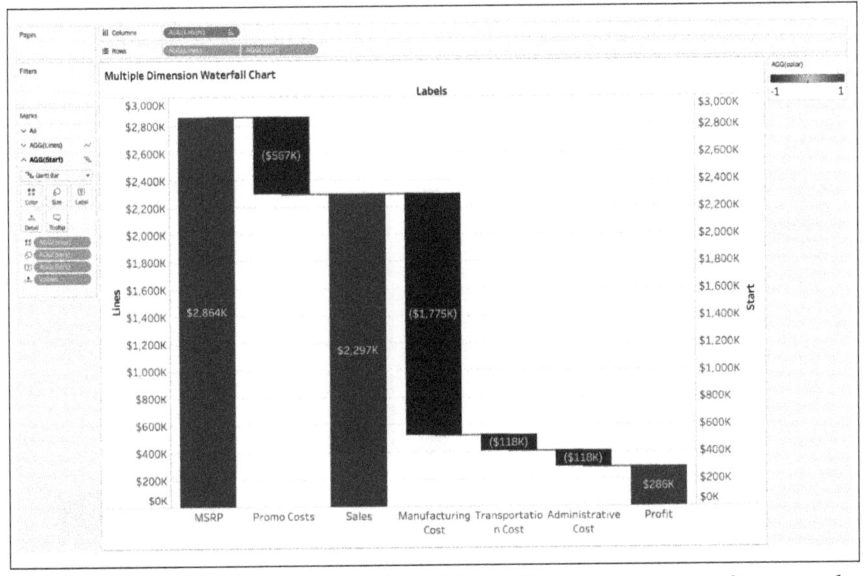

Figure 10-43. The waterfall chart after adding a line to help your audience track its values

g. You are basically done. But it might make sense to clean up the details:

   a. Hide the Start axis on the right.

   b. Remove the [Lines] axis label.

   c. Set the row and column dividers to None.

   d. Set the zero line to None.

   e. Set the columns axis ruler and axis ticks to a very dark gray.

   f. Hide the field labels for Rows.

   The formatted result, shown in Figure 10-44, is a radiant waterfall chart that looks as if it came from a single data source.

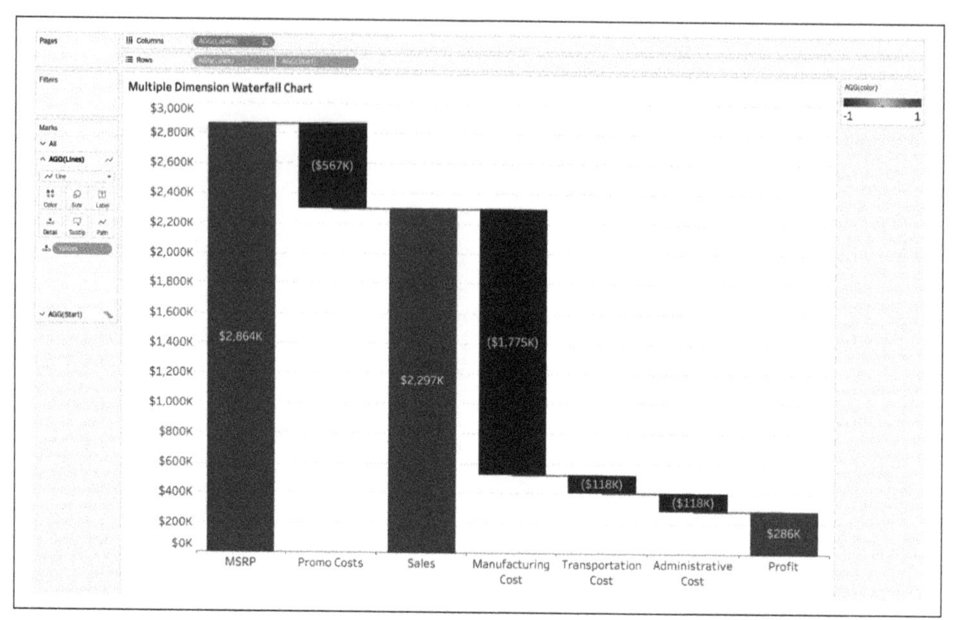

Figure 10-44. Our waterfall chart

While this visualization looks great, you probably will want to know if we can filter this visualization. The answer: yes. There are two solutions to consider. We could use a standard filter or we could build a filter using a parameter.

Let's take a look at the standard filter. Go to the Orders (Sample – Superstore – Modified) dataset. Click and drag [Region] to Filters. Select Central and then show the filter. You will notice in Figure 10-45 that the standard four regions show, but so does a Null value. This Null value shows up because of our blend. If your audience will be able to handle the Null option, then your visualization is complete!

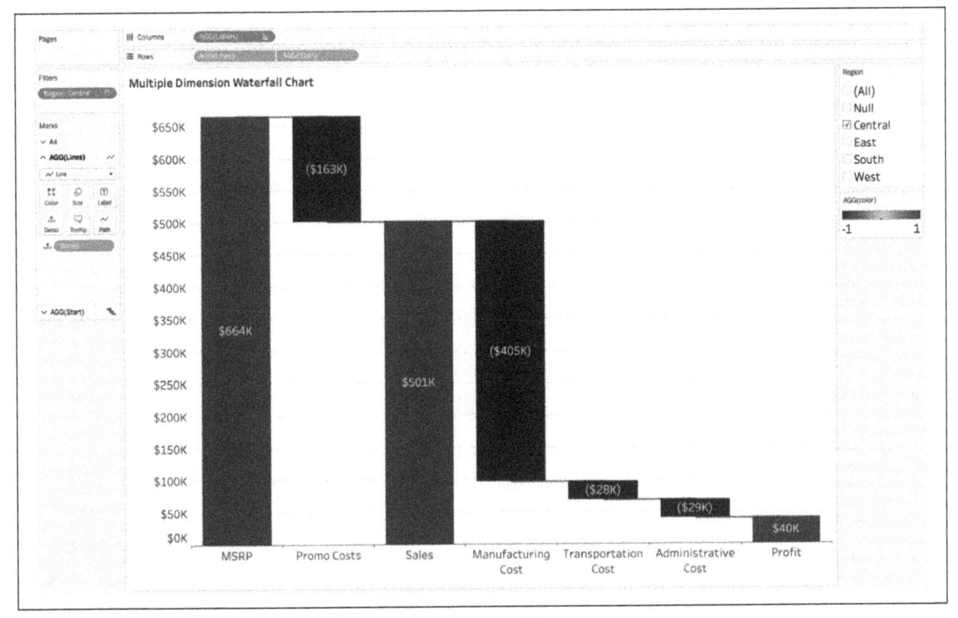

Figure 10-45. Our waterfall chart with a region filter

But if your audience is not going to be satisfied with a Null value being shown, we can use a parameter as a filter. Right-click [Region] and select Create → Parameter. Name the parameter **`[Region Parameter]`**. Add an additional Parameter option called All and place the All member as the first value in the parameter list, as shown in Figure 10-46.

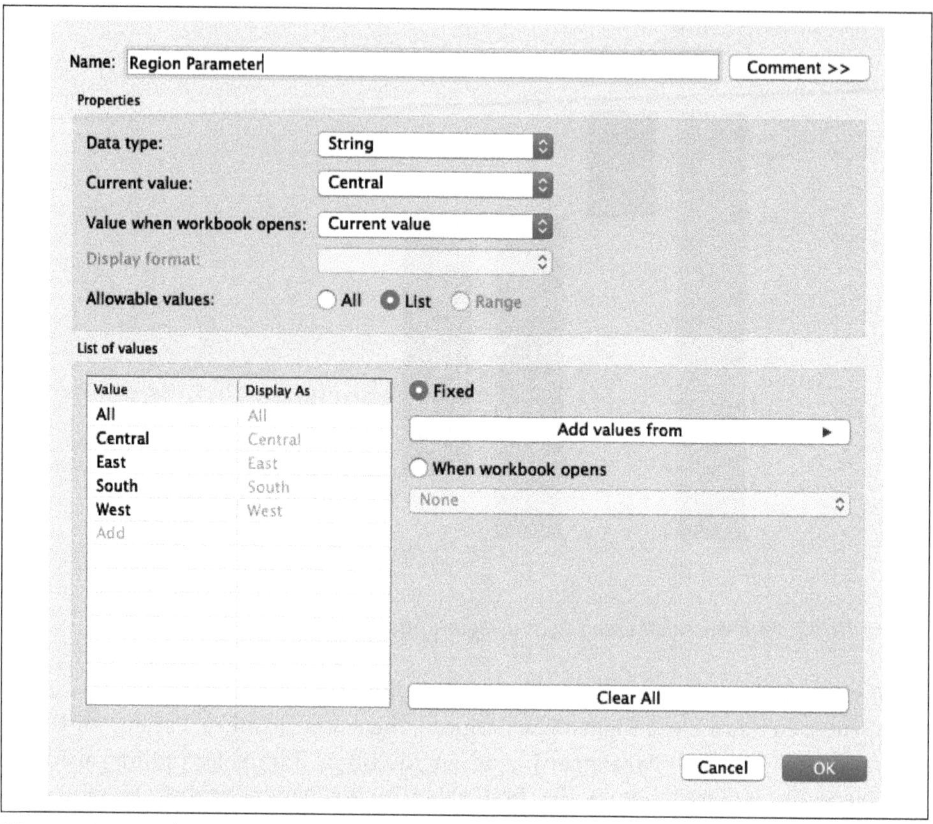

Figure 10-46. When setting up our [Region Parameter], it should look like.

Create a new calculation called **`[Region | TF]`**:

```
CASE [Region Parameter]
WHEN "All" THEN TRUE
ELSE [Region] = [Region Parameter]
END
```

Add [Region | TF] to Filters and select True. Then show [Region Parameter].

Congrats—you now have a filter driven by a parameter that does not include a Null label.

In this strategy, you learned how to use a simple blend to create a visualization using multiple measures. This blend was extremely simple, but required you to use a rarely used method. In our next strategy, we'll show you a new hack that is possible through mapping.

Strategy: Layer Marks with Maps Without Making a Map

In Tableau 2020.4, Tableau introduced map layers. There are some constraints for using map layers: for instance, values need to be between –180 and 180 for Latitude. However, map layers unlock the ability to have more than two mark types layered on the same sheet.

In this strategy, we will show you how to create a multilayered visualization using map layers, as shown in Figure 10-47. This visualization will encompass many of the concepts you've learned throughout this book. You will create a small multiple of sales by sub-category. You will provide a summary by using a text layer. You will visualize prior-year sales with an area chart, and current-year sales with a line and two circle layers. And you will use custom shapes to create axes and labels.

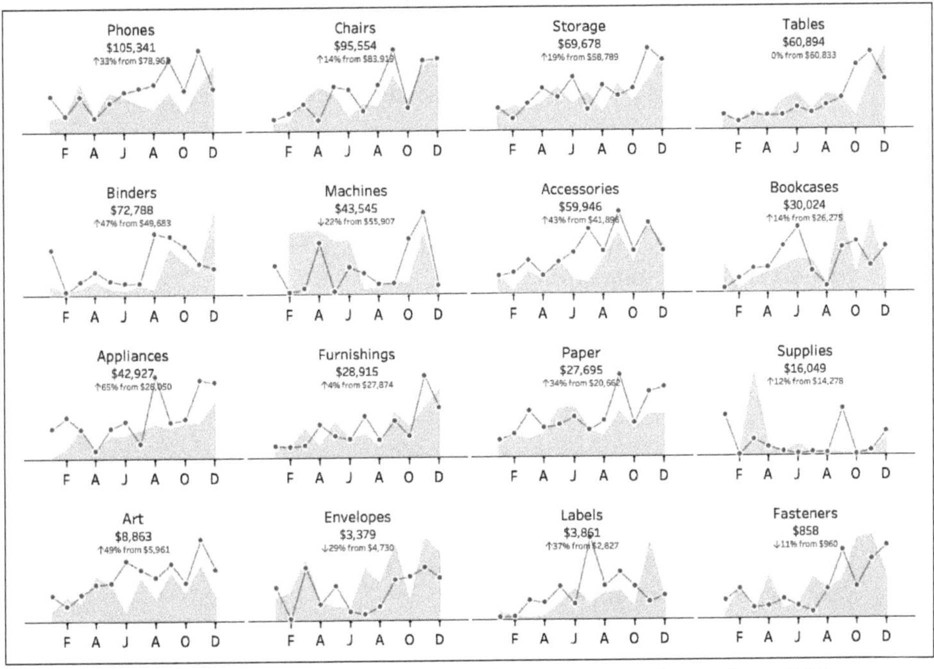

Figure 10-47. A small-multiple area and line chart using map layers from Tableau 2020.4

For this strategy, you will first build a series of calculations; then you will build the visualization from the top layer and work to the bottom. Many of our calculations will be repeated as we will calculate versions for current-year sales and prior-year sales:

1. Build your calculations as follows:

   a. Create a calculation called **[Sales | CY]** to calculate current-year sales:

```
// Sales | CY
IF YEAR([Order Date]) = {MAX(YEAR([Order Date]))}
THEN [Sales]
ELSE 0
END
```

   b. Create a calculation called **[Sales | PY]** to calculate prior-year sales:

```
// Sales | PY
IF YEAR([Order Date]) = {MAX(YEAR([Order Date])) - 1}
THEN [Sales]
ELSE 0
END
```

   c. Create a calculation called **[Sales | Delta]** to calculate year-over-year change in sales as a percent. You will use this as text to provide context:

```
// Sales | Delta
(SUM([Sales | CY]) - SUM([Sales | PY])) / SUM([Sales | PY])
```

   Change the default formatting to a percentage with no decimals. When the values are positive, have an arrow pointing up. When the values are negative, have an arrow pointing down.

   d. Create an LOD calculation called **[Sales | CY | Sub Mo]** to calculate current-year sales by sub-category and month:

```
// Sales | CY | Sub Mo
{
    FIXED [Sub-Category], MONTH([Order Date]):
    SUM([Sales | CY])
}
```

   e. Create an LOD calculation called **[Sales | CY | Sub Mo]** to calculate prior-year sales by sub-category and month:

```
// Sales | PY | Sub Mo
{
    FIXED [Sub-Category], MONTH([Order Date]):
    SUM([Sales | PY])
}
```

   f. Because sales values will find themselves greater than 180, and therefore won't render on our "map," we need to normalize our values. Create a calculation called **[Sales | CY | Sub Mo | Norm]**:

```
// Sales | CY | Sub Mo | Norm
[Sales | CY | Sub Mo]/
{FIXED [Sub-Category] : MAX(MAX([Sales | CY | Sub Mo]),
  MAX([Sales | PY | Sub Mo]))}
```

g. Complete the same normalization, but for prior-year sales. Create a calculation called **[Sales | PY | Sub Mo | Norm]**:

```
// Sales | PY | Sub Mo | Norm
[Sales | PY | Sub Mo]/
{FIXED [Sub-Category] : MAX(MAX([Sales | CY | Sub Mo]),
  MAX([Sales | PY | Sub Mo]))}
```

h. You are going to plot values by sub-category and month. To plot this on our map, we need to use a function called `MAKEPOINT()`. This worker function plots any value to a map. You need to specify two values under this function. Make the map layer by creating a new calculated field called **[Sales | CY | MP]**:

```
// Sales | CY | MP
MAKEPOINT(
        [Sales | CY | Sub Mo | Norm],
        MONTH([Order Date])
)
```

i. Replicate the same process for the prior-year values by creating a new calculated field called **[Sales | PY | MP]**:

```
// Sales | PY | MP
MAKEPOINT(
        [Sales | PY | Sub Mo | Norm],
        MONTH([Order Date])
)
```

j. Create a placeholder for the labels for the small multiples. The `MAKEPOINT()` calculation can be encoded with values in our data source, or you can hardcode the data. You'll hardcode the header at 1.1 (slightly above the plot) and 6.5 (between June and July). Create a calculated field called **[Sub-Category | MP]**:

```
// Sub-Category | MP
MAKEPOINT(1.1, 6.5)
```

k. Create a placeholder for the axis labels. You can actually mix values in the `MAKEPOINT()` function. Specify –.05 so the values show up below the axis. You will also alternate the labels so they show up every other month:

```
// Month | MP
MAKEPOINT(
        -.05,
        IF MONTH([Order Date]) % 2 = 0 THEN MONTH([Order Date]) END
)
```

l. To go with the axis points, you'll need labels, which you can create with the following calculation:

```
// Month Labels
LEFT(DATENAME('month', [Order Date]),1)
```

m. For our last two calculations, we need to specify the grid for our small multiples. You've seen these calculations in Chapter 4:

```
// Cols
(INDEX() - 1) % 4
// Rows
((INDEX() - 1) - [Cols])/4
```

After 14 calculations, we're ready to build this visualization!

2. Build our base (and our summary labels) as follows; many of our calculations will be repeated as we will calculate versions for current-year sales and prior-year sales:

   a. The [Sub-Category] dimension has 17 members. You want only 16 for this visualization. Click and drag [Sub-Category] to the Filters shelf and exclude Copiers. You are filtering copiers because there is some missing data that will give us extra headaches that we don't want to solve in this strategy.

   b. To add your first map layer, click and drag [Sub-Category | MP] to Detail on the Marks card. If everything is going as planned right now, you will see a blue dot just north of the island nation of Sao Tome and Principe. Don't worry, this won't show up on your "map" when you are done.

   Change the mark type from Automatic to Text.

   c. To add labels, add [Sub-Category], sum of [Sales | CY], sum of [Sales | PY], and [Sales | Delta] to Text. Then format by placing [Sub-Category] on the first line with size 15 font and the default font color. On the second line, place sum of [Sales | CY] with the default font and color, but changing the font size to 11. On the third line, place [Sales | Delta] and sum of [Sales | PY] and type `from` (with a space on each side) between the two measures, as shown in Figure 10-48.

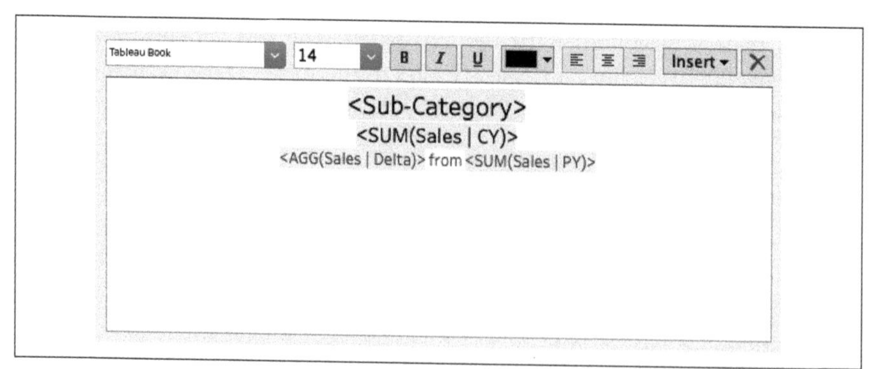

Figure 10-48. Text formatting for the sub-category labels

When you click OK in the text editor, you will immediately notice a bunch of overlapping text.

d. You will now start building out the grid. Double-click to the right of [Longitude (generated)] on the Columns shelf and type the **`[Cols]`** calculation. Press Enter. Change the [Cols] calculation from continuous to discrete.

   Edit the [Cols] table calculation. Select Specific Dimensions and Sub-Category. Then create a custom sort order by descending sum of sales.

e. Repeat the same process for [Rows] by double-clicking the Rows shelf. Edit the table calculation and edit the two nested table calculations of rows, using Specific Dimensions, selecting Sub-Category, and then sort in descending order by sum of sales. Remember, you need to do this for both table calculations in the rows calculation.

   The result, shown in Figure 10-49, is a grid of labels on a map.

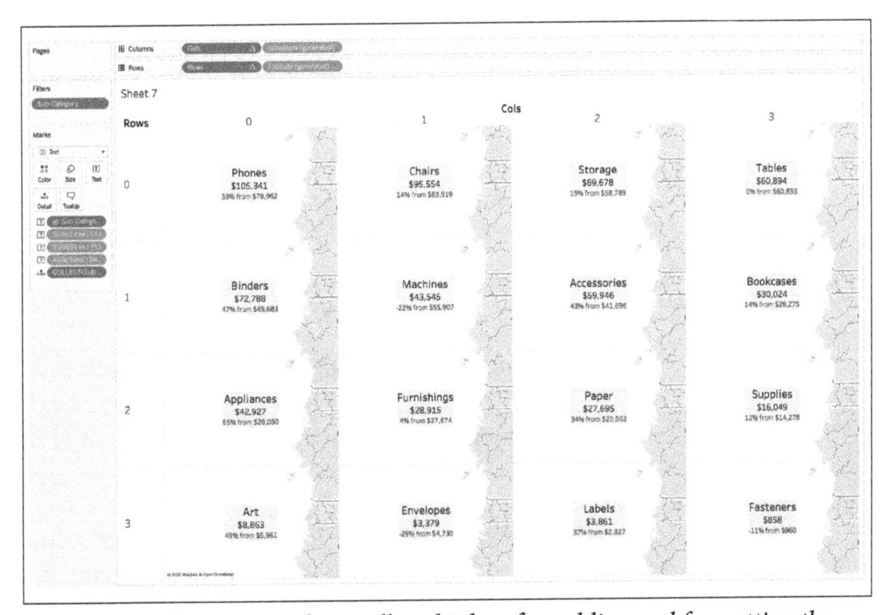

Figure 10-49. Progress on the small multiples after adding and formatting the [Rows] calculation on the Rows shelf

Now let's show you some of the magic. Our final visualization won't actually be a map (even though it's on a map). You can turn off your map from the top menu by choosing Map → Background Maps → None. This turns off your map and shows a standard chart!

f. Edit the [Latitude (generated)] axis and set the range from –.5 to 1.5. Edit the [Longitude (generated)] axis and set the range from –1 to 14. Figure 10-50 now shows the potential of our visualization.

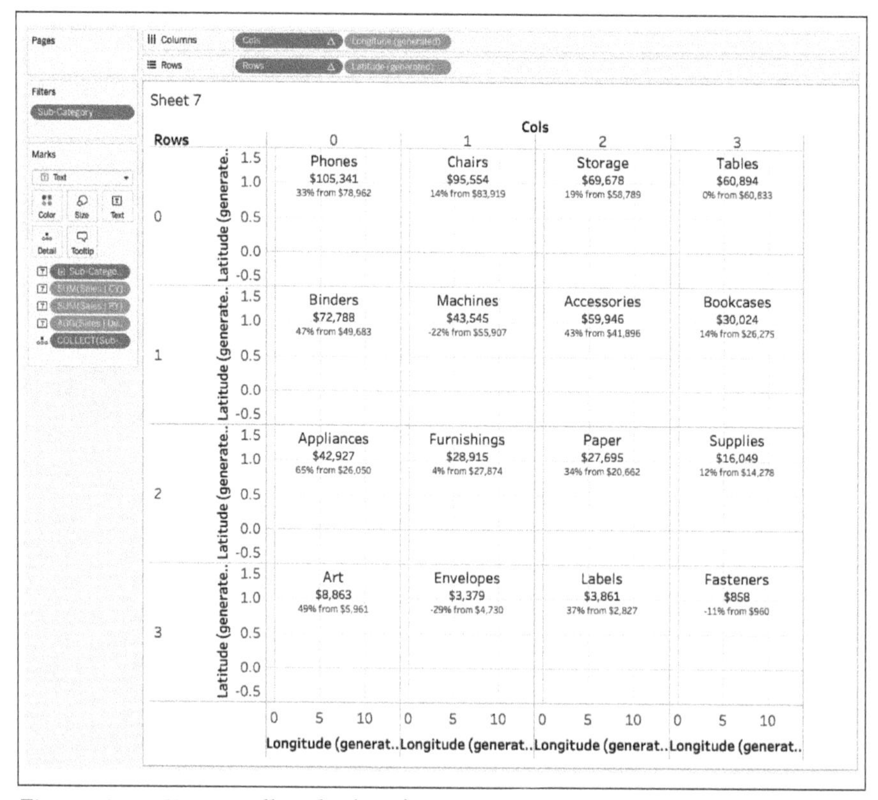

Figure 10-50. Your small multiples after turning off the map layers and editing the axes

You've completed the first layer, but now it is time to add the second layer. Here's the deal: you can't add additional layers without seeing the map view. This means we need to turn back on the map layers. From the top menu, choose Maps → Background Maps → Light.

3. Add a layer for the blue dots for the current year:

   a. After you have turned your map layer back on, click and drag your [Sales | CY | MP] calculation onto your map visualization. As you do this, an option to Add a Marks Layer will appear. Place your field on the callout. This will add another map layer to your visualization. By default, any layer you add is added to the top of your visualization. You can see this by looking at your Marks cards and seeing that the new field is sitting above your [Sub-Category | MP] layer, shown in Figure 10-51.

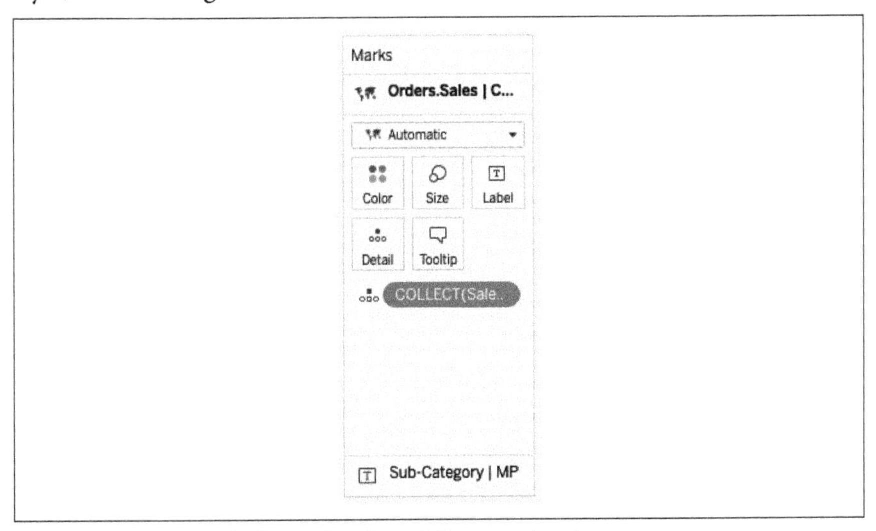

Figure 10-51. Our Marks cards after adding our first layer

You can easily change the layers by clicking the header and dragging the value below the [Sub-Category | MP] layer. You can also rename the layer to Blue Circles by clicking the drop-down, selecting Rename, and typing **`Blue Circles`**.

Change the mark type from Automatic to Circle.

b. Add [Sub-Category] to Detail of the Blue Circles layer. Double-click the [Blue Circles] Marks card to create an ad hoc calculation and type **`MONTH([Order Date])`**.

c. Edit the size of the dots to be approximately 20% of the maximum value. Figure 10-52 shows the visualization at this point.

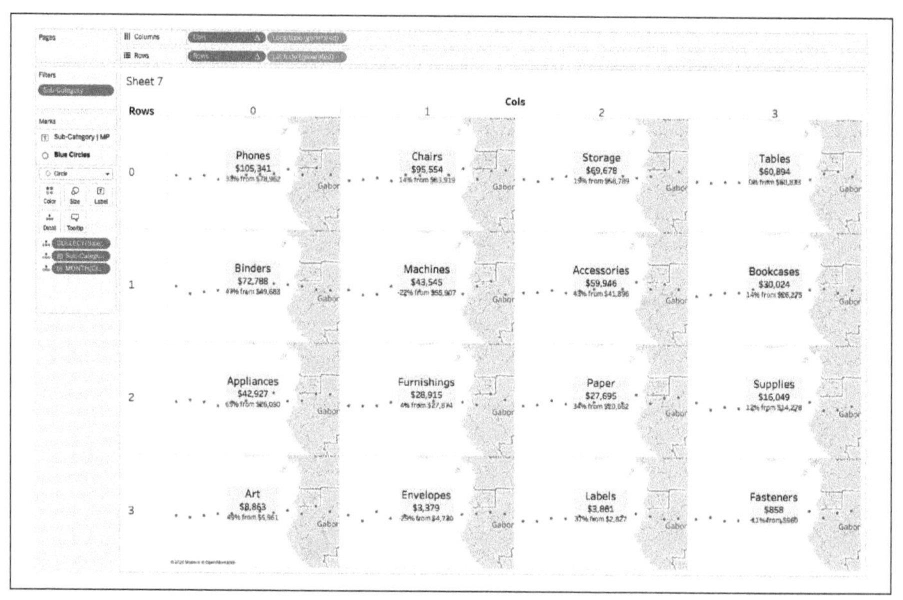

Figure 10-52. Our visualization after step 3

4. To add the white circles behind the blue circles, you'll generally follow the same steps from step 3 for creating the circles. This layer will give the appearance that the blue dots and the gray lines are not touching:

   a. Click and drag your [Sales | CY | MP] calculation onto your map visualization. Move the layer to the bottom. Change the name of the layer to **`White Circles`**.

   b. Add [Sub-Category] and MONTH([Order Date]) to Detail.

   c. Change the mark type to Circle. Change the size of the circle to be about 25% of the maximum value. This should be where the tick is on the left. Change the color of the mark to white and set the opacity to 85%.

5. Add lines that connect the dots; these steps look very similar to steps 3 and 4, but with some small changes:

   a. For the final time, click and drag [Sales | CY | MP] to add a marks layer. Move the layer to the bottom and change the name to **`Gray Lines`**.

   b. Add [Sub-Category] and MONTH([Order Date]) to Detail.

   c. Change the mark type to Line. Change the size of the line to be about 10% of the maximum value. Change the color of the mark to a medium gray and set the opacity to 85%.

Layering two dots and a line chart together gives the effect of a singular, cohesive line chart, which is shown in Figure 10-53. But, as you know, it is not. Let's add context to this line chart with an area chart.

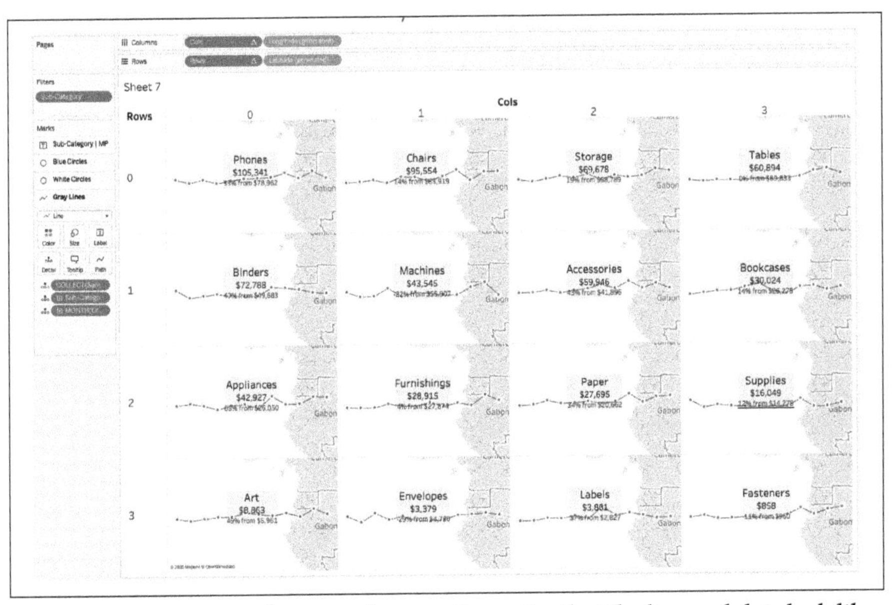

Figure 10-53. Your visualization after step 5—notice that the line and dots look like a singular component

6. When we create our charts, we like to show as much context as possible. This can be done by adding a layer showing last year's sales:
   a. This time, bring in prior-year sales by clicking and dragging [Sales | PY | MP] and adding a marks layer. Move the layer to the bottom and change the name to **`PY Area Chart`**.
   b. Like a broken record, add [Sub-Category] and MONTH([Order Date]) to Detail.
   c. Change the mark type to Area. Change the color to light gray.
7. For our last layer, we are going to create one that looks like a customized axis:
   a. Find [Month | MP] and add a new layer. Move the layer to the bottom and change the name to **`Axis`**.
   b. Add [Sub-Category] and MONTH(Order Date) to Detail. Add our custom calculation, [Months Label], to Text on the layer.

      Change the mark type to Shape. Edit the shape and select an arrow pointing up in the Thin Arrows folder. Format the text label to show at the bottom center. You have now completed the build portion; Figure 10-54 shows the resulting visualization.

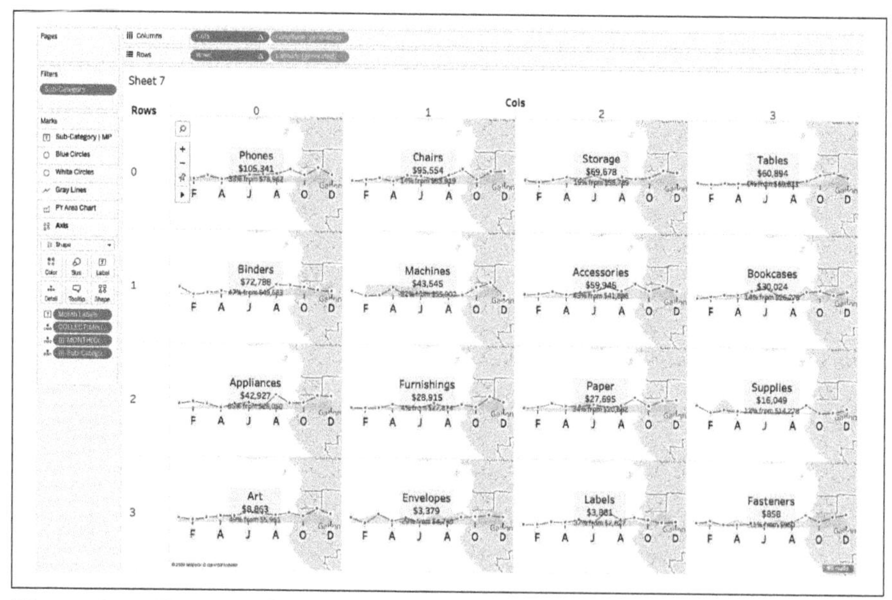

Figure 10-54. Your visualization after completing the build

8. To complete the visualization, turn off the map from the top menu by choosing Map → Background Maps → None.
9. Format the visualization as follows:
   - Hide all the headers.
   - Turn off the grid lines.
   - Remove the zero line on columns.
   - Set the zero line on rows to a black solid line.
   - Change the row and column dividers to the thickest white line possible.
10. Disable selection and format tooltips:
    a. With map layers, you can disable the selection of any values on your visualization by clicking the drop-down on the header of each layer and selecting Disable Selection (Figure 10-55). Disable selection for all of your layers.
    b. For each layer, you can control your tooltips. We suggest turning off tooltips for the Sub-Category | MP and Axis layers. For the Blue Circle, White Circle, Gray Line, and PY Area Chart layers, we suggest showing the month name and the value associated.

After you have formatted the tooltips, your visualization should match that in Figure 10-56.

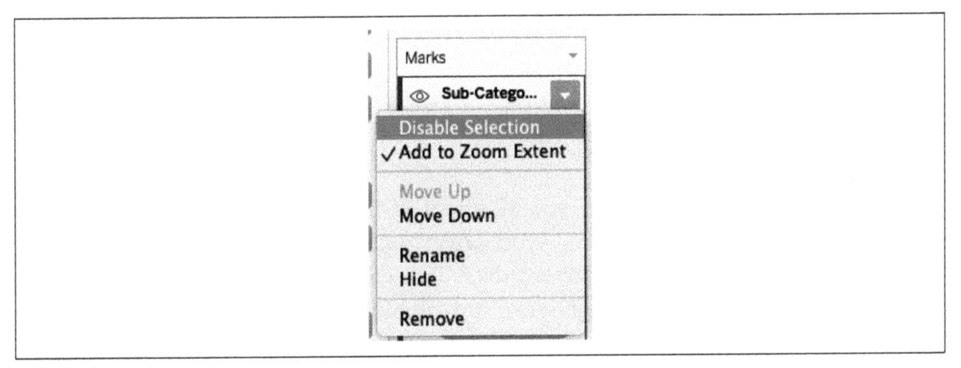

Figure 10-55. Disabling the selection on your map layers

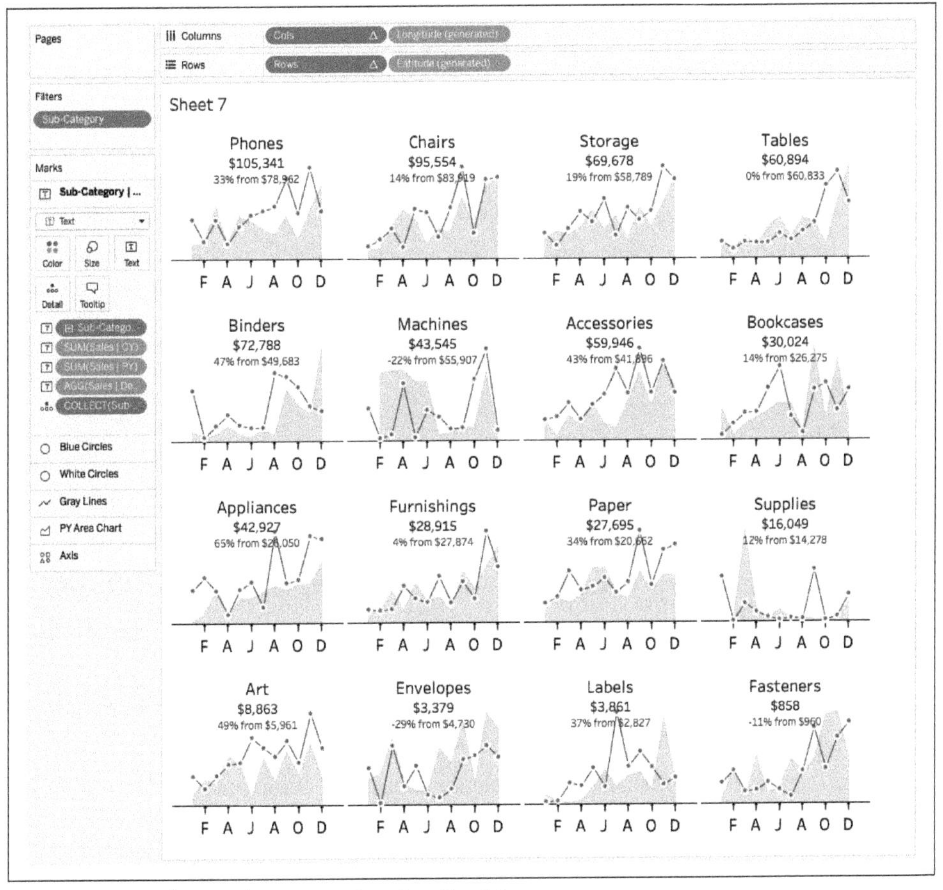

Figure 10-56. The resulting visualization for this strategy

In this strategy, you learned how to build a visualization by using map layers. You specified values on a map by using `MAKEPOINT()`. You learned that you can build layers with the same data used to build a customized component, like the line chart with dots. You also learned that with map layers, you can use multiple mark types without having to do complex data modeling.

Conclusion

As its title indicates, this chapter was all about reshaping your data to create novel visualizations. With data modeling, you aren't necessarily creating aggregations of your data; rather, you are padding your data. This padding can occur in many forms: through table calculations, through joins, or with blends to the same data source. You also learned about scenarios when you might want to join a dataset to itself. Finally, we looked at how map layers can take visualizations to a new level.

All of our examples are just breaking the iceberg with each data modeling type shown. There are so many more possibilities with each data modeling type. We didn't intend to show you each type, but to give you an important introduction to the concepts around data modeling. If you can understand how to model your data, you can build any advanced chart type.

In the next chapter, we will take some of the concepts shown here and in previous chapters and start to use them to build a series of visualizations that, when put together, build interactive visual systems. We stop short of calling these dashboards or even data products because we'll be focusing on the components that will go into dashboard building, not necessarily the dashboards themselves.

Further Reading

- Read Jeff Shaffer's blog post on domain completion and domain padding (*https://oreil.ly/qNgjI*).
- Read Ken Flerlage's blog post on data densification (*https://oreil.ly/KiDdA*).
- Learn more about map layering from Jeff Shaffer's blog (*https://oreil.ly/Drn2n*).
- If you are looking for a challenge, try making the radial stacked bar chart (*https://oreil.ly/lgZzd*) by Toan Hoang.
- Or practice the stepped area chart by Klaus Schulte and reproduced in blog form by Rosario Gauna (*https://oreil.ly/TeoxI*).
- Or re-create an English Premier League table in a blog post by Donna Coles (*https://oreil.ly/4Sbof*).

CHAPTER 11

Advanced Interactivity

Up to this point in the book, we've discussed using actions to build dynamic visualizations (Chapter 9) as well as using data modeling to expand our visual universe (Chapter 10). Dynamic visualizations are what our audiences have come to expect. When they click a value, they expect the visualization to react.

Many times, however, we have not designed our visualizations to react to these actions. We can, as developers, anticipate the actions of our audiences and design around their intuition. We can use data modeling of simple data sources, along with interactions to build visual systems. These systems will look like buttons or dynamic legends or even table headers.

In this chapter, we'll take Tableau to another level by showing you how to build visual systems that will enhance the way your audience works with data. This chapter is not about industry-specific use cases. It is about showing you how to build advanced interaction types by combining the principles from Chapters 9 and 10.

In This Chapter

You'll need the Sample – Superstore dataset. In this chapter, you'll learn how to do the following:

- Execute sheet swapping with a parameter
- Use interactivity to update parameters and build a dynamic UI
- Turn off mark selection
- Use parameters to store multiple values
- Build buttons out of sheets for dynamic interactivity on your dashboards
- Build dynamic pagination
- Drill into maps for more detail

Sheet Swapping with Parameters and Parameter Actions: Superstore Case Study

You are working with the office supply company again. Your end users want to be able to see three chart types: one describing sales, one describing profitability over time, and the other showing profitability by product category. Your users want to see only one chart at a time. You decide to use parameters and sheet swapping to solve the problem.

In the first strategy of Chapter 9, you learned how to use a parameter to update a metric. You can also use the same parameter to change which sheets are showing, by using the parameter inside a filter. Figure 11-1 shows an example that uses a list parameter to show one of three chart types.

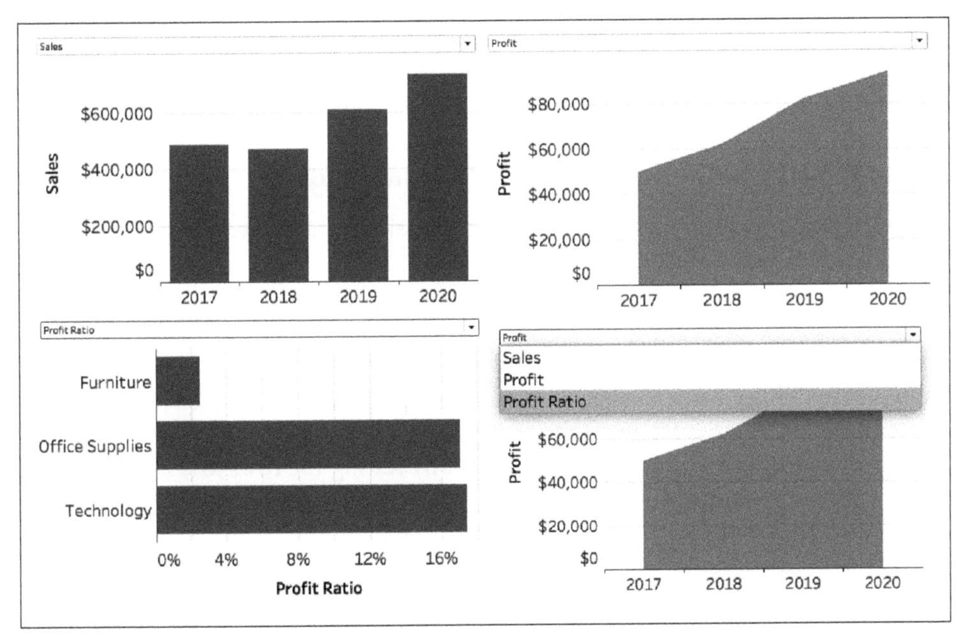

Figure 11-1. Sheet swapping using a parameter. Clockwise from top-left: the Sales sheet activated from the parameter, the Profit sheet activated from the parameter, the options in the parameter, and the Profit Ratio sheet activated from the parameter

Beyond using a drop-down option to update a parameter, as discussed in Chapter 9, you can also use a parameter action on a sheet to drive changes on your dashboard. Figure 11-2 illustrates this option.

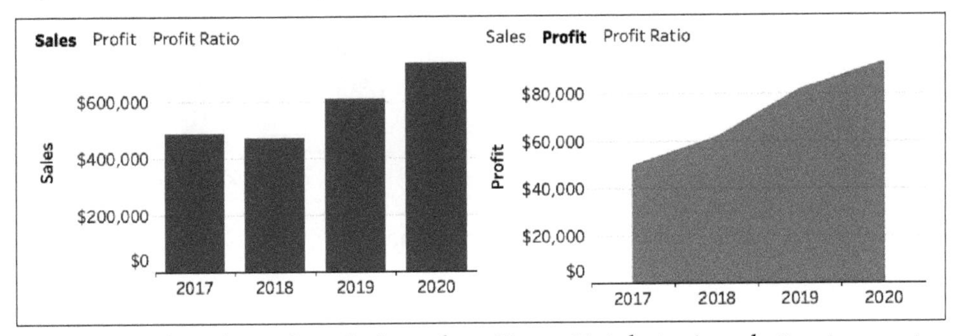

Figure 11-2. The Sales and Profit sheets from Figure 11-1, but using a button to execute the sheet swap

If you are tight for space on your dashboard but still want to somehow fit all of your sheets, we highly suggest you use sheet swapping and that you do this with a metric button.

In the first three strategies, you will learn how to create the example shown in Figure 11-2. Each will use the Sample – Superstore dataset. For the second strategy, you will also need a text editor to build a custom dataset.

Strategy: Use Parameter-Based Sheet Swapping

In this strategy, you will create a sheet swap by using a parameter. You'll expand on the output in subsequent strategies:

1. To build the three visualizations for the sheet swap, connect to the Sample – Superstore dataset and then follow these steps:

   a. To build the Sales visualization (Figure 11-3), add the sum of [Sales] to the Rows shelf. Add discrete year of [Order Date] to the Columns shelf. Change the mark type to Bar. Format the sheet by setting the rows axis ruler to None and the columns axis ruler and ticks to Black.

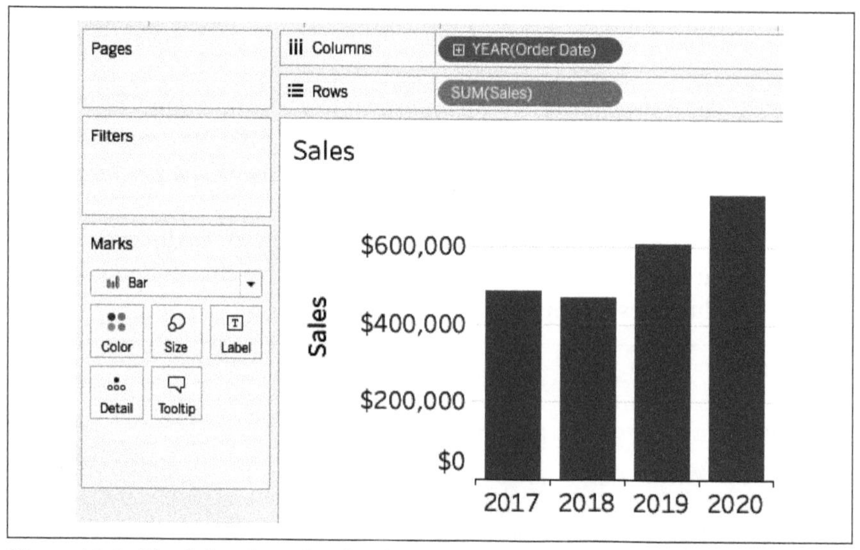

Figure 11-3. The Sales sheet for the sheet swap

b. To build the Profit visualization (Figure 11-4), create a new sheet. Add sum of [Profit] to the Rows shelf. Add discrete year of [Order Date] to the Columns shelf. Change the mark type to Area. Format the sheet by setting the rows axis ruler to None and the columns axis ruler and ticks to Black.

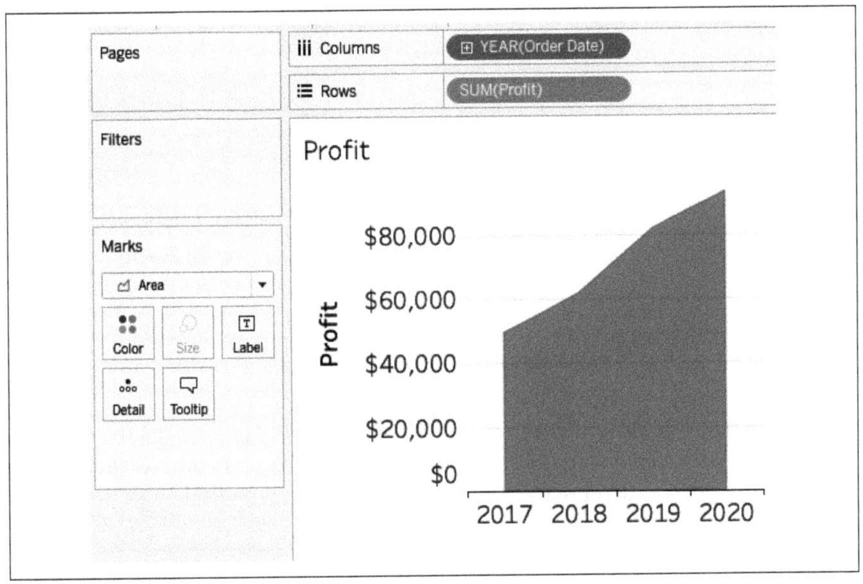

Figure 11-4. The Profit sheet for the sheet swap

c. To build the Profit Ratio visualization (Figure 11-5), create another new sheet. Add [Profit Ratio] to the Columns shelf. If you don't have this calculation, you can create it:

```
// Profit Ratio
SUM([Profit])/SUM([Sales])
```

Add [Category] to the Rows shelf. Change the mark type to Bar. Format the sheet by setting the columns axis ruler to None and the rows axis ruler and ticks to Black.

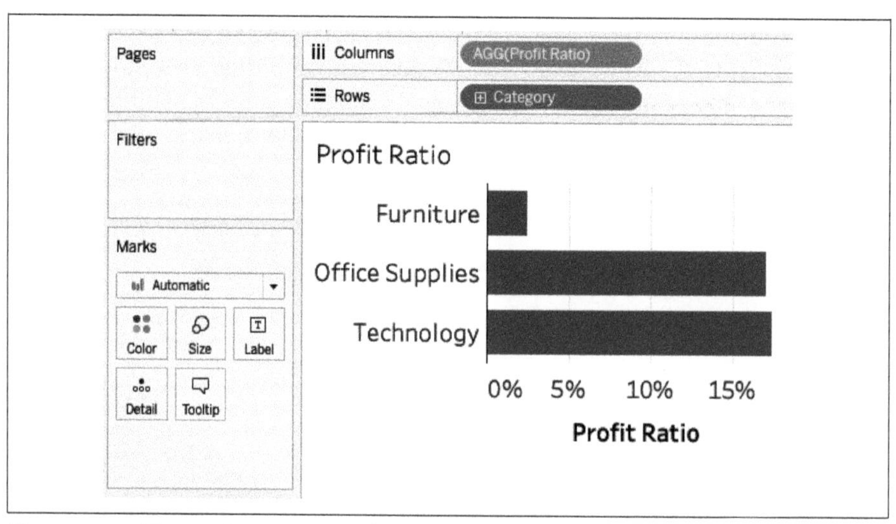

Figure 11-5. The Profit Ratio visualization for the sheet swap

2. Create a string parameter called **`[Names Parameter]`**. Create a list of three values: Sales, Profit, and Profit Ratio, as shown in Figure 11-6. These values align to the three sheets you just created.

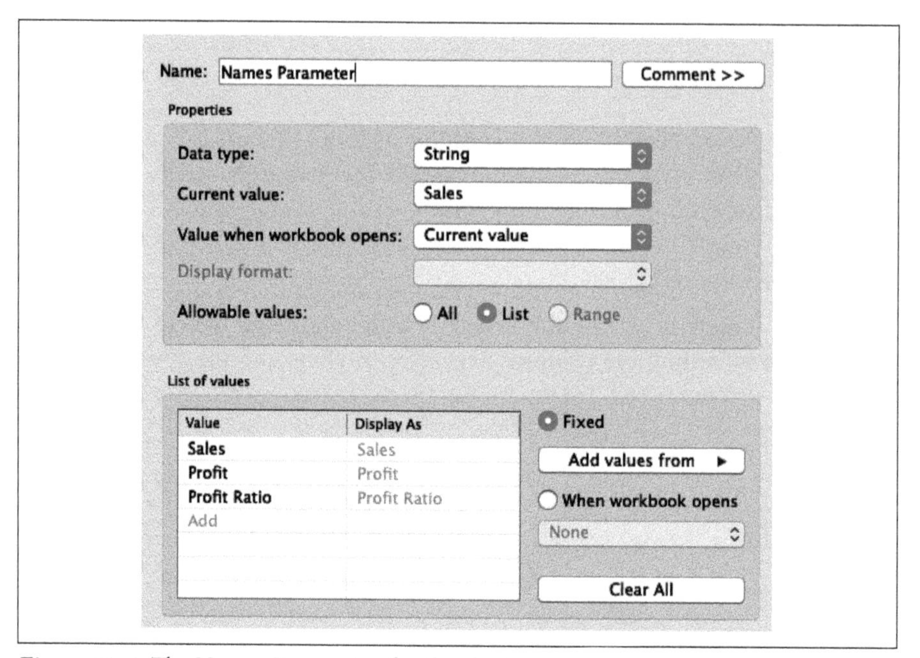

Figure 11-6. The Names Parameter for the sheet swap

3. Place the sheets and parameter on a dashboard:

   a. If you don't already have a dashboard, create a new one.

   b. On the dashboard, add a vertical container.

   c. Add the Names Parameter to the vertical container (Analysis → Parameters → Names Parameter) and hide the title.

   d. Inside the vertical container, below the Name Parameter, add a horizontal container.

   e. Inside the horizontal container, add the Sales, Profit, and Profit Ratio sheets.

   f. Hide the titles of all the sheets. If you do not do this, the titles will all show and the technique will not work.

   Do not adjust the size of the sheets, as it would fix the width of any of the sheets and break our sheet swap. Also, do not distribute the sheets evenly in the container. The result will be similar to Figure 11-7.

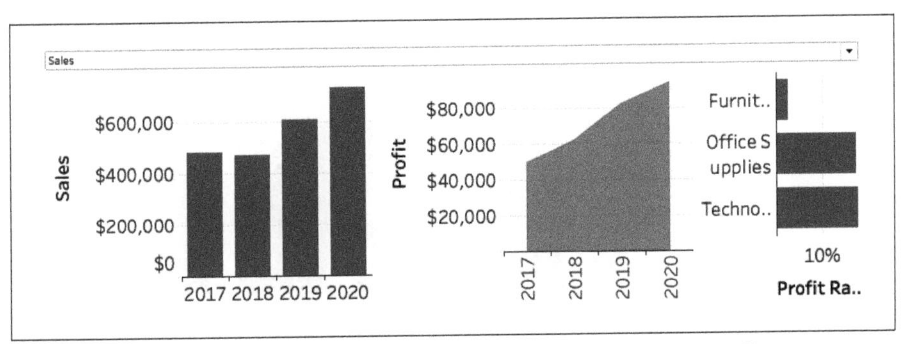

Figure 11-7. In our dashboard for the sheet swap prior to adding any filters, notice that all the visualizations are active

4. Create calculations for filtering each sheet, as follows:

   a. To create the Sales parameter filter, create a calculation called **`[Names Parameter | Sales]`** and type the following:

   ```
   // Names Parameter | Sales
   [Names Parameter] = "Sales"
   ```

   Go to the Sales sheet. Set the parameter to Sales. Add the [Names Parameter | Sales] calculation to the Filters shelf and set the filter to True.

   b. To create the Profit parameter filter, create a calculation called **`[Names Parameter | Profit]`** and type this:

   ```
   // Names Parameter | Profit
   [Names Parameter] = "Profit"
   ```

Go to the Profit sheet. Set the parameter to Profit. Add the [Names Parameter | Profit] calculation to the Filters shelf and set the filter to True.

c. To create the Profit Ratio parameter filter, create a calculation called **`[Names Parameter | Profit Ratio]`** and type this:

```
// Names Parameter | Profit Ratio
[Names Parameter] = "Profit Ratio"
```

Go to the Profit Ratio sheet. Set the parameter to Profit Ratio. Add the [Names Parameter | Profit Ratio] calculation to the Filters shelf and set the filter to True.

When you go back to your dashboard, you will notice that your dashboard now shows only a single sheet—the sheet selected by your parameter (Figure 11-8).

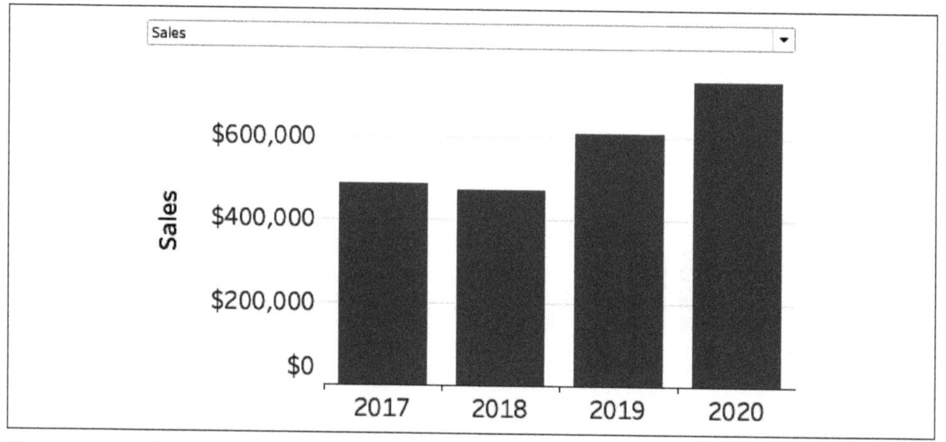

Figure 11-8. The Sales visualization is showing on the sheet swap because the parameter is set to Sales

In this strategy, you learned how to do a sheet swap by using a parameter. A sheet swap allows you to interact with multiple visualizations by hiding the sheets that are not relevant at the time. While this is nice, our audience has to make two clicks every time they want to swap sheets. And since we live in a world where speed and interactivity is so important, it might be better to make this into a single-click interaction.

Strategy: Use a Parameter-Action Sheet Swap

You are going to create a supplementary data source and use this to drive parameter actions. You will then create a sheet using this data source that looks like buttons; these are shown in Figure 11-2. Formatting, plus parameter actions, will then make the visualization feel highly interactive:

1. Open your text editor and create a new data source:

```
Values
1
2
3
4
5
6
7
8
9
10
```

   Using the same techniques as for data modeling in Chapter 10, you will create a single column with 10 data points called Values. Technically, you will need only three points, but we like to add more just in case we need more for future updates—better to future-proof. Save this file as *values.csv*.

2. Connect to the new data source, *values.csv*, you just created.

3. Build the three calculations needed for this visualization:

   a. Using the Values data source, create a calculation called **`[Names]`**. This will convert integers to names:

   ```
   // Names
   CASE [Values]
   WHEN 1 THEN "Sales"
   WHEN 2 THEN "Profit"
   WHEN 3 THEN "Profit Ratio"
   END
   ```

   This calculation will drive all our parameter updates.

   b. Create a calculation called **`[Active Name]`**. This calculation will return a highlighted name after we've formatted it:

   ```
   // Active Name
   IF [Names] = [Names Parameter]
   THEN [Names]
   END
   ```

   c. Create a calculation called **`[Inactive Names]`**. This calculation will return names that are not selected by the parameter name. You'll need to format this value too:

   ```
   // Inactive Names
   IF [Names] != [Names Parameter]
   THEN [Names]
   END
   ```

4. Build the visualization as follows:

   a. Create a new sheet called `Name Swap`.

   b. Change the marks type to Text.

   c. Add [Inactive Names] and [Active Name] to Text.

   d. Add [Values] as a dimension above [Inactive Names] and [Active Name] on the Marks card.

   e. Add [Names] below [Active Name] and [Inactive Names] on the Marks card.

   f. Edit the text. Place [Inactive Name] and [Active Name] on the same line. Change the [Inactive Names] measure to light gray. Change the [Active Name] measure font to Tableau Semibold. After the text, add two underscore values (__) and color the underscores white to match the background of your visualization. See Figure 11-9 for the visual styling.

   g. Turn off the tooltips.

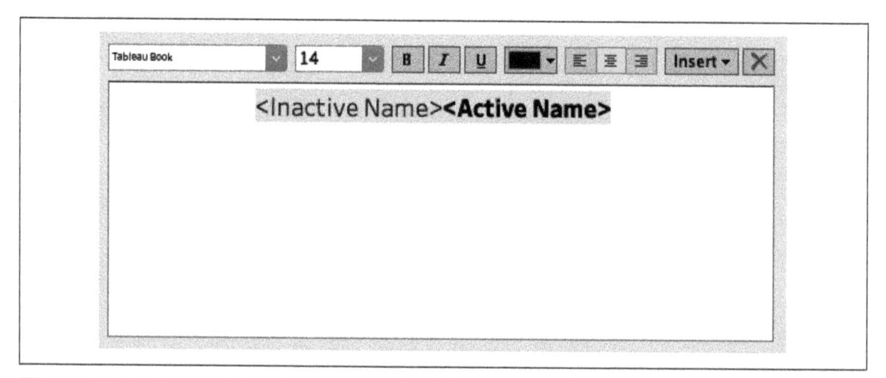

Figure 11-9. Formatting the text on the Name Swap sheet

   h. Add [Values] as a dimension to Filters and set the range from 1 to 3.

   If you change the [Names] parameter to Profit, the result is Figure 11-10. Normally, the text marks would be very close to each other, but we've used the two underscores to add spacing between the text.

Figure 11-10. The Name swap sheet after adding formatting

5. Using the dashboard we created in the first strategy, add the Name Swap sheet in the vertical container above the Names Parameter. Hide the title of the sheet and then remove the Names Parameter from the dashboard. Adjust the height of the sheet so all the values are legible—this is about 60 pixels tall.

6. Add a new parameter action by choosing Dashboard → Actions; then select New Action → Change Parameter. Name the parameter **`Names Update`**. On the dashboard, when your audience clicks a name on the Name Swap sheet, update the Names Parameter with the Names dimension. When they clear the selection, keep the current value. For more detail on the setup, see Figure 11-11.

   Figure 11-12 shows the final result.

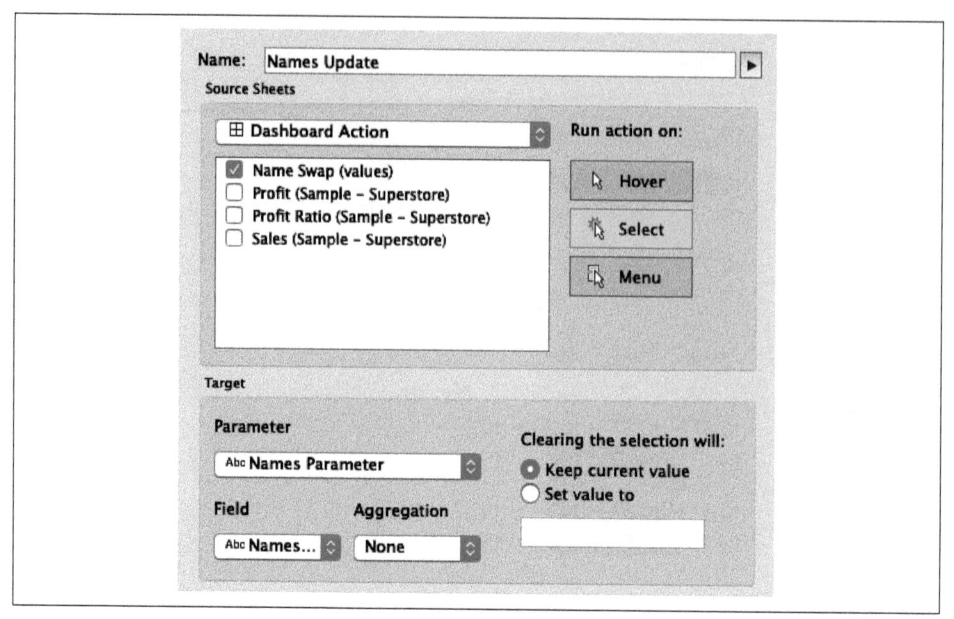

Figure 11-11. The parameter action for the Names Parameter

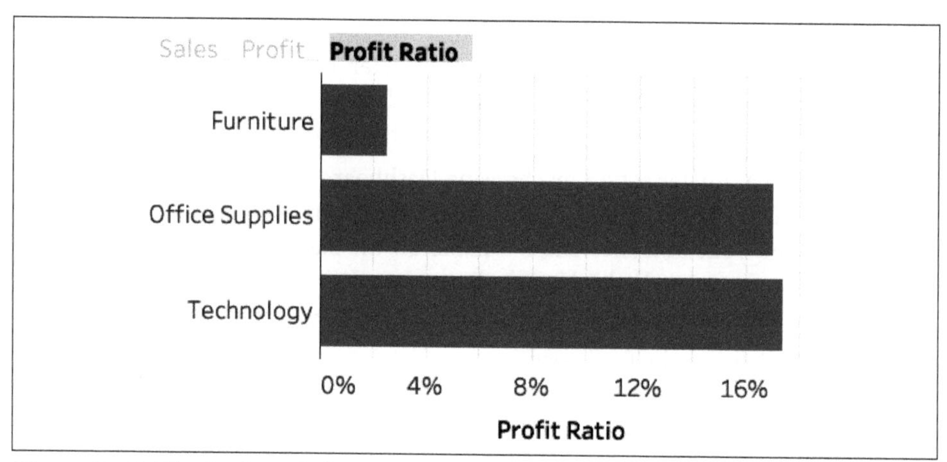

Figure 11-12. The final result for our parameter-action sheet swap

In this strategy, you learned how to create a well-formatted worksheet that looks and feels like a parameter, but—thanks to parameter actions—also acts like a parameter.

Strategy: Automatically Deselect Marks

Your final result is a sheet swap that updates on the selection of a value. But one annoying challenge remains: every time you select a value, the values become highlighted in blue. It would be nice if your sheet automatically turned off this selection. This can be done with a little Tableau hackery, using filter actions across dashboards and sheets.

This hack is something we use all the time. Anytime we create a dashboard, we use this trick, as it makes our dashboards feel more application-like and less business-intelligence-dashboard-like:

1. You need to create two calculations in the *values.csv* data source:

   a. Create a calculation called **[TRUE]** that is equal to the True Boolean:

   ```
   // TRUE
   TRUE
   ```

   b. Create a calculation called **[FALSE]** that is equal to the False Boolean:

   ```
   // FALSE
   FALSE
   ```

   For this strategy, we are going to use the Name Swap sheet you created in the "Strategy: Use a Parameter-Action Sheet Swap" on page 432. Add the [TRUE] and [FALSE] calculations to Detail of the Marks card.

2. Go to the dashboard containing your Name Swap sheet that you are using for the sheet swap action. Choose Dashboard → Actions. In the dialog box, click Add Action → Filter.

   Name your filter action **Names Reset**. Choose the Name Swap sheet from your dashboard. In our example shown in Figure 11-13, our dashboard is called Dashboard Action. Set "Run action on" to Select.

   From there, you will need to change your target sheet from the dashboard to the sheet itself. This is a critical step. If you run the action on the dashboard, the action won't work. Set your filters to show all values when clearing the selection.

   On the last section, Target Filters, choose Add Filter and then set the source field to TRUE and the target field to FALSE. Click OK, and then click OK on the Actions menu too.

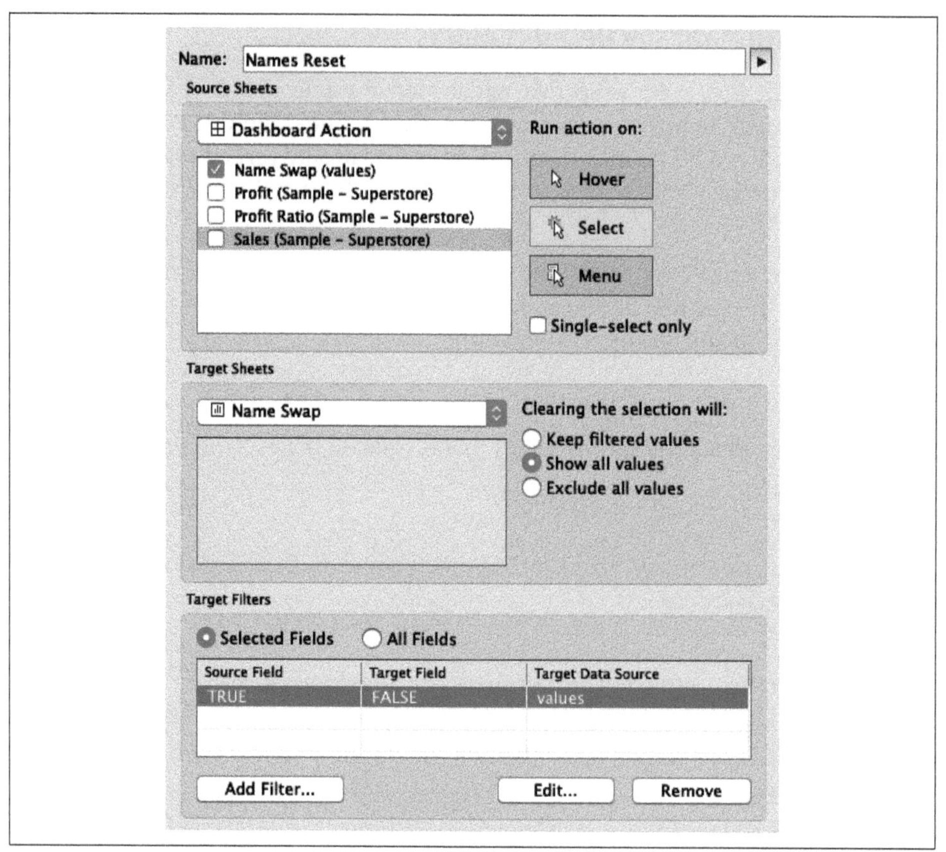

Figure 11-13. The filter action from the Name Swap sheet on the dashboard from the Name Swap sheet itself

The final result is a sheet swap driven by a button. But after you press the button, the sheet is no longer highlighted; see Figure 11-14.

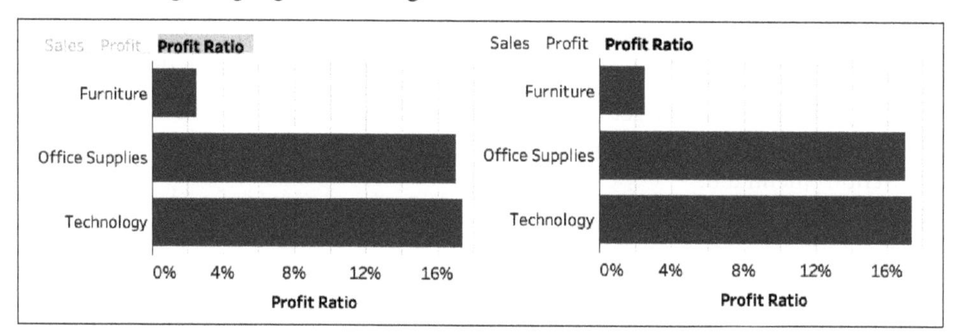

Figure 11-14. The parameter action without the filter action to deselect the mark (left) and the parameter action to automatically deselect the mark (right)

Using Parameters and Parameter Actions to Create Multiple Select Parameters: Superstore Case Study

When we work with parameters and parameter actions, our minds first think that they can be used only to select a value from a list. But a parameter also can be used as an input to store information. One thing we do with parameters is write a parameter plus a dimension on our view back to the same parameter.

Once again, your audience wants to see more charts—this time, a breakdown of sales performance by state. They also want you to allow for analysis to be broken down by segment. Normally, you could use a multiselection filter, but they indicated they wanted something more interactive. In this strategy, you will learn how to use parameters as an input tool.

For instance, consider the visualization on the right side in Figure 11-14. The visualization includes [Category] as a dimension. Imagine we have a parameter with a value that contains Furniture we could update, but we wanted the update to read Furniture, Office Supplies or Furniture, Technology. We could do that by creating a calculation that includes the parameter with the category. This calculation would be something like this:

```
// Hypothetical Calculation
[Parameter] + ", " + [Category]
```

We could then use this hypothetical calculation to feed back into the same parameter. This process could then add information to a parameter without losing the parameter's existing information.

Tableau Hall of Fame Zen Master Jonathan Drummey has written about this process in detail in an amazing post that's linked in the "Further Reading" on page 479.

Why would you want a multiple-select parameter? Well, parameters allow for only a single input, and the format of those inputs is extremely fixed. For parameters, you have the option of a drop-down, an open field, a slider, or radio button selector, as shown in Figure 11-15. But sometimes you need more. Sometimes your audience demands customization.

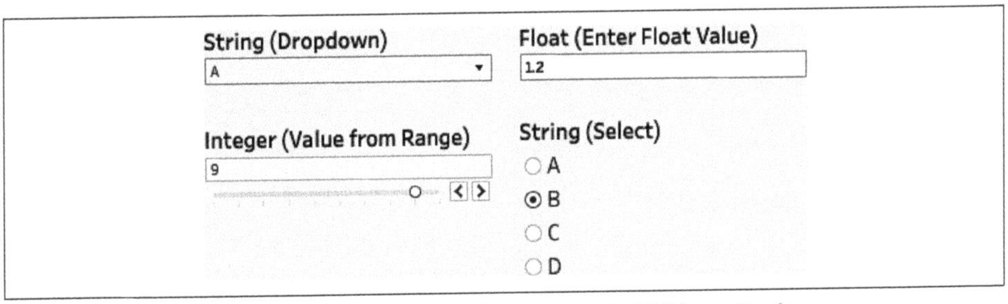

Figure 11-15. The existing parameter display options in Tableau Desktop

Strategy: Create a Multiple-Select Parameter

In this strategy, you are tasked with creating a parameter that allows for multiple selections. On the visualization, you will have on-off toggle values available, and this toggle will filter a separate table that you create. While this is one use-case for parameters as a data source, there are many, many use cases, which we provide links to in "Further Reading" on page 479.

For this strategy, you will be using the Sample – Superstore dataset. Imagine you are an analyst who covers the Southern region and you are tasked with providing a dynamic table reviewing first-quarter sales for an audience of sales reps. Unfortunately, these reps are not very good with technology and have trouble using dropdown menus. They want a quick way to toggle multiple values on and off. Your solution is shown in Figure 11-16.

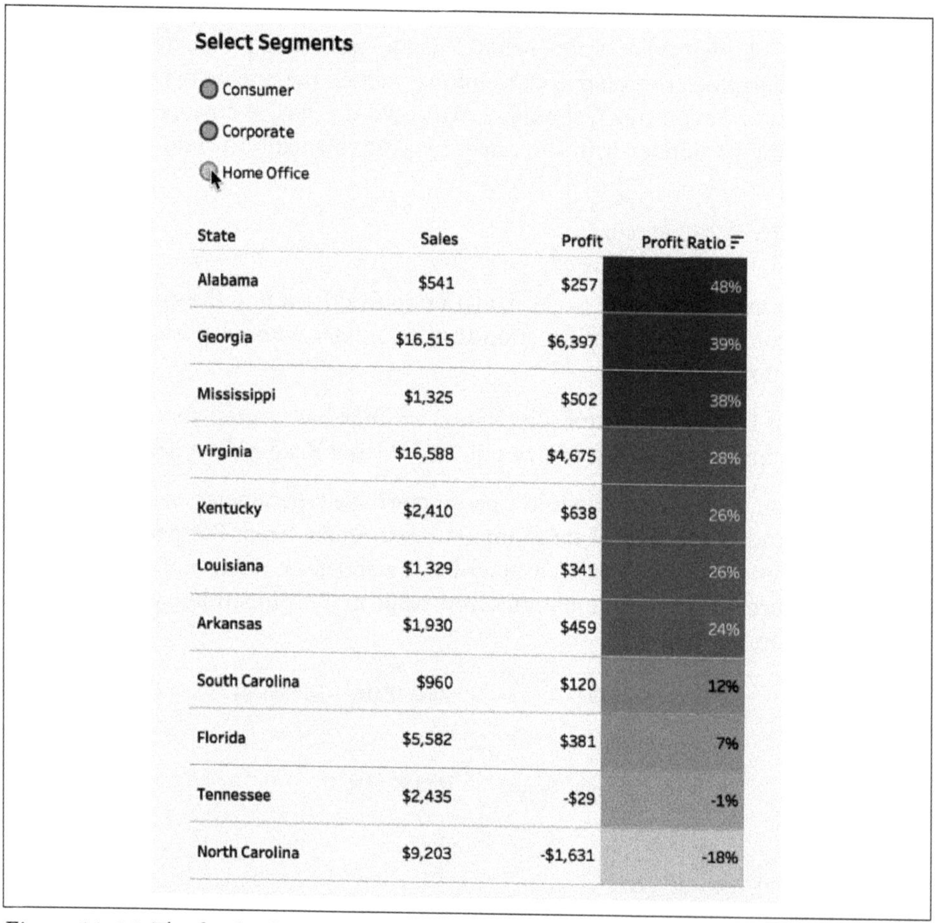

Select Segments

Consumer

Corporate

Home Office

State	Sales	Profit	Profit Ratio
Alabama	$541	$257	48%
Georgia	$16,515	$6,397	39%
Mississippi	$1,325	$502	38%
Virginia	$16,588	$4,675	28%
Kentucky	$2,410	$638	26%
Louisiana	$1,329	$341	26%
Arkansas	$1,930	$459	24%
South Carolina	$960	$120	12%
Florida	$5,582	$381	7%
Tennessee	$2,435	-$29	-1%
North Carolina	$9,203	-$1,631	-18%

Figure 11-16. The final solution for the Automatically Deselect Marks strategy

To build this solution, you'll follow these steps:

1. Build a base visualization (Figure 11-17) as follows:
   a. Add [State] to the Rows shelf.
   b. Add [Measure Names] to Columns.
   c. Add [Measure Values] to Filter. Select sum of [Sales], sum of [Profit], and [Profit Ratio].
   d. Change the mark type to Square.
   e. Add [Measure Values] to Text and Color.
   f. Right-click [Measure Values] used for color and choose to use separate legends.
   g. Set the color of [Sales] and [Profit] to white, using the process outlined in "Strategy: Use Parameter-Based Sheet Swapping" on page 428.
   h. Set the color of [Profit Ratio] to Gold-Green.
   i. Add a filter for the South region.
   j. Add [Order Date] to filters, change to quarter as a date part, and filter to Q1.
   k. Sort [State] descending on [Profit Ratio].

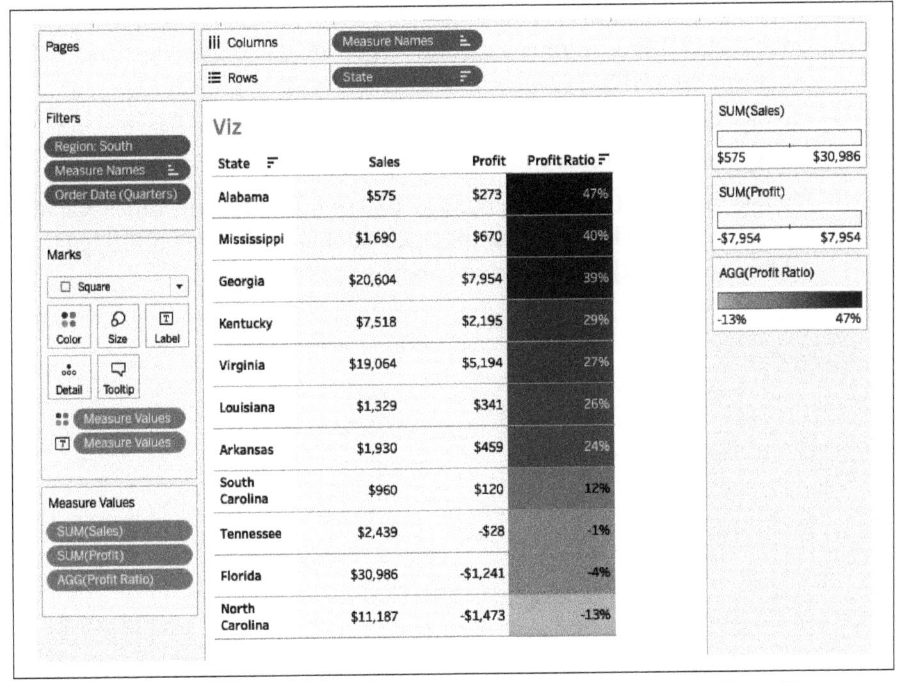

Figure 11-17. Your base visualization before adding any parameter-base filters

2. Create a string parameter, but called **[MultiParam]**. Delete any value in the "Current value" text box and allow all values (Figure 11-18). By leaving the parameter empty, we'll have no values selected—for the moment—from our data source.

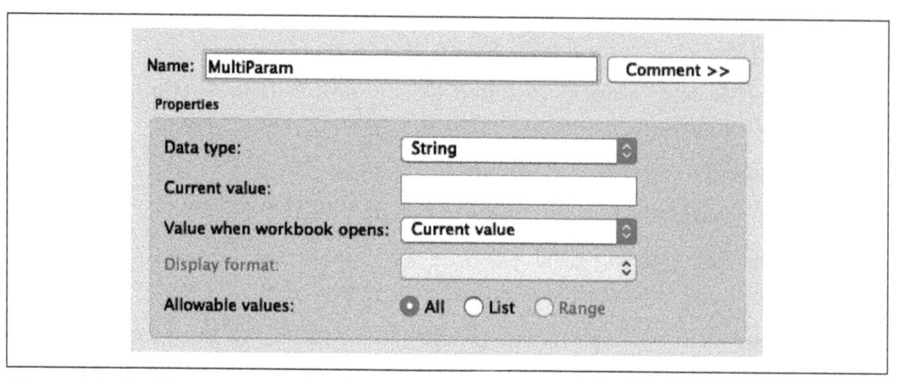

Figure 11-18. The empty multiselect parameter

3. For this example, you'll drive parameter actions with the [Segment] dimension. But to do so, you'll need to build a visualization, which is shown in Figure 11-19. Start by creating a new sheet called **Select Segments**.

   a. Add [Segment] to the Rows shelf.

   b. Double-click the Columns shelf and type **MIN(0.0)**. Double-click again and create another MIN(0.0) ad hoc calculation.

   c. Change the mark type on the left MIN(0.0) Marks card to Custom Shape. Change the custom shape to an open circle.

   d. Change the mark type on the right MIN(0.0) Marks card to Circle. Change the opacity to 40%. This will give the open bubble effect shown in Figure 11-19. Add the [Segment] dimension to Label on the Marks card.

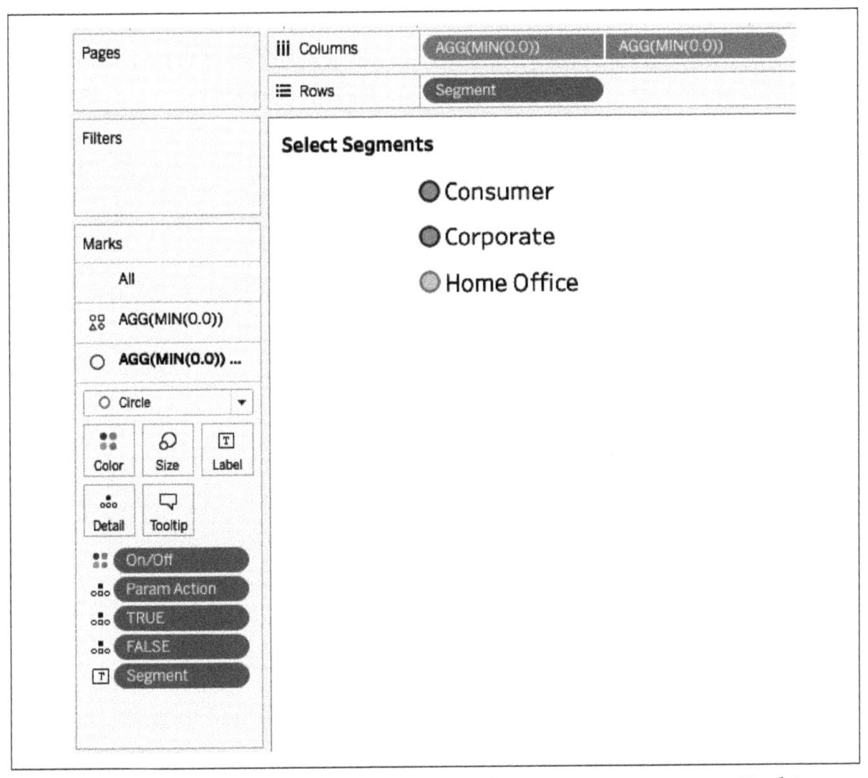

Figure 11-19. The sheet you'll use to drive updates to your parameter. In this example, Consumer and Corporate segments are active.

e. Create a synchronized dual axis. Edit the axis to range from –1 to 10. Then hide the axes.

f. Remove [Measure Names] from Color. Hide the header for [Segment]. Set the row and column dividers to None. Set the zero lines, grid lines, and axis rulers to None. You may need to resize the shapes. Finally, turn off your tooltips.

   Figure 11-20 shows your progress to this point.

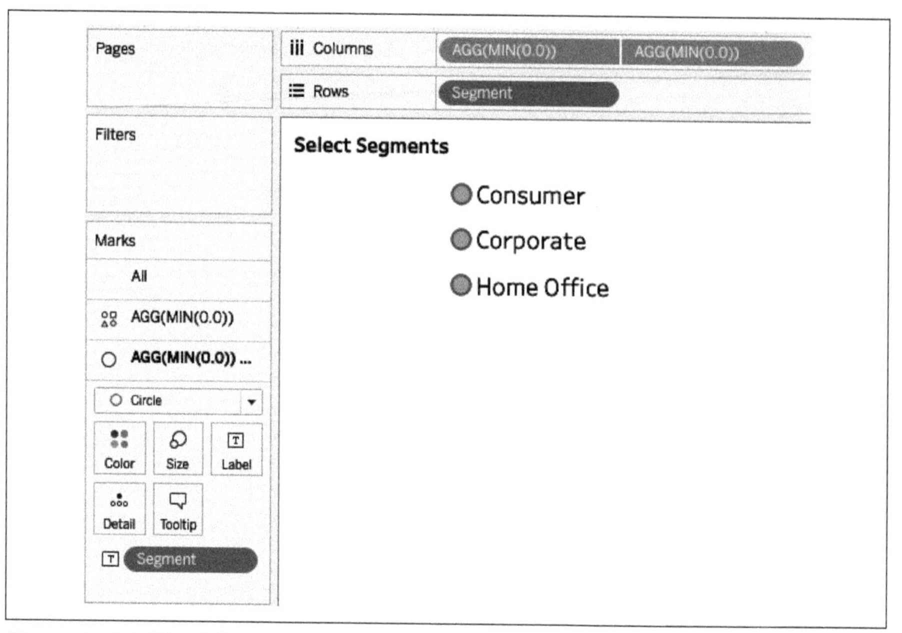

Figure 11-20. The Select Segments sheet after step 3

4. Create a calculation, as follows, that identifies whether a member is in the parameter:

   a. Create a calculation called **`[On/Off]`**. This calculation is a Boolean that uses the `CONTAINS()` function to look for the [Segment] members inside the [MultiParam] parameter—which is currently empty!

   ```
   // On/Off
   CONTAINS([MultiParam], [Segment] + ", ")
   ```

   b. Add this calculation to Color on both Marks cards.

   c. Show the [MultiParam] parameter and type **`Consumer,`** (with a space after the comma). This will activate Consumer and change the color. Add to the parameter **`Corporate,`** (also with a space after the comma). This will activate Consumer and Corporate.

   d. Edit the color by setting the True values to green and the False values to a warm gray.

5. Create the **`[Param Action]`** calculation, which will drive selection/deselection. We will apply this calculation to the parameter action:

   ```
   // Param Action
   IF CONTAINS([MultiParam], [Segment] + ", ")
   THEN REPLACE([MultiParam], [Segment] + ", ", "")
   ```

```
ELSE [MultiParam] + [Segment] + ", "
END
```

If a member of [Segment] is in the [MultiParam] parameter, the calculation will remove the value. If it's not in the parameter, it'll add the value to the MultiParam string.

Add this calculation to Detail on the All Marks card of the Select Segments sheet.

6. Following step 1 from "Strategy: Automatically Deselect Marks" on page 437, create [TRUE] and [FALSE] calculations. Add these calculations to Detail on the Marks cards. Your visualization should now match Figure 11-18.

7. Add the visualization and the Select Segments sheet to the dashboard. Hide the title of both dashboards.

8. Set up the parameter action so that when you select a value on the parameter sheet, the MultiParam sheet will update with the [Param Action] calculation; see Figure 11-21.

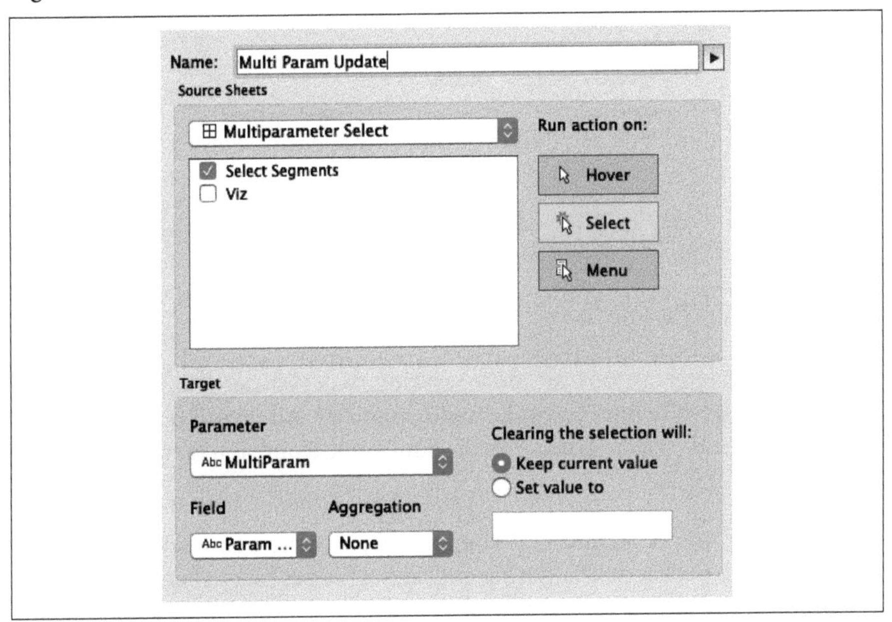

Figure 11-21. The parameter action for the MultiParam parameter

9. Follow step 2 from "Strategy: Automatically Deselect Marks" on page 437 so that with a click, the marks are deselected from the Select Segments sheet; see Figure 11-22.

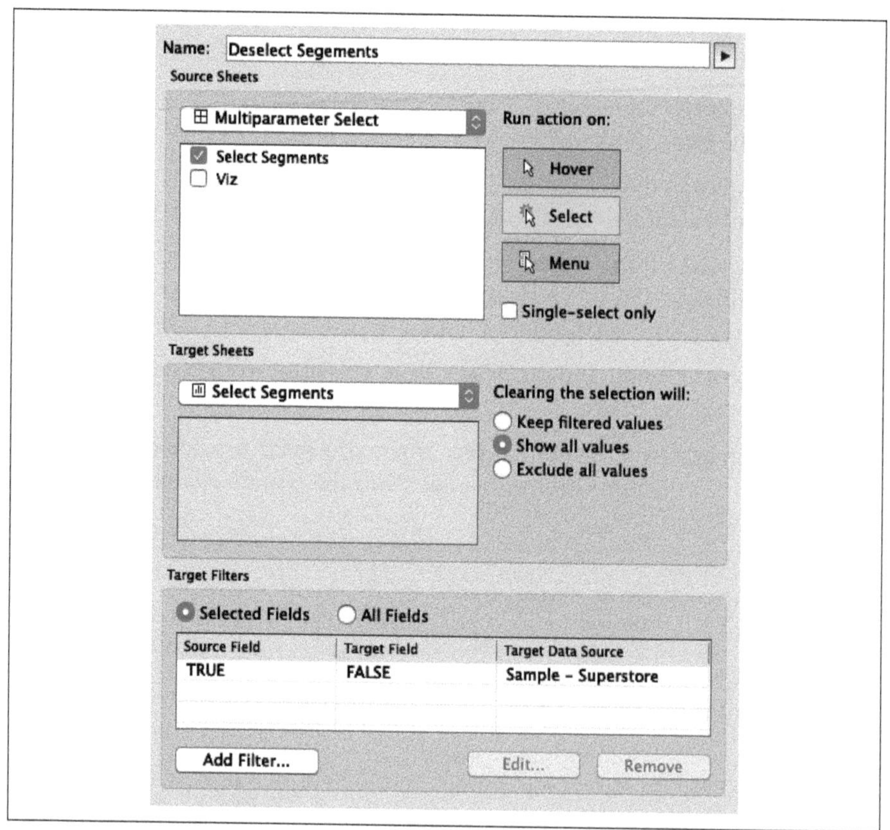

Figure 11-22. The filter action to automatically deselect marks

Your parameter actions are now working, but you need to update your table sheet to apply the filter from the multiparameter. Add On/Off to your table visualization and select True. You've now built your interactive visual system (Figure 11-23).

In this strategy, you've learned that you can use parameters as a data source by applying a parameter back to itself. Storing and adding values to a parameter allows a visualization to have serious interactivity.

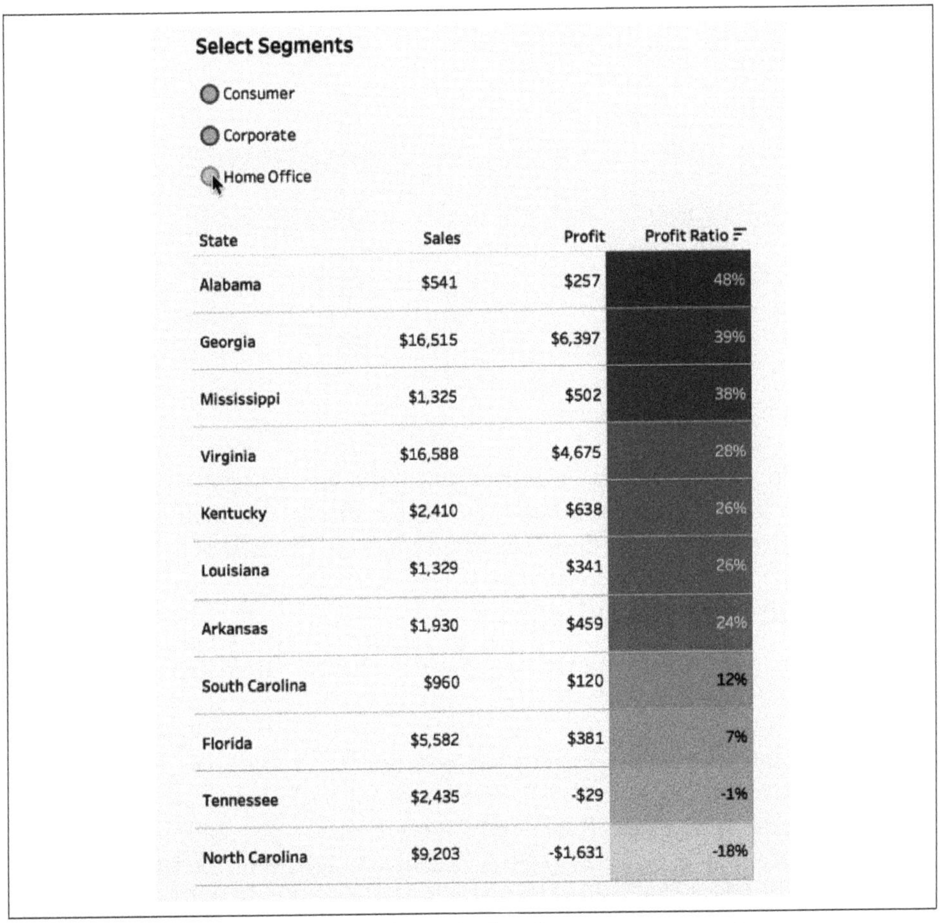

Select Segments

Consumer

Corporate

Home Office

State	Sales	Profit	Profit Ratio
Alabama	$541	$257	48%
Georgia	$16,515	$6,397	39%
Mississippi	$1,325	$502	38%
Virginia	$16,588	$4,675	28%
Kentucky	$2,410	$638	26%
Louisiana	$1,329	$341	26%
Arkansas	$1,930	$459	24%
South Carolina	$960	$120	12%
Florida	$5,582	$381	7%
Tennessee	$2,435	-$29	-1%
North Carolina	$9,203	-$1,631	-18%

Figure 11-23. The final solution for the multiple-select parameter strategy

Swapping Metrics Using Parameter Actions: Office Essentials Case Study

Parameters and parameter actions are so powerful. As you could see in the previous strategy, you can make sheets look like and act like parameters. You can also make sheets look and act like buttons with greater functionality than the base functionalities of Tableau.

Your audience is loving the interactivity that you've shown. As a result, they now want the ability to scroll through various metrics by using buttons. This time, you've decided to use individual sheets to look like buttons and provide the interactivity your audience desires (Figure 11-24).

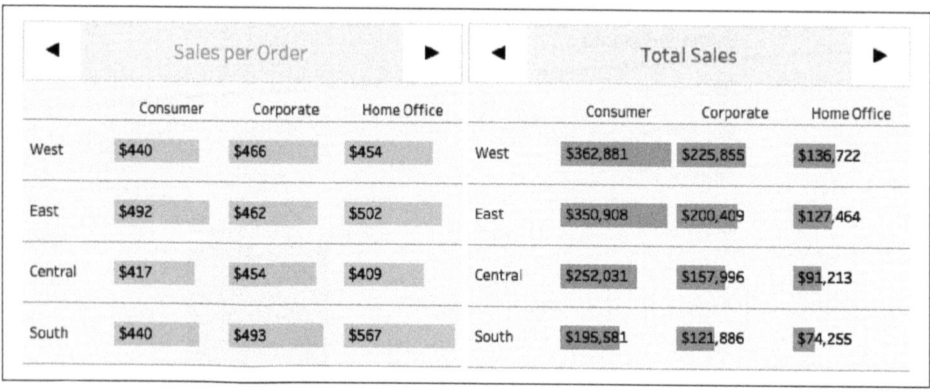

Sales per Order
Consumer
Corporate
Home Office
West
$440
$466
$454
East
$492
$462
$502
Central
$417
$454
$409
South
$440
$493
$567
Total Sales
Consumer
Corporate
Home Office
West
$362,881
$225,855
$136,722
East
$350,908
$200,409
$127,464
Central
$252,031
$157,996
$91,213
South
$195,581
$121,886
$74,255

Figure 11-24. Using sheets as buttons

In this scenario, imagine you are a user-experience designer working on building an advanced interface for your customers. You want to display one metric at a time but have four metrics to show. You decide to create your own carousel of options.

Strategy: Use Sheets as Buttons

In this strategy, you will create buttons that will change parameters and update a measure:

1. Create an integer parameter called **`[Select Metric]`** and choose a list of values. Enter as many values as there will be metrics. In this example, we are swapping four metrics: Total Sales, Total Profit, Percent of Margin, and Total Sales per Order. Under Value, type 1 through 4, as shown in Figure 11-25. Update the Display As list to show the names of the four metrics.

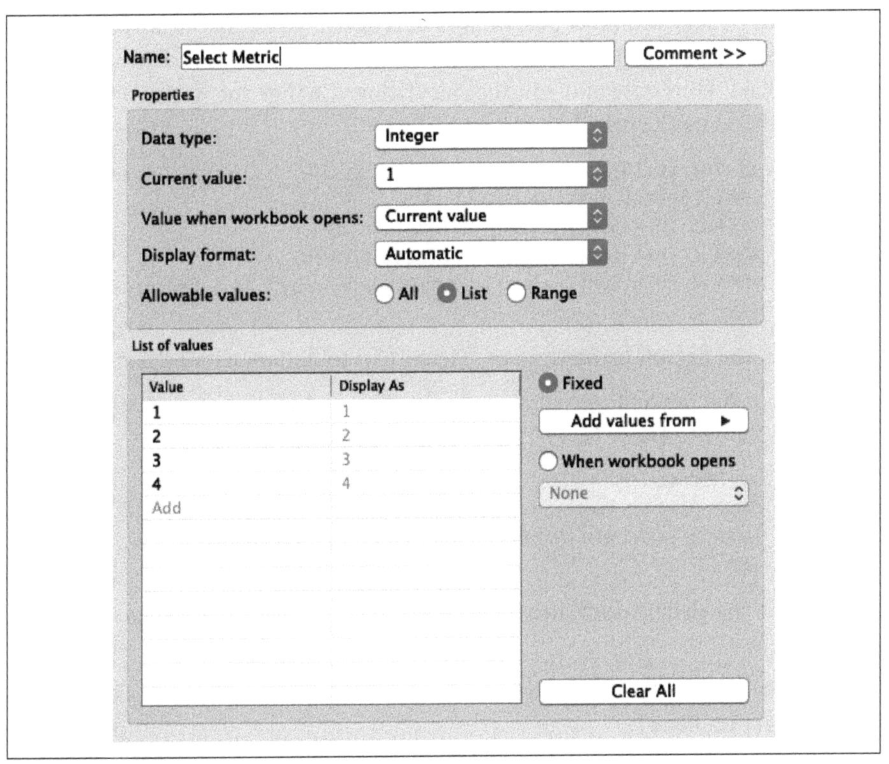

Figure 11-25. The integer parameter for step 1

2. Create the calculations for your base visualization as follows:

   a. Create a calculation called **[Metric]**. We'll use four metrics to match our four parameter values:

```
// Metric
CASE [Select Metric]
WHEN 1 THEN SUM([Sales]) //Total Sales
WHEN 2 THEN SUM([Profit]) //Total Profit
WHEN 3 THEN SUM([Profit])/SUM([Sales]) //% Margin
WHEN 4 THEN SUM([Sales])/COUNTD([Order ID]) // $ per order
END
```

b. Our visualization has percentages and dollar totals. Because you can't change the formatting, you need to make a dimension that's strictly for displaying dollars. Duplicate and edit the calculation. Change the name to **[Metric ($)]** and update the calculation:

```
// Metric ($)
CASE [Select Metric]
WHEN 1 THEN SUM([Sales]) //Total Sales
WHEN 2 THEN SUM([Profit]) //Total Profit
WHEN 4 THEN SUM([Sales])/COUNTD([Order ID]) // $ per order
END
```

Edit the default formatting of [Metric ($)] to dollars with no decimals.

c. Build the percentages label by duplicating and editing the calculation. Change the name to **[Metric (%)]** and update it as follows:

```
// Metric
CASE [Select Metric]
WHEN 3 THEN SUM([Profit])/SUM([Sales]) //% Margin
END
```

Edit the default formatting of [Metric (%)] to be a percentage with no decimals.

d. Create labels that tie metrics to the parameters by creating a calculation called **[Metric Name]**:

```
// Metric Name
CASE [Select Metric]
WHEN 1 THEN "Total Sales"
WHEN 2 THEN "Total Profit"
WHEN 3 THEN "Percent of Margin"
WHEN 4 THEN "Sales per Order"
END
```

3. Build your base visualization, shown in Figure 11-26:

   a. Create a sheet called **Viz**.

   b. Add [Segment] and [Metric] to Columns.

   c. Add [Region] to Rows.

   d. Place [Metric (%)] and [Metric ($)] on Text and left-align the text.

   e. Format your visualization by turning off row banding but adding row dividers.

   f. Add more spacing to the rows and change the size of the bars so they are approximately 60% of the total size of the row height.

   g. Add [Metric Name] to Color and set the opacity to 40%.

   h. Turn off the tooltips.

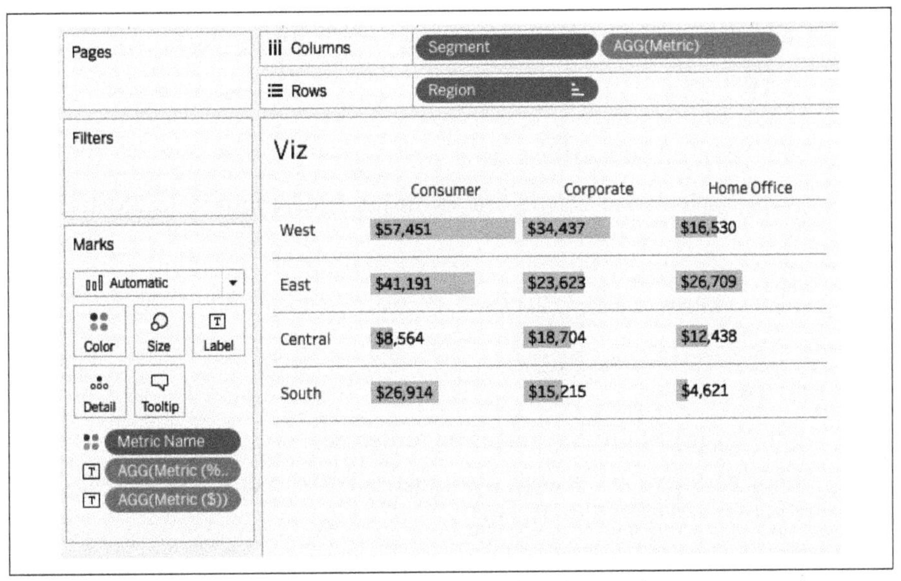

Figure 11-26. The base visualization that we will update with parameter values

4. Create a new left-arrow sheet called **`Left`**.

   a. Create the label with a calculation named **`[Left]`**:

   ```
   // Left
   "◀"
   ```

   Add the calculation as a dimension to Text.

   b. Create a calculation called **`[Subtract Value]`** that will update your [Select Metric] parameter:

   ```
   // Subtract Value
   IF [Select Metric] = 1
   THEN 4
   ELSE [Select Metric] - 1
   END
   ```

   Add the calculation to Detail.

   c. Following step 1 from "Strategy: Automatically Deselect Marks" on page 437, add [TRUE] and [FALSE] to Detail.

   d. Remove the tooltips from your visualization, set axis rulers, grid lines, zero lines, and dividers to None.

5. Create a new right-arrow sheet called **`Right`**.

a. Create the right-arrow label by calling the calculation [Right]:

```
// Right
"►"
```

Add the calculation to Text.

b. Create a calculation called **`[Add Value]`** to update your [Select Metric] parameter:

```
// Add Value
IF [Select Metric] = 4
THEN 1
ELSE [Select Metric] + 1
END
```

Add the calculation as a dimension to Detail.

c. Following step 1 in "Strategy: Automatically Deselect Marks" on page 437, add [TRUE] and [FALSE] to Detail.

d. Remove the tooltips from your visualization, and set axis rulers, grid lines, zero lines, and dividers to None.

The results of the Left and Right sheet are shown in Figure 11-27.

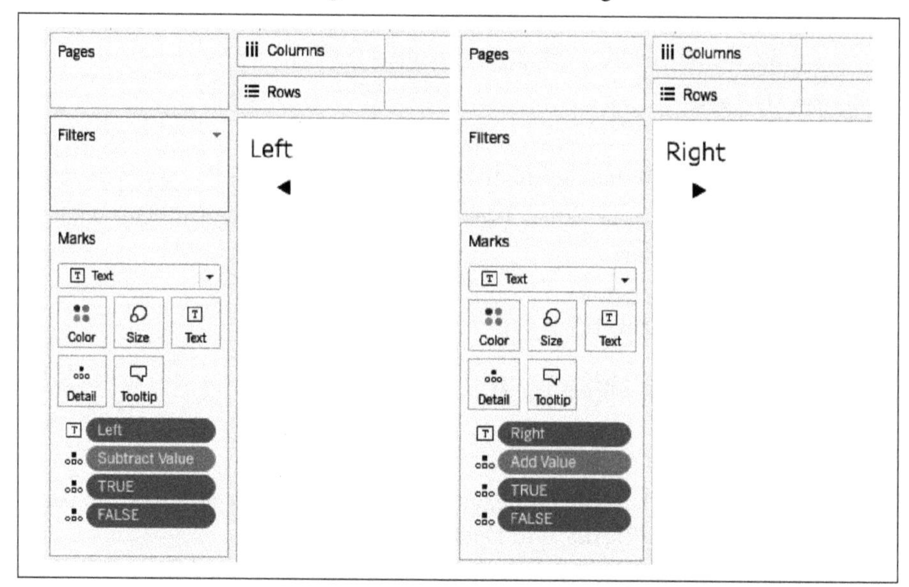

Figure 11-27. The Left and Right sheets for the table pagination strategy

6. Create the text display sheet, shown in Figure 11-28, as follows:

a. Create a sheet called **`[Metric Name]`**.

b. Add [Metric Name] to Text and align to the middle center.

c. Add [Metric Name] to Color.

d. Turn off your tooltips.

e. Format your visualization by removing any lines or dividers.

f. Update the color (optional).

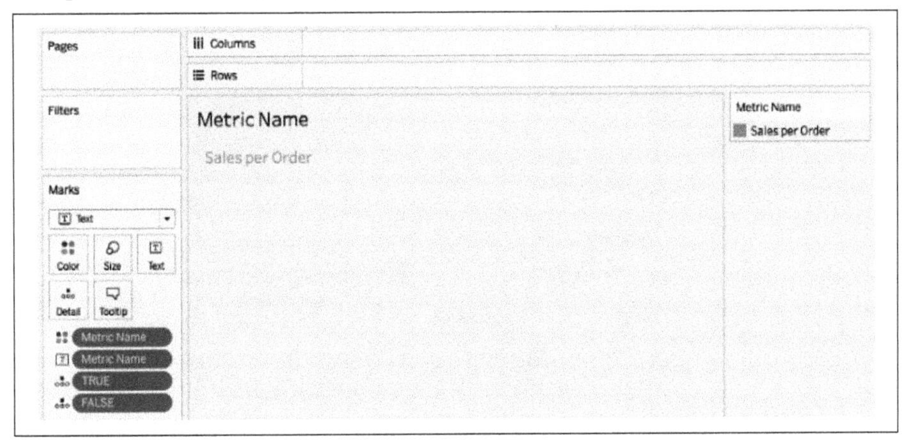

Figure 11-28. The Metric Names sheet prior to adding to a sheet

7. Build your dashboard:

   a. Create a new dashboard; for this example, we are using 400 × 300.

   b. Add a vertical container to the dashboard.

   c. Add Viz to the dashboard inside the vertical container. Remove any legends or parameters that are automatically added as well.

   d. In the vertical container, above Viz, add a horizontal container. Add a light gray border around the container. Set the height of the container to 40 pixels.

   e. Add the Left, Metric Name, and Right sheets into the horizontal container. Hide all the titles. Order appropriately and remove padding from each sheet.

   f. Fix the width of the Left and Right sheets to 40 pixels.

   g. Format the background of [Metric Name] to a slightly darker color than the background.

8. You need to add a parameter action for both the left and right arrow sheets:

   a. On the Left sheet, update the [Select Metric] parameter by using [Subtract Value]; see Figure 11-29.

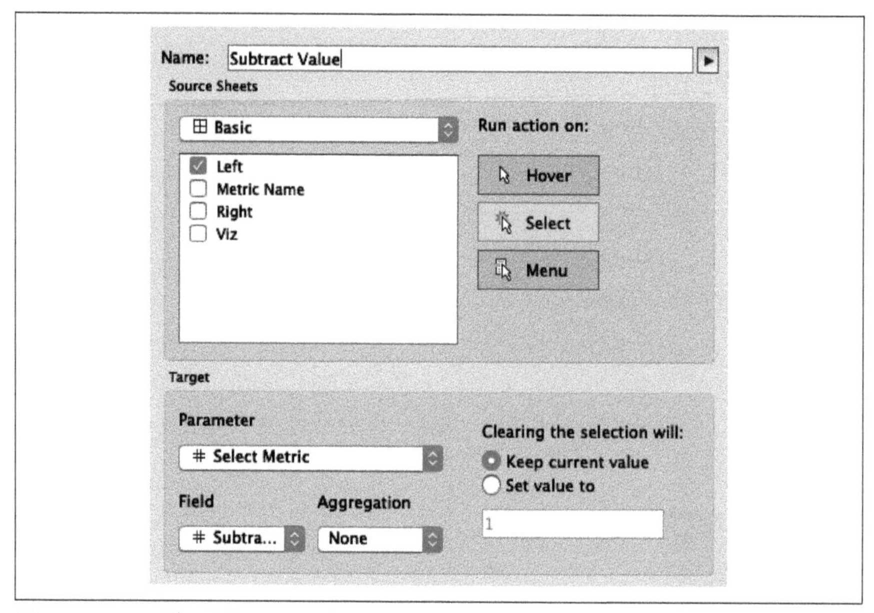

Figure 11-29. The Subtract Value parameter action

b. On the Right sheet, update the [Select Metric] parameter by using the [Add Value] calculation; see Figure 11-30.

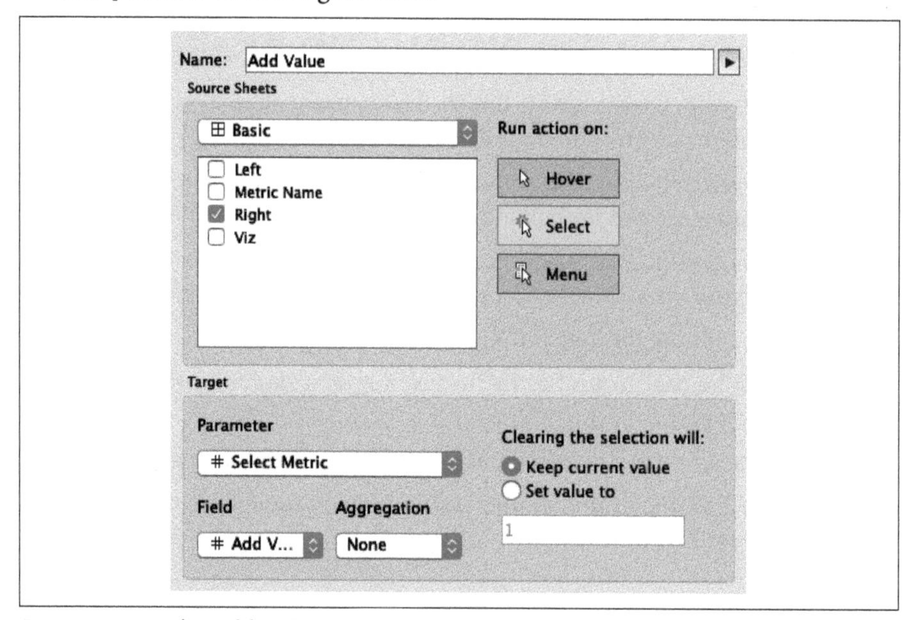

Figure 11-30. The Add Value parameter action

9. For the Left and Right sheets, follow step 2 in "Strategy: Automatically Deselect Marks" on page 437 to turn off mark selection.

   Figure 11-31 shows the final visualization.

◄ Percent of Margin ►				◄ Total Profit ►			
	Consumer	Corporate	Home Office		Consumer	Corporate	Home Office
West	16%	15%	12%	West	$57,451	$34,437	$16,530
East	12%	12%	21%	East	$41,191	$23,623	$26,709
Central	3%	12%	14%	Central	$8,564	$18,704	$12,438
South	14%	12%	6%	South	$26,914	$15,215	$4,621

Figure 11-31. The final visual for the sheets-as-buttons strategy

In this strategy, you learned how to turn sheets into interactive components to help drive changes on your dashboard. This is done with parameters and parameter actions. When you layer in the mark deselection functionality, dashboards move from business-intelligence tool to full-fledged application.

Adding Pagination to Tables with Parameters, Parameter Actions: Superstore Case Study

In the previous strategy, we introduced the concept of using pages as buttons. While this might be a small thing, we see this as unlocking lots of amazing possibilities for our dashboard designs. In this next strategy, you will tackle adding table pagination and building a suitable UI to go with that pagination.

There is no doubt, you've won over your audience with the amazing interactivity you've added to dashboards. But now you've been tasked with your most difficult task to date: the CEO is asking you to build tables that can perform at the speeds she wants. The challenge is that the dashboards are rendering slowly because the dashboard has too many marks. The solution: create pagination with a UI that any web developer would be jealous of.

Sometimes our tables are too long in Tableau, and sometimes we want to combine sheets to make a single table. We can do this with pagination! And when you think about it, most tables we interact with on the web have pagination; this is a subtle hint that we, as developers, should be designing tables with pagination. Figure 11-32 shows an example.

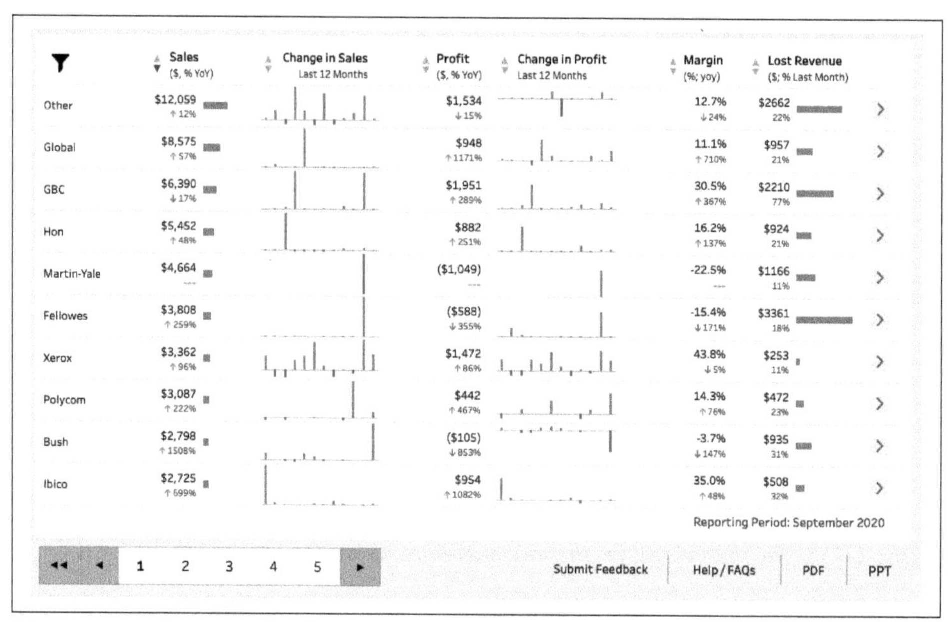

Sales ($, % YoY)
Change in Sales Last 12 Months
Profit ($, % YoY)
Change in Profit Last 12 Months
Margin (%; yoy)
Lost Revenue ($; % Last Month)
Other $12,059 ↑12% $1,534 ↓15% 12.7% ↓24% $2662 22%
Global $8,575 ↑57% $948 ↑1171% 11.1% ↑710% $957 21%
GBC $6,390 ↓17% $1,951 ↑289% 30.5% ↑367% $2210 77%
Hon $5,452 ↑48% $882 ↑251% 16.2% ↑137% $924 21%
Martin-Yale $4,664 ($1,049) -22.5% $1166 11%
Fellowes $3,808 ↑259% ($588) ↓355% -15.4% ↓171% $3361 18%
Xerox $3,362 ↑96% $1,472 ↑86% 43.8% ↓5% $253 11%
Polycom $3,087 ↑222% $442 ↑467% 14.3% ↑76% $472 23%
Bush $2,798 ↑1508% ($105) ↓853% -3.7% ↓147% $935 31%
Ibico $2,725 ↑699% $954 ↑1082% 35.0% ↑48% $508 32%
Reporting Period: September 2020
1 2 3 4 5
Submit Feedback
Help/FAQs
PDF
PPT

Figure 11-32. An example of pagination—and many other custom-built features for a table

In the next strategy, you are tasked with building a table that has pagination. You will need to show total sales, total profit, profit ratio, and total orders for each customer. Let's also stretch our imagination—your team has agreed that the only scroll bar that shows on your dashboard should be the one on a web browser. Therefore, your dashboard should have no scroll bars.

You are tasked with building a visual interface that shows details of 15 customers at a time, arrows for navigation, and an interactive page counter. You will do this using the Sample - Superstore dataset. You will re-create the table and visual components in Figure 11-33.

Customer Name	Sales	Profit	Profit Ratio	Total Orders
Sean Miller	$25,043	-$1,981	-8%	5
Tamara Chand	$19,052	$8,981	47%	5
Raymond Buch	$15,117	$6,976	46%	6
Tom Ashbrook	$14,596	$4,704	32%	4
Adrian Barton	$14,474	$5,445	38%	10
Ken Lonsdale	$14,175	$807	6%	12
Sanjit Chand	$14,142	$5,757	41%	9
Hunter Lopez	$12,873	$5,622	44%	6
Sanjit Engle	$12,209	$2,651	22%	11
Christopher Cona..	$12,129	$2,177	18%	5
Todd Sumrall	$11,892	$2,372	20%	6
Greg Tran	$11,820	$2,163	18%	11
Becky Martin	$11,790	-$1,660	-14%	4
Seth Vernon	$11,471	$1,199	10%	10
Caroline Jumper	$11,165	$859	8%	8

◄ 1 2 3 4 5 ►

Figure 11-33. The output of the table pagination strategy

Strategy: Add Table Pagination

This strategy requires four sheets: one for the table, one for the numerical display, one for the left arrow, and one for the right arrow. You'll start with the table, get it set up, then transition to the arrows, and finally create the display:

1. Connect to the Sample – Superstore dataset.
2. Build the base table:
   a. Add [Customer Name] to Rows.
   b. Sort [Customer Name] in descending order using sum of [Sales].
   c. If you don't have it, create a calculation called **`[Profit Ratio]`**:

   ```
   // Profit Ratio
   SUM([Profit])/SUM([Sales])
   ```

   d. Create a calculation called **`[Total Orders]`**:

   ```
   // Total Orders
   COUNTD([Order ID])
   ```

e. Add [Measure Names] to Columns.

f. Add [Measure Values] to Text. Place SUM(Sales), SUM(Profit), Profit Ratio, and Total Orders on the [Measure Values] Marks card.

g. Remove row banding and all lines except for the row dividers.

h. Format the customer names to match the font style of the text in the table. Figure 11-34 shows the table at this stage.

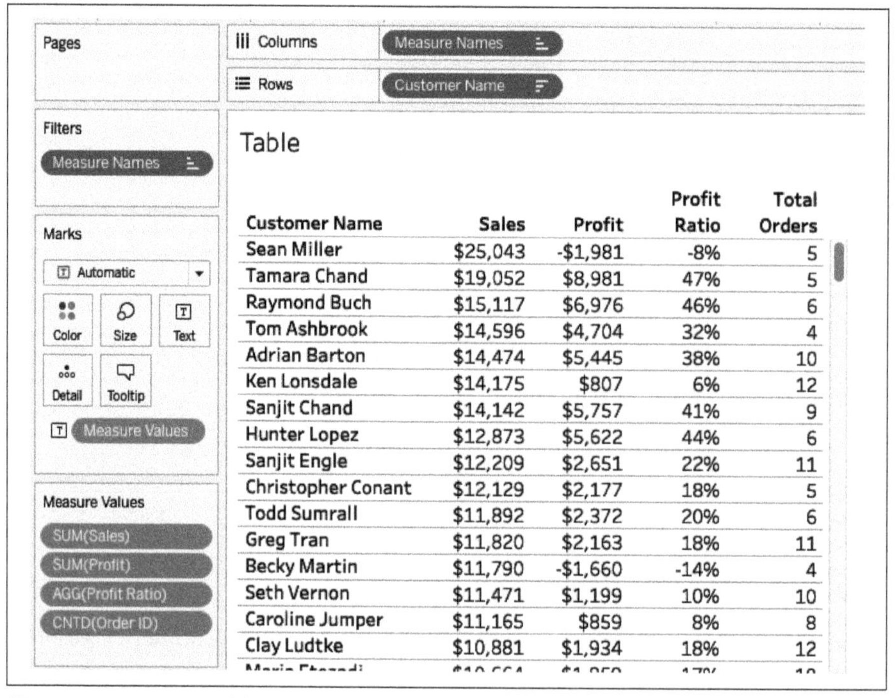

Figure 11-34. Your table prior to adding pagination

3. Create a series of parameters that will be used to drive changes on your table:

   a. Create an integer parameter called **`[rows to show]`** that will indicate the number of rows to show (Figure 11-35). For this example, you'll specify 15, but you can change it later.

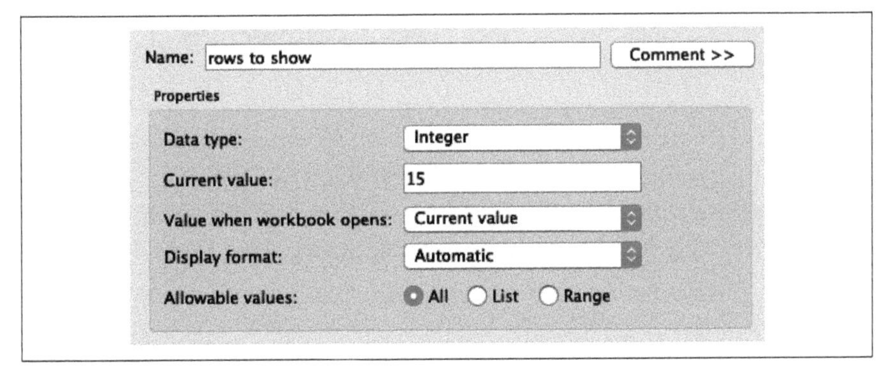

Figure 11-35. The setup of the rows to show parameter

b. While you are at it, create a second parameter called **`[page number]`** that will indicate the page number. Set the integer in the parameter to 1 and be sure to allow all values, as shown in Figure 11-36.

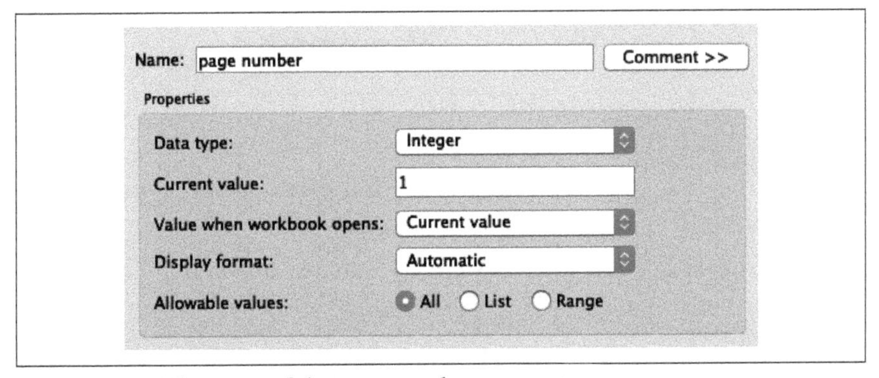

Figure 11-36. The setup of the page number parameter

c. Create a calculation called **`[Customer Name | Index]`**. This function just applies the `INDEX()` function to sort and rank your customers:

```
// Customer Name | Index
INDEX()
```

This calculation will come in handy for building your other calculations.

Instead of hardcoding values in calculations (like the total rows to show), use a parameter. This will allow you to quickly change the values without having to open the calculation.

d. Your next calculation is called **[Page | Customer Name | Index]**. This will calculate which rows to show on which page:

```
// Page | Customer Name | Index
((([Customer Name | Index] - 1) - (([Customer Name | Index] - 1)
  % [rows to show]))/[rows to show]) +1
```

e. Now you simply need to create a calculation that shows which page to show. Create a Boolean based on the page parameter called **[Page Number | TF]**:

```
// Page Number | TF
[Page | Customer Name | Index] = [page number]
```

f. Add [Page Number | TF] to Filters on the table sheet and select True. You will now see 15 rows from page 1. This is based on the rows to show and page number parameters. This leaves you with the visualization shown in Figure 11-37.

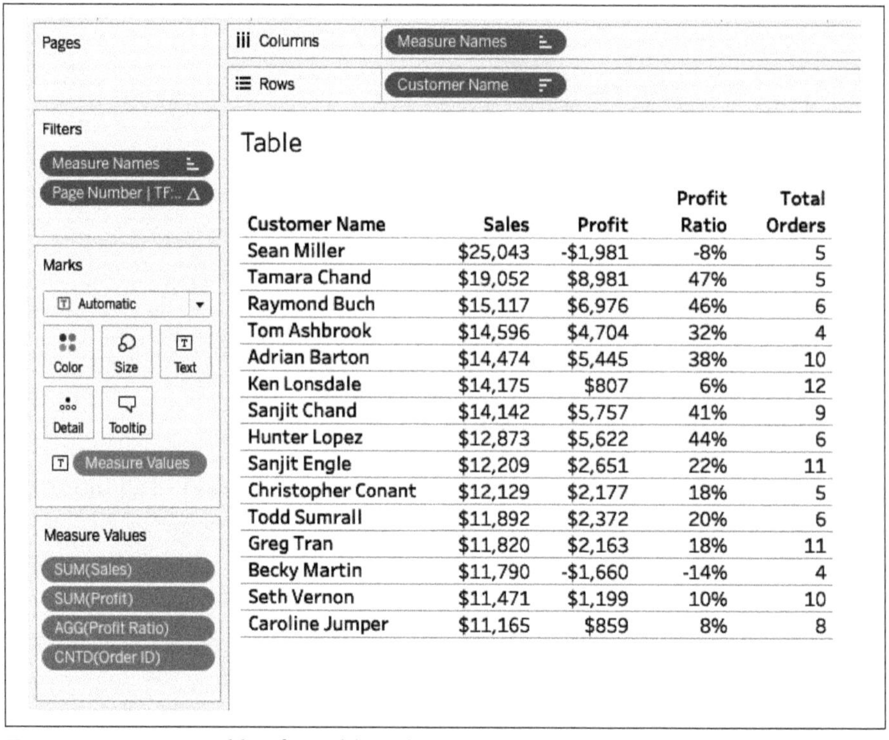

Customer Name	Sales	Profit	Profit Ratio	Total Orders
Sean Miller	$25,043	-$1,981	-8%	5
Tamara Chand	$19,052	$8,981	47%	5
Raymond Buch	$15,117	$6,976	46%	6
Tom Ashbrook	$14,596	$4,704	32%	4
Adrian Barton	$14,474	$5,445	38%	10
Ken Lonsdale	$14,175	$807	6%	12
Sanjit Chand	$14,142	$5,757	41%	9
Hunter Lopez	$12,873	$5,622	44%	6
Sanjit Engle	$12,209	$2,651	22%	11
Christopher Conant	$12,129	$2,177	18%	5
Todd Sumrall	$11,892	$2,372	20%	6
Greg Tran	$11,820	$2,163	18%	11
Becky Martin	$11,790	-$1,660	-14%	4
Seth Vernon	$11,471	$1,199	10%	10
Caroline Jumper	$11,165	$859	8%	8

Figure 11-37. Your table after adding the page number filter

Now that you have your filter working on your table, the only thing you need to do is update the parameter by using the arrows or page display!

4. Start by creating a new sheet called **`Left Arrow`**:

   a. Create an ad hoc calculation and add ◀ to Text. Align the text to the center-middle. Set the font to Arial and change the size to 12.

   b. Create a calculation called **`[Page | -]`**. This calculation will subtract one value of the page number unless you are at the lowest number:

      ```
      // Page | -
      IF [page number] = 1
      THEN 1
      ELSE [page number] - 1
      END
      ```

      Add [Page | -] to Detail on the Marks card.

   c. From step 1 of "Strategy: Automatically Deselect Marks" on page 437, add [TRUE] and [FALSE] calculations to Detail.

   d. Turn off the tooltips for the Left Arrow worksheet. Set any lines or row dividers to None.

5. Build the right arrow by creating a new sheet called **`Right Arrow`**:

   a. Create an ad hoc calculation and add ▶ to Text. Align the text to the center-middle. Set the font to Arial and change the size to 12.

   b. Create a calculation called **`[Page | +]`**. This calculation will add 1 to the value of the page number unless you are at the maximum page:

      ```
      // Page | +
      IF [page number] = FLOOR({COUNTD([Customer Name])}/[rows to show]) +1
      THEN FLOOR({COUNTD([Customer Name])}/[rows to show]) + 1
      ELSE [page number] + 1
      END
      ```

      Add [Page | +] to Detail on the Marks card.

   c. Following step 1 from "Strategy: Automatically Deselect Marks" on page 437, strategy, add [TRUE] and [FALSE] calculations to Detail.

   d. Turn off the tooltips for the Right Arrow worksheet. Set any lines or row dividers to None. The result is Figure 11-38.

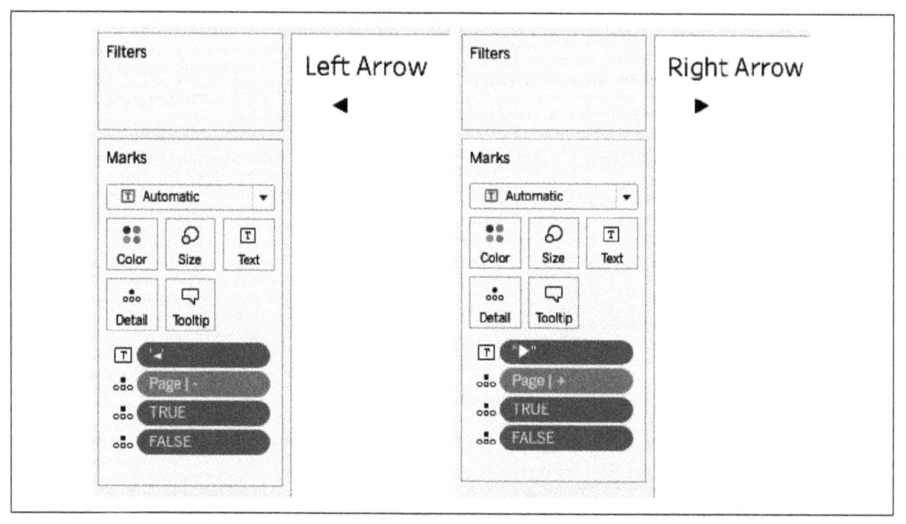

Figure 11-38. Your Left Arrow and Right Arrow sheets after completing steps 4 and 5

6. The page navigator is the most complicated component on the dashboard. You need to display the selected page, the surrounding pages, and make it interactive. Also, the selected page might not always be in the middle of the list of five, so we need to make sure we can highlight the correct page at the correct time:

   a. Start a new sheet called **`Display`**. Add [Customer Name] to Detail on the Marks card. Sort in descending order by sum of sales.

   b. Add [Customer Name | Index] to Columns. Change the measure to Discrete. Edit the table calculation, select Specific Dimensions, and select Customer Name. Use a custom sort and select descending for sum of [Sales].

      This means our pagination will be based on the sum of sales! Each index value is a [Customer Name] and is sorted on the sum of [Sales].

   c. Create a calculation called **`[Label Page Number Active]`**:

```
// Label Page Number Active
IF [Page Number | TF]
THEN [page number]
END
```

      Add this calculation to Label. Make sure the calculation is discrete. This will display only the active page. We will eventually format this text. Edit the table calculations for the nested table calculations; choose Specific Dimensions and select Customer Name. Be sure to do this for both calculations.

d. Create a calculation called **`[Label Page Number Inactive]`**:

```
// Label Page Number Inactive
IF NOT [Page Number | TF]
THEN [Page | Customer Name | Index]
END
```

Add this calculation to Text. As in step 6c, edit the table calculations for the nested table calculations; choose Specific Dimensions and select Customer Name. Be sure to do this for both calculations.

e. Click to edit the text. Place the two dimensions on the same line. For [Label Page Number Not], select a lighter gray, Tableau Book, Size 12. For [Label Page Number], select black, Tableau Semibold, Size 12. The text should look similar to Figure 11-39.

Format [Label Page Number Active] and [Label Page Number Inactive] to display as integers.

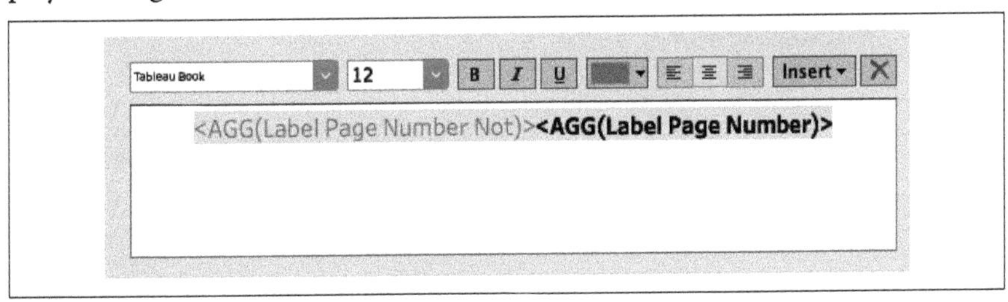

Figure 11-39. The text for the page numbers

That leaves you with the results shown in Figure 11-40.

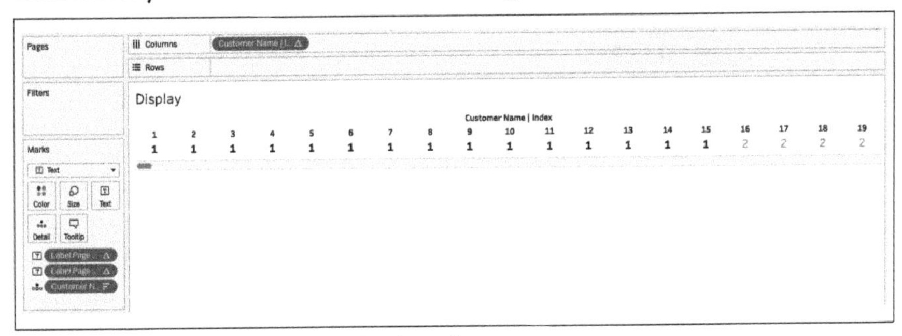

Figure 11-40. Your pagination display after step 6e

f. From step 1 of "Strategy: Automatically Deselect Marks" on page 437, add [TRUE] and [FALSE] calculations to Detail.

g. [Customer Name | Index] shows all customers, but we don't really need to show all the customers. We need to show just one customer from each page.

The best thing we can do is show the first customer from each page—since the last page might consist of only that single customer.

Write the following formula called **[Page n | 0]**:

```
// Page n | 0
([Customer Name | Index] % [rows to show]) - 1
```

Convert [Page n | 0] to a discrete value and place it on Filters. Edit the table calculation of the filter. Select Specific Dimensions and Customer Name. Edit the filter to select 0 from the filter options. This results in the visualization in Figure 11-41.

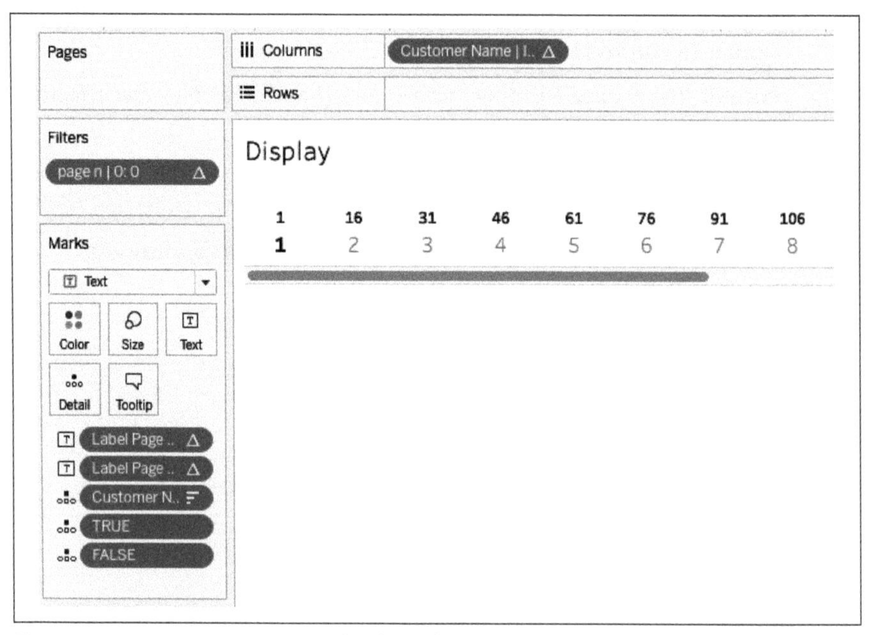

Figure 11-41. Your pagination display after adding a filter to show only the first member of each page

h. This last bit is the hardest part. You need to know which pages to show. This is about setting a range: a lower bound and an upper bound. You will do that with two calculations:

Create the **[threshold | bottom]** calculation:

```
// threshold | bottom
IF [page number] < 3
THEN 1
ELSEIF [page number] > FLOOR({COUNTD([Customer Name])}/[rows to show]) - 1
THEN FLOOR({COUNTD([Customer Name])}/[rows to show]) - 3
ELSE [page number] - 2
END
```

And create a **[threshold | top]** calculation:

```
// threshold | top
IF [page number] > FLOOR({COUNTD([Customer Name])}/[rows to show]) - 1
THEN FLOOR({COUNTD([Customer Name])}/[rows to show]) + 1
ELSEIF [page number] < 3
THEN 5
ELSE [page number] + 2
END
```

To make easy for yourself, build a single calculation that will define the upper and lower bounds, and call that calculation **[threshold | tf]**:

```
// threshold | tf
[Page | Customer Name | Index] >= MIN([threshold | bottom])
AND
[Page | Customer Name | Index] <= MIN([threshold | top])
```

Add [threshold | tf] to the Filters shelf. Edit the table calculation, choose Specific Dimensions, and select Customer Name. Then edit the filter and select True.

i. Now you just need to format by removing row dividers and hiding headers. Align text to the center-middle. Turn off your tooltips. Your visualization should resemble Figure 11-42.

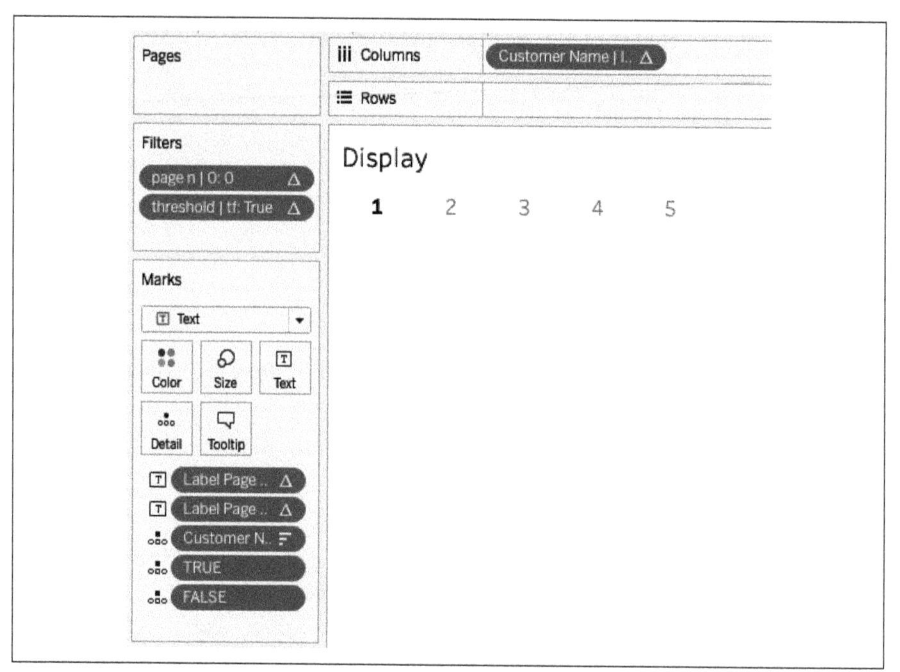

Figure 11-42. Our pagination display after completing step 6

7. Build the dashboard:
    a. Create a new dashboard and set the size to 400 wide and 500 tall.
    b. Add a vertical container to the dashboard.
    c. Add the table to the container and fit the sheet to the entire view.
    d. Add a horizontal container below the table in the vertical container.
    e. Add the left arrow, display, and right arrow into the container. Set all the objects to fit the entire view. Hide the titles of the sheets.
    f. Remove all the inner and outer padding associated with each of the three sheets.
    g. Set a border around the container.
    h. Fix the height of the container to 40.
    i. Fix the width of both the left and right arrows to 40.
    j. Add a background color to each of the arrows. Your dashboard should look similar to Figure 11-43.

Customer Name	Sales	Profit	Profit Ratio	Total Orders
Sean Miller	$25,043	-$1,981	-8%	5
Tamara Chand	$19,052	$8,981	47%	5
Raymond Buch	$15,117	$6,976	46%	6
Tom Ashbrook	$14,596	$4,704	32%	4
Adrian Barton	$14,474	$5,445	38%	10
Ken Lonsdale	$14,175	$807	6%	12
Sanjit Chand	$14,142	$5,757	41%	9
Hunter Lopez	$12,873	$5,622	44%	6
Sanjit Engle	$12,209	$2,651	22%	11
Christopher Cona..	$12,129	$2,177	18%	5
Todd Sumrall	$11,892	$2,372	20%	6
Greg Tran	$11,820	$2,163	18%	11
Becky Martin	$11,790	-$1,660	-14%	4
Seth Vernon	$11,471	$1,199	10%	10
Caroline Jumper	$11,165	$859	8%	8

◀ **1** 2 3 4 5 ▶

Figure 11-43. Your pagination dashboard after step 7

8. Add dashboard actions to the left arrow, display, and right arrow:

   a. From your dashboard, choose Dashboard → Actions → Add Actions → Change Parameter. Select the Left Arrow sheet from the Pagination dashboard, and change the page number parameter by using the [Page | -] calculation with no aggregation on select, as shown in Figure 11-44.

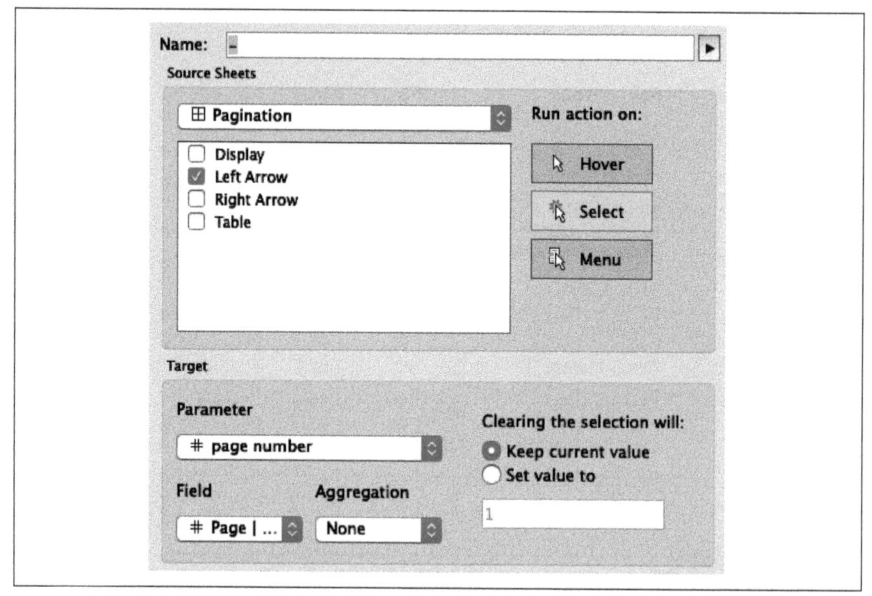

Figure 11-44. The left arrow parameter action for table pagination

   b. Select the Right Arrow sheet from the Pagination dashboard, and change the page number parameter by using the [Page | +] calculation with no aggregation on select, as shown in Figure 11-45.

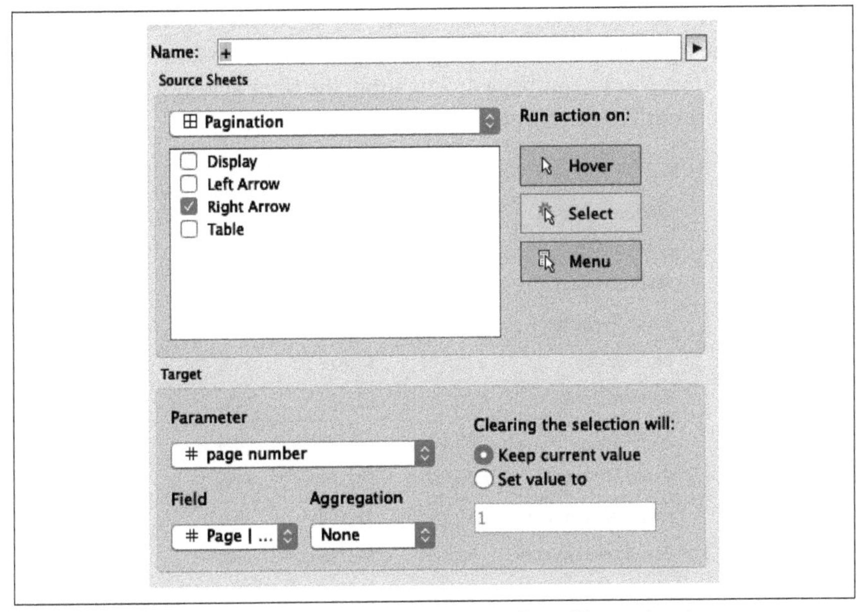

Figure 11-45. The right arrow parameter action for table pagination

c. The parameter action will be for the display. Update the page number parameter by using [Label Page Number Inactive] with no aggregation on select, as shown in Figure 11-46.

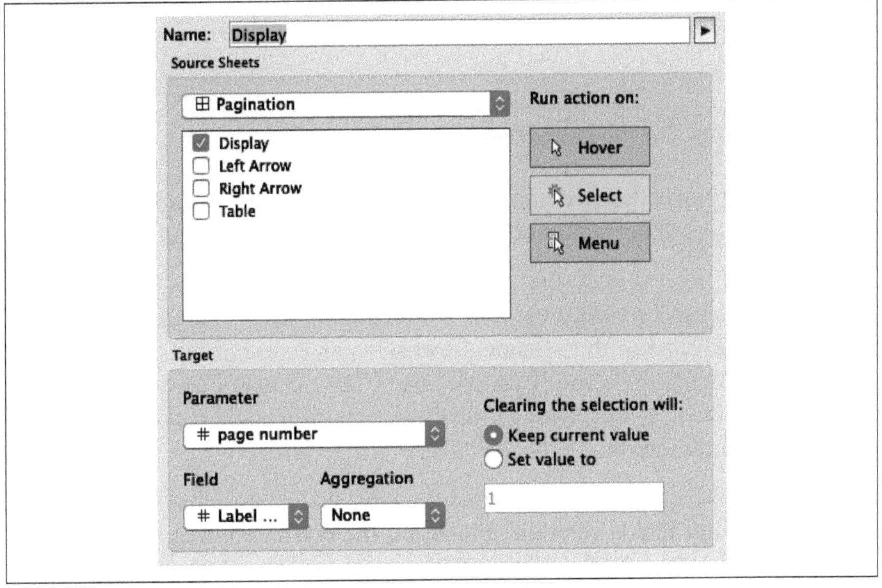

Figure 11-46. The page display action for table pagination

9. Make three separate filter actions, one for each sheet with an action: the left arrow, the right arrow, and the page display. Follow step 2 in "Strategy: Automatically Deselect Marks" on page 437 to apply the filter action for each of the sheets. The final result is shown in Figure 11-47.

Customer Name	Sales	Profit	Profit Ratio	Total Orders
Clay Ludtke	$10,881	$1,934	18%	12
Maria Etezadi	$10,664	$1,859	17%	10
Karen Ferguson	$10,604	$1,660	16%	7
Bill Shonely	$10,502	$2,616	25%	5
Edward Hooks	$10,311	$1,394	14%	12
John Lee	$9,800	$229	2%	11
Grant Thornton	$9,351	-$4,109	-44%	3
Helen Wasserman	$9,300	$2,164	23%	8
Tom Boeckenhauer	$9,134	$2,798	31%	7
Peter Fuller	$9,063	-$614	-7%	4
Christopher Mart..	$8,954	$3,900	44%	4
Justin Deggeller	$8,828	$1,620	18%	8
Joe Elijah	$8,698	$1,262	15%	10
Laura Armstrong	$8,673	$2,059	24%	11
Pete Kriz	$8,647	$2,038	24%	12

◀ 1 **2** 3 4 5 ▶

Figure 11-47. Page 2 of the table pagination strategy

In this strategy, you learned more about connecting sheets and parameters with parameter actions. It helped unlock an enhanced UI that you wouldn't have with base Tableau.

One brilliant extension of this use case is applying the same calculations you just created to a second sheet with the same dimensions on rows but different visualizations being shown. Using the same [Page Number | TF] calculation, you can synchronize values being shown. In fact, with a little formatting, you could make it look like a single, cohesive visualization.

For example, in Figure 11-48, we added the Sales by Category bar chart. By applying the same sort and filters, we were able to put the two sheets in a single container to show what looks like a single cohesive table.

This method has some drawbacks: you have to leave the sort fixed as is. And that's fine as long as your users know. This means you'll want to turn off our sort controls. You can do this sheet-by-sheet from the top menu by choosing Worksheet and then unchecking Show Sort Controls.

Customer Name	Sales	Profit	Profit Ratio	Total Orders	Sales by Category		
Sean Miller	$25,043	-$1,981	-8%	5	$680	$882	$23,482
Tamara Chand	$19,052	$8,981	47%	5		$1,054	$17,998
Raymond Buch	$15,117	$6,976	46%	6	$76	$776	$14,265
Tom Ashbrook	$14,596	$4,704	32%	4	$125	$761	$13,710
Adrian Barton	$14,474	$5,445	38%	10	$1,280	$11,489	$1,704
Ken Lonsdale	$14,175	$807	6%	12	$1,263	$9,655	$3,257
Sanjit Chand	$14,142	$5,757	41%	9	$2,031	$12,081	$30
Hunter Lopez	$12,873	$5,622	44%	6	$1,141	$91	$11,641
Sanjit Engle	$12,209	$2,651	22%	11	$634	$2,245	$9,330
Christopher Conant	$12,129	$2,177	18%	5	$701	$106	$11,322
Todd Sumrall	$11,892	$2,372	20%	6	$2,225	$863	$8,804
Greg Tran	$11,820	$2,163	18%	11	$6,219	$3,624	$1,977
Becky Martin	$11,790	-$1,660	-14%	4	$2,070	$1,338	$8,382
Seth Vernon	$11,471	$1,199	10%	10	$8,332	$202	$2,937
Caroline Jumper	$11,165	$859	8%	8	$6,267	$658	$4,240

◄ 1 2 3 4 5 ►

Figure 11-48. Two well-formatted sheets can create what looks like a single visualization when you use pagination

This is the power of pagination. When you can synchronize several sheets, you can begin to layer in multiple visual components that are shaped like a table, but really are complex dashboards driving concise and informative insights.

Creating Dynamic, Single-Click Drill-Throughs: Superstore Case Study

For much of this chapter, we have used parameters and parameter actions to drive change on a dashboard. Parameters are just one of several dashboard actions that are available for you to build interactive, dynamic visual systems. For this chapter's final strategy, you will build dynamic drill-throughs that allow your audiences to filter and navigate simultaneously. For instance, in your table in Figure 11-47, your audience might have questions about the specific products ordered by Clay Ludtke.

You've designed some amazing data products in this chapter, but now your audience wants to be able to drill into order details for individual customers. In this strategy, we'll show you a technique we use when working with tables to drill into the details your audiences want.

The typical analyst might say: go to a different dashboard and then select Clay Ludtke from a separate filter. But you can drive this action from a single click of your existing dashboard so your audience maintains a seamless experience. Figure 11-49 shows an example.

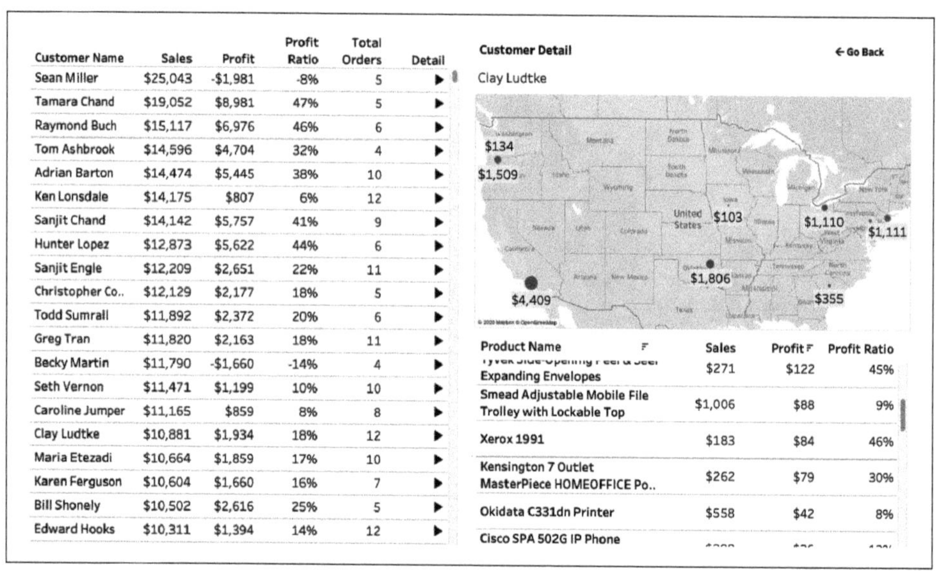

Figure 11-49. The top-level visualization (left) and the product-level detail after clicking through to a customer with the name of Clay Ludtke (right)

Strategy: Build Drill-Through Interactivity

In this strategy, you will build two dashboards: one showing customer-level information that serves as the top level of our dashboard, and a second one showing details about customer orders. Both dashboards will have no visible filters, but we will use a series of actions that will allow your audience to traverse to and automatically filter the product-level dashboard and then return to the main dashboard with just two clicks:

1. Connect to the Sample – Superstore dataset.
2. Build the top-level visualization by creating a new sheet called **`Table`**.
   a. Following step 2 in "Strategy: Add Table Pagination" on page 457, create [TRUE] and [FALSE] calculations.
   b. Add the arrows by creating a new calculation called **`[Detail]`**:

      ```
      // Detail
      1
      ```

   c. Drag [Detail] into the [Measure Values] Marks card. Right-click SUM(Detail) and select Format. Under the numbers formatting, choose Custom and then type "►". This will turn all the values to the right arrow.
   d. Following step 1 from "Strategy: Automatically Select Marks" on page 444, create **`[TRUE]`** and **`[FALSE]`** calculations. Add these calculations to Detail on the Marks cards.
3. Build your top-level dashboard by creating a 700 × 800 dashboard called **`Top Level`**. Add Table to your dashboard. Your dashboard should look like the image on the left in Figure 11-49.
4. Build your detail-level visualizations. Create two sheets, one called **`Map`** and the other called **`Product`**:
   a. For the Map sheet, shown in Figure 11-50, do the following:
      a. Add [City] and [State] to Detail. This should generate latitude and longitude and create a map.
      b. Add [Sales] to Size and Label on the Marks card.
      c. Add [Customer Name] to Detail on the Marks card.
      d. Edit the title of the Map sheet and replace it with the [Customer Name] dimension.

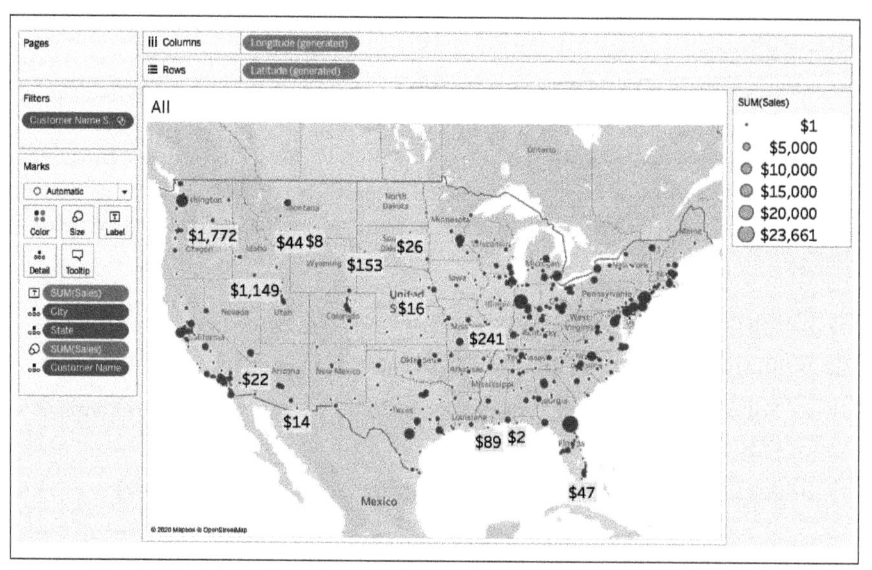

Figure 11-50. The Map sheet for our detailed view

b. For the Product sheet, shown in Figure 11-51, do the following:

  a. Add [Product Name] to Rows.

  b. Add [Measure Names] to Columns.

  c. Add [Measure Values] to Text.

  d. Filter [Measure Names] to display only SUM(Sales), SUM(Profit), and Profit Ratio.

  e. Edit the title of the Map sheet and replace it with the [Customer Name] dimension.

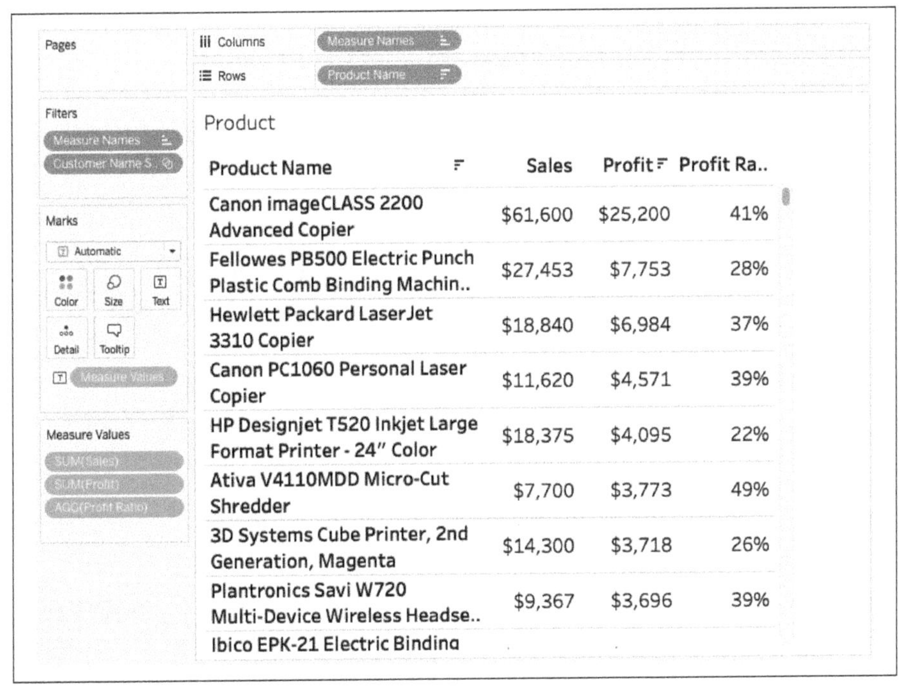

Figure 11-51. The Product sheet for our detailed view

5. Build your top-level dashboard:

   a. Create a 700 × 800 dashboard called **`Detailed Level`**.

   b. Add a vertical container.

   c. Add your Map and Product sheets into the vertical container.

   d. Set the height of the map to 400.

   e. Add a horizontal container above the map.

   f. In the horizontal container, add a text box and type **`Customer Detail`**. Change the font to Tableau Bold, Size 15.

   g. In the horizontal container, to the right of the customer detail text, add a Navigation object.

   h. Edit the Navigation button. Navigate to the top-level dashboard you created in step 3. Edit the title to **`Go Back`**. Change the text color to black and the background to None. This will give your audience the interactivity back to the top level.

   i. Edit the height of the horizontal container and set it to 50. At this point your dashboard should resemble Figure 11-52.

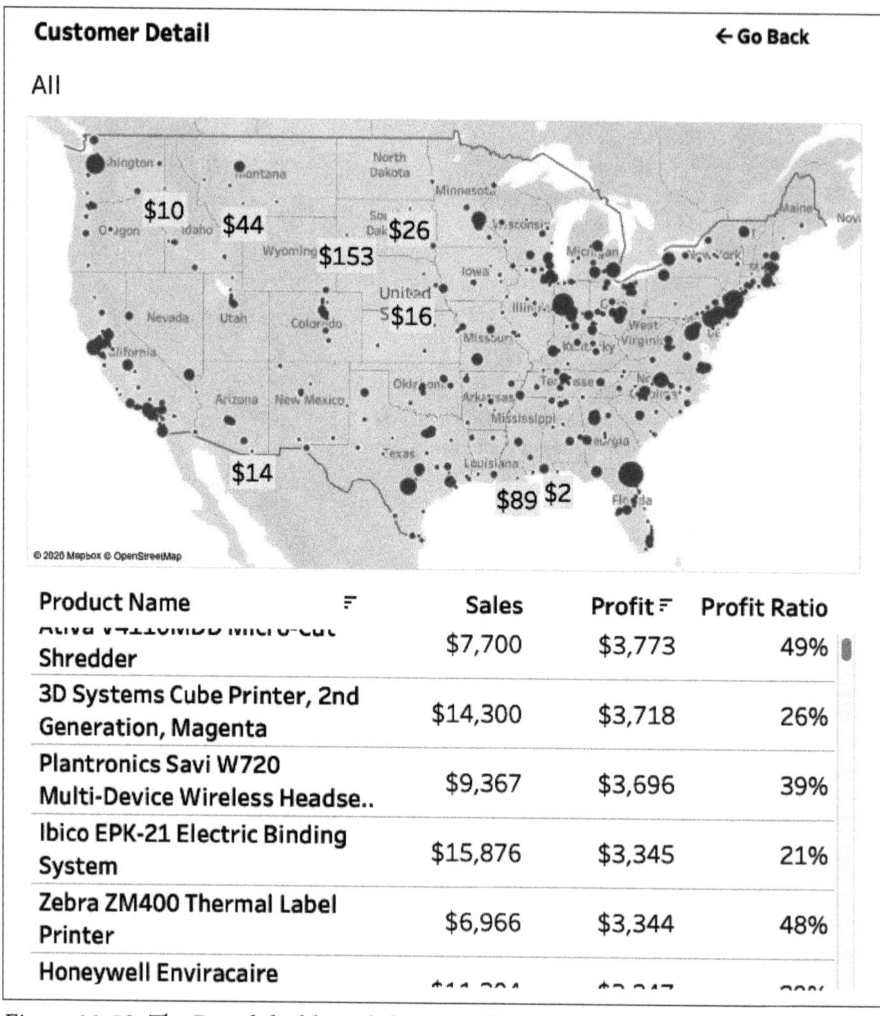

Customer Detail
← Go Back
All
North Dakota
Minnesota
$10
$44
$26
Wyoming
$153
Iowa
Nevada
Utah
United
$16
Missouri
West Virginia
Arizona
New Mexico
Arkansas
Mississippi
Texas
Louisiana
$14
$89
$2
© 2020 Mapbox © OpenStreetMap
Product Name
Sales
Profit
Profit Ratio
Shredder
$7,700
$3,773
49%
3D Systems Cube Printer, 2nd Generation, Magenta
$14,300
$3,718
26%
Plantronics Savi W720 Multi-Device Wireless Headse..
$9,367
$3,696
39%
Ibico EPK-21 Electric Binding System
$15,876
$3,345
21%
Zebra ZM400 Thermal Label Printer
$6,966
$3,344
48%
Honeywell Enviracaire

Figure 11-52. The Detail dashboard showing all customers

6. Right now your two visualizations on the Detail dashboard do not filter by customer. You need to add this functionality by using a set parameter:

   a. Go to the Map sheet. On the Data tab, right-click Customer Name and choose Create → Set. Name the set **`Customer Name Set`** and select the first customer, Aaron Bergman. Then click OK.

   b. Go to the Map worksheet. Click and drag [Customer Name] Set to the Filters shelf. Repeat with the Product worksheet.

7. Add interactivity to your top-level dashboard by adding three actions to the sheet: a set action to filter to a specific customer, a go-to action to drill into the detail sheet, and a filter action to automatically deselect a mark, which makes for a clean user experience for your audience. On the top-level dashboard, from the top menu, choose Dashboard → Actions.

   a. From the Actions menu, choose Add Action → Change Set Values. Following Figure 11-53, from the top-level dashboard, select the Table sheet. On select, update the Customer Name set. When you select, have it assign values to the set. And when it clears, add all values to the set. Click OK.

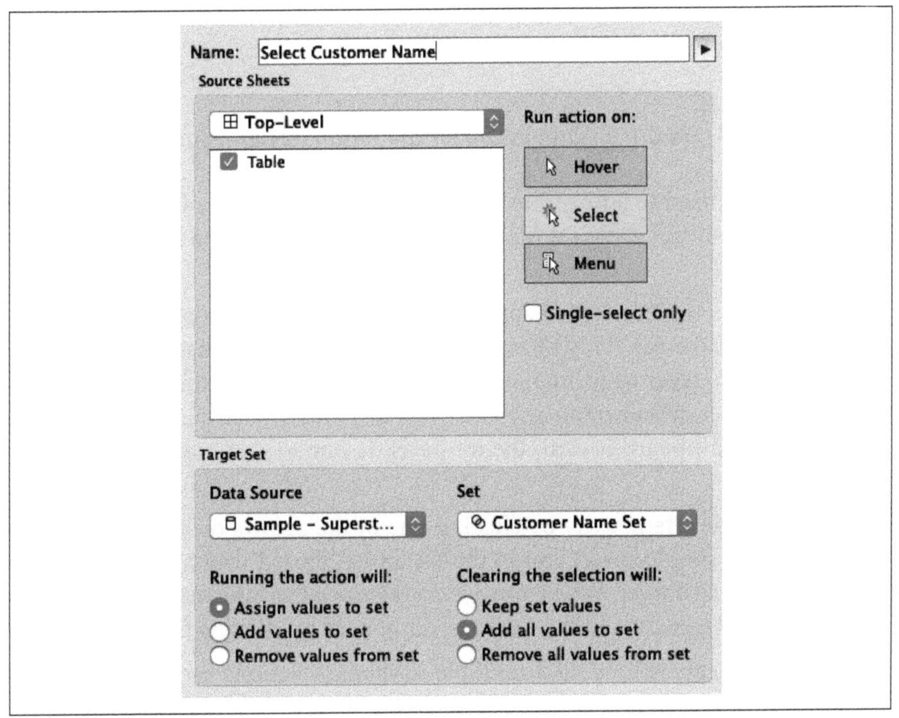

Figure 11-53. The set action for Customer Name on the top-level dashboard

   a. From the Actions menu, select Add Action → Go to Sheet. From the top-level dashboard, select the Table sheet. Then select the Detail dashboard as the target sheet (Figure 11-54).

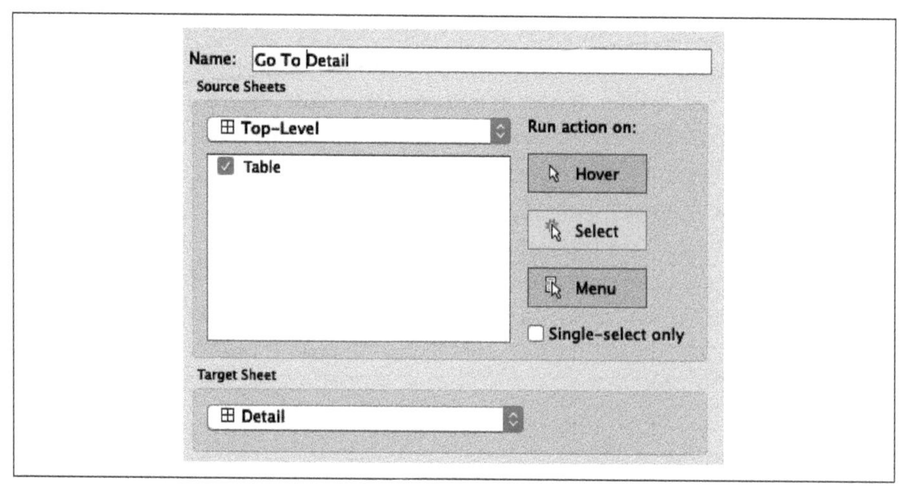

Figure 11-54. The Go to Sheet action for the Top-Level dashboard

Finally, you need to add a filter action on the Table sheet. Follow step 2 in "Strategy: Automatically Deselect Marks" on page 437 to apply the filter action for each of the sheets.

The result is three actions happening simultaneously. The set action selects the customer in the set. This filters the sheets on the Detail dashboard. The Go to Sheet action takes your audience—in a single click—to the Detail dashboard. And the filter action resets the marks so that when your audience returns from the Detail dashboard back to the top-level dashboard, no marks will be highlighted.

Test out your interactivity by clicking one of the arrows on your top-level dashboard. This should direct you to the Detail dashboard with specific details about a particular customer. When you are done inspecting information about that customer, you can return to the top-level dashboard by clicking Go Back.

In this strategy, you applied multiple dashboard actions on the same sheet. By doing this, you were able to build an intuitive data product that your audiences can easily navigate.

Conclusion

In this chapter, you used parameters in several ways. In the first strategy, you used parameters and filters to create an action known as a sheet swap.

Whether presenting a very subtle trick that connects "buttons" to a supplementary data source that drives parameter selection and sheet swapping, or showing how to use a parameter as a data source and store values, this chapter showcased some of Tableau's best features to build.

This chapter, perhaps more than any other, brought together several of Tableau's key features. These features required you to use multiple sheets and format them in a way that makes them look like a single component. And while they look simple, you learned that building these interactive systems takes time. Despite that time, they offer your audiences a streamlined experience working with your dashboards. And that's important; as developers, it is our job to put in the extra work so that our audiences can have a streamlined work experience. The more we are able to streamline, the faster they will be able to get to critical insights, engage with the data, and continue to use the product.

In the next chapter, we continue to build on interactive components by exploring dashboard design.

Further Reading

- Read "Parameter Actions: Using Parameters as a Data Source" (*https://oreil.ly/4yhsu*) by Jonathan Drummey.
- To learn how to create tables like the example shown in Figure 11-47, read the compilation by Luke Stanke (*https://oreil.ly/ynlWM*).
- If you want to create searchable parameters and filters, read Lindsey Poulter's post (*https://oreil.ly/DCp1X*).
- To learn about creating interactive legends, read the post by Luke Stanke (*https://oreil.ly/8oGtX*).
- For information about synchronized scrolling, read the post by Luke Stanke (*https://oreil.ly/4qb6B*).

CHAPTER 12

Building Dashboards and Data Products

For much of the book, we've focused on creating various chart types. In this chapter, we will change our focus from creating chart types to building and developing dashboards that are useful for our audiences. In this chapter, you will learn how to bring together worksheets and components for visual systems and learn how to format them to give your audience the best experience possible.

In This Chapter

Unlike in previous chapters, in which you've been asked to create dashboards from scratch, this time you will start with an existing Tableau workbook. In this chapter, you'll learn how to do the following:

- Ask your audience questions when gathering information for dashboard development
- Consider four dashboard types when developing a dashboard
- Create a curated dashboard so your audience feels like they are in charge
- Build unique desktop and mobile views on the same dashboard by using device designer
- Design for those with cognitive challenges like dyslexia
- Design for individuals with color impairments
- Ensure that your dashboard designs are consistent

Dashboard Design

Good *dashboard design* starts with understanding your audiences, the context in which they use the data, and the consequences that come with the decisions made. For novice developers, the typical dashboard creation process starts by creating a few sheets and then deciding to place the sheets on a dashboard. When the developer creates a dashboard, they add a new dashboard, drag the sheets onto the dashboard, do a little manual sizing, add a title, show some filters, and call it a wrap.

If you can't relate to this process, you've probably been developing data products for so long that you forgot what it was like when you started. So, what went wrong in the dashboard creation process with that novice developer?

Well, here's what happened. That novice developer did the following:

- Thought about the data from only their perspective so the information in each worksheet is intuitive only to the developer
- Used the default sizing for the dashboard of 1000 × 800 and did not consider whether this product is being used by an executive steering committee or salespeople who have access to only a mobile device
- Used the default spacing and formatting, which made it tougher for their audiences to understand the information provided
- Failed to consider the order in which information should be presented and how the filters should be applied

How do you avoid making these mistakes? It's simple: get to know your audience, and work with and develop with them in an iterative fashion. And don't forget to remind yourself and your audience that it's not just a dashboard—it's a data-centric product that is constantly evolving.

That's right, you have to think of your dashboard as a product. The failure of any developer is considering a dashboard "done." To us, there is only one type of done dashboard, and that's a dashboard that's done being used and can be scrapped. We don't like done dashboards because that means a whole lot of effort being thrown in the trash!

Know Your Audience

So how do you avoid your data products being lost forever? Again, get to know your audience. The way we do this is by asking questions. Here's a sample of the types of questions we ask our audiences:

People-focused

- How big (in terms of total people) is the audience for this product?
- Tell me about their role in the organization.
- What insights do they most frequently discuss?
- Where and when are they accessing this product?
- How frequently will users interact with this product?
- What type of device(s) will your audience use to interact with the product?
- What type of decisions will be made after working with this data?

Metrics-focused

- What key questions are you trying to answer?
- What metrics are your audience focused on?
- Do the metrics behave/look different when you start to compare them across different dimensions?

Data-focused

- At what level of granularity are the metrics most frequently viewed?
  - Current period versus prior full period
  - Day, week, month, year
- How deep do they go into the data?
  - Are they focused on top-level indicators?
  - Do they drill deep into row-level information?
- Where does decision making often stop with the data (and transition more into a data investigation)?

Security-focused

- Of the data presented in the data product, how much can any single audience member see?
  - All of the dashboard?
  - Certain parts?
  - Customized to an individual?

If you can get a full understanding of how your audience will interact with your product, it has significantly better odds of sustaining itself over a longer period of time.

Dashboard Types

When we create dashboards, we group the components required for each dashboard into four categories: static, curated, exploratory, and mobile. A single dashboard is not necessarily exclusive to a single category. One dashboard might be only static because it's used in PowerPoint slides for a steering committee. Another dashboard might rely on static monthly KPIs, but users have the ability to drill into the row-level data to explore and understand the makeup of the KPIs.

Static

Components of a *static visualization* require no additional filtering or slicing. These visuals are what-you-see-is-what-you-get. Although KPIs are most associated with static components, these components are not only KPIs. If you are reviewing information at a weekly or monthly cadence, and these reviews use the same data with no interactivity, then you have a static dashboard. Figures 12-1 and 12-2 show examples.

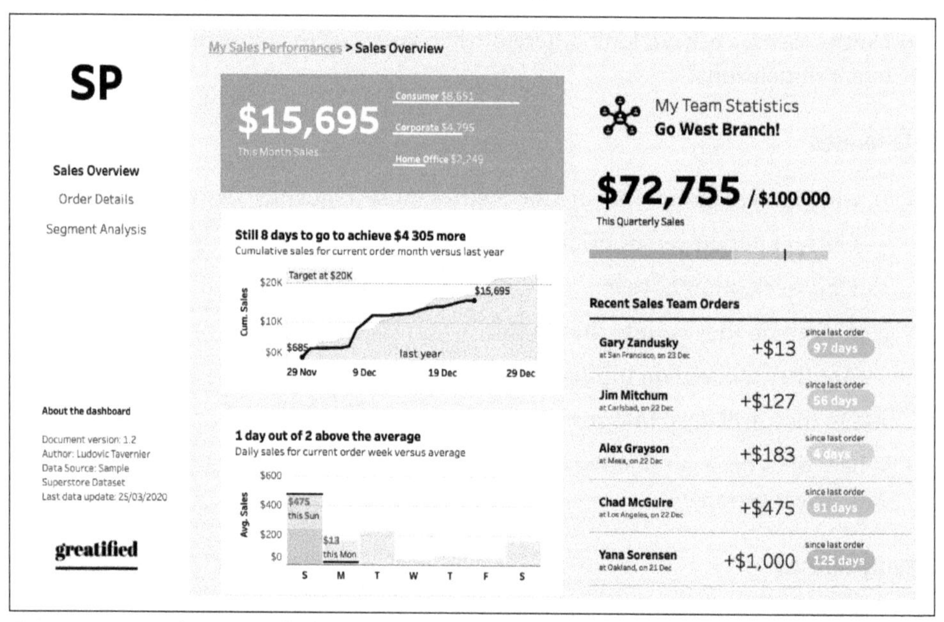

Figure 12-1. In this static dashboard, a great design highlights progress to target for a sales team. There are no filters, only a prescriptive view for the audience. (Dashboard courtesy of Ludovic Tavernier.)

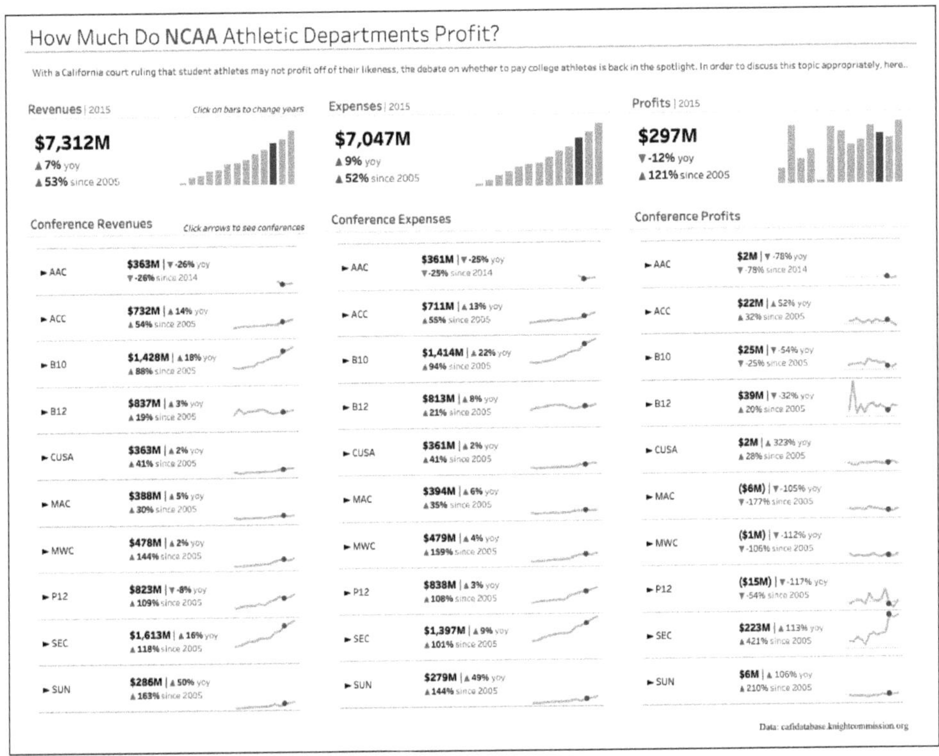

Figure 12-2. This static dashboard uses small multiples to highlight key metrics and the patterns in the metrics for the audience very quickly. (Dashboard courtesy of Spencer Baucke.)

In the world of data visualization, there is an aversion to using static dashboards because they are essentially reports. However, we've worked with lots of clients that utilize Tableau to create custom PowerPoint decks or reports. We should embrace the way our audiences interact with data. Our jobs are first to meet our audience where they are and then to slowly change their habits.

Creating static data products does come with benefits. When you create an exclusively static data product, your audience very likely already trusts the data being presented. Whether it's presenting financials to a C-suite of a Fortune 100 company or a social media review of top blog posts by a small consulting company, static comes with implicit trust.

Exploratory

Exploratory dashboards are essentially the polar opposite of a static dashboard. These types of data products typically include data reported at the most granular level and contain lots of filters. The purpose of an exploratory dashboard is less insights-driven and more search-driven. Figure 12-3 shows an example.

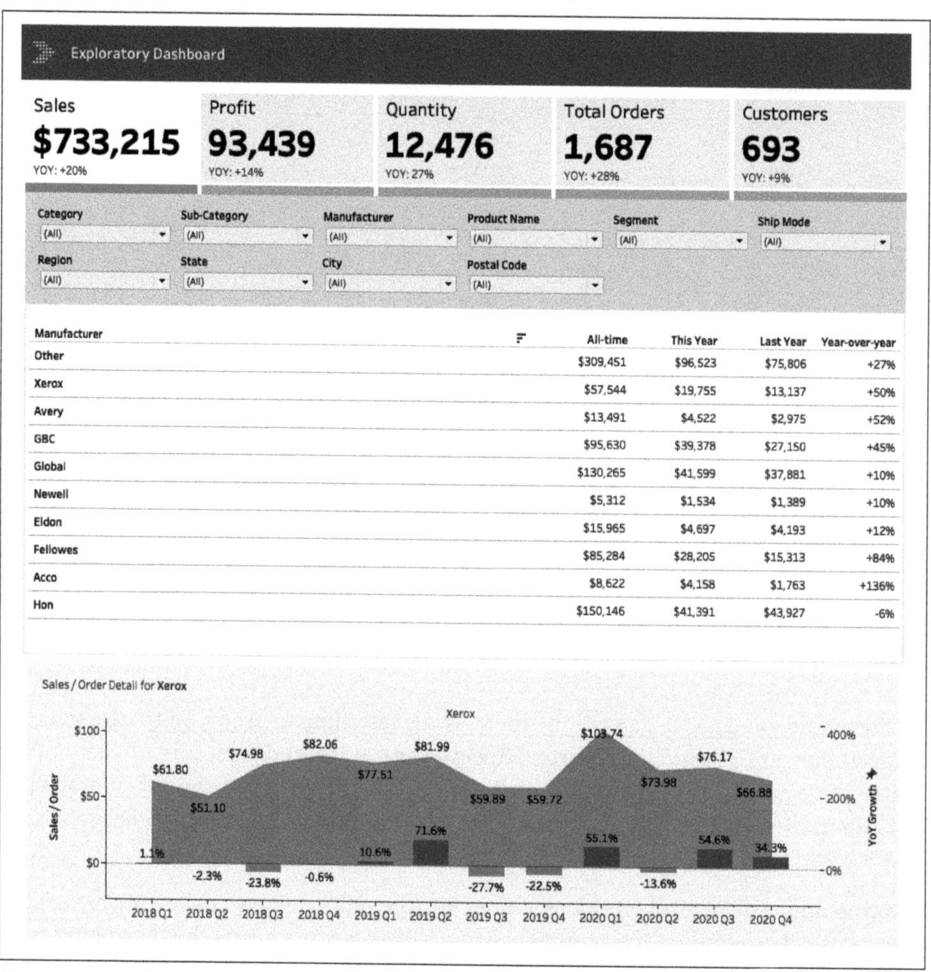

Figure 12-3. This exploratory dashboard uses multiple filters and drill-downs to showcase insights

In most instances, your audience is interested in transactional information like, "What products did Sean Miller order on October 12th?" Who would be interested in such a detailed question? A customer service representative who is about to call Sean Miller about the support ticket he just submitted!

The most common complaints about creating exploratory dashboards is that they have too many filters or are too slow (because they display a lot of information). You could tell your audience this is a trade-off of having granular access to the data, but there are solutions for speeding things up! Exploratory information is typically presented in a very large table. This means Tableau has a lot of marks to render! To speed this up, we suggest using table pagination to build drill-through interactivity that was covered in Chapter 11.

Exploratory dashboards are created to either provide transactional information or help audiences build trust with any static or curated data products. As you build trust, you will find your audience spends less time in the exploratory portions of your dashboard.

There is one more reason your users might use exploratory dashboards: you are not providing relevant insights for them to perform their jobs. This worst-case scenario can be avoided by getting to know your audience and building around their needs.

Curated

If we had to rank dashboard types, we would rank *curated dashboards* as the most important. Curated data products are interactive and allow any user—to an extent—to explore and learn from the information being presented.

Curated dashboards (Figure 12-4) represent the output that occurs when you put your user at the center of your designs. As the developer of this data product, you've learned how your users interact with the data and the types of decisions that are made with the information presented. Since you interact with the data more regularly than your audience, you know the stories that exist in the data and you share those stories in visual form. This also allows you to establish appropriate guardrails.

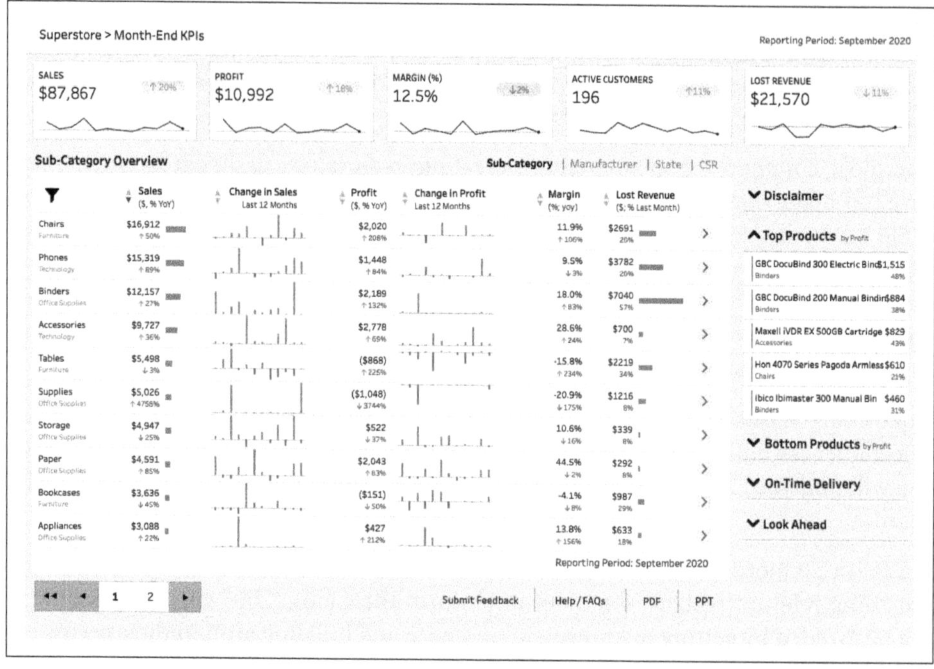

Figure 12-4. This curated dashboard provides insights for the audience to examine, but also directs the user throughout

A well-designed curated data product is no different from a well-curated museum exhibit. What do we mean by this? Think about your experience; when you walk through a museum exhibit, do you look at every component, read every card, and watch every video? No, you don't have time for that! Instead, you might walk into an exhibit and observe the overall theme of the room; you might look at seven to ten visual components, and dive deep by reading the details of two or three. Does everyone else follow the exact same path you took? No! Does everyone else leave an exhibit with a similar understanding? Yes!

This is what you need to do with your curated dashboards. You need to let the data speak, but it's your job to help guide your audience through the data, highlighting key stories. If someone in your audience wants to dive deep, let them, but provide guardrails in the form of allowing them to a specific level of granularity in the data or limiting the number of filters they can interact with.

If you fail to put up guardrails, individuals will *p-hack*. This term, popular in statistics, is just a simple way of describing the misuse of data to find patterns that might look significant while understating that the correlations might not actually exist in the data. You've probably seen some of these spurious correlations on TV. For instance, there's a significant relationship between per capita cheese consumption and the number of people who die by becoming tangled in their bedsheets. For real! Look at Figure 12-5.

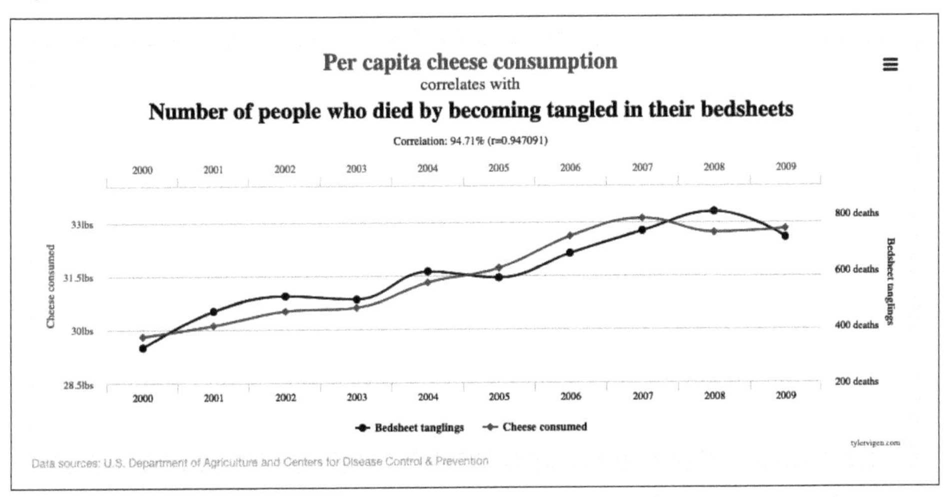

Figure 12-5. A spurious correlation example from Tyler Vigen's Spurious Correlations project (https://tylervigen.com/spurious-correlations), with data from the US Department of Agriculture and the Centers for Disease Control and Prevention

It's your job to make sure that everyone interacting with the data product is leaving with the same story. The last thing you want is someone p-hacking your data products and creating unnecessary discord about actions from the insights.

Mobile

Our last dashboard type is the *mobile dashboard*. As we mentioned from the outset, these dashboard types are not mutually exclusive. But when we think of mobile dashboards and data products, they take on fundamentally different designs from all other media types. Therefore, we've separated them out in this discussion to delve a little deeper in the following section.

Mobile Dashboard Design

One of the hardest things to get right is developing a mobile dashboard. Our first challenge is that our viewport is going to be extremely narrow. We can meet some of our audience's needs by building a long-form dashboard that is navigated with scrolling. If there is one thing we've learned about humans in the last 10 years, it's that they love to scroll and scan on their phones. Billion-dollar social media organizations have been established by sticking to simplistic user interfaces that allow their audiences to scan and scroll forever. You should embrace this design style with your mobile dashboards.

Of course, enabling our users to use mobile dashboards means you can't show three to five KPIs side by side. You can stack the KPIs or include a horizontal scroll. In your tables, this means showing just three to five columns of information (probably fewer).

While this might constrain the information you present, it actually opens up your designs to be more direct and forces your users to tell you exactly what they need to see. If you ask, "What do you really need to see?" you are likely to hear, "Well, I'm only really interested in this column," or "I don't even look at all of this."

Design Details

When it comes to the design of your mobile dashboard, you need to keep spacing of your dashboard components at the top of your mind. You want to place them so you can leverage the scan-and-scroll design model. If your visuals lack appropriate spacing, users will quickly disengage. In addition to spacing, we suggest using a simplified color palette. This means using four or five discrete colors and one gradient color palette.

Mobile dashboards also mean losing a lot of detail compared to your desktop versions. For instance, if you are working with line charts that have labels on each value, you probably aren't going to have that sort of space available. Another loss of context comes in the form of tooltips. While you can easily use a mouse to hover over a value, this isn't as easy on mobile devices. If you are using a scatter plot or line chart, getting the value you are looking for might require a few taps. That's not to say you can't use tooltips, but you certainly can't rely on your audience to find the information as easily as they would by hovering with a mouse. In fact, if you do use tooltips, it may be worth adding text, explicitly telling your audience that tooltips are available via tapping.

In this section, we'll give you three great examples for building mobile dashboards. But first, it's important to understand when Tableau will render an available mobile dashboard over a desktop or tablet view. If you have created a mobile view for your audience, the dashboard will render as mobile when your screen size is 500 pixels or smaller.

Setting up a mobile dashboard is easy. If you are working with the newest version of Tableau Desktop, Tableau automatically adds a mobile dashboard for you. The design of this mobile view is very hit-and-miss. If you are a novice developer who is just putting sheets on a view, it works perfectly fine. But if you've put effort into designing a desktop dashboard, it's very unlikely to match exactly what you are looking for. This can best be embodied by looking at how the simple, but formatted desktop dashboard in Figure 12-6 looks after it is automatically converted into a mobile dashboard in Figure 12-7.

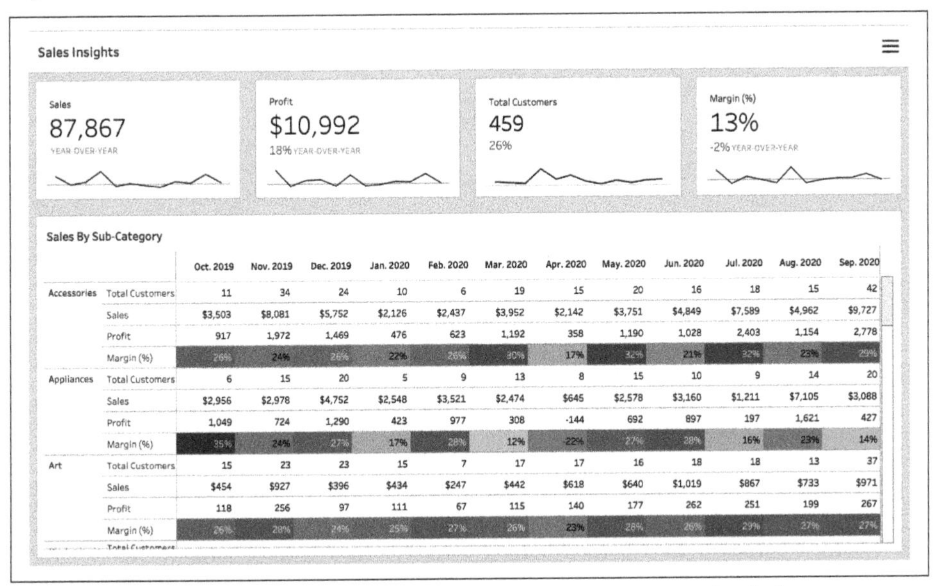

Figure 12-6. A formatted dashboard with four KPIs and a table on the desktop

Needless to say, if you are looking to maintain any formatting you've put into the desktop version, you'll need to put that same effort into reformatting the mobile version. One easy thing to do is to stack our KPIs rather than placing them side by side, as shown in Figure 12-8. If you really want to save space, you could use the button technique to scroll through the four KPIs.

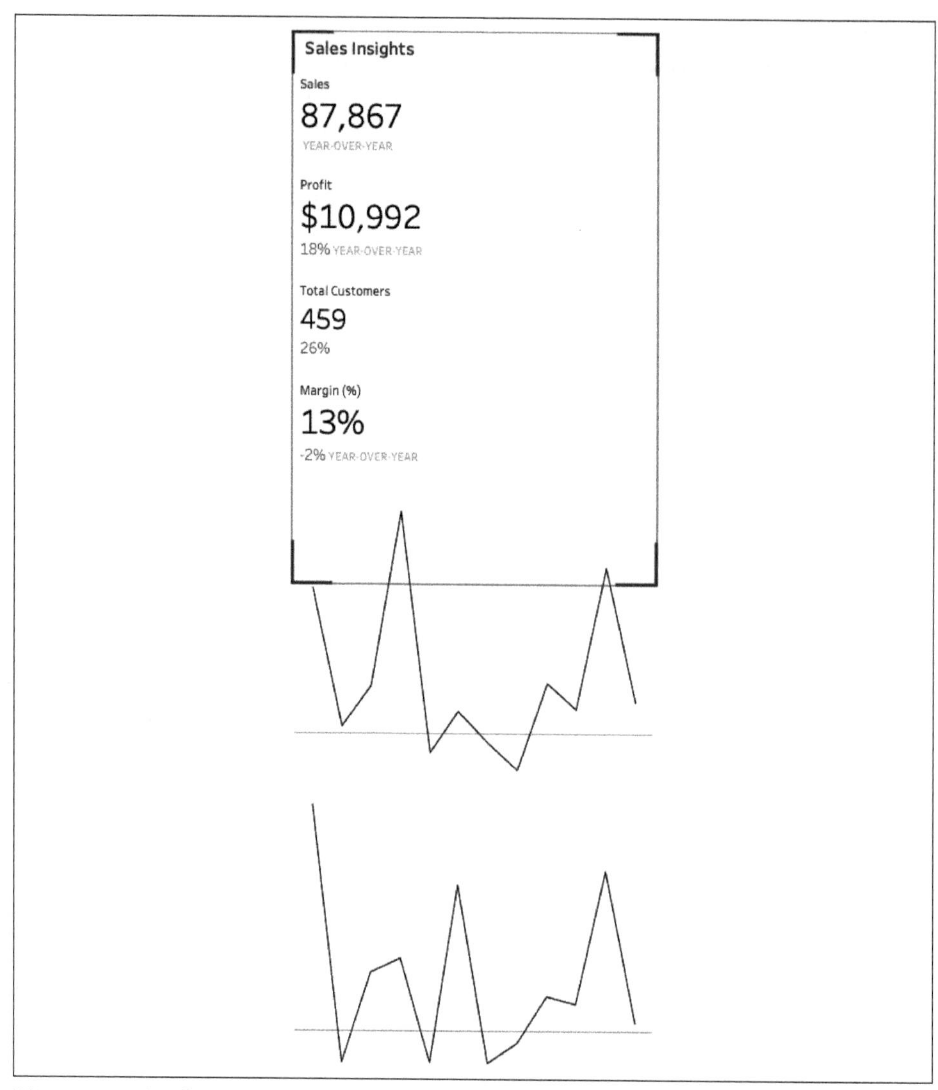

Figure 12-7. In this automatically generated mobile dashboard, the gray outlined box represents the active viewport

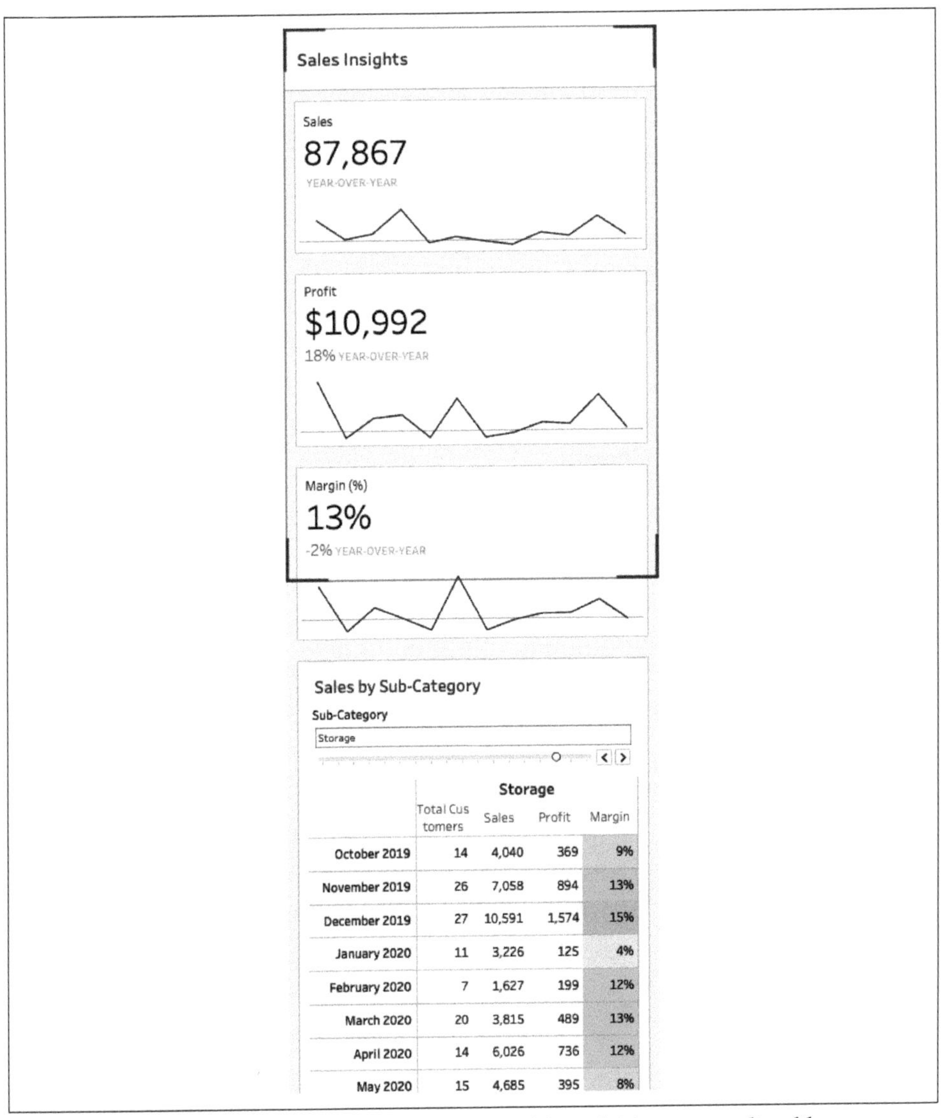

Figure 12-8. This is a fully customized mobile dashboard. The gray outlined box represents the active viewport.

If you work with tables—which is very likely—you aren't going to be able to fit your standard tables onto a mobile dashboard. Take a look back at the table in Figure 12-6. There's no way you're making that table effective for your mobile audience. In reality, you have room for three to five columns of information, depending on the complexity and width of those columns.

An easy solution is to swap your columns and rows and then place Filter on your new column values. If you are using the table in Figure 12-6, this means dates move to rows, the sub-categories move to columns, and you filter your data by individual sub-categories. You technically don't need the sub-category on columns if you are going to use it as a filter, but we (Ann and Luke) prefer this because it provides a label for the data being shown (in addition to the filter). Figure 12-9 shows the result.

While we haven't said this, you are creating two different tables and—depending on the device type—having one or the other display. This might seem straightforward, but one challenge is that all sheets on your dashboard must be on the default (desktop) dashboard view.

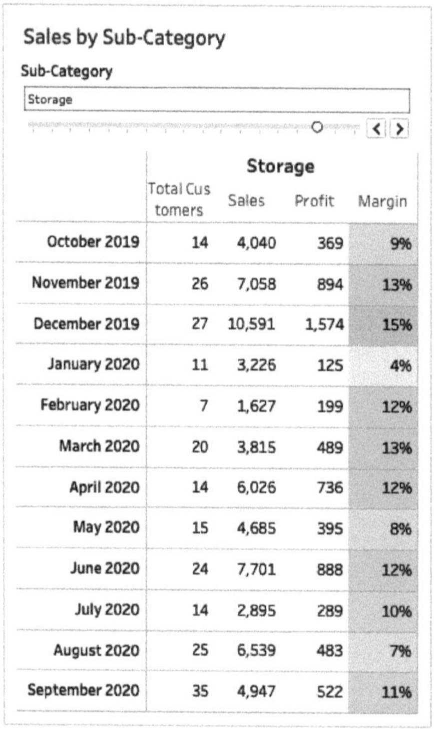

Sales by Sub-Category

Sub-Category

Storage

	Storage			
	Total Customers	Sales	Profit	Margin
October 2019	14	4,040	369	9%
November 2019	26	7,058	894	13%
December 2019	27	10,591	1,574	15%
January 2020	11	3,226	125	4%
February 2020	7	1,627	199	12%
March 2020	20	3,815	489	13%
April 2020	14	6,026	736	12%
May 2020	15	4,685	395	8%
June 2020	24	7,701	888	12%
July 2020	14	2,895	289	10%
August 2020	25	6,539	483	7%
September 2020	35	4,947	522	11%

Figure 12-9. In this table for mobile dashboards, the total number of columns is limited

Strategy: Display Different Visuals for Desktop and Mobile

In this strategy, you will learn how to display different tables on different devices. You will work with the Sample – Superstore dataset to build this interactivity:

Part 1: Connect

1. Connect to the Sample – Superstore dataset.
2. Add an extract filter using [Order Date]. Set the range to between January 1, 2017 and October 27, 2020. This will allow us to simulate the data as if today is October 27.
3. If the calculation does not exist, create a calculation called **`[Margin]`**:

```
// Margin
SUM([Profit])/SUM([Sales])
```

4. Change the default number formatting of Margin to a percentage with no decimals.
5. Change the default color to the Red-Green-White Diverging palette. Use the full color range, set the start to –0.25, the end to be 0.40, and the center to 0, as shown in Figure 12-10.

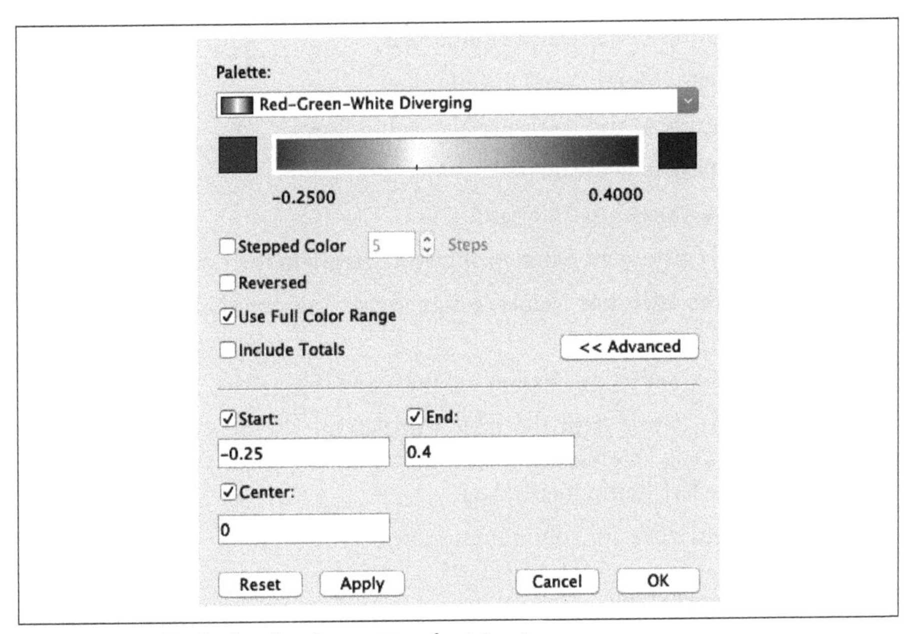

Figure 12-10. Default color formatting for Margin

6. If the calculation does not exist, create a calculation called **`[Total Customers]`**:

```
// Total Customers
COUNTD([Customer Name])
```

7. Set the default colors to custom diverging between #FFFFFF and #FFFFFF with two steps. This will make all values appear white.
8. Repeat color formatting for [Sales] and [Profit] measures.
9. Create a Boolean calculation called **`[Full Months]`**. This will return the most recent month only if the last date reported is the last day of the month:

```
// Full Months
[Order Date] < DATETRUNC("month", {MAX([Order Date]) + 1})
```

10. Create a Boolean calculation called **`[Last 12]`**. This is a table calculation we'll use to return the most recent 12 months of data:

```
// Last 12
LAST() < 12
```

Part 2: Create the desktop table

1. Create a new sheet called **`Desktop Table`**.
2. Create a custom date from Order Date. Choose Month/Year as the level of detail. Leave the name as Order Date (Month/Year).
3. Add Order Date (Month/Year) to Columns.
4. Add [Sub-Category] and [Measure Names] to Rows.
5. Change the mark type to Square
6. Add [Measure Values] to Color and Text.
7. Include Total Customers, Sales, Profit, and Margin in the [Measure Values] card.
8. Right-click the Measure Values assigned to Color and choose Use Separate Legends.
9. Add the [Full Months] calculation to Filters and select True.
10. Add [Last] to Filters. Edit the table calculation. Use Specific Dimensions and select Order Date (Month/Year). Edit the filter and select True. You should now show only the last 12 months of data.
11. Hide your field labels for columns and rows.

Part 3: Create the mobile table

1. Duplicate the Desktop Table sheet. Call the new sheet **`Mobile Table`**.
2. Swap the rows and columns.

3. Right-click [Sub-Category] on Columns and select Show Filter.
4. Change your filter type to a single-value slider.

Part 4: Create the desktop dashboard

1. Create a new dashboard. Keep the default size settings this time.
2. Add the Desktop Table sheet to the dashboard. Fit the width of the sheet to the dashboard.
3. Remove legends from the dashboard.
4. Change the title to **`Sub-Category Details`**.

   After this step, your visualization will be similar to Figure 12-11.

Sub-Category Details

		Oct. 2019	Nov. 2019	Dec. 2019	Jan. 2020	Feb. 2020	Mar. 2020	Apr. 2020	May. 2020	Jun. 2020	Jul. 2020	Aug. 2020	Sep. 2020
Accessories	Total Customers	11	34	24	10	6	19	15	20	16	18	15	42
	Sales	$3,503	$8,081	$5,752	$2,126	$2,437	$3,952	$2,142	$3,751	$4,849	$7,589	$4,962	$9,727
	Profit	$917	$1,972	$1,469	$476	$623	$1,192	$358	$1,190	$1,028	$2,403	$1,154	$2,778
	Margin	26%	24%	26%	22%	26%	30%	17%	32%	21%	32%	23%	29%
Appliances	Total Customers	6	15	20	5	9	13	8	15	10	9	14	20
	Sales	$2,956	$2,978	$4,752	$2,548	$3,521	$2,474	$645	$2,578	$3,160	$1,211	$7,105	$3,088
	Profit	$1,049	$724	$1,290	$423	$977	$308	($144)	$692	$897	$197	$1,621	$427
	Margin	35%	24%	27%	17%	28%	12%	-22%	27%	28%	16%	23%	14%
Art	Total Customers	15	23	23	15	7	17	17	16	18	18	13	37
	Sales	$454	$927	$396	$434	$247	$442	$618	$640	$1,019	$867	$733	$971
	Profit	$118	$256	$97	$111	$67	$115	$140	$177	$262	$251	$199	$267
	Margin	26%	28%	24%	25%	27%	26%	23%	28%	26%	29%	27%	27%
Binders	Total Customers	31	52	43	20	11	32	30	32	31	27	30	54
	Sales	$6,695	$4,929	$17,225	$9,437	$627	$2,785	$4,916	$2,929	$2,427	$2,472	$12,873	$12,157
	Profit	$2,290	$1,028	$7,170	$1,916	($824)	$578	$1,699	$782	$239	$697	$2,593	$2,189
	Margin	34%	21%	42%	20%	-132%	21%	35%	27%	10%	28%	20%	18%
Bookcases	Total Customers	4	8	5	2	5	4	7	9	10	4	2	7
	Sales	$1,376	$5,797	$1,617	$298	$1,126	$1,928	$1,979	$3,732	$5,309	$1,700	$460	$3,636
	Profit	$114	$15	($111)	$72	$115	$6	$28	($232)	$313	$261	($102)	($151)
	Margin	8%	0%	-7%	24%	10%	0%	1%	-6%	6%	15%	-22%	-4%
Chairs	Total Customers	13	24	23	6	5	8	5	15	14	14	10	27
	Sales	$5,911	$13,173	$14,951	$2,248	$3,552	$5,525	$2,099	$9,137	$8,495	$3,970	$9,419	$16,912
	Profit	$494	$369	$2,019	($97)	$154	$1,021	$133	$325	$814	$196	$206	$2,020
	Margin	8%	3%	14%	-4%	4%	18%	6%	4%	10%	5%	2%	12%
Copiers	Total Customers	2	1	2	1	---	6	---	1	---	3	2	1

Figure 12-11. Your very simple desktop dashboard with a table

5. Since you need to include your Mobile Table somewhere on your desktop dashboard in order for it to render on your mobile dashboard, you need to get creative. There are two popular methods:

- Float each mobile sheet onto your dashboard (in this case, just the Mobile Table) and set the *x* and *y* locations to –1 and the height and width to 1. Although this works, it's hard to control the dashboard.
- Add a floating vertical container onto the dashboard, as shown in Figure 12-12. Then add all the mobile-only sheets to the container (in this case, the Mobile Table sheet and the Sub-Category filter). Then adjust the vertical container so it's located outside of the desktop viewport by setting the *x* location to 1200, the *y* location to 0, the width to 375, and the height to 800.

We prefer the second method, as it allows us to track which sheets will be added to the mobile view. As you add sheets, you will not need to format in this container, as much of the formatting would be reset.

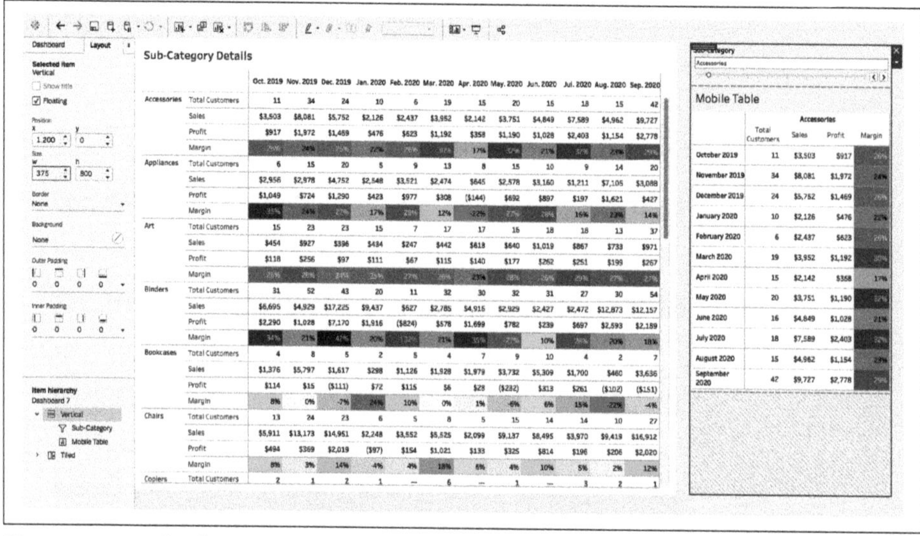

Figure 12-12. The floating container off the dashboard on the desktop dashboard

Part 5: Create the mobile dashboard

1. If your version of Tableau hasn't added the mobile dashboard, add a Phone device. By default, Tableau has brought the Desktop Table onto the mobile view.
2. Click the unlock icon next to Phone to edit the mobile dashboard.
3. Remove the Desktop Table from the phone device dashboard.
4. Adjust the size of the Mobile Table worksheet and hide the title. Your dashboard will now be similar to Figure 12-13.

Sub-Category

Accessories

	Accessories			
	Total Customers	Sales	Profit	Margin
October 2019	11	$3,503	$917	26%
November 2019	34	$8,081	$1,972	24%
December 2019	24	$5,752	$1,469	26%
January 2020	10	$2,126	$476	22%
February 2020	6	$2,437	$623	26%
March 2020	19	$3,952	$1,192	30%
April 2020	15	$2,142	$358	17%
May 2020	20	$3,751	$1,190	32%
June 2020	16	$4,849	$1,028	21%
July 2020	18	$7,589	$2,403	32%
August 2020	15	$4,962	$1,154	23%
September 2020	42	$9,727	$2,778	29%

Figure 12-13. The mobile view of your very simple dashboard with a table. Notice that the two dashboards are different from mobile to desktop.

In this strategy, you learned that sometimes you need to build completely different views for your mobile dashboard. When you are building these views, all the sheets need to go on your default dashboard. Otherwise, you will not be able to add them to your mobile view. If you want to improve the interactivity of your slider's filter on your mobile view, you could re-create the action. Finally, remember that any formatting necessary must happen on the dashboard of interest.

If you force us to describe in two words what mobile dashboards are all about, we'd pick "audience" and "simplicity." First, know your audience and the information they really need. Know why and how they are going to use the data! Second, keep your mobile dashboards simple. There's only so much information an individual can take. Narrow your focus to the key information. If you keep the interface simple, it should allow your audience to easily scan and scroll for insights.

Accessibility

Good dashboard design is accessible design. In this section, we discuss two common challenges for the novice developer: designing to include individuals with visual impairment (in this example, color blindness) and with learning disorders (in this example, dyslexia).

We know the challenges all too well when it comes to dyslexia, as one of the authors of this book deals with mild dyslexia. We try to stay positive when reading and writing, staying lighthearted about the mistakes—but, no doubt, dyslexia is extremely frustrating when trying to communicate.

While we've classified this portion of this chapter as "Accessibility," many of the concepts covered here are industry standards for design in general. They became standards out of the needs for all users, but we'll unpack these standards so you know where they came from.

Dyslexia

Let's start by understanding what *dyslexia* is. Someone with this learning disorder has trouble decoding letters and words when reading. Individuals with dyslexia have challenges identifying parts of speech. What's more, approximately 5% to 10% of the population has dyslexia (*https://oreil.ly/68s43*).

What's important to remember is that dyslexia isn't an on/off learning disorder. There's a spectrum. For some people, words and letters are in a constant state of flux. For milder cases, words or parts of words flip. This can look like an individual spending more time reading than others or trouble spelling, but it can be more extreme and manifest as trouble pronouncing words or completing math problems.

So how can we temper dyslexia in our data visualizations? Let's start with fonts.

Types of fonts

While dyslexia is usually stereotyped as mixing up letters and words, it also covers any symbology being used on a visualization, so it's important to understand that it's more than just words. But words do matter and are a great place to start if you are trying to figure out how to create accessible designs for people with dyslexia.

We have many types of fonts to choose from when it comes to creating a dashboard. These fonts are typically broken into five categories: sans serif, serif, display, handwriting, and monospace.

If you are looking for more resources, check out "Good Fonts for Dyslexia" (*https://oreil.ly/0thvy*) by Luz Rello and Ricardo Baeza-Yates and the British Dyslexia Association website (*https://oreil.ly/7h98D*).

Serif fonts have the small features at the ends of strokes. Serif is better to use for your average reader when the font size is smaller than 12. The most common serif fonts for dashboards are Times New Roman, Garamond, Georgia, Palatino, Merriweather, and Source Serif Pro, as shown in Figure 12-14.

Times New Roman	Beware; for I am fearless, and therefore powerful.
Garamond	Beware; for I am fearless, and therefore powerful.
Georgia	Beware; for I am fearless, and therefore powerful.
Palatino	Beware; for I am fearless, and therefore powerful.
Merriweather	Beware; for I am fearless, and therefore powerful.
Source Serif Pro	Beware; for I am fearless, and therefore powerful.

Figure 12-14. Several of the most common serif fonts

Sans serif fonts do not have the small features at the ends of strokes. They have been shown to be easier for people with dyslexia to read than serif fonts. The most common sans serif fonts for dashboards, as shown in Figure 12-15, are Arial, Verdana, Century Gothic, Trebuchet, Calibri, Open Sans, Helvetica, and Tableau font. Even Comic Sans is acceptable (yes, you read that right).

Arial	Beware; for I am fearless, and therefore powerful.
Verdana	Beware; for I am fearless, and therefore powerful.
Century Gothic	Beware; for I am fearless, and therefore powerful.
Trebuchet	Beware; for I am fearless, and therefore powerful.
Calibri	Beware; for I am fearless, and therefore powerful.
Open Sans	Beware; for I am fearless, and therefore powerful.
Comic Sans	Beware; for I am fearless, and therefore powerful.
Helvetica	Beware; for I am fearless, and therefore powerful.

Figure 12-15. Several of the most common sans serif fonts

Display fonts like those in Figure 12-16 are often designed for short-form uses like titles, logos, or headings. These fonts usually include unique decorative features. Display fonts can contain serif or sans serif elements. While these fonts are often graphic in nature, they typically don't work for people with dyslexia, so we recommend avoiding display fonts.

Lobster	Beware; for I am fearless, and therefore powerful.
Comfortaa	Beware; for I am fearless, and therefore powerful.
Bebas Neue	Beware; for I am fearless, and therefore powerful.
Abril Fatface	**Beware; for I am fearless, and therefore powerful.**

Figure 12-16. A sampling of decorative display fonts

Interestingly, many people with dyslexia find it easier to read fonts that look like *handwritten fonts* versus serif fonts. Those in Figure 12-17 are some examples.

Dancing Script	Beware; for I am fearless, and therefore powerful.
Indie Flower	Beware; for I am fearless, and therefore powerful.
Pacifico	Beware; for I am fearless, and therefore powerful.
Shadows Into Light	Beware; for I am fearless, and therefore powerful.
Architects Daughter	Beware; for I am fearless, and therefore powerful.
PERMANENT MARKET	BEWARE; FOR I AM FEARLESS, AND THEREFORE POWERFUL.

Figure 12-17. A sampling of handwritten fonts

Monospaced fonts use characters that have the same width. Monospaced fonts were used with typewriters because every time you typed a character, the register needed to move the same distance. These fonts were also used on early computers because they required less graphics memory. Monospaced fonts are best when aligning numbers vertically (though most fonts set numerical values to be monospaced). Some of the most popular monospaced fonts are Courier New, Consolas, Fixed Sys, Menlo, and Source Code Pro.

While many of these fonts are straightforward, many display fonts use monospacing to enhance their graphic nature. If you look at the following three fonts—Courier New, Arial, and Times New Roman—only Courier New, the monospaced font, perfectly aligns the characters vertically. Samplings of monospaced fonts are shown in Figure 12-18

Courier New	Arial	Times New Roman
12,345	12,345	12,345
+ 6,789	+ 6,789	+ 6,789
19,134	**19,134**	**19,134**

Courier New	Beware; for I am fearless, and therefore powerful.
Consolas	Beware; for I am fearless, and therefore powerful.
Source Code Pro	Beware; for I am fearless, and therefore powerful.
Roboto Mono	Beware; for I am fearless, and therefore powerful.
Inconsolata	Beware; for I am fearless, and therefore powerful.

Figure 12-18. A sampling of monospaced fonts

When given the choice for your users, you should choose sans serif fonts over serif fonts—but there are things you can do to make serif fonts work.

Choosing the best font

If we had to pick one font, based on research—including eye tracking—we suggest that you use Arial font for your users. Does that mean we use only this font? Absolutely not. And we have other formatting tips you can use to make your fonts more friendly for your audience.

Regardless of the font you select, a couple of additional tips can help enhance the readability of your dashboard: make sure your font is big enough and pay attention to font styling.

First, use at least 12-point font. If you have no problem reading size 8 fonts (or smaller), remember that your audience doesn't experience text the same way as you. Many times, we pick our font sizes after we've built our dashboards and data products. You should consider the opposite: stick with larger fonts when you build a worksheet and then develop your designs around larger text.

Second, consider font styling. Just because you've selected a font that's accessible doesn't mean you've made an accessible visualization. The styling you select with your font matters too. Bold fonts can be useful for readers with dyslexia, but avoid other formatting like italics, underlining, and all caps. Italics and underlines make it harder to track the text because they tend to crowd characters. All Caps tends to simplify the characters into a block form, making it more difficult to read.

Readers with dyslexia prefer a darker font, but not black. The contrast makes words easier to read. This is also a pro-tip that we use on our dashboards: instead of making all of our fonts the same color, we sometimes make our headers a slightly darker version of the base text. For instance, our base font might be the hex code #666666 (RGB 102, 102, 102), which is actually closer to a middle-gray tone. And our headers will be #494949 (RGB 73, 73, 73), a darker, but not super-dark gray.

Another interesting thing happens when you change your font color, as seen in Figure 12-19. The use of lighter tints will make your fonts look thinner!

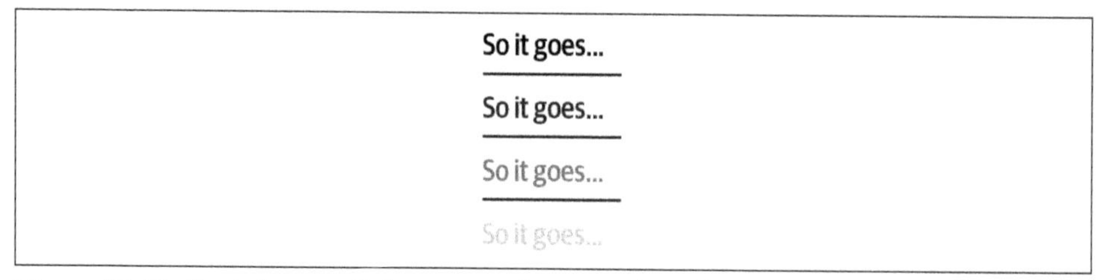

Figure 12-19. Fonts become lighter upon changing the color

Each line uses Chivo font, size 16. But the font colors are different. In many cases, you could use the top line as a header and the bottom line as the standard text. Users would treat them differently in your font hierarchy.

Text alignment

In addition to making smart font choices, you must also align your text. *Left-aligned text* is often easiest to read, though there are exceptions for charts like horizontal bar charts. We align text (whether at the left, right, or center) so that it's easy to scan it to find the insights we're looking for. If text is not aligned, scanning it and parsing details becomes more difficult.

In addition, do your best to keep as much of the text horizontally aligned rather than vertically aligned. Vertically oriented text is much more difficult to read. Even for expert readers, some of us need to tilt our heads to read the labels! While you won't be able to accomplish this on every one of your visualizations, it should at least be a consideration.

Let's take a look at Figure 12-20. At the top left, the text is left-aligned, and you can quickly read the sub-categories as a result. But the distance from the end of the text to the bars varies. This is OK unless the space becomes excessive for any one value. For instance, the distance from the Art label to the bar might be too much, given that the whitespace takes up more space than the actual text.

At the top right, you'll see that the text is right-aligned. This is also a good option because the space between the text and the bars is consistent. And when you are developing, the goal is consistency!

The chart at the bottom left, however, uses text that is center-aligned. This is the worst of both worlds. You have inconsistency at the start and inconsistency in the spacing between the text and bars. This styling should be avoided.

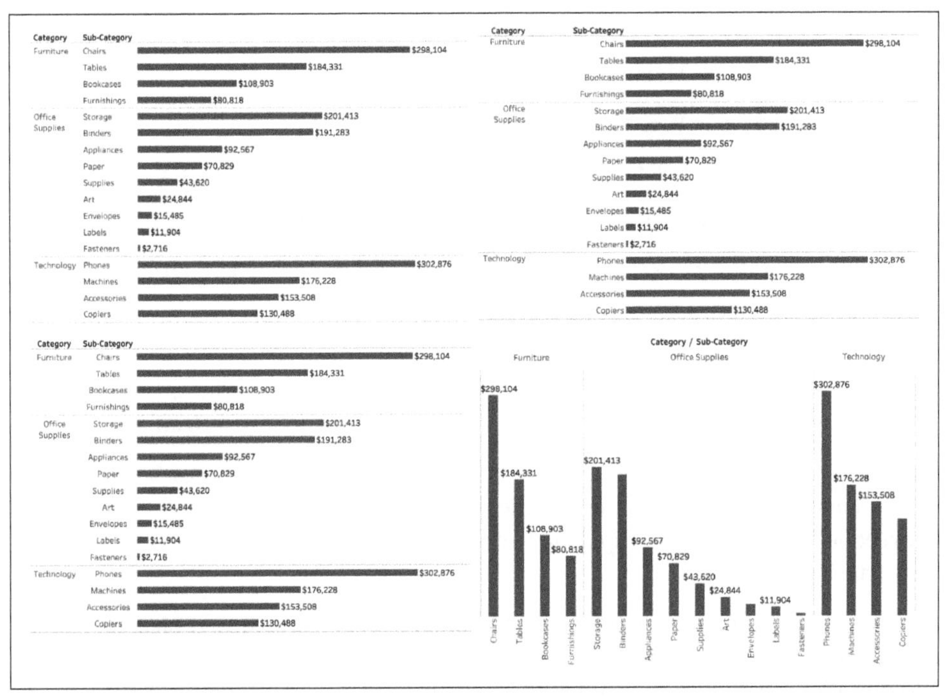

Figure 12-20. Four versions of a bar chart with different text alignment: left-aligned (top left), right-aligned (top right), center-aligned (bottom left), and vertically aligned (bottom right)

Finally, at the bottom right, we are using vertical text for the sub-category labels. This text is much more difficult to read than the values at the top of each bar. We did what we could to make the headers as readable as possible. Had we rotated them horizontally, they would not have shown up at all. This is one reason horizontal bar charts are much more valuable than vertical bar charts.

Next, let's consider spacing.

Interword spacing

In general, *interword spacing* (the space between words) should be at least 3.5 times the interletter spacing (the space between letters). While there is very little you can do in Tableau to control the kerning (the distance between characters), you can add spaces between letters.

But if you choose to do this, you have to be very thoughtful about your spacing between words. Novice developers often use one space between letters and two spaces between words. This is incorrect. You should have three or four spaces between words, depending on the font, especially if you are choosing to use All Caps for titles.

Consider this example:

Line 1: The Best Stock Market in US History

Line 2: THE BEST STOCK MARKET IN US HISTORY

Line 3: T H E B E S T S T O C K M A R K E T I N U S H I S T O R Y

Line 4: T H E   B E S T   S T O C K   M A R K E T   I N   U S   H I S T O R Y

Line 5: T H E   B E S T   S T O C K   M A R K E T   I N   U S   H I S T O R Y

Line 1 is a bold Arial font, and this is actually the easiest to read for users with dyslexia. Line 2, in all caps and with standard spacing, is probably the second-hardest font to read. By using all caps, your audience loses the teeth (all the small, imperfect spaces) that are between each letter, which make the words easier to read.

On line 3, we've added a space between each letter, and two spaces between each word. This is arguably the most difficult line to read. While we've added edges between letters by using all caps, the two spaces between each word aren't enough to distinguish the words.

Here's a test for you: rewrite line 2. Time yourself to see how long it takes to add an additional space between each word. Then type line 3 and see how long it takes to add one additional space between each word. It's harder than it looks!

Lines 4 and 5 are better options. While both have one space between the letters of each word, both lines have three and four spaces, respectively, between each word. If we revisit the key principle here, interword spacing, it's clear why we should use 3.5 times the interletter spacing: it keeps our text readable for all users.

Headings

For your headings, use a font size that is at least 20% larger than the base font. If you need to emphasize the text further, use bold.

Line length

One extremely underrated characteristic of text design is the number of characters that should occur on a single line. This point is easily lost on the novice developer. If they add text to a dashboard and it spans the length of the dashboard, they just add text! It doesn't matter if the dashboard is 300 pixels wide or 1400 pixels wide. They add text and click OK! The reality is that any line of text should be about 60 to 70 characters in length. Think about a newspaper—with those tiny fonts—typesetters don't text-wrap the entire length of the paper; they add columns.

Think about that. If we go back to the start of the previous paragraph, the line should be no longer than this:

"One extremely underrated characteristic of text design is the number"

And that's 69 characters!

Our advice on the topic: it's really hard to track how many characters are on a single line. In general, our guidance is to format so you have eight to twelve words on a single line. Instead of hitting Return in your text editor, use left and right padding. This will make the text wrapping much easier than trying to do it all by hand. This is especially useful if you are designing long-form executive dashboards.

Take a look at the example in Figure 12-21. You'll notice the following:

- Despite using an All Caps font, we added three spaces between each word.
- All the text displayed is about 80 characters wide.

MLS GOALKEEPERS IN 2018

This shouldn't be suprising but MLS goalkeepers, for the most part, block shots at the center of the goal. Also not suprising, keepers save fewer shots near the posts. Overall, MLS goalkeepers save about 68% of the 4.5 shots on goal every game.

LEAGUE AVERAGES | 2018

SHOTS FACED	SAVE %	ADJ. SAVE %
7.0 /96 MIN	68%	43%

The Adjusted Save Percentage is similar to save percentage but also weights the quality of the save. Across the league, the Adjusted Save Percentage average is 43% – meaning a high quality shot will be saved 43% of the time. This is lower than the actual save percentage of 68% because of the number of lower quality shots that have been taken.

Figure 12-21. This dashboard text uses lines fewer than 80 characters long

Line spacing

Line spacing is very difficult to control in Tableau. Ideally, spacing between lines is about 1.5 times the height of a particular line. Right now, the only way to control line spacing in Tableau is to add your own Return breaks for each line, add a single character on the next line (we use an underscore), and match the color to the background. Set the font size of the single character to 50% of the height of the standard text line. This is extremely tedious but can yield extremely strong results. The following is an example of what you would do to create the 1.5 spacing.

This line is size eighteen font and the font size of

_

the underscore is size nine. Set the color of the

_

underscore to match the background.

Hey, not all of our tips are time-savers! Sometimes it's just good to know what you can and cannot do in Tableau.

Padding

As we mentioned earlier, dyslexia doesn't affect just text. Numbers and symbols also apply. This means your visualizations are likely to affect an audience member with dyslexia. If there is one feature that is significantly underused by all users of Tableau, it's padding. As you've seen in previous chapters, removing padding can make multiple visualizations look like a single chart type. But we can also use padding to provide appropriate spacing between our visuals to increase readability. Padding provides the space your visualization needs to let key insights stand out.

The padding options are available on the Layout tab, shown in Figure 12-22, after you've added a worksheet to the dashboard. You have the option to synchronize your inner and outer padding or change the top, left, right, or bottom individually. If you want a worksheet to stand alone, you can use the All Sides Equal option as you change the padding. If you want to format two objects in a container, you can uncheck All Sides Equal and adjust accordingly. You can also adjust the padding of containers.

Outer padding adds space outside the worksheet border, while inner padding adds space between the worksheet border and the details in the worksheet. If you are using both inner and outer padding and change the background color of your worksheet, the inner padding will match the color selected, while the outer padding will remain transparent and display whatever is behind the sheet.

By default, Tableau adds a 4-pixel outer padding to visualizations. This is not enough. In general, we add 10 or 20 pixels of outer padding and 5 or 10 pixels of inner padding (yes, that much!). Our general rule for padding is to add too much padding to start and then work backward. For instance, you might want to start with padding of 50 pixels. If you choose to slowly add padding, go until it feels uncomfortable, and then go a little farther just to be safe.

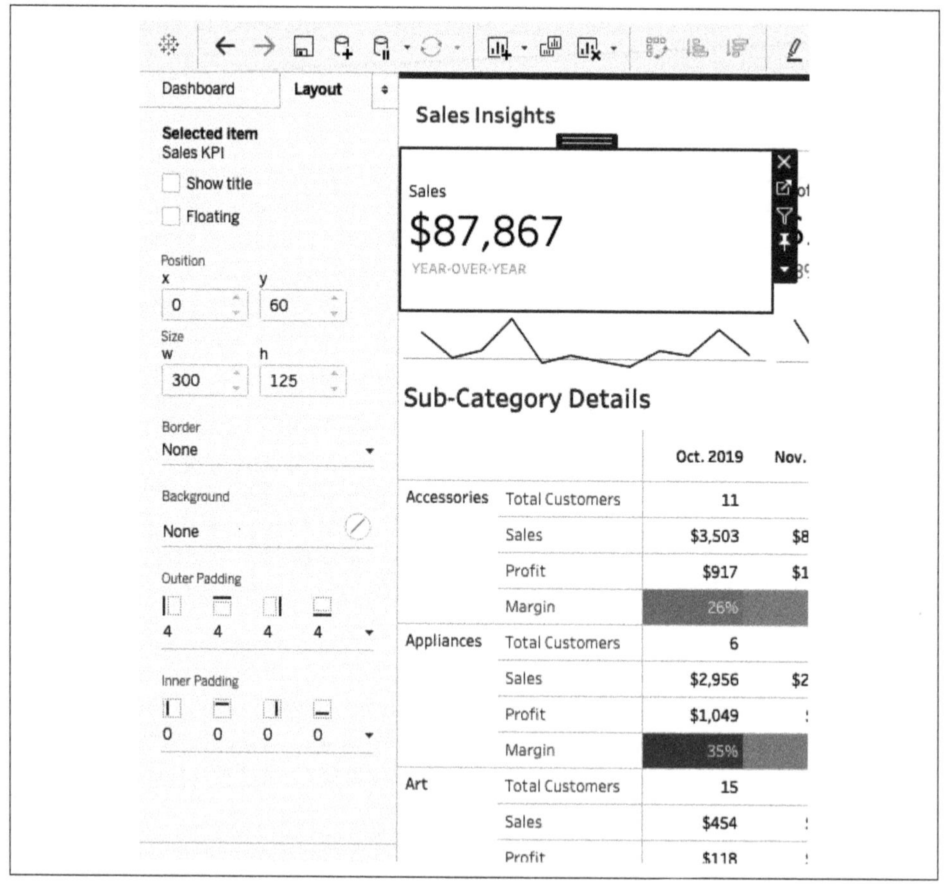

Figure 12-22. The Layout tab showing the outer and inner padding for a sales KPI. Notice that Tableau uses a default Outer Padding of 4.

As you develop, be sure to pair text alignment with padding. The two combined provide a powerful tool to increase the readability of your dashboards.

Let's take a look at Figure 12-23. This example keeps the standard 4-pixel padding on each visualization, but there is no spacing between the components. We've done a good job left-aligning the KPIs and right-aligning the numerical values, but it still feels a little cramped.

Sales Insights

Sales
$87,867
YEAR-OVER-YEAR

Profit
$10,992
18% YEAR-OVER-YEAR

Total Customers
459
26%

Margin (%)
13%
-2% YEAR-OVER-YEAR

Sub-Category Details

		Oct. 2019	Nov. 2019	Dec. 2019	Jan. 2020	Feb. 2020	Mar. 2020	Apr. 2020	May. 2020	Jun. 2020	Jul. 2020	Aug. 2020	Sep. 2020
Accessories	Total Customers	11	34	24	10	6	19	15	20	16	18	15	42
	Sales	$3,503	$8,081	$5,752	$2,126	$2,437	$3,952	$2,142	$3,751	$4,849	$7,589	$4,962	$9,727
	Profit	$917	$1,972	$1,469	$476	$623	$1,192	$358	$1,190	$1,028	$2,403	$1,154	$2,778
	Margin	26%	24%	26%	22%	26%	30%	17%	32%	21%	32%	23%	29%
Appliances	Total Customers	6	15	20	5	9	13	8	15	10	9	14	20
	Sales	$2,956	$2,978	$4,752	$2,548	$3,521	$2,474	$645	$2,578	$3,160	$1,211	$7,105	$3,088
	Profit	$1,049	$724	$1,290	$423	$977	$308	($144)	$692	$897	$197	$1,621	$427
	Margin	35%	24%	27%	17%	28%	12%	-22%	27%	28%	16%	23%	14%
Art	Total Customers	15	23	23	15	7	17	17	16	18	18	13	37
	Sales	$454	$927	$396	$434	$247	$442	$618	$640	$1,019	$867	$733	$971
	Profit	$118	$256	$97	$111	$67	$115	$140	$177	$262	$251	$199	$267
	Margin	26%	28%	24%	25%	27%	26%	23%	28%	26%	29%	27%	27%
Binders	Total Customers	31	52	43	20	11	32	30	32	31	27	30	54
	Sales	$6,695	$4,929	$17,225	$9,437	$627	$2,785	$4,916	$2,929	$2,427	$2,472	$12,873	$12,157

Figure 12-23. A dashboard with no additional inner or outer padding–only the defaults

Let's add some outer padding to this visualization. This will help separate the KPIs from each other and the table. Additionally, we'll change the background color to light gray. Figure 12-24 shows the result of this manipulation.

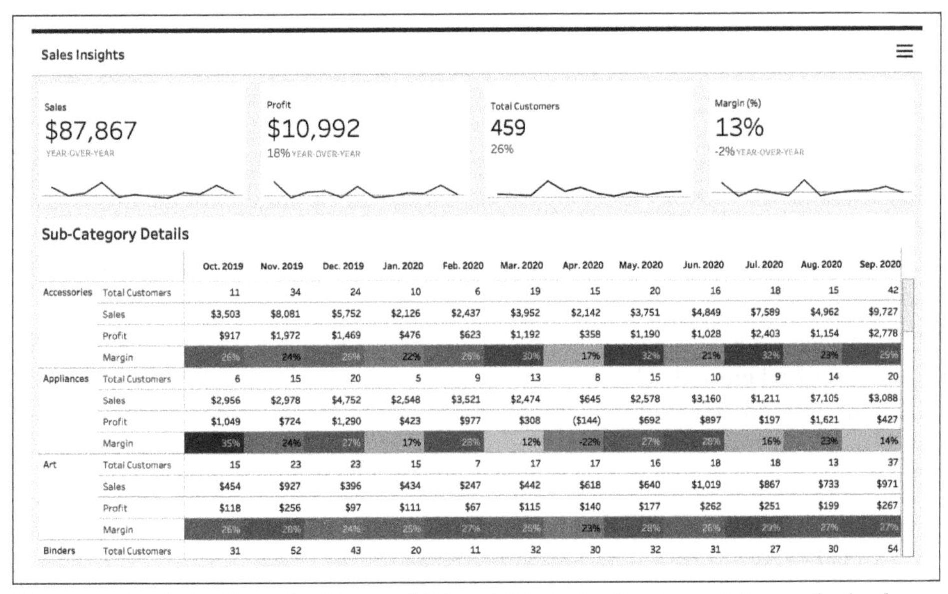

		Oct. 2019	Nov. 2019	Dec. 2019	Jan. 2020	Feb. 2020	Mar. 2020	Apr. 2020	May. 2020	Jun. 2020	Jul. 2020	Aug. 2020	Sep. 2020
Accessories	Total Customers	11	34	24	10	6	19	15	20	16	18	15	42
	Sales	$3,503	$8,081	$5,752	$2,126	$2,437	$3,952	$2,142	$3,751	$4,849	$7,589	$4,962	$9,727
	Profit	$917	$1,972	$1,469	$476	$623	$1,192	$358	$1,190	$1,028	$2,403	$1,154	$2,778
	Margin	26%	24%	26%	22%	26%	30%	17%	32%	21%	32%	23%	29%
Appliances	Total Customers	6	15	20	5	9	13	8	15	10	9	14	20
	Sales	$2,956	$2,978	$4,752	$2,548	$3,521	$2,474	$645	$2,578	$3,160	$1,211	$7,105	$3,088
	Profit	$1,049	$724	$1,290	$423	$977	$308	($144)	$692	$897	$197	$1,621	$427
	Margin	35%	24%	27%	17%	28%	12%	-22%	27%	28%	16%	23%	14%
Art	Total Customers	15	23	23	15	7	17	17	16	18	18	13	37
	Sales	$454	$927	$396	$434	$247	$442	$618	$640	$1,019	$867	$733	$971
	Profit	$118	$256	$97	$111	$67	$115	$140	$177	$262	$251	$199	$267
	Margin	26%	28%	24%	25%	27%	26%	23%	28%	26%	29%	27%	27%
Binders	Total Customers	31	52	43	20	11	32	30	32	31	27	30	54

Figure 12-24. A dashboard with an additional 10 pixels of outer padding and a background color change

The KPIs are now distinctly tied to their sparklines. However, the text for each KPI and for the table is sitting directly next to their borders, and this makes the text difficult to read. If you add 10 more pixels of inner padding to the KPIs and table, the text will be separated from the borders, increasing the readability. We'll also add a border around the KPIs and table. This border is just the next tint darker than the background color. The result is Figure 12-25.

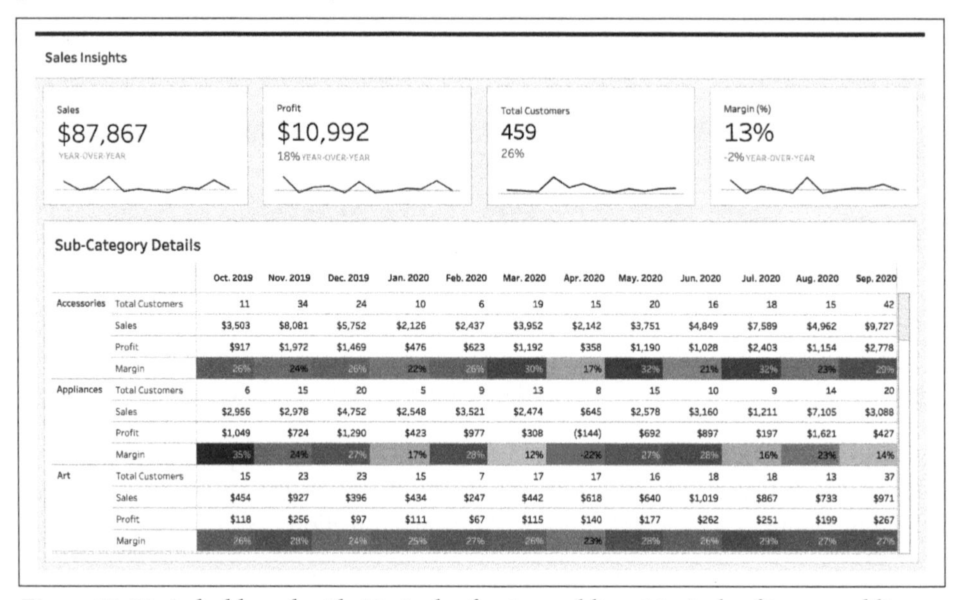

Figure 12-25. A dashboard with 10 pixels of outer padding, 10 pixels of inner padding, a background color change, and an outline around the KPIs and table

Using these techniques, we have added appropriate spacing to enhance the readability of our dashboard. One last note on this dashboard worth noting: the color of all the fonts is a medium-gray, #666666, specifically.

Background color

We've already discussed how to format your font colors, but it's also important to consider the background colors you are using for your visualizations. It's best to avoid white backgrounds for audience members with dyslexia. Instead, use an off-white background. In general, you want your data products to have contrast—just not extreme contrast like black text on a white background (or vice versa).

Let's take our preceding example one step further by changing the background of the text to a light gray. Figure 12-26 shows the change.

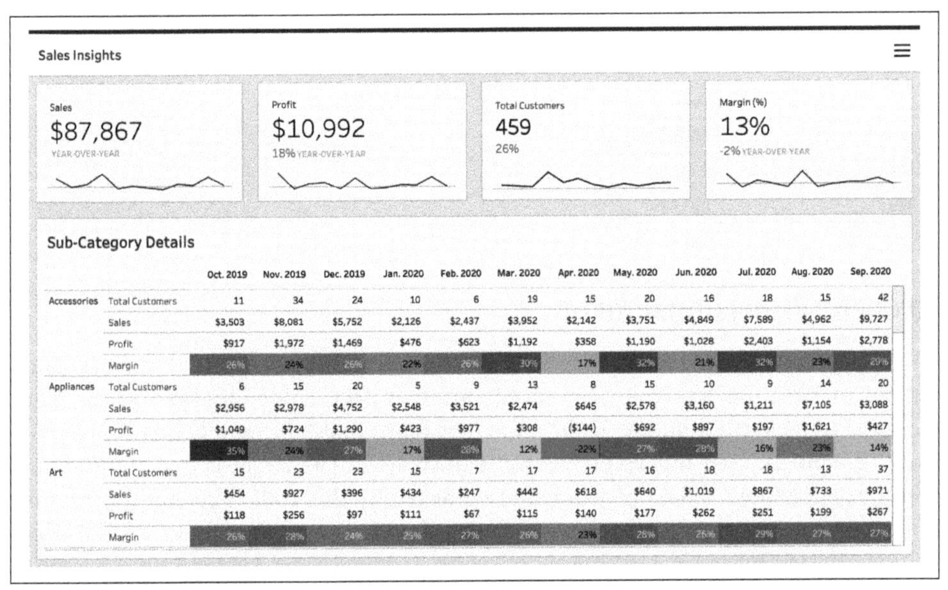

Figure 12-26. Changing the background colors to a light gray enhances readability

Writing style

If you must include text on your dashboard, just remember the following:

- Be concise with your text.
- Give clear instructions and narratives.
- Avoid double negatives.
- Use consistent language across products.

If you follow the guardrails discussed in this section for users with dyslexia, you're not designing a dashboard that is accessible for one group—rather, you are improving the readability of your product for all users.

Lists

Our last tip for font styling: when in doubt, make a list for your end users. Lists are highly effective and make content easier to digest because you've broken it into chunks. Just think, had we narrowed all these paragraphs into something this easy to read, it would have been much easier to understand!

Table 12-1. Tips for Font Styling

Do	Avoid
- Use at least size 12 font	- Underlining
- Dark, but not black, fonts	- Italics
- Bold	- All Caps
- Make lists	

Color and Visual Impairment

Of the population with vision, 90% to 95% have normal color vision. For the remaining 5% to 10%, the way color is viewed varies. In fact, color blindness arises from recessive traits located on the X chromosome. Since males have only one X chromosome, the presence of this recessive trait is displayed. The result is that about 8% of men but only about 0.64% of women are affected by this trait. This has a profound impact on the vision of males versus females. And knowing that about 1 in 20 users will be affected by the color choices on your dashboard, understanding a bit more about color selection is worthwhile.

The most common type of color vision impairment is deuteranomaly. This impairment type affects 6% of males and 0.4% of females. Audiences with this impairment are considered green-weak, which means many red, orange, yellow, and greens will be muted as compared to typical vision.

The most common form of color blindness is red–green color blindness. Those affected have difficulty discriminating red and green hues because of an absence or mutation of the red or green retinal photoreceptors. The protanomaly impairment affects 1% of males and 0.01% of females. This affects the ability of the individual to distinguish colors in the red-yellow-green spectrum. Like protanomaly, deuteranopia affects the red-yellow-green spectrum, but the average color they perceive is more cyan.

Other types of color impairments exist, as shown in Figure 12-27, but the majority of people affected struggle with green. So, what does that mean for developers? The red-white-green and red-yellow-green diverging color palettes are the most likely to be viewed incorrectly. That means the dashboard we've shared since the beginning of the chapter isn't the best choice (of course, about 5% of you knew that through your own experiences, and another 10% were aware of this color blindness concept).

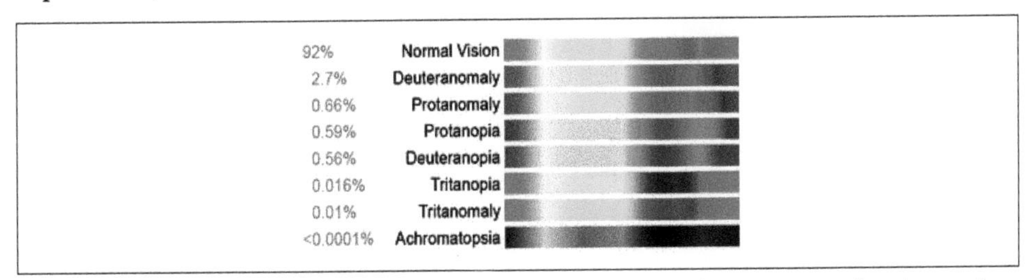

Figure 12-27. A breakdown of vision types and the estimated percentage of each population; source: Wikipedia (https://en.wikipedia.org/wiki/File:Color_blindness.png)

So, what can you do to make better color selections? This is fairly simple: check to see whether your color palettes and visualizations are accessible, and if they are not, change the palette to colors that are. In fact, let's take a look at Figure 12-28 to see how our mobile dashboard looks across the four vision types.

If you have typical vision, you will perceive that the reds and greens are still noticeable for deuteranomaly and protanomaly, but the deuteranopia view combines red and greens to make views that look more like warm pink and olive green tones.

As we said previously, other forms of color blindness exist, beyond those affecting red-green. Tritanopia and tritanomaly cause blue-yellow anomalies, for example. This affects less than 1% of the male population and less than 0.01% of females. Figure 12-29 shows our mobile dashboard through the eyes of people who see this way.

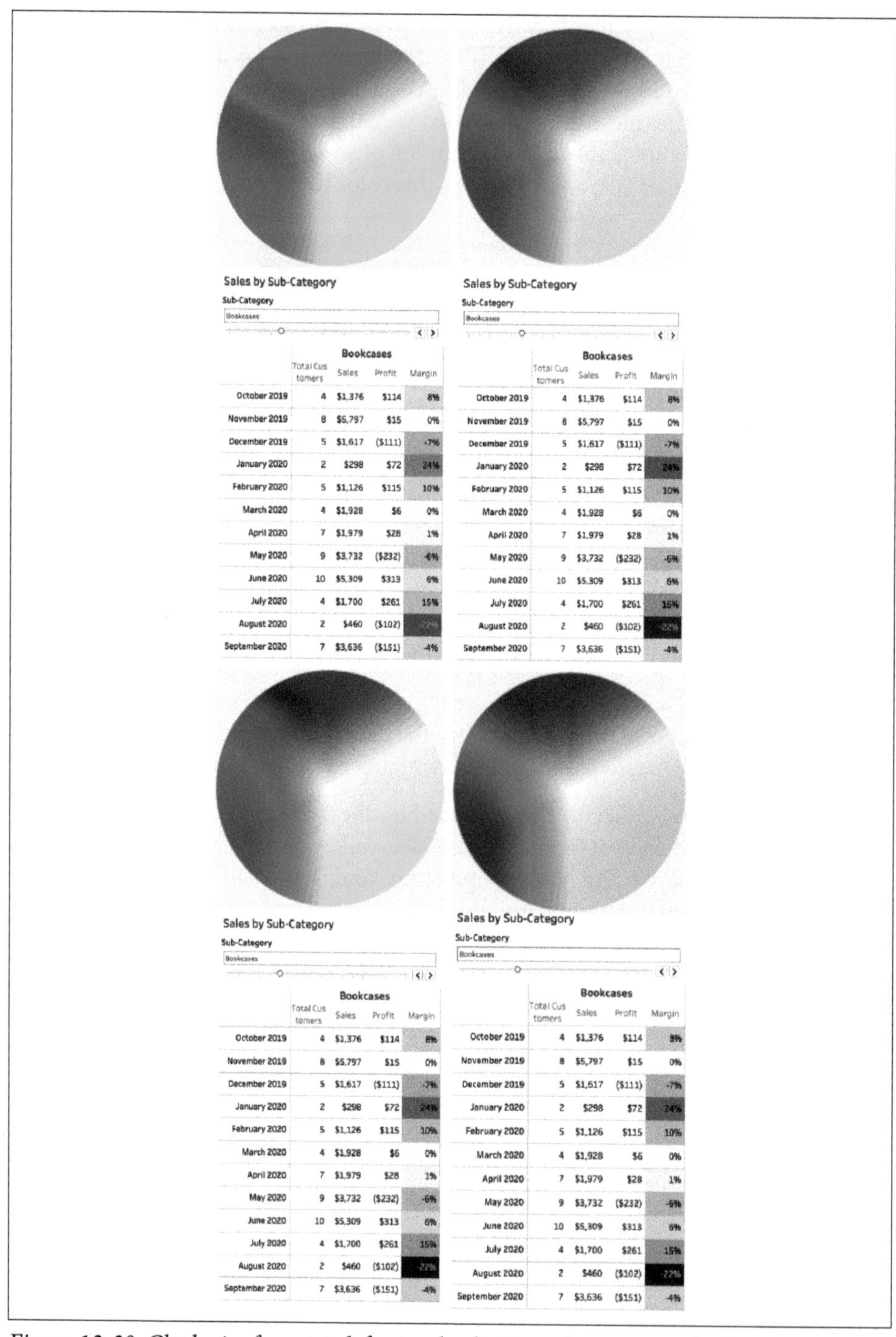

Figure 12-28. Clockwise from top-left: standard-vision, deuteranomaly, deuteranopia, and protanomaly views of the mobile dashboard

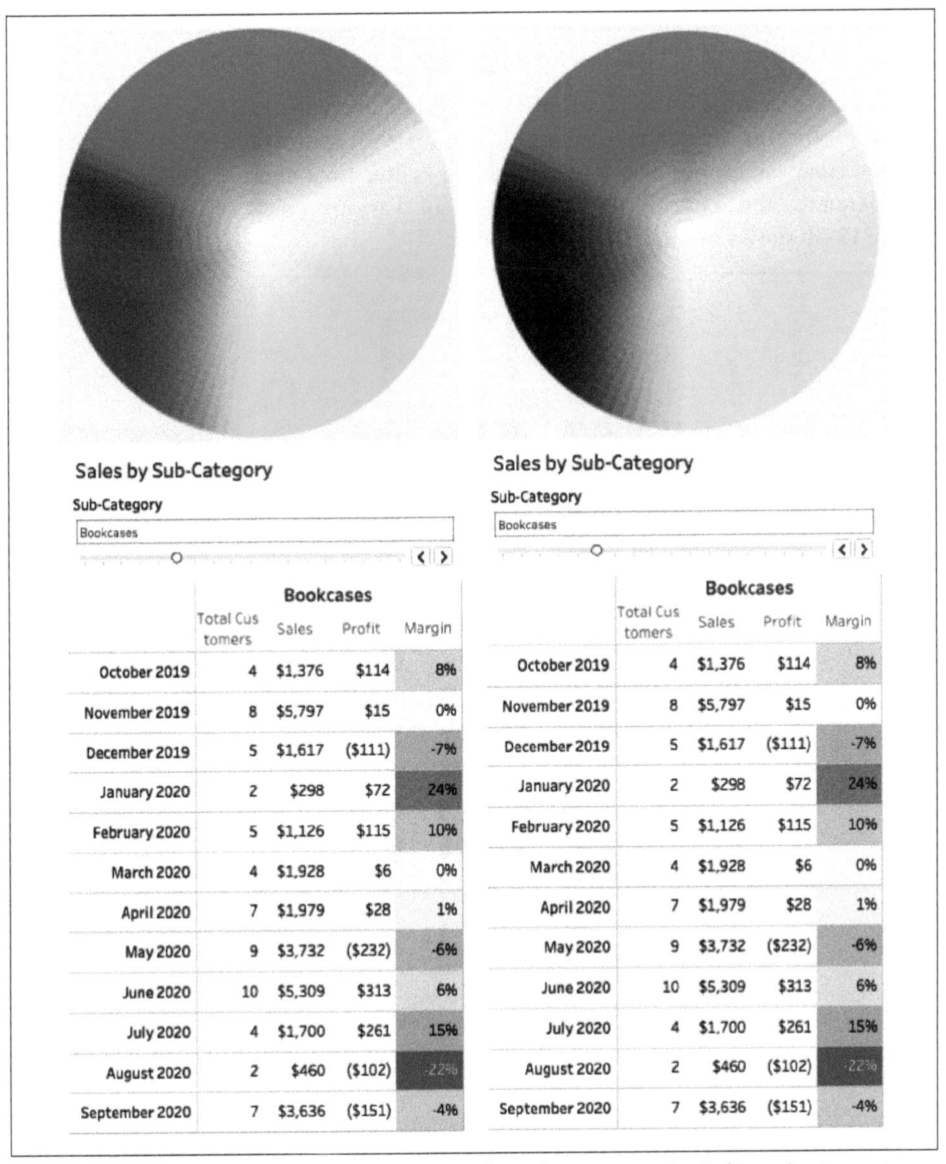

	Bookcases			
	Total Customers	Sales	Profit	Margin
October 2019	4	$1,376	$114	8%
November 2019	8	$5,797	$15	0%
December 2019	5	$1,617	($111)	-7%
January 2020	2	$298	$72	24%
February 2020	5	$1,126	$115	10%
March 2020	4	$1,928	$6	0%
April 2020	7	$1,979	$28	1%
May 2020	9	$3,732	($232)	-6%
June 2020	10	$5,309	$313	6%
July 2020	4	$1,700	$261	15%
August 2020	2	$460	($102)	-22%
September 2020	7	$3,636	($151)	-4%

	Bookcases			
	Total Customers	Sales	Profit	Margin
October 2019	4	$1,376	$114	8%
November 2019	8	$5,797	$15	0%
December 2019	5	$1,617	($111)	-7%
January 2020	2	$298	$72	24%
February 2020	5	$1,126	$115	10%
March 2020	4	$1,928	$6	0%
April 2020	7	$1,979	$28	1%
May 2020	9	$3,732	($232)	-6%
June 2020	10	$5,309	$313	6%
July 2020	4	$1,700	$261	15%
August 2020	2	$460	($102)	-22%
September 2020	7	$3,636	($151)	-4%

Figure 12-29. Our dashboard as seen by people with tritanomaly (left) and tritanopia (right)

As you develop your dashboards, we suggest you analyze color selections by using a tool like Coblis (*https://oreil.ly/hexEb*) that can simulate color blindness!

Since many companies love to use the stoplight approach for color palettes, what are alternatives? Tableau's default orange-blue color palette concurs with almost all forms of color blindness, but most companies want red for negative. We've found most success in converting red to blue and keeping green.

When we are creating our own palettes, we do a detailed analysis of each color. When we were selecting the color palettes to use for Workout Wednesday, we not only selected base values, but also looked at using lighter tints and darker shades, adding transparency, and considering the effects of various color vision impairments. Figure 12-30 shows our analysis.

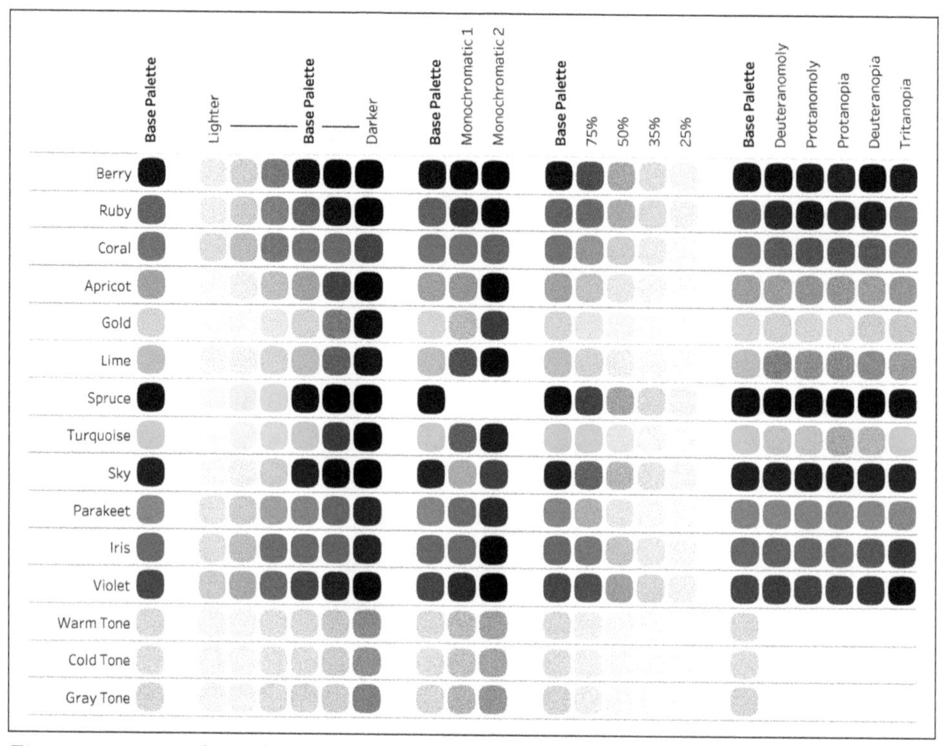

Figure 12-30. A color palette created for Workout Wednesday, highlighting how the colors could be used and how colors would display to different vision types

If you work with the same palettes regularly, we highly recommend creating something similar to Figure 12-30 to assist in color analysis. It's impossible to select the perfect palette that eliminates issues with all colors. But creating a table like this one would allow developers to have a large palette to select from while also being able to quickly analyze color selections.

If you are starting to develop your own palette, we recommend using *coolers.co* or the Palette Creator (*https://oreil.ly/bYOP5*) tool Luke developed for creating and analyzing color palettes on Tableau public. This tool allows you to select colors and see what they will look like as you change color values–so you can avoid mistakes after you've chosen values.

Formatting Consistency

In the last section of this chapter, we want to talk about the little things: the details that are necessary to get correct on a dashboard. This list does not encompass all the details that we sweat when creating our dashboards, but highlights top priorities.

Rounding Numbers

Rather than sticking to the default formatting of dashboards, we are always concerned about getting the formatting of our numbers correct and consistent. The level of precision needed for an axis will be different from the level of precision needed for a tooltip. Consider the example in Figure 12-31.

Look at the Profit (%) axis, and you will see that the label shows precision to the third decimal. This is an incredibly precise value for this axis. There is no difference in meaning between 50.000% and 50%. Both articulate the same value. Additionally, the extra digits are redundant information that is potentially distracting your end users from key insights.

When you are creating any numerical label, ask yourself one question: can my audience clearly articulate the difference between 0.001 and 0.002? If the answer is no, change the level of precision and ask the same question. In this case: can my audience clearly articulate the difference between 0.01 and 0.02? Repeat the process until you find yourself at an appropriate level of detail.

If you find yourself trying to straddle providing the seemingly unnecessary precision required by some of your audience and displaying values so your audience quickly finds insights, consider leaving the highly precise values in the tooltips and your less, but more appropriate, level of detail articulated on the view itself. You can see this in Figure 12-31: the level of detail for Discount is to the nearest percent on the axis but to the nearest tenth of a percent in the tooltips.

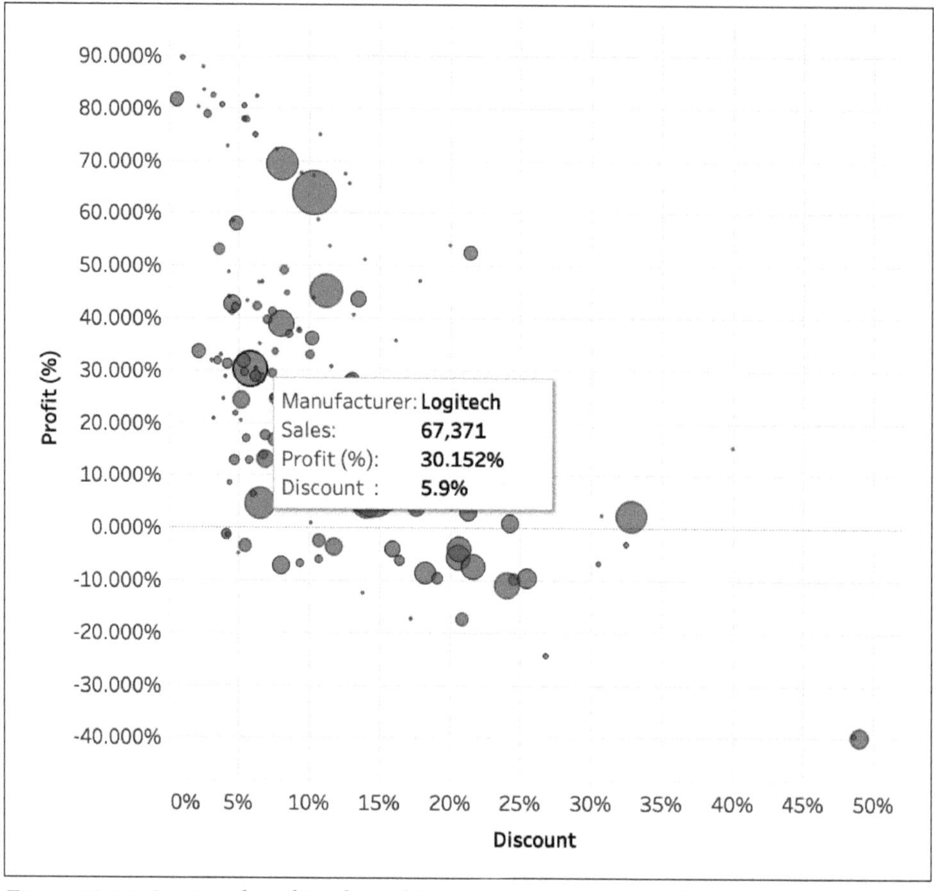

Figure 12-31. Scatter plot of Profit and Discount as a percentage by manufacturer in the superstore dataset, highlighting proper axis labels

To display more-precise values on tooltips instead of the view itself, sometimes you will need to duplicate the same measure and place it on Detail.

Axes Versus Labels

As you develop data products, your goal is to reduce *redundancy*—information that is presented more than once. One redundancy that we often see on our visualizations is the labeling (and encoding of the scale of a visualization).

Let's consider a simple bar chart that shows sales by sub-category (Figure 12-32). This visualization provides three levels of encoding for the same sales value.

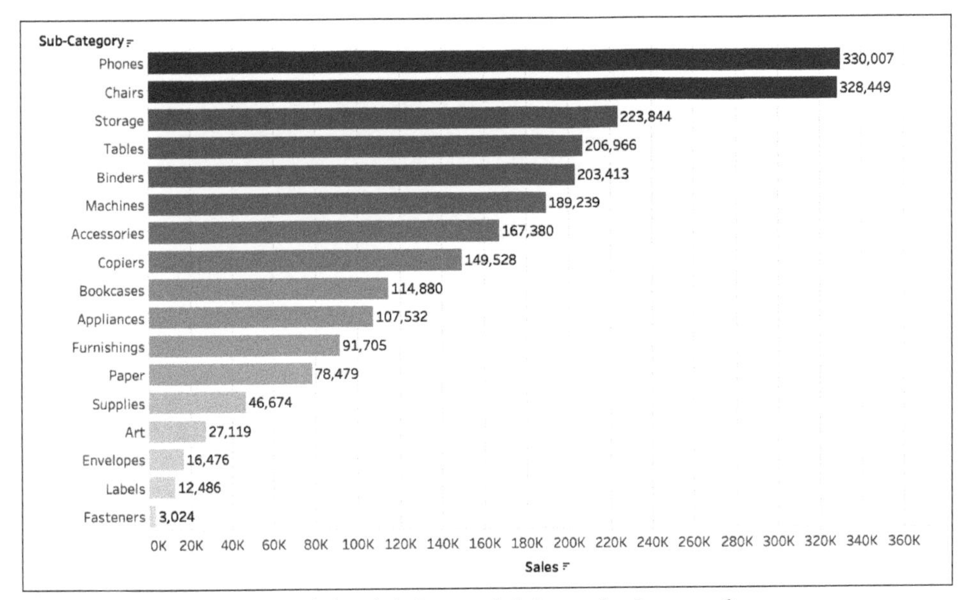

Figure 12-32. Bar chart with bar labels, axis labels, and color encoding

First, we have our bar length—which is actually quite useful. We added an axis to help provide the bars' scale. But we then added a second encoding by adding a physical label with more precision. On top of that, we added a third encoding of sales with color. We have too many encodings for the same value (and it could have been worse —we could have placed [Sales] on Size too).

Our goal is to reduce redundancy (particularly with our scale). To do so, you should first remove [Sales] from Color. From there, you are presented with two ways of encoding scale: you can either choose to have an axis or you can choose to have labels. Many novice developers use both, but you don't need both as they are doing the same thing: justifying the length of the bars!

We personally recommend having the labels on the bars rather than axes, as shown in Figure 12-33, because your audience will likely want to know the actual values of the bars rather than the relative values. What we mean by this is that your audience would rather know the exact values of phones and chairs rather than saying both are around $330,000 but chairs are slightly less than phones.

If you choose to have labels on your bars, you won't need the axis (and the axis ruler and grid lines that come with the axis).

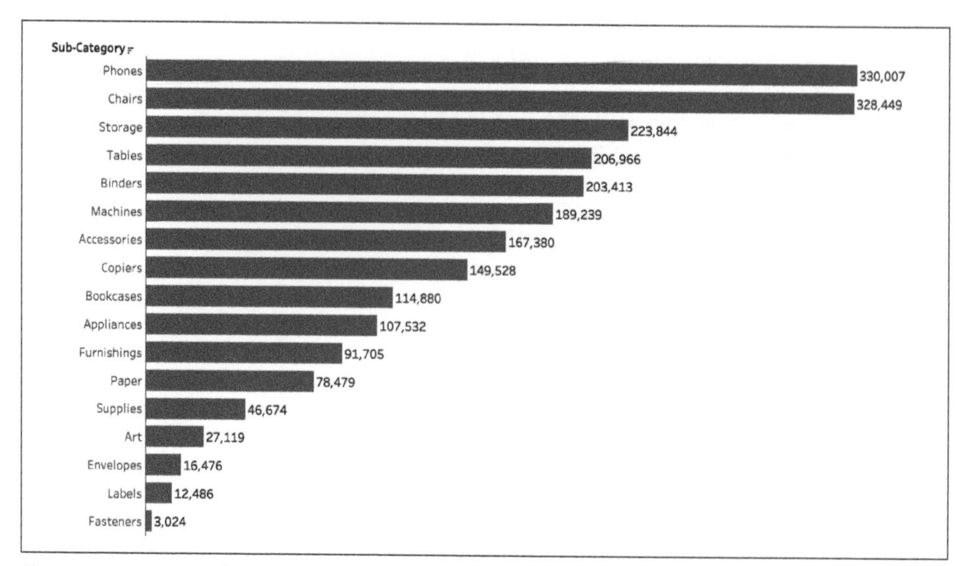

Figure 12-33. Bar chart with mark labels only

If you choose the axis, you'll need grid lines; see Figure 12-34. And for both, you should have a well-defined baseline that showcases the start of the bars!

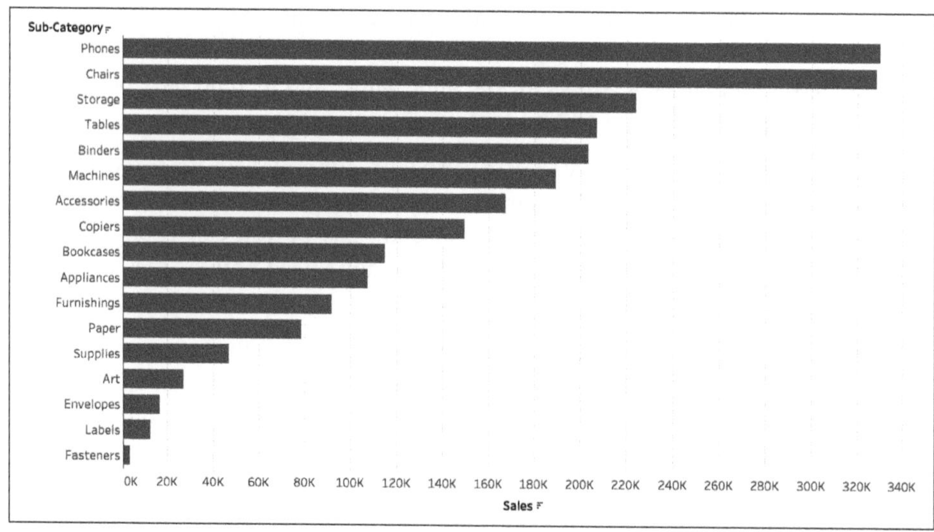

Figure 12-34. Bar chart with axis labels only

In the end, you must ask yourself, what do you want to showcase: an axis or a label? This will help you reduce the redundancy on your charts.

Font Styling

One of the greatest challenges (in our minds) is making sure font styling is consistent from sheet to sheet and from dashboard to dashboard. We're always asking ourselves: Are the fonts identical on my axes? What about the titles? And those marks labels? There are so many labels to check!

Your options for typography are practically unlimited and can feel overwhelming. If you've dabbled in typography, you'll know that you might want to consider dozens of factors when selecting font styling. However, the goal is to not to become overwhelmed in the minutiae of your dashboard. Instead, you should focus on high-level scenarios that will help drive the selection of your font types.

When we select fonts, we narrow our selection to the following categories:

- Title
- Subtitle
- Heading
- Subheading
- General text
- Axis label
- Axis title
- KPI
- Primary mark
- Secondary mark
- Row headers and table text
- Column headers

For each of these options, you must decide which font you'll use, and which size, and which color. Generally speaking, we suggest that you stick with the Tableau font. You can use numerous font stylings. If you do find yourself using a different set of fonts, use one or two fonts at most—and always begin by looking at your organization's existing brand standards.

Once you've created your style guide, stick to it as much as possible, but remember: this is a guide to help you. Exceptions to the rules will exist.

If you need to have two sheets identically formatted, right-click the tab that *is* formatted and select Copy Formatting (Figure 12-35). Then right-click the tab that *needs* the formatting and select Paste Formatting.

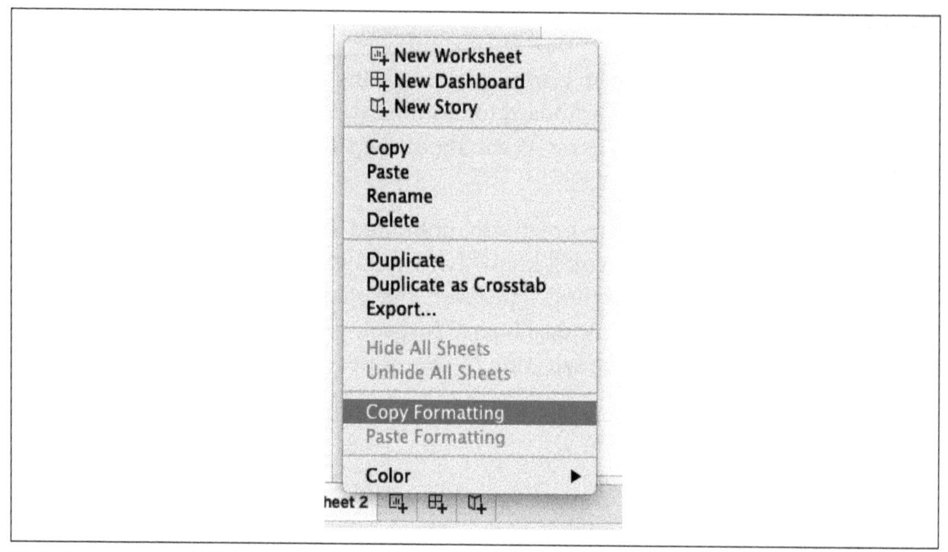

Figure 12-35. Right-click a tab to copy the formatting

Throughout this chapter, we've had one glaring issue shown on our dashboard. Some of you knew it right away because you are color-blind and immediately recognize poor color selection. Some of you are aware of the challenges of using red and green. For the remainder, you probably went about things without hesitating.

Conclusion

The goal of this chapter was not to provide a comprehensive list of to-dos when constructing a dashboard, but for you to start thinking about some of the components that must be considered when constructing a data product like a dashboard. You need to not only create a visual for your audience, but also construct it in a way that makes sense for your audience given their context.

In the first part of this chapter, you explored dashboard types. The type of dashboard you create will depend on your audience. The types of visuals you use will be dictated by the desired dashboard type. To determine the type of dashboard you create, you should first be empathetic to your audience by asking questions, and see how they currently work with the data that will power your data product.

Depending on their needs, they may want a static, exploratory, curated, or mobile dashboard. Static dashboards provide little interactivity and are often focused on KPIs. Exploratory dashboards are often loaded with filters but allow a user to ask specific questions of the data. Curated dashboards allow your users to interact with the dashboard, but the interactivity is controlled by you, the author.

Mobile dashboards are not mutually exclusive to the previous dashboard options but often need to be considered their own type, given that your audience will interact with them completely differently than the other options. Given that they present less information than a desktop dashboard, mobile dashboards will most likely mirror static or curated dashboards. Although you must present less information on a mobile dashboard, you often can create your best work because space restrictions lead your users to tell you what's most important. This allows you to create a dashboard focused on the important things, and not on the fluff that rarely gets used!

In addition, we discussed just two accessibility considerations: font selection for those who may have dyslexia, and color selection for those who don't see color the same as the majority of the population. While font and color selection is useful for those who have dyslexia or have color impairments, the principles that apply to make your data products more accessible are actually just universal practices regardless of impairment.

In the final part of the chapter, we discussed three examples of focusing on the little things to make your dashboards better: displaying appropriate precision on chart elements, reducing redundancy of labels, and consistently formatting text. While these three approaches do not make up the full extent of design elements that need to be considered when formatting your dashboard, they are three of the most critical for cleaning up your designs for your audience.

While this chapter discussed what you can do your dashboards right now, the next chapter covers trends you can expect in the future in regard to dashboard development.

CHAPTER 13

The Broader Tableau Ecosystem

The bulk of the book up to this point has focused on both how and why to create various charts and visualizations. We've built up your visualization skills, so you now know how to work with a variety of data and types of analysis. You've also learned how to think about your visualizations as data products by focusing on the end-user experience and interactivity.

This chapter is different. Instead of focusing on case studies and strategies, we now want you to take a step back and think more holistically about working with data for visual analysis. Specifically, we'll provide guidance on where and when to prep your data and how to craft a best-in-class analytics platform. Finally, we will end this chapter by sharing some trends that are on the horizon.

In This Chapter

This chapter has no hands-on strategies. In this chapter, you'll learn the following:

- Where and when to clean and prepare your data for analysis
- Key components to building a thriving data analytics platform
- Our thoughts on where visualization and analytics are heading in the future

Data Preparation: Where and When

Throughout this book, you've been spoiled. We've provided relatively clean datsets that already contain key pieces of information that are ready for analysis. While that was intentional to provide learning opportunities to you, it's not always feasible or realistic. Analysts often must work to collect important data, clean it to make it intelligible, and reshape it for visualization. In Chapter 11, we provided strategies for keeping that workflow inside Tableau Desktop, but that's not always possible.

When the task of cleaning data becomes a job of its own, knowing the right time to transition to a different tool, and how, is important. Here are some examples of when it is necessary or recommended to remove data cleaning from Tableau Desktop:

- Combining many data sources with multiple data grains. This scenario often rears itself in marketing analytics. Analysts must rationalize data from different platforms that have varying levels of dimensions. For the analyst working with this kind of data, preparing the data sources separately and identifying a common level of aggregation and then combining them by using a union is crucial.
- Generating row IDs or ranks. While you can use Tableau to generate ranks and indices, this can be done only via table calculations and relies on data within a worksheet. If you need to add a row ID to every row in your dataset, or create multiple rankings, this should be done before visualization.
- Joining two vastly different source systems, like data from an API and data from a spreadsheet. While Tableau Desktop has the capability to do cross-database joins, that functionality is not available with every data source. A good example is data from an API, like the built-in Google Analytics connector, containing information on web traffic, and data from an Excel spreadsheet with goals associated with the traffic objectives.
- Complex data cleansing or the usage of advanced statistical/mathematical analysis on your data. In Chapter 9, we outlined examples of when to do mathematical analysis outside Tableau Desktop. If your data falls into this scenario, you can most likely fold in those processes with the preparation tools listed next.

Spreadsheets: Microsoft Excel

It's impossible to write a book on business analytics without mentioning Microsoft Excel. Because of the proliferation and robustness of this tool, the data professional is likely to have used this for data preparation and/or data storage in the past. It is not uncommon for professionals to start with Excel and then work through the data ecosystem to more sturdy tools like Tableau. Spreadsheets work well for analysts because they provide full control over the underlying data. Cleaning can be done with formulas, macros, or simply by hand. Similarly, analysts can construct pivot tables to preaggregate their data, or copy and paste data from multiple systems into one source.

So, when should you use spreadsheets as a data preparation tool? There are two real use cases for using Excel: when that's the only tool you have or when you are doing a super-quick one-off analysis. If Excel is the only tool at your disposal, we definitely recommend using it to help shape your data. It has several useful functions and formulas that can clean up your data before analysis. We have found it especially useful for niche financial functions that aren't covered within Tableau Desktop and would take more than few lines of code in Python or R to construct. As stated, you can also copy and paste data from numerous sources into a single spreadsheet, so if you have no other way to combine, this becomes an obvious solution. We're also OK with it in an extreme ad hoc fashion. If you're constructing subset tables for a static analysis or know that you won't need to refer back to how the data was cleaned later, it may be a good solution. Most people are functional in Excel, and the popularity of the tool makes it relatively easy to search for solutions or tutorials on how to complete a task.

Beyond those two use cases, we highly recommend and encourage you to try the other tools mentioned next. Excel currently cannot handle large datasets (one million plus) and encourages behavior we try to avoid—modifying source data. There's also no real way to programmatically repeat data-cleaning tasks in Excel, so it is unrealistic to deploy a cleaned dataset via Excel to a large audience.

Traditional Data Preparation: SQL

Arguably the most popular programming language, and certainly the most popular query language, is *Structured Query Language* (*SQL*), which has been the predominant way to combine and query data housed inside relational databases, as shown in Figure 13-1. In fact, Tableau is built off of SQL; it translates your drag-and-drop interactions (which Tableau calls *VizQL*) into SQL queries to your data sources. At the start of our analytics careers, SQL was often thought of as the most crucial and necessary skill for business intelligence developers and analysts to have in order to be effective in their roles.

Practitioners with high skill in SQL benefit from the ability to create datasets that push back aggregations and data cleaning to the source. In many large enterprises, it is still the primary tool used to create datasets (often called *reporting tables*). From the perspective of working with Tableau, it can abstract many complicated calculated fields or pure cleaning tasks to the "backend," freeing up the analyst to work on constructing analysis on top of the datasets.

When should you consider using SQL? You should use SQL if your data is already housed in a relational database and your organization has adopted the idea of having a data warehouse and/or reporting databases. If your data is already housed inside these types of databases, using SQL will allow you to create new data objects (tables or views) that Tableau visualizations can feed off of. In these environments, it is also not uncommon for the task of preparing data to be separated into two roles: the visualization designer and the data engineer. So, the necessity to have SQL skill can often be offloaded to the data engineer.

```
8  SELECT
9  ACCOUNT,
10 BRANCH_ID,
11 REGION,
12 SUB_REGION,
13 REGION_SERVICE,
14 BRANDED_STATUS,
15 SERVICE,
16 DATE,
17 DEVICE,
18 KEYWORD_MATCH_TYPE,
19 KEYWORDS,
20 INTERFACE,
21 IMPRESSIONS,
22 CLICKS,
23 CONVERSIONS,
24 COST,
25 REVENUE,
26 NULL AS REVENUE_CONVERSIONS
27
28 FROM
29
30 (SELECT DATE,
31 ACCOUNT_NAME AS G_ACCOUNT_NAME,
32 CAMPAIGN_ID,
33 CAMPAIGN_NAME AS G_CAMPAIGN_NAME,
34 DEVICE,
35 KEYWORD_MATCH_TYPE,
36 KEYWORDS,
37 'Google' AS SYSTEM_INTERFACE,
38 SUM(ALL_CONVERSION_VALUE) AS REVENUE,
39 SUM(CLICKS) AS CLICKS,
40 SUM(CONVERSIONS) AS CONVERSIONS,
41 SUM(COST) AS COST,
42 SUM(IMPRESSIONS) AS IMPRESSIONS
43
44 FROM "MKT_DB"."GOOGLE_ADS"."KEYWORD_PERFORMANCE_REPORT"
45   WHERE DATE>='05/22/2019'
46
47 GROUP BY
48 DATE,
49 ACCOUNT_NAME,
50 CAMPAIGN_ID,
51 CAMPAIGN_NAME,
52 DEVICE,
53 KEYWORD_MATCH_TYPE,
54 KEYWORDS,
55 SYSTEM_INTERFACE) AS KWRD
56
57 INNER JOIN "MKT_DB"."MASTER_LOOKUPS"."SEARCH_PLATFORM_CAMPAIGN_KEY" AS MKEY
58 ON KWRD.G_ACCOUNT_NAME = MKEY.ACCOUNT_NAME
59 AND KWRD.G_CAMPAIGN_NAME=MKEY.CAMPAIGN_NAME
60 AND KWRD.SYSTEM_INTERFACE=MKEY.INTERFACE
61
```

Figure 13-1. An example SQL query used to add dimensions to data from Google

What are the drawbacks of using SQL to prepare your data? Most often it is time—both in gathering requirements and in developing the correct queries. In organizations that have separate workflows for data preparation and analysis, visual designers are often left waiting for data engineering to construct datasets to use. This waterfall approach also makes it rigid to changing analytical requirements and slows the process of iterative analysis, where working with the data can actually influence identifying the most important facets to analyze. Programming, or code, can also have a black-box effect on your process. It can be extremely difficult to unravel the queries constructed and to explain them to business end users. We have found that this process often can lead to datasets that lack trust.

Modern Data Preparation: ELT Tools

Another approach to preparing data is working with *ELT tools*. *ELT* is a play on a data engineering acronym that stands for *extract, transform, load*—the process of extracting data from a source system, transforming it through programming (SQL), and loading it into a database or warehouse for analysis. ELT flips the load and transform steps, by leveraging modern computer hardware to temporarily load data into memory (RAM) and then use that hardware to push transformed data into its final home. Additionally, most ELT tools have a more friendly UI, which relies less on coding, and more on tools and widgets (as illustrated in Figure 13-2). Tableau itself has an ELT tool called Prep Builder, born out of the need of Tableau Desktop users to have more capabilities in preparing their data.

When is the best time to use ELT tools? There are two compelling reasons to lean on ELT tools when working with your data: lack of a centralized database, and speed to develop and iterate. If you don't have access to a database, ELT tools can serve as an interim solution or virtual database, where you can comingle, combine, and reshape data. We find this on two extreme sides in the business realm: smaller businesses that rely heavily on cloud applications with interoperability, and extremely large organizations in which analysts are separated by many organizational layers from source applications or databases and must instead rely on flat-file exports from systems.

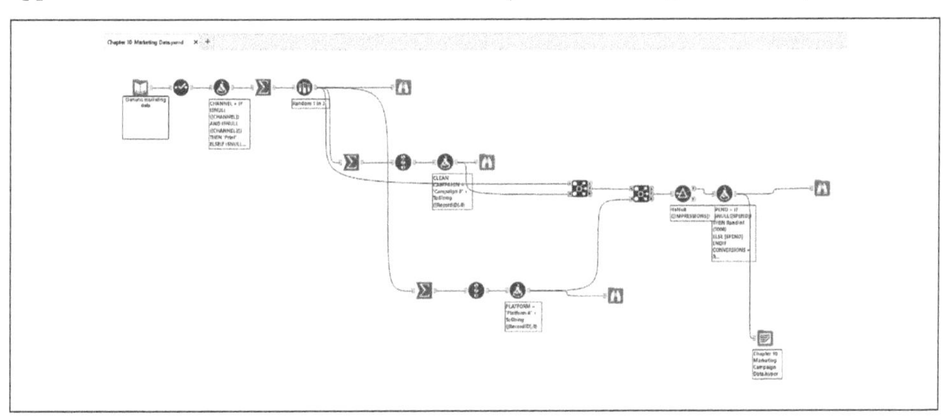

Figure 13-2. A data preparation workflow in Alteryx used to construct the data for Chapter 10

ELT tools also benefit from faster development time. Because users can drag and drop tools, test their outputs, and then reengineer, solutions are often quick to come by. ELT tools also often have data profiling, a way to summarize and display ranges of data, built in. Therefore, when working with a net-new data source, analysts who bring data into their ELT tools can gain a firm understanding and do initial exploration before they visualize. Additionally, what would be a complex or long SQL query instead becomes a series of tools for the analyst. And because data is preprocessed and relies on commodity hardware, no performance burden or impact occurs on the source systems. These tools also allow analysts to annotate on individual steps and end up looking like process diagrams, which are easier to communicate to their end-business users. And finally, analysts can benefit from exporting repeated steps for use in other workflows, eliminating the need for repetitive development or coding.

We've just made ELT tools sound amazing, so what's the downfall? These tools are earlier in their life cycle than other data engineering and development strategies, so they are currently missing version control and robust collaboration solutions. One person often must own the data-cleaning process from start to finish and ends up being responsible for maintaining the final workflow. With the breakneck pace of iteration, this can lead to friction within the organization. For organizations with dedicated enterprise data teams, using ELT tools can be seen as "shadow IT," or bypassing the process. It may also conflict with broader enterprise strategies that are in place. We recommend offsetting the tension this may bring by positioning ELT tools as great for prototyping usage. Positioning these as prototype tools has great benefit; the business users get the analysis they desire, and the data engineering teams have strong requirements.

Review

The three options we outlined are not all inclusive but serve to broadly cover the most common approaches to cleaning and preparing your data outside Tableau Desktop. Since each strategy has pros and cons, we encourage you to explore the options that may work best within your environment. Also think of these strategies as working in concert with each other—a blended approach that puts the need of people to have access to data at the forefront.

Building a Thriving Analytics Platform

If we haven't said it enough, it's important to reiterate that analysis and visualization are one piece of a larger picture. While they are wildly important because that's where stakeholders will go to understand the data available and make decisions, those decisions can't be made if data isn't commonly available, ready for analysis, and consumable in a scalable way. At its heart, Tableau Desktop is a data reader; it consumes whatever data you feed in. It is not a cure-all solution to having good analytics at any organization.

Instead, to have a thriving analytics platform, data practitioners, and particularly data leaders, need to be conscious of the entire pipeline of data and consider complementary technologies needed at each major step. When we think about the data analytics pipeline, we break it into six pieces, with visualization at the tail end (Figure 13-3).

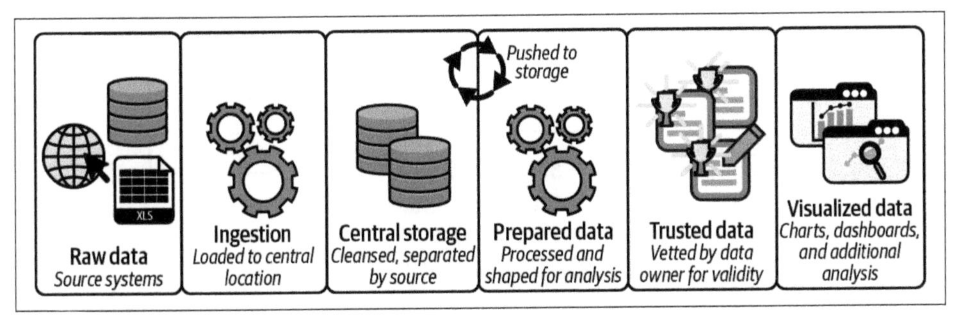

Figure 13-3. The data analytics pipeline, from raw data to visualization

First, it is important to consider where your data is coming from. For many organizations, this is a blend of many disparate applications, databases, and external data sources. All pertinent information rarely resides natively in one system of record. Think about your own personal health or financial records: do they all exist in one portal? Most likely they don't, because you've changed healthcare providers or banks over time. Organizations have the same issue: changing technologies, relying on third parties, and organization-agnostic data (like demographic information) mean that data lives in disparate sources.

To create a thriving culture of analytics, the most common practice typically is centralizing all of those data sources into a single location. Two competing theories indicate how to manage centralized data: the data lake and the data warehouse. For the *data lake*, the goal is to ingest as much data from source systems as possible, dumping it directly into the data lake and worrying about cleaning it and organizing it later in the pipeline. Data lakes are usually a form of unstructured data store, like Apache Hadoop, which can handle many file types and does not require data to be stored in rows and columns. In a *data warehouse*, data is cataloged, organized, and goes through an initial cleansing to be neatly stored and accessed. Traditionally, warehoused data is stored in a relational format (rows and columns), but recent technologies allow unstructured or semistructured data (like JSON files) to be stored as well.

Depending on whether you choose a data lake or a warehouse, the next steps are likely to be similar. In a data lake, the lack of up-front cleansing means that your organization will most likely be cleaning data as needed for analysis and reporting and probably has another layer of data storage (often called a *data mart*). Figure 13-4 shows a common practice for data lakes.

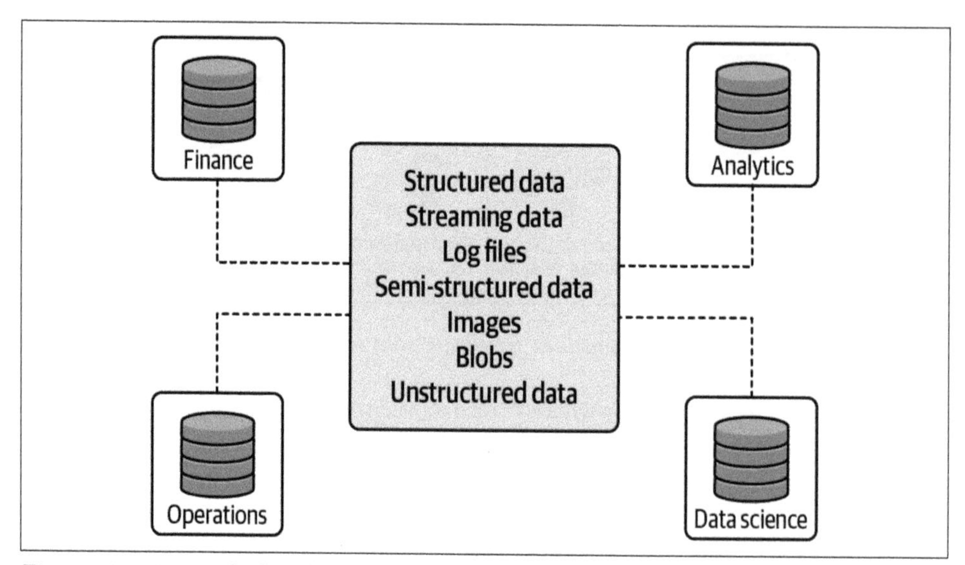

Figure 13-4. Example data lake with data marts that feed off a central source of data

In the data lake scenario, preparing data (step 4, shown previously in Figure 13-3) is managed by the owners of the data mart. Data is cleansed and prepared relative to the analysis that needs to be done.

If your organization instead has a data warehouse, data will most likely go through a base cleaning and rationalization process upon storage. Most data sources conform to a schema, data will be separated and normalized into tables that relate to one another, and data fields will be assigned types (such as numbers, text, or Boolean values). Here, more up-front work is required to ingest and store the data, but the payoff is that all consumers are working from the same quality of base data. Additionally, data warehouses are likely to have some tables developed for reporting and analysis, which can include combining separate data sources and data tables into highly denormalized tables designed for querying and analysis. That said, newer database technologies (Snowflake is a good example) are blurring the lines between data lakes and data warehouses, allowing organizations to dump data into relational databases without requiring an extreme amount of formal processing and organizing.

Independent of how your organization chooses to centralize its data, the next step is functionally the same for both scenarios. Data is further cleaned, processed, and refined for query and analysis. This is where both SQL and ELT tools come into play. Data engineers add in business logic, create robust datasets, and do data reshaping for visual analysis. In the end, independent of where the data is then stored, it is nearing completion and availability for visual analysis.

On to step 5, which is the creation of trusted data. During this step, no technologies are used, but instead datasets are given back to stakeholders and domain experts to validate accuracy. During this step, data dictionaries may be generated to formally define fields and calculations. This step is often neglected in the process as everyone is eager to get to step 6, visualization and analysis, but we caution you here. This step is critical in avoiding data source proliferation and having a strong and stable data platform centered around trust. A thriving data analytics platform includes this step, ensuring that end consumers, both experienced analysts and designers, and casual business users can trust the data and truly take action off of it.

After data certification, we're at the final step, visualization and analysis of data. And with this step, we've covered how to create compelling and impactful visualizations, but not how to share them. It is critically important at this step to have some sort of platform to house and catalog them (Figure 13-5). When working with Tableau, that means Tableau Server or Tableau Online—a software application that allows for Tableau visualizations and data sources to be stored, and accessed via a web browser (or mobile application), and that includes user permissions and security. And while we don't delve into the details, it is important to note that when standing up this type of environment, it is critical to ensure that it can handle both the volume of content and users who may be accessing data products. Fortunately, Tableau Server is a highly scalable environment capable of managing thousands of users and dashboards.

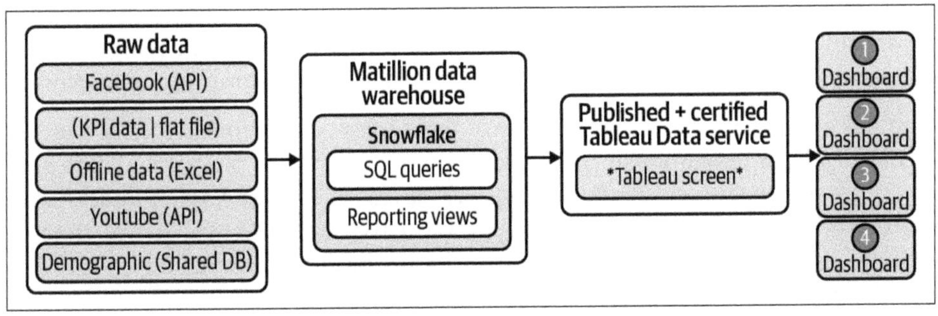

Figure 13-5. An example data platform for a marketing agency

As you've just witnessed, creating a holistic and best-in-class analytics platform takes a blend of the right technologies, processes, and mindsets. If you're just starting on your analytics journey, we encourage you to interrogate your own data ecosystem to find any opportunities to improve or add more transparency. And while there's no right answer, having pieces or concepts like those mentioned previously will lead you down the path of success.

Future Trends

We want to wrap up this chapter by exploring trends in visualization and analytics that are on the horizon. While visualization has long been a concept in supporting data analysis, in the past 10 to 15 years, the practice of visualization with a business organization has exploded in popularity—mostly due to commodity hardware and the proliferation of digitized data. Given that it is in its relative infancy, a lot of exciting opportunities lie ahead.

Data Journalism and the Rise of Data Storytelling

First is the idea of integrating storytelling, or narrative, into business analytics. If you look beyond business intelligence, you'll find that *data journalism* occupies a significant portion of the data visualization space. Here, the emphasis on visuals isn't on communicating business KPIs, but on using charts to support a compelling narrative. Journalists are using animation, annotated text, and whitespace (scrolling) to better emphasize and communicate their topics.

These techniques are primed to bleed into the business world in the future. As audiences become more data literate (able to understand visualization and data), designers will be able to further incorporate narrative into their own visualizations. Additionally, with advancements in machine learning and artificial intelligence, automatically generated narratives are also on the horizon.

We don't, however, see the advancement of technologies in this area removing the need for a visualization designer and/or someone well versed in interpreting charts for an audience. Those are likely to remain critical roles, with the shift being from the manual development or generation of insights, to a more programmatic approach.

And moreover, we think that popularity of these roles will continue to drive another recent trend: the development and deployment of data literacy programs. These programs are designed to ensure that all business users have a base level of understanding of both data foundations: chart types and basic statistical analysis. At time of writing, Tableau has already started to make a big push at advocacy of this concept and providing free information to its customers.

In the coming years, data and analysis likely will be fully recognized in the organization for nearly every role and expand to include broader enablement and training programs. The idea that data, business intelligence, and analytics are "technical" in nature and belong to IT arms of an organization will be completely stripped away.

Analytics Within Human Workflows

The next trend on the horizon is the concept of *analytics anywhere*, and particularly in the flow of operations. For nontechnical roles such as a sales manager, customer service representative, product owner, or project manager, the need for data is ever growing. With data and insights at their fingertips, each party can make more-informed decisions and often service their stakeholders (internal or external) more favorably. Currently, we see that most users go to a separate designated area (like

Tableau Server) to access the information they need. In the future, we see those lines blurring such that the necessary analyses exist within their current workflow.

No example of this trend is clearer than the acquisition of Tableau Software in 2019 by Salesforce, a major software organization with a prolific cloud-based customer relationship manager (CRM) application. Salesforce's huge acquisition demonstrates that a robust analytics and visualization platform, embedded directly where customer account information is stored, is critical for the organizations it supports day to day. As with the growth of its Einstein products (which are focused on predictive analytics and machine learning), the acquisition of Tableau has led to further and more complete integration of highly analytical and data-oriented tools into human workflows.

In the coming years, we think this trend will continue to grow, and the lines will further blur between a business intelligence hub or siloed-off repository where all reports and dashboards are housed, and daily operational applications. In the not-too-distant future, employees of all types likely will start leveraging interactive analytics to drive better decision making and information gathering at the place they spend most of their time.

Analytics as a Product

The final trend, which has shown major growth in the past three years, is the idea of *analytics as a product*. This is a fully owned and managed product that is sold to customers in a variety of industries, both as a value-added premium service to existing enterprises and as a standalone product. Think of a hospital with an electronic health record (EHR) application. Currently, this is used to store and access patient health information, but we also see it containing analytics to help inform medical practitioners—from providing operational insights like capacity and efficiency, to more holistic perspectives, like total care of a patient.

We believe that the role of measuring these types of metrics and performing these types of analyses will soon be separated from an organization and recaptured by the software that supports the task. And we can certainly say that this transformation is already starting to take place.

In addition to enhancing services for clients to include analytics, visual analytics will (and already has) become a standalone product for purchase. Many subject areas are already regularly receiving this type of treatment, from retail marketing trends, to advertising insights and consumer behaviors from large social media and technology firms such as Facebook.

Along the same vein is the idea of integrated analytics offered back as a service. Organizations like TrackMaven (used for marketing analytics) offer the promise of integrating all related data sources into a unified platform and constructing reporting and analysis deemed best in class for the variety of channels and objectives. These organizations also have an extreme advantage—the mass consumption of data from a variety of *their* customers, meaning they can also serve up aggregated insights on performance and construct comparison benchmarking in their respective disciplines. This task is something that a single organization would unlikely be able to achieve on its own, as it requires the cooperation of competitors and broad data-sharing. Instead, by virtue of opting into these services, these organizations gain this data sharing as an extreme benefit.

Conclusion

This chapter focused on three major topics: where and when to prepare data, building a best-in-class data platform, and future trends in analytics.

You've learned three approaches to preparing data. We've provided you with examples of when that function may need to be separated from the visualization process. You should now feel confident in knowing the tools at your disposal and have the ability to articulate the pros and cons of choosing each. And most importantly, we hope you take away that there isn't one best approach. Quite often a blend of all three exists within an organization to help achieve the analytical results you're hoping for.

You've also learned the layers of a best-in-class data analytics platform—one that starts with raw data and works its way all the way up to the point of visualization. You've also learned that data can be stored in one of two major ways to help facilitate this type of platform—the data lake and the data warehouse—and the benefits associated with both.

We've discussed three trends within data visualization and analytics: data storytelling, analysis within the human workflow, and analytics as a product. As you expand and grow your own visualization skills, we encourage you to look at the sister discipline of data journalism as a way to innovate and evolve your own data products. Additionally, we've challenged you to think hard about where dashboards and data products should reside, and shared the conclusive thought that in the future (for efficiency and maximum gains), we see them residing in the same application as an office worker's day-to-day operations.

And finally, we've said that analytics itself is becoming a product—one that is separating itself from analysts and business intelligence professionals to a completely separate product offering. It can be added to existing applications to enhance their value or work as a standalone product from data aggregators, allowing organizations to do necessary comparisons and benchmarks with their competitors and to further understand their own industry.

CHAPTER 14

Industry Frameworks

In this final chapter of the book, we want to touch on some core industries that rely on analytics as well as the key strategies and techniques to mastering each. Throughout the course of this book, we've consciously exposed you to case studies and strategies covering an array of industries. However, we want to now take a step back and share insights that we have both learned throughout our careers on successful ways to approach each industry.

Our goal with this chapter is to provide you a framework for working with your stakeholders and to share some of the key tenets necessary to construct meaningful analytics and data products. We hope this knowledge serves as a guide and assists you with navigating your role. In particular, we will discuss frameworks and strategies for the following industries:

- Healthcare
- Education
- Logistics
- Marketing
- Sales
- Retail
- Finance

As we guide you through each of these industries, we encourage you to see the parallels among them and the techniques that you may be able to glean to apply to your own scenarios.

In This Chapter

This chapter does not present any hands-on strategies. In this chapter, you'll learn the following:

- Strategies for approaching various types of industry analysis
- Techniques on how to work with various stakeholders
- The key foundational elements for crafting data products within each domain

Healthcare

Healthcare is a fascinating industry that covers a variety of organization types, from the providers that administer care, to the insurance carriers (payers) who manage claims and process benefits, to pharmaceutical and medical equipment manufacturers. At the heart of the healthcare industry is the drive and necessity to improve health and health outcomes. The byproduct of this guiding light is emphasis on patient care as well as on continued operational efficiency (reducing cost) and increased innovation. Knowledge of these concepts is key in providing insights to stakeholders.

With regards to patient care, analytics often takes the form of segmenting populations of care, by health markers such as chronic medical conditions, socioeconomic distinctions such as federal or state insurance coverage, or by stage of life. Here it is typical to find different types of ratios used to express metrics.

One is a *per thousand patient (PP) ratio*. The PP ratio is a way to express frequency in layman's terms, one of the most common being the existence of a disease among 1,000 people or the number of available hospital beds per 1,000 people. The uniqueness of this ratio is that by computing against 1,000, the number becomes analogous to the number of impacted individuals. It also serves to express smaller numbers with more prominence; for example a disease that affects 1% of a population becomes 10 patients per 1,000.

Often metrics of this nature are trended over time with goals or thresholds added to help gauge their performance. Take a concrete example of a PP ratio for hospital beds available. There can be both a goal to have 100 beds available at all times per 1,000 people, aimed at ensuring access to care, and simultaneously a tracking of current capacity with a threshold so that if the number goes below 10 PP, additional measures are to be taken to ensure that the population can be served.

The other type of metric you are bound to come across is a *performance measurement*. This measurement has a numerator, indicating those impacted by the performance measurement's goal, and a denominator, indicating the population of patients applicable to the measurement. These are common measurements for assessing and ensuring that quality metrics are upheld.

As an example, the Healthcare Effectiveness Data and Information Set (HEDIS) is an extremely popular way to assess the performance of healthcare plans; it constructs ratios across domains of care and benchmarks of comparison. These measures are usually expressed on a numerical scale from 0 to 100, representing the compliance of a measure. One example is the Adult BMI Assessment (ABA) (*https://www.ncqa.org/hedis/measures/adult-bmi-assessment*), which measures the percentage of members from 18 to 74 years of age who had an outpatient visit and whose body mass index (BMI) was measured. These measures are critically important because they are typically reported back to authorities on an annual basis and demonstrate the KPIs with which payers and providers operate against. When analyzing these, it is important to compare against available benchmarks, changes among different populations served, and to express them in clear "hit or miss" terms.

Since most of the metrics you are going to encounter are ratios or rates, they lend well to line charts for time-based comparisons. Often these time horizons are broader (monthly, quarterly, annually), to allow for enough data collection and occurrence. They also often lag because of the insurance industry and may not be finalized for three or more months while claims are finalized and payments are made. So, in this realm, it is important to display a long range of data, typically at least two years. It is also important to express percentage point difference and percentage differences (rate of change)—the first used to easily communicate the magnitude of the change, and the second to express the velocity of that change.

Lastly within the healthcare industry, it is important to master the concept of *protected health information* (*PHI*). PHI is anything that refers to individually identifiable health information and is often a major source of concern. It must be considered at all times to ensure that data isn't disaggregated enough to identify an individual data point as a person and that a data breach could also not provide individual information about a person. To that end, data storage has extra precautions and security guidelines, and analyses are often expressed only in aggregates. When working with stakeholders, it is important to identify what is and isn't PHI and to identify the necessary levels of aggregation to allow for insight and anonymity.

Education

We like to separate the industry of education into two practices: primary/secondary and higher education. *Primary and secondary* is anything from elementary through high-school graduation, and *higher education* encompasses universities and colleges. We separate them because they have a very different set of data points and measurements that are important.

Elementary, Middle, and High School

For everything up through high school, the most important analyses often are related to *assessments* (or standardized tests). Assessments are done on a semiannual or annual scale and measure the knowledge of a student against state and/or national standards of mastery. These values are used to not only communicate the mastery level of individual students, but also, aggregated together, to show effectiveness of curriculum, faculty, and school district. They also often become a critical KPI for the amount of funding a public school may receive.

Assessment data presents its own set of challenges. First, because it is typically administered by a body other than a school district, it is not integrated with other key applications (like a student information system) within a district. Therefore, demographic elements, attendance, and grade history are not easily and readily tied to the final score. Second, the data is often available as overly aggregated exports with poor structure. The poor data structure means that data must be preprocessed and typically pivoted (columns turned into rows) before thorough analysis can be done. Additionally, overaggregation takes place in the form of providing outcomes only at a grade level (versus individual student) or an individual school. This overaggregation makes it difficult for faculty and administration to compile a holistic view for their district while simultaneously identifying areas of true improvement.

To unlock the power of assessment data, we recommend obtaining results at an individual student level, per assessment. Then pivot the data so that metrics that were previously columns present as dimensions (a good example is assessment subject). And finally, use the one key piece of information known to be available, a student ID, to marry up information from other systems of record. District administrators will want to be able to see results at a district, school, grade, and teacher level—and want the ability to compare peers among the lower levels of aggregation to pinpoint improvement opportunities. Pivoting the data also allows demographic student information to be married up with assessments more easily.

The last step in unlocking the value of assessment data is to form cohorts of students (often based on graduation years), allowing the entire course of a student population's educational career to be seen and analyzed. This form of insight is powerful in identifying any gaps in curriculum and to track the long-term impacts and variation of socio-economic factors (for example, students on free or reduced lunch as compared to those who are not).

College and University

Data and analytics within secondary institutions center on something completely different from the first 12 years. Here, the important factors are primarily on enrollment and graduation. *Enrollment* is at the forefront because it informs how much tuition will be received and how it should be dispersed. *Graduation* is the key indicator of successful instruction and how the institution may be perceived among its peers.

The challenge within this realm is working with unique academic calendars that vary in length and starting dates. To work with this type of data, we often normalize against the first date of a semester's enrollment and express all charts in "days since enrollment start." These can be valuable in doing comparisons across time and academic years. Because enrollment is predictable, prior years are usually used for goal setting.

Logistics

No industry is more obsessed with time than *logistics*. Here, stakeholders will ask for all kinds of analysis related to time, knowing that their ultimate objective is to make sure things flow through a process seamlessly and, typically, that goods arrive on time. Because this industry runs on the notion of keeping everything within a time service quota, intra-day results, and comparisons of smaller time components (minutes and seconds) come to the forefront.

Additionally, with the invention of the Internet of Things (IoT), logistics is usually inundated with a massive amount of sensor and telemetry data. This data can also be useful for another critical element: optimization. Optimization has several facets within this industry, from the optimization of a fleet and effectiveness of utilizing staff members, to the optimization of routes themselves to reduce costs and loss due to inefficiencies.

Effective analysis in the realm of logistics means allowing for the vast amount of time analyses and providing many modes of comparison. Comparing performance against the same day of a prior week, or the same day of the prior year, and, in particular, up through the same point in time of those comparison periods, is critical. Instead of trailing toward a goal or the end of a workday, stakeholders often will want to pace against past performance. By pacing, stakeholders can more easily identify any immediate problems and fine-tune adjustments and reactions.

Marketing

Since the mass introduction of digital marketing, no other industry is more likely to face multiple and disparate data sources than *marketing*. Additionally, since most marketing channels are externally managed, like social media or search engine–based ads, marketers have the least amount of control over how and what data is initially collected for analysis. The most critical part for a marketing stakeholder to get right is to combine all the data for a full and holistic picture. Because of this, the biggest challenge present in marketing analytics is data engineering. A strong data pipeline and engineering element must be involved to ensure that all data can be rationalized and consumed.

Once data is rationalized, the next problem that presents itself is the massive number of KPIs in this industry. These metrics fall into three broad categories: one based on *volume*, such as total number of clicks or conversions; one based on *engagement* or *impact*, such as a conversion rate; and the final category related to overall *effectiveness* of marketing activities, such as return on advertising spend (ROAS). Stakeholders often want to see all these metrics at once, resulting in a cluttered and dense data display. We recommend breaking these up by providing two types of analysis: one that is focused on deep diving into a single metric, and the other that aims to explore the relationship among two or more metrics (think scatter plot).

The last aspect of marketing that can be a challenge is relating offline data or institution data to marketing activities. As stated earlier, one of the most compelling metrics in the advertising realm may be ROAS, so that means being able to equate revenue back directly or indirectly to advertising spending. This can present a challenge in multiple ways. First, it may limit how granular digital marketing activities can be compared to revenue; simply said, you may not be able to directly equate a marketing activity to a particular revenue event. Additionally, some lag is bound to occur in how revenue relates to advertising, so it can be important to measure and monitor how a change an advertising is reflected in a lagged revenue metric.

Sales

We broadly define *sales* as any activities that happen up to and before a purchase of a product by a customer. For an initial purchase, it is the identifying and tracking of potential customers, the engagement activities that happen with the customer along the way, and ultimately whether the interaction ends with a sale. For ongoing customers, it is the frequency with which they make additional purchases and keeping the sales relationship intact.

Within sales analytics, performance almost always has an external quota or goal, one that is often tied to a quarterly (or monthly) financial calendar. Common analyses relate to measuring the gap to reach a goal, or distance from goal. There is also a strong emphasis on using the past to inform the trends of the future. Historical data also becomes critically useful for adding a metric to purchasing habits or effectiveness of individual sales managers or divisions. You'll find a lot of comparison and ranking among peer groups, which aligns with the competitive and fast-paced environment.

Stakeholders will be looking to measure effectiveness through their sales pipeline and through their individual team members. They'll want to keep an extra close eye on this, so it is common to build out simple mobile dashboards to satisfy these metrics.

Retail

Retail and *consumer goods* is another industry that has a long history of leveraging analytics to measure performance and operational effectiveness. Transaction-level data has long been available and forms the cornerstone of most retail analytics. In this industry, the focus for stakeholders is both maintaining or growing revenue and ensuring predicable customer growth and activity. Because of the highly seasonal nature of retail, organizations commonly use prior-year performance as the baseline of goals and expected outcomes. There's also an emphasis on being able to aggregate the data to provide a top-level performance summary, but also to have filtering abilities to pinpoint performance by product lines or geographies.

More-sophisticated analysis in retail includes customer information to derive deeper understanding of their purchasing behaviors and determine effective marketing strategies. Here a full historical database of customer transactions, along with more personal information (likely obtained from a rewards program) can be mixed to generate customer personas and define key metrics like lifetime customer value.

Additionally, from the operational side of retail, inventory data is used heavily to monitor the inflow and outflow of products, measuring the speed with which products are leaving shelves and providing insight into both hot items (those that are selling quickly) and stale items. Because cash is often tied up in inventory, it is critical for these organizations to have a clear understanding of the balance.

Finance

Finance, the final industry we'll discuss, deals with everything associated with the movement of money throughout an organization. Because this arm of an organization deals so significantly with money, common accounting concepts like balance statements, income statements, statements of cash flow, and a profit-and-loss sheet are often the foundation of analytics that need to be available. These provide financial executives the high-level information they crave to assess the financial health of an organization, to assess any risk, and undertake financial planning.

In organizations with more maturity around financial analytics, this extends to understanding at a deeper level how various cost centers impact these numbers. In this capacity, they can provide strategic guidance on decisions being made as new products or new departments are developed.

Additionally, finance serves as the hub of understanding any outstanding payments and debt that exist. They are often responsible for marrying up invoices and payments, which can often reside in different applications, and capturing key metrics like time to pay and funding sources.

Conclusion

In this chapter, you have learned about frameworks associated with key industries and the methods of approach for working with their respective stakeholders:

- For healthcare, key considerations relate to data privacy. This industry relies on metrics that describe quality of care and effectiveness of care through multiple types of ratios.
- For education, you now know the differences between primary/secondary and higher education priorities, and how it is important for both groups to be able to aggregate and disaggregate data to gauge performance.
- For logistics, you know now it's all about time. This industry focuses on intra-day operations and ensuring that everything is operating smoothly and service deadlines are being met. This often means working with historical data as a baseline for pacing operations and as a way to quickly identify any processes that may be out of control.
- For sales, the goal is to understand the entire life cycle of a customer, from initial touchpoint to a sale, and to be able to categorize and quantify what happens along the path. Because sales is often a competitive environment, it is common to do peer-to-peer comparisons to rank performance of individuals and groups.
- For retail, transactional data is the basis of most analysis, serving to describe performance or revenue and customer growth, which are both highly seasonal. More-nuanced analysis in this industry extends to include inventory, to understand the velocity of various products, identifying the right levels of products that are selling well and those that are becoming stagnant.
- For finance, almost every analysis can be related back to accounting. You'll speak the language of this industry by focusing on basic accounting principles to initially describe the flow of money within an organization. Additionally, you'll support stakeholders in gaining broader understanding by providing more risk- and planning-related metrics focused on timing of payments, distribution of money, and types of payment inflow.

Index

C

D

E

F

G

L

M

N

O

P

Q

R

S

T

U

V

W

Y

About the Authors

Ann Jackson is founder and managing director at Jackson Two, a boutique consulting firm specializing in data analytics and visualization. In her practice, she empowers businesses to fully utilize their data assets. Ann has been recognized as a Tableau Zen Master for her proficiency with the product, her participation and leadership within the global community, and her contagious passion for data analytics.

Luke Stanke coleads Tessellation, an analytics consultancy firm that focuses on data visualization and advanced analytics, collaborating with many Fortune 500 companies. Luke has been repeatedly recognized as a Tableau Zen Master for his proficiency with and passion for the product, as well as his support of the Tableau community in building innovative solutions.

Colophon

The animal on the cover of *Tableau Strategies* is Stephanie's astrapia (*Astrapia stephaniae*), a bird of paradise.

In 1884, this colorful bird was named in honor of Princess Stéphanie of Belgium, the wife of Crown Prince Rudolf of Austria. Male astrapias have an iridescent blue-green and purple head with long central tail feathers up to 84 cm in length and purplish-black in color. Conversely, females are darker and subtler in color, with dark brown feathers, a bluish-black head, and a coppery brown belly. These iridescent colors lend themselves to the genus name, *Astrapia*, which means "flash of lightning."

This species is native to Papua New Guinea, through eastern and western highlands. You can often find Stephanie's astrapia in an umbrella tree, where it feeds on the fruit as well as insects and spiders that crawl up the branches.

Courtship peaks during the dry season, with males "advertising" themselves to females in packs of two to five (or more) birds. When a female displays interest, the male pursues her in a flighty chase; in these groups, males will hop between perches, hopping more and more as other males find their own mates.

The species is not threatened, but local poachers do target the birds for their vibrant plumes. Many of the animals on O'Reilly covers are endangered; all of them are important to the world.

The cover illustration is by Karen Montgomery, based on a black and white engraving from *Shaw's Zoology*. The cover fonts are Gilroy Semibold and Guardian Sans. The text font is Adobe Minion Pro; the heading font is Adobe Myriad Condensed; and the code font is Dalton Maag's Ubuntu Mono.

www.ingramcontent.com/pod-product-compliance
Ingram Content Group UK Ltd.
Pitfield, Milton Keynes, MK11 3LW, UK
UKHW050950120325
456138UK00005B/192
9781492080084